Lecture Notes in Computer Science 1798

Edited by G. Goos, J. Hartmanis and J. van Leeuwen

Springer
Berlin
Heidelberg
New York
Barcelona
Hong Kong
London
Milan
Paris
Singapore
Tokyo

Franz Pichler Roberto Moreno-Díaz
Peter Kopacek (Eds.)

Computer Aided Systems Theory – EUROCAST'99

A Selection of Papers from the
7th International Workshop on
Computer Aided Systems Theory
Vienna, Austria, September 29 - October 2, 1999
Proceedings

Springer

Series Editors

Gerhard Goos, Karlsruhe University, Germany
Juris Hartmanis, Cornell University, NY, USA
Jan van Leeuwen, Utrecht University, The Netherlands

Volume Editors

Franz Pichler
Johannes Kepler University
Institute of Systems Science
Altenbergerstr. 69, 4040 Linz, Austria
E-mail:pichler@cast.uni-linz.ac.at

Roberto Moreno-Díaz
University of Las Palmas de Gran Canaria
P.O. Box 550, 35080 Las Palmas, Spain
E-mail: moreno@ciic.ulpgc.es

Peter Kopacek
Vienna University of Technology
Institute for Handling Devices and Robotics
Favoritenstr. 9-11, 1040 Vienna, Austria
E-mail: kopacek@ihrt1.ihrt.tuwien.ac.at

Cataloging-in-Publication Data

Die Deutsche Bibliothek - CIP-Einheitsaufnahme

Computer aided systems theory : a selection of papers from the 7th International Workshop on Computer Aided Systems Theory, Vienna, Austria, September 29 - October 2, 1999 ; proceedings / EUROCAST '99. Franz Pichler ... (ed.). - Berlin ; Heidelberg ; New York ; Barcelona ; Hong Kong ; London ; Milan ; Paris ; Singapore ; Tokyo : Springer, 2000
(Lecture notes in computer science ; Vol. 1798)
ISBN 3-540-67822-0

CR Subject Classification (1998): J.6, I.6, I.2, J.7, J.3, C1.m, C.3

ISSN 0302-9743
ISBN 3-540-67822-0 Springer-Verlag Berlin Heidelberg New York

Springer-Verlag Berlin Heidelberg New York
a member of BertelsmannSpringer Science+Business Media GmbH

Printed in Germany
Typesetting: Camera-ready by author, data conversion by PTP Berlin, Stefan Sossna
Printed on acid-free paper SPIN 10720123 06/3142 5 4 3 2 1 0

Preface

Computer Aided Systems Theory (CAST) deals with the task of contributing to the creation and implementation of tools for the support of usual CAD tools for design and simulation by formal mathematical or logical means in modeling. Naturally, the basis for the construction and implementation of CAST software is provided by the existing current knowledge in modeling and by the experience of practitioners in engineering design. Systems Theory, as seen from the viewpoint of CAST research and CAST tool development, has the role of providing formal frameworks and related theoretical knowledge for model-construction and model analysis. We purposely do not distinguish sharply between systems theory and CAST and other similar fields of research and tool development such as for example in applied numerical analysis or other computational sciences.

The here documented EUROCAST conference which took place at the Vienna University of Technology reflects current mainstreams in CAST. As in the previous conferences new topics, both theoretical and application oriented, have been addressed.

The presented papers show that the field is widespread and that new developments in computer science and in information technology are the driving forces.

The editors would like to thank the authors for providing their manuscripts in hard copy and in electronic form on time. The staff of Springer-Verlag Heidelberg gave, as in previous CAST publications, valuable support in editing this volume.

March 2000

Franz Pichler
Roberto Moreno-Díaz
Peter Kopacek

Table of Contents

1 Survey Papers

2 Conceptual Frameworks, Methods and Tools

3 Intelligent Robots

4 Modeling and Simulation

1 SURVEY PAPERS

The CAST Project: Experiences and Future Perspectives

Franz Pichler

Johannes Kepler University Linz
Institute of System Sciences
Systems Theory and Information Engineering
Altenbergerstraße 69, A-4040 Linz
pichler@cast.uni-linz.ac.at

1 Introduction

The origin of Systems Theory lays in the kind of complex problems experienced by engineers (specifically in the field of communications and control) and scientists (in biology and ecology) in the mid of the 20th century. Then it became obvious, that the usual mathematical modeling concepts based on analysis (differential equations) and linear algebra (linear equations and matrices) were not any more appropriate. New mathematical methods in dealing with actual problems in formal modeling tasks were required.

In Communications Engineering Karl Küpfmüller, well known as engineer and as Professor, suggested to model transmission lines and related components not on the level of electrical networks (and related differential equation systems) but on the higher level of functional frequency descriptions (by the transfer function) and look for ("top down") means of computational determination of the physical realization on the level below.

Such an approach in modeling he called "systemstheoretical". His book "Die Systemtheorie der Elektrischen Nachrichtenübertragung" of 1949 can be considered as the "birth" of Systems Theory for Information Technology. From that it should be clear, that Systems Theory should not be considered as a "theory of systems" but rather as a collection of useful concepts, methods and tools for the support of "top down" multi-level modeling in engineering and science. By this definition it is obvious that formal concepts and methods will in a Systems Theory have an important role. It is likewise quite clear that in order to deal with complex design and simulation problems in engineering and science a computer assistance for the effective application of systemstheoretical methods is a necessity. The field of Computer Aided Systems Theory (CAST), the field which should provide the proper software tools for the application of Systems Theory, deserves therefor the vital interest of engineers and scientists working on complex design and simulation projects.

2 CAST/CAD/CAM Tools: The Beginnings

In the past most attention has been given to the development of computer assistance tools for practical engineering tasks. This resulted in CAD tools for structuring of

F. Pichler, R. Moreno-Díaz, and P. Kopacek (Eds.): EUROCAST'99, LNCS 1798, pp. 3–7, 2000.

engineering tasks and in CAM tools for physical layouting the design to prepare for implementation. The functional design of engineering systems, which needs the application of formal methods, however, has not received the proper attention which is deserves.

In the development of CAD/CAM tools the domain of microelectronics and there the design of VLSI circuits has reached a high standard both in applied methodology and in efficient implementation. This resulted in university education that VLSI design became teachable and in industry that also non-specialized firms were finally able to design customized circuits. Formal methods, which support functional design steps on a higher level of description, however, do not dominate in such tools.

This fact gave us in the mid of the 80's the idea to implement some of the formal methods which are available in systems theory for such tasks and to integrate them into existing VLSI-design tools. We considered this activity as part of "Computer Aided Systems Theory" (CAST). The first "CAST-tool" to be integrated to VLSI CAD/CAM tools (to became a "CAST/CAD/CAM/tool) was CAST.FSM, a Lisp-software supporting the application of finite state machine theory in VLSI design.

3 CAST Conferences: The History

For traditional areas of engineering, especially in control engineering and in communication engineering tools which support formal (mathematical) modeling exist already for a long time (see for example Jamshidi-Herget (eds.) 1985). For (classical) mathematical systems theory, as defined by Küpfmüller, Zadeh, Kalman, Mesarovic, Wunsch and others, which centers on the concept of a "dynamical systems with input and output" (in different degree of abstraction and specialization) "CAST tools" in comparable quality did not exist.

An exception constitute the reported activities of George Klir and his systems group at SUNY-Binghamton to develop and implement the "General Systems Problem Solver" (GSPS) for the support of problems in general systems (as defined by G. Klir).

In 1984 we started at the University Linz, Institute of Systems Science, the "CAST project" with the goal to develop software tools which support systems theoretical methods in formal modeling of systems in the domain of information technology. Furthermore to integrate such tools (CAST-tools) into existing CAD/CAM tools and to show its practical applicability. This project got full support by Siemens AG ZT Munich (research laboratories) specifically there by its leader Prof. Heinz Schwärtzel and his group. As a result different versions of prototype-CAST tools, such as CAST.FSM (a finite state machine problem solver), CAST.FOURIER (an abstract harmonic analysis problem solver) and CAST.LISAS (for modeling and simulation of cellular systems) were developed (see Pichler-Schwärtzel (eds.) 1992). In addition international conferences on the topic of "CAST" were started and organized (CAST workshop'88 (Linz), EUROCAST'89 (Las Palmas), EUROCAST'91 (Krems), EUROCAST'93 (Las Palmas), CAST'94 (Ottawa), EUROCAST'95 (Innsbruck), EUROCAST'97 (Las Palmas)). These CAST-conferences received international interest and participation, most of the papers delivered by the speakers have been published (Lecture Notes in Computer Science, Springer-Verlag Berlin-Heidelberg).

The topics covered at the CAST conferences have purposively a wide spectrum. Besides of papers which are in the core of CAST research the organizers accepted also papers which are potentially close to CAST or which give promise of new areas for the development or of possible applications of CAST-tools. Of specific interest was here the connection of Systems Theory and CAST to the field of Artificial Intelligence and related areas. In this direction the engagement of Roberto Moreno-Diaz and the research groups at the University of Las Palmas (Gran Canaria, Spain) and at different other Spanish universities deserves to be mentioned here. All in all we would like to consider the past activities in the "CAST project" as satisfying since many researchers and practitioners in engineering became aware of the importance of formal models and associated algorithms to structure and optimize design. They follow the rule that design has to be done "top down" using models of different levels of abstraction and specialization and acknowledge the availability of formal models to help in verification the fulfillment of requirements. Furthermore they have trust in deductive methods for systems analysis to approve the desired quality of the designed system. This facts are satisfying to all researchers working in theoretical fields such as applied mathematics or mathematical systems theory. This satisfaction is independent to what degree CAST tools (in the strict sense of definition) might have migrated into existing CAD/CAM tools until today.

4 Directions for Future Research

After ten years of activities in CAST matters there is the question in what direction future research should go. I will try to point out three possible directions:

(1) Systems Theory and CAST tools for Macro Architecting,
(2) Systems Theory and CAST tools for complex hierarchically distributed intelligent systems,
(3) Investigations of dynamical systems in "Bourbaki style" and development of associated CAST tools.

To (1): The engineering systems of today a very often complex systems build up by reusable components which consists of conventional engineering systems. To assure the quality of the design of the overall system it is advisable to have formal models for higher architectural levels available. In consequence this requires associated methods for structuring and optimization of such formal models. Furthermore appropriate CAST tools for the application of such methods to support "Macro-Architecting" (see Pichler 1998) are necessary. Examples of such complex engineering systems are manifold. The internet and the quality assurance of it with regard to the security of private data or the existing cellular nets for mobile-telephony provide fashionable examples.

To (2): A hierarchical structure of a complex system gives, as we know, an analyst often the chance to apply formal methods in a recursive manner going from one level of the hierarchy to the other. This means that the complexity is reducible and effective deductive methods can be applied. The situation is more complex in the case that the components of a hierarchy are multifaceted and therefor inhomogen. This is the case of intelligent components which have a certain autonomy. Arthur Koestler (1967, 1968, 1978) introduced for such hierarchies the concept of "holons" (describing the components) and also the proper relations between components. He called such a hierarchy a "holarchy". Furthermore he defined by a canon of rules for important properties which a holarchy (Koestler calls it a "SOHO" structure - SOHO stands for Self Organizing Hierarchical Order) is required to have.

We believe that in the future specific systemstheoretical methods for holarchies are needed. With associated CAST tools the engineering quality of the design of "complex distributed intelligent systems" can be improved. We do, however, not believe that evolution mechanisms - comparable to such which are intuitively proposed for a free capitalistic market - will finally sort out bad species in proper time. We rather would like to emphasize the need of rational proofs for the assured quality of such design tasks.

Many activities for a methodological framework and associated tools for such systems are under way (we refer here to the topic of "Multi-Agent Systems", see J. Ferber 1999).

To (3): In the introduction to this paper we pointed out that Systems Theory is essentially a collection of concepts and methods for dealing with multilevel modeling and we emphasized the need for a top down approach. Following the classification as given by Mesarovic we can distinguish between two main kinds of multi-level models: multi-strata models and multi-layer models. Multi-strata models are , as we know, characterized by the fact that the different levels represent the real system which is in discussion by different levels of abstraction. In contrast, in a multi-layer model the different layers model components of the real system and reflect their order in the hierarchy with respect to tasks such as partition of the work load or decision making.

By investigating dynamical systems in "Bourbaki style", as we propose, we understand the aquiring of knowledge to represent dynamical systems in different levels of abstraction and to relate such representations "top down" by proper morphisms (dynamorphisms in the sense of Michael Arbib).

Although the theory of dynamical systems is in mathematics highly developed and ranges from such abstract fields as topological dynamics to rather specialized topics such as dynamical systems generated by linear constant differential equations or finite state machines, the study of dynamorphisms to relate the different possible representations seems to be not developed in what we would like to call "Bourbaki style". There are, however, some important partial results available (e.g. the algebraic theory of linear systems as developed by Rudolf Kalman). However much research work is left to be done.

To stress the importance of this direction of research in Systems Theory and CAST, we would like to give an example for possible applications.

Microsystems, as it is well known, are highly integrated circuits which are based on different technologies such as microelectronics, micromechanics, microacoustics, microhydraulics, micro-vacuum tubes, micro optics and possible others. By these different technologies it is possible to implement different kind of "machines" and to integrate them by coupling to a complex system on a single silicon chip. For modeling a microsystem we need, depending on the different realization technologies for its components and their couplings, a variety of different modeling concepts and tools. The complexity of the modeling process is certainly a new challenge in systems design. The paradigms which are existing in modeling and tool making, as for example, used in microelectronics, mechanical engineering or control engineering, have to be revised and adapted to new needs (DeMan, 1990). In addition to the development of CAD tools for the engineering support of the designer formal mathematical methods and related tools have to be investigated and elaborated. The related research constitutes for the near future an important task in applied mathematics and mathematical systems theory.

The development of a theory of dynamical systems (in "Bourbaki style") and associated CAST-tools can be considered as an important part of research in this direction. It would provide mathematical means to "lift" models which are represented by dynamical systems from a lower (refined) level to a higher (coarse) level by appropriate morphisms. Furthermore such a theory and associated CAST tools would also allow to investigate the possibilities of decompositions of dynamical systems to achieve a refinement transformation from a higher level to a lower level of dynamical systems representation. Both kind of transformations are needed for giving theoretical support to microsystems design. More arguments on this topic may be found in a recent paper of the author (Pichler (1998)).

References

Jamshidi M., Herget C.J. (1985) Computer-Aided Control Systems Engineering. North-Holland, Amsterdam-New York-Oxford

Pichler F., Schwärtzel H. (eds.) (1992) CAST Methods in Modelling - Computer Aided Systems Theory for the Design of Intelligent Machines. Springer-Verlag, Berlin-Heidelberg

Pichler F.(1998) Systems Theory for Macro-Architecting in the Computer- and Information Sciences. In: Cybernetics and Systems'98, ISBN 3-85206-139-3, Austrian Society for Cybernetic Studies, Vienna, R. Trappl (ed.), pp. 50-53

Koestler A. (1967): The Ghost in the Machine. Hutchinson&Co Ltd., London

Koestler A., Smythies J. R. (eds.) (1969) Beyond Reductionism. New perspectives in the life sciences. The Alpbach Symposium 1968. Macmillan Company, New York

Koestler A. (1978) Janus. A Summing Up. Hutchinson&Co Ltd., London

Ferber J. (1999) Multi-Agent Systems: An Introduction to distributed artificial intelligence. Addison-Wesley, Reading Massachusetts

Pichler F.(1999) Arthur Koestler's Holarchical Networks: A Systems-theoretical Approach. In: Publikation des Schweizerischen Wissenschaftsrates, Thomas Bernold (ed.) (in press)

DeMan, H. (1990): Microsystems: A Challenge for CAD Development. In: Microsystems Technologies 90, H. Reichl ed., Springer Verlag Berlin, pp. 3-8

Pichler F. (1998) Design of Microsystems: Systems-theoretical Aspects. In: Systems: Theory and Practice, R. Albrecht (ed.), Springer Verlag Wien, pp. 107-115

Cast Methods in Biocybernetics

R. Moreno-Díaz

Instituto Universitario de Ciencias y Tecnologías Cibernéticas
Universidad de Las Palmas de Gran Canaria. Campus de Tafira
E-35017 Las Palmas de Gran Canaria. Spain
Tel. +34 928 458751, Fax: +34 928 458785
rmoreno@ciicc.ulpgc.es

1 The Classics

The systematic use of what we now call Systems Theory in the description of biological systems, and more precisely, the nervous system, took off in the Forties although many of the basic ideas had been being managed in philosophic and scientific circles almost since the Ancient Greeks. From 1943 to 1945, a kind of synergetic process was started up, triggered as the result of three basic works. First, Norbert Wiener, Arthur Rosemblueth and Julian Bigelow's study (1943) on the nature of teleological processes where the crucial idea was that what was relevant in a homeostatic process was the *information* return and not the *energy* return via the feedback links. It is representative of the analytical approach.

Following this, came the work of the young British philosopher, Kenneth Craick, published in the form of a small book called *On the Nature of Explanation* in 1943, a pursuit of a Theory of Knowledge which would be contrastable like any other Natural science. Craick offered a clear and powerful frame work within which to express the acquisition, processing, storage, communication and use of knowledge, through a type of behavioural approach.

Third, the work of Warren McCulloch and Walter Pitts, *A Logical Calculus of the Ideas Immanent in Nervous Activity*, which was published in 1943. They elaborated the concept of a "formal neuron" and came to the final conclusion that a network of formal neurons, with an effective infinite memory tape, can compute any number, which is computable by a Turing Machine. It corresponds to the logical approach.

From these studies, the systems theoretical approach evolved with powerful input from the Theory of Communication of Shannon and key figures in the field of Computer Science such as Von Neuman, in the case of the latter, with application to questions of computability, performability, capacity for reproduction and reliability of functioning.

2 The Logical Approach

The original paper of McCulloch and Pitts in 1943 is the first illustration of how logical tools can be used to describe properties of the nervous system. Strictly, by the introduction of a very simple counterpart of a neuron the first explanation of automata in terms of neurons was achieved.

F. Pichler, R. Moreno-Díaz, and P. Kopacek (Eds.): EUROCAST'99, LNCS 1798, pp. 8–13, 2000.

The formal theory was completed by the introduction of the interaction of afferents (logical counterpart of the so-called presynaptic inhibition) and later by structuring the nets in a hierarchy, to account for the neuronal synthesis of arbitrary probabilistic automata.

From the propositional logic realization point of view, a formal neuron is a threshold unit, which fires or not at discrete time t+1 if the weighted addition of input signal values is equal or larger than a threshold at time t. That is, is strictly a linear logical circuit component, and two layers of formal neurons are necessary for the realization of any arbitrary logical function, or in McCullochs words, to realize an arbitrary proposition of the input arguments.

In the fifties, there was a preoccupation with the questions about reliable computation and computation in the presence of noise, which gave as result the introduction of the probabilistic formal neuron, in which each input configuration has only a certain probability to fire the neuron. Also, fluctuating thresholds were considered. The "logical stability" of these networks was the addressed problem to find a model for reliable computation with unreliable units.

In the early 60´s a new biological paradigm, the so-called presynaptic inhibition, served as biological counterpart to introduce the interaction of afferents in formal neurons, so that a single of these revised unit is then able to compute any logical function of the input arguments. Formal neurons came then to be logical universal units.

The theory developed, so that a much more transparent relation between automata theory and formal neural nets was obtained by the late Sixties. As a by-product of the analysis of stability and oscillations in formal neural nets, it was found that a fixed structure (anatomy) can compute all the possible patterns of oscillation for N formal neurons, and that each pattern can be evoked by an external input, serving as an example of dynamic, associative memory. It was obtained that any deterministic automaton can be realized by a neural net and vice versa, by relatively simple constructive rules. The extension to arbitrary probabilistic automata or Markov chains was worked out in the Eighties which was based on the 'axon axonal interaction'. By then, the theory of formal neurons, as it came from formulations of prepositional logic, was practically finished, except for new formulations to introduce fuzzy concepts and the problem of "contingency", which appears also in artificial vision systems. Considering the 'automata-neural net duality', this problem appears because small changes in the network structure may provoke large changes in the state-transition matrices of the probabilistic or deterministic automaton, that is, in the automata behaviour, and vice versa. There is still a challenge in appropriately introducing learning in Formal Neurons, and to find equivalent analysis and synthesis constructive theorems to go from continuous state variable automata to networks and vice versa, all from the point of view of the biological counterparts.

Formal neural nets in their different formulations, are of particular interest to find connectivistic or granular representations of behaviour_formulated at the automata language level, to find potential explanatory diagrams for natural neural nets and to provide for preliminary connectivistic mechanisms when modeling natural neural nets

A different trend is that rooted in Rosemblatt's concept of Perceptron which had an enormous upraise in the 80´s, the so called artificial neural networks. This went, however, away from neurosystems descriptions to form the parallel field of distributed granular computation.

3 Analytical Approaches

Analytical models and theories of parts of the nervous system can be developed when treating the front (sensory and effector) ends of the nervous system, where the level of symbolic complexity is still low. That is the case for the visual processing in the vertebrate retina, where more or less classical systems approach gives good description.

Probably the most provocative single paper on the neurophysiology of a vertebrate visual pathway, that fired a legion of modelling efforts, has been that on the frog by Maturana, Lettvin, McCulloch and Pitts, in its two versions, engineering and anatomico-physiological . In fact, the behavioural connotations of the interpretations and conclusions of the authors inspired models and disquisitions far beyond from the crude "facts", that is, the nature of the neurophysiological signals as recorded from single fibers.

Soon after, an program was started at MIT to develop computer programs which processed visual data from a vidicon camera (actually, the images were simulated, not real) with the aim to duplicate the described behavior of the retinal cells. The modelling was been developed, however, practically with no detailed anatomical or physiological support, and only driven by the desired results. A completely satisfactory model was not possible until the analytical approach was taken, in which all the proposed mechanisms underlying the behavior of the model had anatomical and physiological justifications.

Lettvin, Maturana, and co-workers distinguished four major groups of retinal ganglion cells which report to the tectum. These have been designated as follows: Group 1 -edge detector, Group 2 -Bug detector, Group 3 -dimming detector, and Group 4 -event detector ganglion cells. Of these, relatively simple explanations can be given to the operations of the Group 1, 3 and 4 ganglion cells. The Group 2 or bug detector ganglion cell, however, is more intricate and the most exciting cell to model because it is sensitive to small dark convex objects which move towards the centre of the responsive retinal field (RRF) of this cell. In essence, it is the most specialized pattern recognition cell of the frog's retina. The models provide in fact for a fair illustration of how to proceed in analytical and neurophysiologically based models, by starting with a list of properties, or a list of specifications which the model should meet, and a list of the anatomical and neurophysiological restrictions in order to minimally converge in an optimum model.

The analytical approach can be extended to higher vertebrate retina, by the introduction of the concepts of fast and delayed local signals at different retinal levels, lateral non linear spatial interaction, local non linear operations and integration at the cell body. The typical tools of systems engineering can in this way be successfully applied to models of cat's avian and amphibian retina, leading to a generalized model for vertebrate retinal processing.

The most recent illustration of the analytical techniques cames from the explanation of the microstructure of receptive fields of retinal cells by means of discrete Newton filters and Hermite weighting functions, which provide for a very appealing system representation of dendro-dendritic interaction.

4 The Multilevel Structure of Systems Tools

Models, theories and formal representations and descriptions of real nerves and neurons have been dependant on the type of formal tools available at each time, so that even Descartes explained nerve action in terms of fluidodynamics. Only in the 1930's have strong enough system tools become available to attempt the system's description of sensory nets.

Besides formulations on the biophysics of membranes, the description of functions on neural nets took off significantly after the 1950's. One of the first models of a neural net soundly proposed is precisely by Pitts and McCulloch, to recognize auditory and visual forms independent of position and size. Later in the 60's, models of the action of the reticular formation neural nets of the vertebrates were proposed, where the neurons operate as rather complex pools, governed by probabilistic action.

The most successful formal descriptions of real, more or less complicated neural nets are those of the retinal neural nets, which range from very complex operation in the case of specialized lower vertebrate retina to the more simple system description applicable to simple neurons in higher vertebrates. These nets take into account the layered structure of the system, with forward propagation of signals being processed, and very strong lateral interconnections.

In the 80's some rather sophisticated neuron models were proposed, with neuronal action already using a conditional, rule-based action. Later, the concept of a frame came up from Artificial Intelligence methods. One peculiarity of these nets is that the data processed by the units are taken-- as corresponding to anatomy--from certain "receptive fields", which, in general, transform into a "data field" within the "input space" and "output space" to produce new output-processed data. These conceptual models had been of use for models of higher level visual processing, association like neural processing, and reliability of neural tissue.

Naturally, different levels of description of a real neural net require different levels of tools (see Table 1). One of the main problems in the formal description of real nets is that deeper layers of neurons, i.e. more central nets receive inputs the code of which is unknown, and which carry a heavy load of semantics. Also, the degree of cooperativity and the subsequent degree of reliability increases. That is, when going from receptors to cortex, the semantic complexity increases, as well as it decreases when going from cortex to motor and other effector ends.

Thus, at the level of neurotransmitters, membrane phenomena and action potentials, the appropriate tools are those of biochemistry and biophysics. At the level of simple sensory neural codes and multiple coding, signal processing tools are appropriate. For visual coding in the retina and decoding in effectors (motor, glandular), the level tool is that of the classical systems theory. Finally, the level of neural input-output interaction and coordination, central neural code, generation of universals (invariants) and social-like interaction of neural nets, the level tools of Algorithmic, Symbolic and other from Artificial Intelligence became necessary.

Table 1.

LEVEL	TOOLS OF
Neurotransmitters, membrane phenomena, action potentials	Biochemistry-Biophysics
Biophysics of neural codes and multiple codes	Biophysics-Signal Processing
Sensorial codes, decoding in effectors (motor and glandular), coding in retinal ganglion cells	Classical System Theory
Neural nets, input –output interaction and coordination	Algorithmic (Logic, Symbolic)
Central neural code, cooperative processes, generation of universals, social-like interaction of pools of neurons	Symbolic, A.I. Techniques

Concepts derived from the engineering of complex robotic systems are of help in understanding the control and command mechanisms of the nervous system. For high level modeling and description of neural nets it seems nowadays that the most sophisticated tools of intelligent, goal-based agents must be used to obtain relevant non-trivial neural net formal descriptions.

5 Selected References

Lettvin, J.Y.; Maturana, H.R.; McCulloch, W.S.; Pitts, W.H. (1959): "What the frog's eye tells the frog's brain". Proc IRE 27, pp 400-415.

McCulloch, W.S. and Pitts, W.H. (1943): "A logical Calculus of the Ideas Immanent in Nervous Activity". Bull. Math. Biophysics, 5, pp 115-133.

McCulloch, W.S. and Moreno-Díaz, R. (1967): "On a Calculus for Triadas", in "Neural Networks", Caianiello, ed. Springer-Verlag Berlín, Heidelberg. New York, pp 78-86.

McCulloch, W.S. (1969): "Regenerative Loop". The Journal of Nervous and Mental Disease, 149, pp 54-58.

Mira, J.; Moreno-Díaz, R.; Cabestany, J. eds. (1997): Biological and Artificial Computation: From Neuroscience to Technology. Springer Verlag. Berlin. Lecture Notes in Computer Science no.1240.

Mira-Mira, J., Sánchez-Andrés, V. (Eds.) (1999). Foundations and Tools for Neural Modeling Lecture Notes in Computer Science No. 1606 and Engineering Applications of Bio-Inspired Artificial Networks. No. 1607. Berlin: Springer.

Moreno-Díaz, R. and McCulloch, W.S. (1968): "Circularities in Nets and the Concept of Functional Matrices" in L. Proctor, ed. Biocybernetics of the C.N.S. Little and Brown. Massachusetts., pp145-150

Moreno-Díaz, R. (1971): "Deterministic and Probabilistic Neural Nets with Loops". Mathematical Biosciences 11, pp 129-136.

Moreno-Díaz, R. (1995). "Natural and Artificial Neural Nets". in Genetic Algorithms in Engineering and Computer Science (Winter, Périaux, Galán, Cuesta eds.), Chichester, England: John Wiley and Sons, pp83-110

Moreno-Díaz, R.; Mira-Mira, eds. (1995): Brain Processes, Theories and Models. The MIT Press. Cambridge MASS. USA.

Neumann, J. von (1956): "Probabilistic Logics and the Synthesis of Reliable organisms form Unreliable Components". in Automata Studies, Shannon and McCarthy, eds. Princenton University Press, Princenton, New Jersey.

Sutro, L. (1966): Sensory Decision and Control Systems. Instrumentation Laboratory R-548. MIT, Cambridge, MA.

Wiener, N. (1948): Cybernetics. The Technology Press. John Wiley and Sons, Inc. New York.

On the Way to the Next Generation of Robots

Peter Kopacek
Institute for Handling Devices and Robotics
Vienna University of Technology
Favoritenstraße 9-11, A-1040 Vienna, Austria
Tel: +43-1-58801-31800, FAX: +43-1-58801-31899
kopacek@ihrt1.ihrt.tuwien.ac.at
http://www.ihrt.tuwien.ac.at/IHRT/

Abstract. The field of robotics is one of the most innovative in the last decade. Conventional industrial robots from the late 70´s are now only a tool on the production level. One of the oldest dreams of the robotic community – intelligent, mobile and humanoid robots – starts to become a reality because of the rapid development of "external" sensors.

External sensors (e.g. visual, auditive, force-torque) offer intelligent robots the possibility to see, hear, speak, feel, smell like humans. Compared with conventional, unintelligent, industrial robots, intelligent robots can fulfill new, innovative tasks in new application areas.

1 Introduction

Industrial robots have been widely applied in many fields to increase productivity and flexibility and to help workers from physically heavy and dangerous tasks. From similar aspects the need on robots in service sectors - like robots in hospitals, in households, in amusement parks - is rapidly increasing. Cheap and accurate sensors with a high reliability are the basis for „intelligent" robots. For these intelligent robots known, conventional but complex applications are now possible to be accomplished as well as new applications are available not only in industry.

There are three "starting" points for the development of intelligent robots:
(a) Conventional, stationary industrial robots;
(b) Mobile, unintelligent platforms (robots);
(c) Walking machines.

In all three cases unintelligent robots were equipped with external sensors step-by-step. This was possible because of the rapidly decreasing prices of the hard- and software for these sensors during the last years. Furthermore it was necessary to create new user interfaces. Users of unintelligent, industrial robots were persons with some technical knowledge. This is and will be only partially the case for intelligent robots for unconventional applications. Therefore modern human machine interfaces (e.g. speech programming) must be developed.

F. Pichler, R. Moreno-Díaz, and P. Kopacek (Eds.): EUROCAST'99, LNCS 1798, pp. 14–19, 2000.

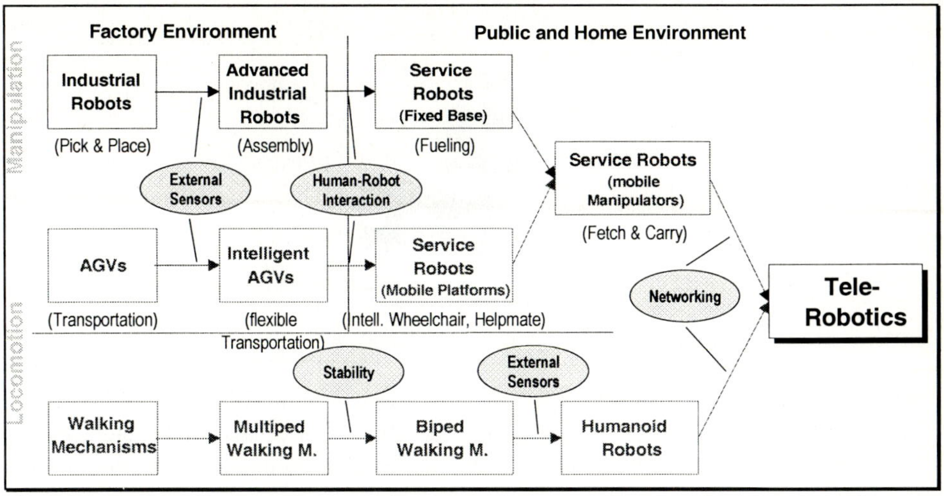

Fig. 1. From Industrial to Service Robotics [4]

2 Intelligent Robot Systems

2.1 Intelligent Stationary Robot Systems

First ideas to create intelligent robots were to attach external sensors on conventional, stationary, classical robots. One examples for some first trials are assembly robots equipped with vision systems for part recognition. Such robots with additional force-torque sensors are also necessary for disassembly operations. One of the newest applications of stationary, intelligent robots is automated car fuelling. The robot is responsible for opening the tank and filling in the petrol.

2.2 Intelligent Mobile Robot Systems

Mobility is one of the main features of an advanced robot. Mobility can be carried out by wheels, chains, vacuum sucks, and others. Movable unintelligent robots are known since some years mostly used for transportation tasks in factories on the shop floor level. But they were guided by wires in the floor and therefore only partially "mobile"; like a car on a street. Mobile robots today are equipped with various external sensors (visual, auditive, proximity, ...). This development was economically possible because of the decreasing prices of hard- and software of such sensors. Therefore mobile robots today are able to find an optimal way, in an unknown environment, by means of sensors. The main parts of such a mobile robot are: body, drives, power supply, moving devices, sensors, on-board computer, operating panel / communication module, safety devices (Fig. 2).

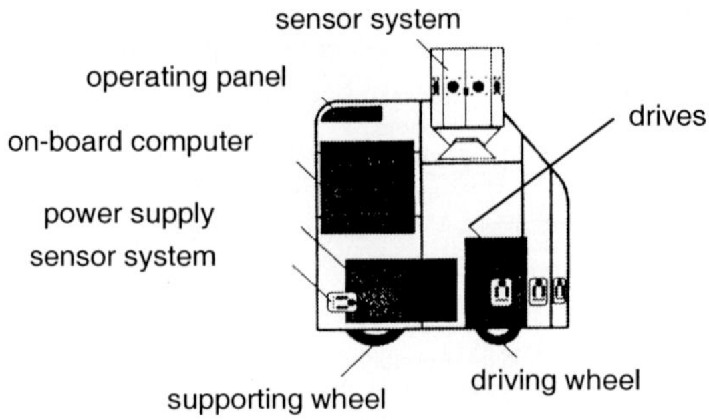

Fig. 2. Main components of a mobile robot platform.

A common problem of all these robots is power consumption. The batteries available today are very heavy (approximately 40 – 50% of the total weight of the robot) and have only a limited capacity (5min – 1h operating time; depending on the tasks to be carried out).

For control tasks, such mobile robot systems are usually equipped with an on-board computer (fast 586 or similar) and connected by wire or wireless with a stationary host computer. Connection by wire reduces dramatically the mobility of the robot. The drives are usually AC/DC motors. Safety devices (e.g. bumpers, ...) are necessary in cases of sensor failures.

Intelligent mobile robots are commercially available today from different producers in various dimensions, for various purposes to reasonable prices [3]. They are in most cases single purpose devices – dedicated for distinct tasks in distinct application fields. This is a similar development to industrial robots in the late 70ies to "single purpose robots". Because of the relatively high development costs of such a robot only few companies worldwide are willing to take this risk. That is one of the main reasons for the relatively low number of intelligent, mobile robots applied in industry. In the future it will be necessary to create intelligent, mobile, multi-purpose robots. One possibility could be a modular hard- and software system to combine these modules for special applications. First steps in this direction are in progress now.

2.3 Walking Machines

Walking mechanisms are a classical research field in kinematics since some decades. A lot of suggestions were made to imitate animal and human legs. First developments were multi-legged robots with 4, 6 or more legs. But walking of humans on two legs is - from the viewpoint of control theory - a complex, nonlinear stability problem. Now we have the necessary computer hard- and software available to solve this problem on line. We are on the way to so-called humanoid robots.

2.4 Robots in Entertainment

One of the newest application areas of service robots is the field of entertainment, leisure and hobby because people have more and more free time. In addition modern information technologies lead to loneliness of the humans (telebanking, teleshopping, and others). Therefore service robots will become a real "partner" of humans in the nearest future. One dream of the scientists is the "personal" robot. In 5, 10 or 15 years everybody should have at least one of such a robot. Because the term personal robot is derived from personal computer the prices should be equal.

Robot Soccer. From the viewpoint of multi-agent systems, a soccer game is an appropriate example of the problems in real world, which can be moderately abstracted. Multi-agent systems deal with research subjects such as cooperation protocol by distributed control, effective communication and fault tolerance, while exhibiting efficiency of cooperation, adaptation, robustness and being in real-time – playing robot soccer offers a perfect testbed for research in these fields.

Depending on the category, a robot soccer team consists of one or three robots. For the current "major league" – MiroSot - each player's mechanism and control must be packed into a cube no larger than 7.5 centimeters on a side (Fig. 3). The 130 x 90 centimeters playing field is bounded on all sides to prevent the ball – an orange painted golf ball – from going out of play.

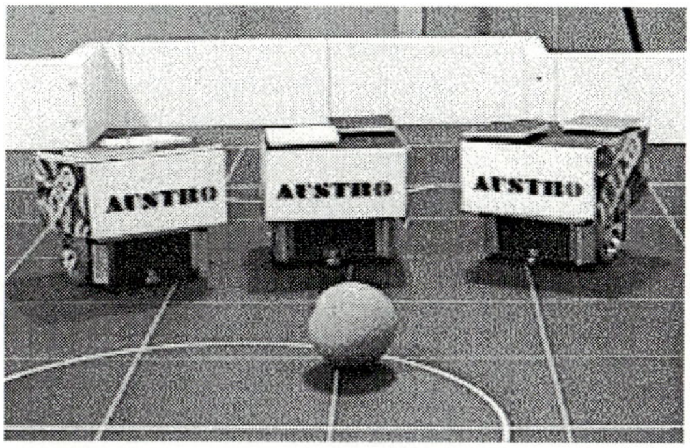

Fig. 3. IHRT Robot Soccer Team "AUSTRO" – European Champion 1999

The teams gather information about the location of the players and of the ball from small video cameras suspended above the playing field. That information goes first to a central control unit ("Host") and then – as a set of command data – by radio to the robots themselves. For such a "Vision-Based Robot Soccer System" described above, the Host is responsible for the calculation of the robots behavior. Nevertheless, development is going in direction of a "Robot-Based Robot Soccer System", where each robot hosts all the functions for fully autonomous behavior. All calculations are done locally in each of the robots – the host computer (if any!) processes vision data and forwards position information to the robots. This can be considered as a distributed control system, where each robot has its own intelligence.

3 Mobile Robot Systems at IHRT

3.1 MaxiFander (Denning Branch International Robotics)

This robot platform has been designed especially for research and education. The option of mounting heavy equipment (up to 25 kg) on the robot such as robot arms or even a PC chassis as well as possible handling at outdoor conditions, like rough, uneven surfaces, gravel, etc., together with the very simple structure of the robot makes it to an excellent tool for education in robotics. Together with the wide array of sensors provided by the robot (rotating sonar transducer, infrared proximity detectors, touch sensors, microphones, optical line followers), MaxiFander allows a comprehensive view of mobile robot control techniques. The robot comes with onboard 386DX microprocessor (486DX upgrade is also available) and source code is written in programming language C++ - download of the application programs is possible by means of floppy disk or serial interface. For input/output, the robot is equipped with a handheld numeric keypad with 16x2 alphanumeric LCD display.

3.2 Nomad 200 (Nomadic Technologies, Inc.)

Nomad 200 is an integrated mobile robot system with different sensing options. Besides tactile (sensor ring with 20 independent pressure sensitive sensors), 16 channel infrared, 16 channel ultrasonic, and 2D laser sensor systems the platform is being controlled by means of an on-board multiprocessor system, which performs sensor and motor control, host computer communication, and supports on-board programming. The drive system of the platform utilizes a three servo, three wheel synchronous drive mechanical system which provides a non-holonomic (with zero gyro-radius) motion of the base and an independent rotation of the upper structure.

Primarily, both robots are used for the laboratory courses in robotics. In addition, the platforms are the basis of some research and development work, among them:

Development of a Task Oriented Language for Mobile Robots. The general idea is to provide with a programming language interpreter which should be able to deal with unexpected events that are normal to occur in the execution of a robot task, such as meeting an obstacle. Also the interpreter must be able to provide priority based scheduling of surveillance monitors when several events take place simultaneously.

Intelligent path planning and collision avoiding algorithms using Neuro-Fuzzy approach. Two different strategies, the optimization of Fuzzy systems by means of Neural Networks as well as training of Neural Networks with Fuzzy mechanism, are being implemented to MaxiFander platform.

Integrated Robot Navigation in CAD-Environment. This navigation system is being designed for both application of mobile systems in unknown environment as well as for pre-defined surroundings. A special Graphical User Interface (GUI) - following the specifications of the chosen CAD package AutoCAD - should provide a convenient access to the real as well as to the simulated robot, and to the representation of the environment. Through this GUI, the user can send commands to the robot, monitor command execution by seeing the robot actually moving on the screen, visualize instantaneous and cumulated sensor data. The user should also be able to create and modify a simulated environment by means of standard CAD functions, and use it to test robot programs. Programming of the robot platforms is accomplished by means of a common Meta-Language – an Interpreter kernel executed by the on-board control system on each platform transforms the commands to the particular robot programming language.

4 Conclusions

Future application oriented research in robotics is dominated by two main directions. Robots for classical applications have to be equipped with additional features like combined force and position control, external sensors based on microsystems, flexible and lightweight robots. Because of the decreasing number of installed robots new application fields will be recognized. One of these fields are the service robots. Service robots look quite different than conventional ones and therefore research have going on in additional directions such as external sensors, new grippers and gripping devices, new kinematic structures. Efforts have to be undertaken to further development of key components of these robots towards efficiency, performance, miniaturization and cost. Here the collaboration of research institutions, service industry and robot and component manufacturers has the potential to create valuable synergies.

Last developments tends towards humanoid robots and "Multi-Agent-Systems – MAS". Agents are similar to "holons" offering the possibility to apply MAS in product automation.

References

1. Dillmann, R., Rembold, U., Lueth, T. (ed.): Autonome Mobile Systeme 1995. Springer-Verlag, Berlin Heidelberg New York (1995)
2. Kopacek, P.: Developing Trends in Manufacturing Automation. Proceedings of the 5th Symposium on Automatic Control and Computer Science, Vol. 1. Iasi, Romania (1995) 43-53
3. Kronreif, G., Probst, R., Kopacek, P.: Modular Service Robots - State of the Art and Future Trends. Proceedings of the 8th International Conference on Advanced Robotics ICAR'97. Monterey, USA (1997) 51-56
4. Schmidt G.: Unpublished transparency (1996)

2 CONCEPTUAL FRAMEWORKS, METHODS AND TOOLS

Representation of the RCS Reference Model Architecture Using an Architectural Description Language

Elena Messina, Christopher Dabrowski, Hui-Min Huang, and John Horst

National Institute of Standards and Technology,
Gaithersburg MD 20899, USA
{emessina | cdabrowski | horst | hhuang}@nist.gov

Abstract. The Real-Time Control System (RCS) Reference Model Architecture provides a well-defined strategy for development of software components for applications in robotics, automated manufacturing, and autonomous vehicles. ADLs are formally defined languages for specification of software system's designs. In this report, we describe the results of an investigation into the use of an ADL to specify RCS software systems, and assess the potential value of ADLs as specification and development tools for RCS domain experts. The report also discusses potential influence of ADLs for commercial software development tools and component-based development.

1 Introduction

Architectural Description Languages (ADLs) are specification languages for rigorously describing and analyzing software system designs. This report provides the results of an investigation into the use of ADLs for formally defining the National Institute of Standards and Technology (NIST) RCS Reference Model Architecture [5].

The RCS Reference Model Architecture provides well-defined guidelines for construction of control software for autonomous real-time systems. We are studying means of formally representing architectures such as RCS in order to facilitate development, understanding, and analysis of complex systems. Communicating RCS in an unambiguous manner was our initial motivation. Beyond that, several potential benefits may accrue. ADLs may provide means of guiding construction of complex systems according to a given reference architecture. This can enhance productivity, reliability, and help ensure conformance to a specified architecture. Analysis tools associated with ADLs may enable developers to study the behaviors and performance of their system design. Although our study focussed on the RCS architecture, we believe that our results are applicable to other architectural models for complex systems.

All significant ADLs were reviewed in a literature search that identified key language features relevant to RCS. Individual ADLs were examined in detail to assess their suitability as specification languages for capturing the structure and

F. Pichler, R. Moreno-Díaz, and P. Kopacek (Eds.): EUROCAST'99, LNCS 1798, pp. 23–37, 2000.

function of the RCS Control Node. A detailed, comparative analysis of ADL features was not the scope of this study; readers interested in such an analysis should consult [16]. A single ADL—*Rapide* [15]—was selected to construct a prototype specification of a significant portion of the RCS Intelligent Control Node. The conclusions from out experiment provide a basis for both identifying requirements for ADLs to specify RCS software architectures and components as well as recommending future research directions for ADLs. We believe that our findings can be relevant to the application of ADLs to other complex, real-time architectures.

2 The RCS Reference Architecture

2.1 Overview of RCS Concepts

Developed over the course of two decades at the National Institute of Standards and Technology and elsewhere, RCS has been applied to multiple and diverse systems [4]. RCS application examples include coal mining automation [11], the NBS/NASA Standard Reference Model Architecture for the Space Station Telerobotic Servicer (NASREM) [1], and a control system for a U.S. Postal Service Automated Stamp Distribution Center [26]. Manufacturing applications include the Open Architecture Enhanced Machine Controller [3]and an Inspection Workstation [18].

RCS provides a reference architecture and an engineering methodology to aid designers of complex control systems. Guidelines are provided for decomposition of the problem into a set of control nodes, which are arranged hierarchically. The decomposition into levels of the hierarchy is guided by control theory, taking into account system response times and other factors, such as planning horizons. RCS is focussed primarily on the real-time control domain. It can be further specialized into application-specific versions. 4-D/RCS [5] is one such version, which is aimed at the design and implementation of control systems for intelligent autonomous vehicles for military scout missions. 4-D/RCS has been selected as the software architecture for Department of Defense Demo III eXperimental Unmanned Vehicle (XUV) Program, managed by the Army Research Laboratories [23]. This particular flavor of RCS was studied with respect to ADLs. Figure 1 is a high-level diagram depicting a portion of an RCS hierarchy for an autonomous vehicle. Each node in the hierarchy is built upon the SP-WM-BG-VJ internal elements shown in Figure 2. RCS prescribes a building-block approach to designing and implementing systems. The building blocks are control nodes that contain the same basic elements. The elements, shown in Figure 2, are behavior generation (BG), world modeling (WM), value judgement(VJ), and sensory processing (SP). Associated with WM is a Knowledge Base (KB), which contains longer-term information. Each node receives goals from its superior and, through the orchestration of BG, WM, JV, and SP, generates a finer resolution set of goals for its subordinate nodes. The RCS control node uses an estimated model of the world, generated via SP and WM, to assess its progress with respect to the

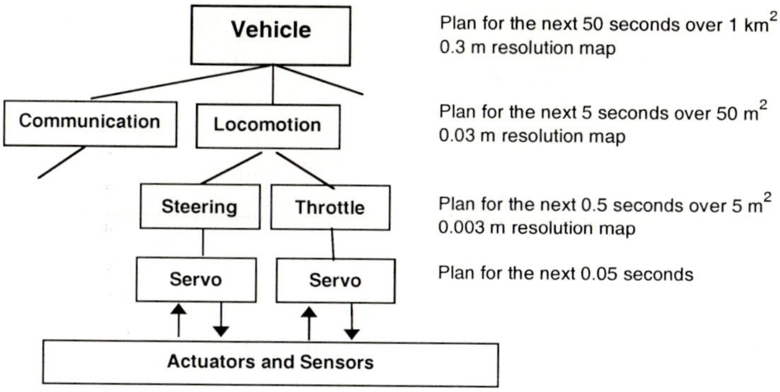

Fig. 1. Example RCS Hierarchy for an Autonomous Scout Vehicle

goals it was given and to make necessary adjustments to its behavior. BG's submodules are the job assigner (JA), a set of plan schedulers (SC), a plan selector (PS), and a set of executors (EX). One SC and EX exist for each subordinate controlled by a particular RCS node. JA decomposes incoming commands into job assignments for each of its subordinates. Each SC computes a schedule for its given assignment. JA and SC produce tentative plans based on available resources. PS selects from the candidate plans by using WM to simulate the execution of the plans and VJ to evaluate the outcomes of the tentative plans. The corresponding EX executes the selected plan, coordinate actions among subordinates and correcting for errors between the plan and the state of the world estimated by the WM.

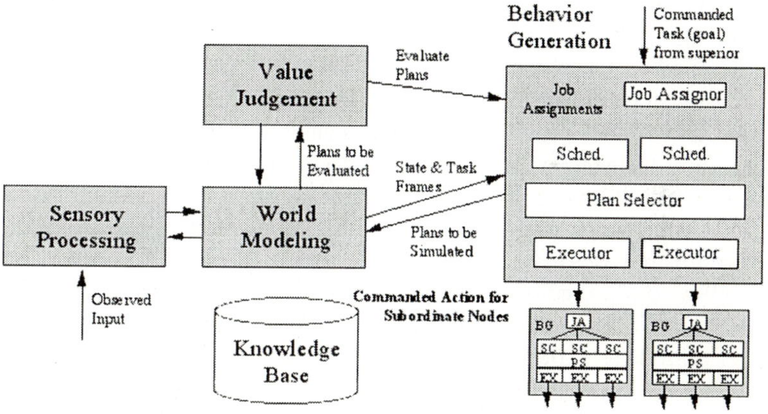

Fig. 2. Model for an RCS Control Node.

3 Architectural Description Languages

3.1 Overview of Architectural Description Languages (ADLs)

Garlan and Perry define a software architecture as consisting of the "structure of the components of a program/system, their interrelationships, and principles and guidelines governing their design and evolution over time" [9]. An ADL is "a language that provides features for modeling a software system's conceptual architecture" [16].

ADLs provide language constructs for specifying the essential elements of a system's software architecture. A generic set of ADL capabilities has been identified in [10][16][28]. ADLs commonly describe software components by defining their interfaces and behavior in response to externally or internally generated stimuli. An interface definition may include a *signature*, i.e., messages and commands received and sent, as well as constraints on the signature.

Some ADLs support specification of computations performed by a system, referred to as system behavior. Usually, an ADL employs a formally defined descriptive method or underlying *computational* model to provide the necessary semantics. Constraints on behavior are also defined in terms of the computational model. Examples of computational models are Finite State Machines (FSM), Communicating Sequential Processes (CSP) and Partially-Ordered Sets of Events (POSETS). An ADL may allow description of the behavior of component interfaces, component internals, and component connections.

Shaw [22] and Allen [6] provide rules or constraints that place limitations on how components may be connected and what system topologies may be described. One example of an *architectural style* is a top-down hierarchical architecture. Some ADLs allow explicit declaration of architectural styles.

The use of a well defined, rigorous specification language provides a basis for formal analysis of a specification and the verification of software system designs. Some ADLs employ formal proof techniques to determine whether desirable properties, such as internal consistency, hold within a specification. Analysis of ADL specifications may also take place through simulation support tools, which allow the specification to be executed and a result to be computed, thus simulating the computations to be performed by the system being specified.

Gaps in the support provided by object-oriented tools and methodologies [13] further stimulated interest in the potential of ADLs. Object-oriented methods in general are data-centric, providing only for some generic behavior description capabilities. Analysis of the architecture and simulation of the execution of an architecture are not possible in most object-oriented tools. In response to these gaps, the Object Management Group recently issued a Request for Proposals under the title "UML Profile for Scheduling, Performance and Time" [21]. This proposal is aimed at expanding the UML to include support for modeling of time-related paradigms, which are essential for the design and specification of real-time systems.

3.2 The *Rapide* ADL

Rapide [15] is an ADL and supporting tool set developed at Stanford University in the mid-1990s. This ADL was chosen as the primary focus of this study because of its well-developed capability for representing and simulating real-time system designs.

Rapide supports most of the features described above that are common to ADLs. *Rapide* permits definition of a set of component *interface* types, each of which has a signature that includes events generated and received by components of that type and a description of the component's behavior. An interface may also define *constraints* that require dependencies between events, place limitations on the order of events, constrain parameter values, or make other limitations. The internal details of the components themselves—known as *modules*—may also be specified. A module description specifies internal behavior and supporting data structures that allow the module to conform to its *interface*.

The software architecture is formally described by connecting types of events generated in one *interface* specification to events received by another *interface*. A *module* conforming to an *interface* may be decomposed into a sub-architecture consisting of a set of connected component *interfaces*.

Connections between types of events of different *interfaces* and the specification of a component's behavior define causal dependencies of the events. During the simulated execution of a software architecture, these dependencies can be aggregated to form POSETs, or partially ordered set of events.

An event is said to be causally dependent on all events that either directly result in its generation or in the generation of its predecessors. It is independent of all other events. In actual *Rapide* specifications of architectures, very large causal sequences of event types and event constraints can be defined both in interface definitions and as part of connections between interfaces. The causal sequences serve as a basis for "executing" a specification using *Rapide* software support tools to produce simulations.

In *Rapide* the POSET is the basis for automated analysis conducted by an associated toolset. A *Rapide* specification may be defined using the RAPARCH tool, which has a graphical front-end, to specify interfaces and interface connections in a software architecture. When compiled and executed, the *Rapide* specification produces a POSET for the defined architecture. *Rapide* provides a simulation tool called RAPTOR for producing an interactive graphical animation of the execution of the specification in which interfaces and connections are depicted as icons while event icons move between interfaces. The POSET Viewer gives a static picture of a POSET with events and causal arrows between events. Query functions can be used to select interesting subsets of the POSET and provide detailed information. A method is provided for verifying system designs against a more general Reference Model architecture based on comparison of POSETs.

4 The Experiment

In order to help answer the questions about the applicability of ADLs to RCS, the *Rapide* ADL was used to specify a large piece of the RCS Intelligent Control Node. The specification was developed by two of the coauthors: one focusing on the study of ADLs; and the other, a domain expert in design of RCS systems who regularly reviewed the model and guided its evolution. The specification was reviewed and verified by a larger group of experts in RCS. In addition, the use of an ADL to ascertain conformance of individual system designs to the Reference Model Architecture was examined. A detailed description of the experiment, including the *Rapide* source can be found in [7].

Overview of the Prototype Specification Component interfaces were defined for each 4-D/RCS Intelligent Control Node together with the events handled, sent, and received and applicable constraints for the module. The specification provided the decomposition of the Control Node into its major subcomponents. Behavior Generation was further decomposed into Job_Assigner (JA), a set of Schedulers (SC), a set of Executors (EX), and a Plan Selector (PS). World Modeling (WM) was decomposed into Simulation and Knowledge Base components. The architecture specification included the connections between the interfaces defined for the modules. A sufficient amount of behavior was included to allow the architecture to be simulated using the *Rapide* toolset. The entire specification encompassed more that 1000 lines of *Rapide* code.

Details of Job_Assigner and Scheduler Functions The use of ADLs to specify RCS is illustrated in a sample *Rapide* description of the interaction of two subcomponents of the RCS Behavior Generation Module: The Job_Assigner and a Scheduler (of which there may be several instances). The conceptual design for this representative fragment of the Reference Model functionality is shown graphically in Figure 3. The fragment contains only a subset of the actual events and behavior defined for these components. The specifications of algorithms for computing schedules and selecting plans in underlying modules are omitted from the Reference Model Architecture because they are application-specific.

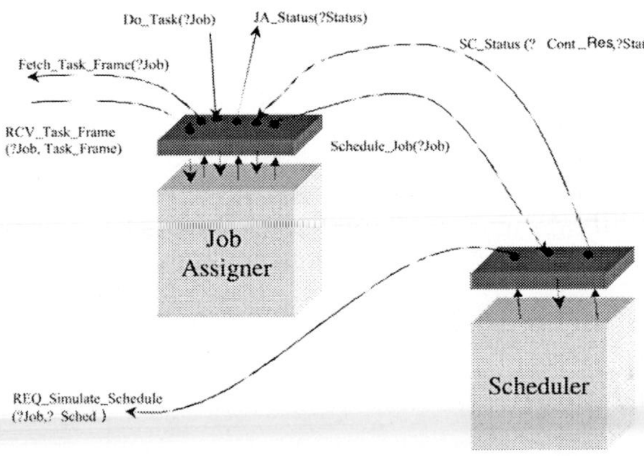

Fig. 3. Job_Assigner and Schedulers in the Behavior Generation Module

Figure 3 shows a *Job_Assigner* component defined as a *Rapide* interface. The *Job_Assigner* interface signature receives a Do_Task event representing an input task in which ?Task is the argument variable for a task name. The *Job_Assigner* generates a *Fetch_task_frame* event with the job name as an argument that is passed to the *World_Modeling* module. *World_Modeling* returns a task frame data structure containing information necessary to perform the task received by *Job_Assigner* as a *RCV_Task_Frame* event. The underlying module for *Job_ Assigner* decomposes the task frame into job assignments (not shown) for the schedulers. Figure 3 depicts the generation of a *Schedule_Job* event, representing a job assignment to the *Scheduler* interface. The *Scheduler* receives the *Schedule_Job* event. Its underlying module computes a schedule, which is transmitted as an event through the interface outside of *Behavior_Generation* to the *World_Modeling* plan simulator. This is depicted as a plan in Figure 2. Ultimately, the simulated plans are evaluated by *Value Judgement* and returned to the *Plan Selector* in the *Behavior Generation* module (not shown in the example). The *Scheduler* interface is also shown as returning a Status event with a ?Status variable. Values for specific status events would be generated in underlying modules that conform to the interface.

Specification of the Interfaces, Behavior, and Constraints A partial *Rapide* specification of the *Job_Assigner* interface is given below. The *Job_Assigner* is declared to be of type Interface. The signatures for the events received by, and sent from this interface are provided including variable arguments and their types.

```
TYPE Job_Assigner_Interface IS INTERFACE;
ACTION
    IN
        Do_Task (Task : Task_Command_Frame),
        RCV_task_frame (Task : Task_Command_Frame; TF : Task_Frame),
        SC_Status (CR : Controlled_Resources; ST : String);
    OUT
        Schedule_Job (Job : Task_Command_Frame),
        Fetch_task_frame (Task : Task_Command_Frame),
        Decompose_task_frame (TF : Task_Frame),
        JA_Status (?status);
BEHAVIOR
    (?Task : Task_Command_Frame)
     Do_Task (?Task) ||> Fetch_task_frame (?Task);
    (?Task : Task_Command_Frame; ?TF : Task_Frame)
     RCV_task_frame (?Task, ?TF) ||> Decompose_task_frame(?TF);;
END;
```

A portion of the behavior depicted in Figure 3 is also specified. The receipt of a Do_Task command to perform a task triggers a request for a task frame containing essential information needed to perform the task. A causal connection is defined between these two events. The receipt of a RCV_task_frame command results in a Decompose_task_frame in which (?TF) denotes the variable placeholder for the task frame which is transferred. The Schedule_Job and Status events are generated through the interface by underlying conforming modules which also instantiate the necessary arguments. These are omitted from this portion of the specification example.

The specification of the Job_Assigner is supplemented by the declaration of constraints shown below.

```
CONSTRAINT
– (C1) Do not allow causally independent Do_Task and Schedule_Job events
NEVER (?Task : String; ?Job : String)
            Do_Task (?Task) || Schedule_Job (?Job);
– (C2) Do not allow causally independent Do_Task and Status Message events
NEVER (?Task : String; ?status : String)
            Do_Task (?Task) || JA_Status (?status);
```

Constraint C1 prohibits the independence of Do_Task and Schedule_Job events, while constraint C2 prohibits independence of Do_Task and Status_Events. These constraints require that that these events *must always be* related in a causal sequence.

Specification of the Architecture The specification of the portion of the Behavior Generation architecture from Figure 3 is given below. This specification shows the connection of the events between the Job_Assigner, an array of Schedulers and the Plan Selector.

```
ARCHITECTURE BG_Module_Arch () ...
IS
      JA : Job_Assigner_Interface IS Job_Assigner_Module();
      SC : array [integer] of Scheduler_Interface IS
            (1..$Num_Controlled_Resources,...)
      PS : Plan_Selector_Interface IS Plan_Selector_Module();...
CONNECT
      (?Job : Task_Command_Frame)
      JA.Schedule_Job(?Job) ||> SC i.RCV_Schedule_Job(?Job);
      (?CR : Controlled_Resources; ?ST : String)
      SC[ i].SC_Status (?CR, ?ST) ||> JA.SC_Status (?CR, ?ST);
      ...
      (?CR : Controlled_Resources; ?Job : Task_Command_Frame;
      ?Sched : Schedule; ?ST :string)
      PS.SND_PS_Status (?CR, ?Job, ?Sched, ?ST) ||>
            SC[i].RCV_PS_Status (?Job, ?Sched, ?ST);
```

Note that each of these components is declared as an instance of one of the types defined above. This is followed by explicit connections between OUT events in the interface of one component and IN events in another interface via the CONNECT keyword. The *Rapide* symbol "||>" is used to indicate a causal connection between these events.

Execution of the RCS Intelligent Control Node Architecture The declaration of causal connections between events in *Rapide interfaces* and in the declaration of the architectures defines a causal sequence of events. The execution of this architecture produces a POSET, which can be used to analyze sequences of events and causality. A portion of the POSET generated by the execution of the RCS control node is shown in Figure 4, which omits intervening events not described in the partial specifications given above. The figure shows the causal connection between the Do_Task event and a Fetch_Task_Frame event that retrieves information necessary to initiate scheduling activity. When the Job_Assigner receives the Task_Frame, this triggers the Decompose_Task_Frame event followed by the Schedule_Job event that is forwarded to a set of Schedulers. POSETs such as the one shown were useful for analyzing sequences of events and communicating behavior of the architecture.

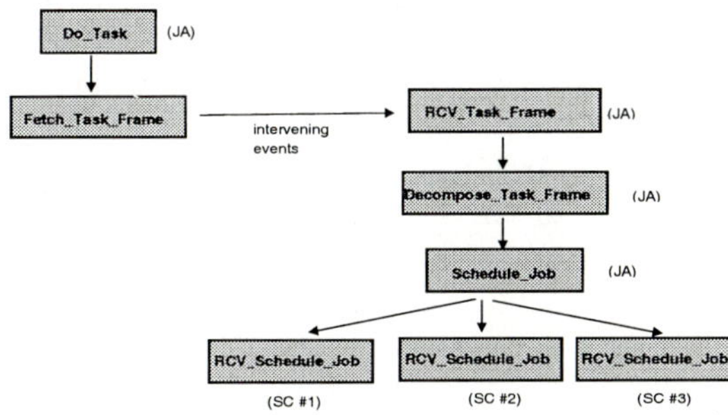

Fig. 4. Event trace of *Rapide* Reference Model Specification

Verification of Individual System Designs Against the Reference Model Architecture *Rapide* provides a capability for verifying that the behavior of a system design, or *concrete* architecture, conforms to that of a more *abstract* architecture, such as the RCS Reference Model Architecture. This is accomplished by first declaring a set of constraints in the abstract architecture and then declaring an equivalence, or mapping, of events from the *concrete* to the *abstract* architecture. The *abstract* architecture is then executed with the "mapped" events of the *concrete* architecture replacing events originally defined in the *abstract* architecture to create a POSET event trace similar to Figure 4. Conformance to constraints of the *abstract* architecture is tested. If constraints are violated, error messages appear in the POSET as events, indicating that the concrete architecture is non-conformant. The conformance verification feature was exercised and found to be useful. One of the challenges facing RCS developers is ensuring that their systems comply with the RCS reference model architecture.

5 Conclusions and Recommendations

5.1 Specifying and Analyzing RCS

Based on informal review by RCS experts, the Intelligent Control Node specification was successful in capturing and representing major RCS architectural concepts. To date, it is the most rigorous representation of the reference model architecture. However, the specification had to be simplified and modified to allow the application of specific RCS keywords, and supplemented by the use of graphical support. The simulated execution of the Control Node Architecture reinforced the specification and proved to be a valuable aid in communicating the architecture by enabling reviewers to visualize the topology and high-level execution of the Intelligent Control Node.

The structure and behavior of the RCS Reference Model Architecture were captured by *Rapide*. Aided by the simplification of the specification and the use of graphics, the Control Node module Interfaces and signatures were clearly defined. Module connections, even though not definable in *Rapide* as explicit types–or first-class objects–were also easily communicated. The successful representation of the RCS Control Node hierarchy indicates that representation of other parts of the multi-level architecture described in Section 2 should be possible.

The ability to create a precise, communicable specification of the RCS Reference Model Architecture led to potential improvements to the architecture itself. Two possible changes to the Architecture as described in [5] were identified, one of which will be described. In the model described in Section 4, Job_Assigner applies a Fetch_Task_Frame operation to retrieve the task knowledge necessary for task decomposition. Although this operation is not explicitly stated in the RCS Reference Model, we found it consistent with the usage of task frames and found it effective in our experiment. Therefore, this operation may be proposed as one of the accepted Job_Assigner functions in its specification. This illustrates the potential of ADLs as practical tools for development of the Reference Model Architecture and software designs in general.

The use of an ADL to verify the behavior of an application system design against the Reference Model Architecture was demonstrated as a proof-of-concept in the Control Node prototype. However, RCS domain experts maintain that verification of the system topology is at least equally important for the Reference Model Architecture. This form of verification involves showing that the application system contains the same basic structure including components, event connections, and data structures as the Reference Model. As a result, two kinds of verification are important from the standpoint of RCS. The first is verification to the structure of the Reference Architecture including existence of specific components, events, and control flows. The second is verification of behavior, including behavior within components and behavior across component connections and an entire architecture. Verification of behavior is the focus of *Rapide.*

Further research is necessary to define techniques for demonstrating consistency with system topology. Work in extending *Rapide*'s POSET model to verification of system structure has been reported in [27]. In SADL [19], Moriconi describes a general approach, called *architectural refinement*, that utilizes theorem proving techniques. In this approach, proofs are constructed to show that in the case when a more general or abstract architecture is applied to produce a more detailed design, that any system that correctly implements the more detailed design also correctly implements the abstract architecture. Refinement is used to demonstrate correctness with respect to the connectivity of events between modules at different levels of abstraction; and this approach may be applicable to the problem of verifying application system designs.

5.2 Appropriate Abstractions for RCS Architectures

Owing to evidence in biological systems and theory of control science, RCS prescribes rules for decomposing the control hierarchy for a system. In his "Outline for a Theory of Intelligence," [2], Albus proposed that:

"In a hierarchically structured, goal-driven sensory interactive, intelligent control system architecture, control bandwidth decreases about an order of magnitude at each higher level, perceptual resolution of spatial and temporal patterns decreases about an order of magnitude at each higher level, goals expand in scope and planning horizons expand in space and time about an order of magnitude at each higher level, and models of the world and memories of events decrease in resolution and expand in spatial and temporal range by about an order of magnitude at each higher level."

These English language rules must be encoded into ADLs in order to represent fully the semantics of an RCS system. Temporal scales and spatial extents relative to other levels of the hierarchy must be represented and validated. While existing ADLs can meet some of these requirements, further work on ADLs adding methods to define these measures and to express constraints among them is needed to allow specifications such as those quoted to be stated and applied.

The syntactic description was simplified and altered to conform to the descriptive forms familiar to RCS experts. RCS experts found specifications much easier to understand when RCS terminology was used. As an example of this approach, instead of declaring an RCS module such as SCHEDULER as a *component* or *interface type*, it should be possible to introduce a higher-level language type called *RCS_Module* in a specification that could serve as a "meta type" for the definition of interface types that are specific to RCS such as SCHEDULER. As a long-term goal, ADLs should allow specifications to be stated at a sufficiently high level of abstraction for non-computer scientists so that they are easier to understand than a program written in C++. This argues for the development of either a flexible ADL with an extensible syntax that can be specialized for RCS or a domain-specific ADL that utilizes RCS terminology.

To facilitate communication of RCS system behavior, an ADL must provide an effective means for abstractly specifying algorithms, component behavior, and performance. While some ADLs may allow representation of all or most of the behavior needed for RCS, this requirement may lead to defining additional language constructs to more directly represent specific RCS behavior. It may also require additional facilities for guiding developers in generating their component specifications, through for example, templates that they can fill in, as proposed in [12] and [17]. As with system structure, such capabilities would allow ADLs to specify essential aspects of behavior at a higher level of abstraction than for programming languages. These capabilities could be part of a domain-specific ADL with a syntax that is customized for RCS systems.

5.3 General Software Development Support

It is often important to be able to divide the processing into atomic processing components that can be executed serially or in parallel, which will facilitate process cessation and make it deterministic. Therefore, it is important that an ADL be able to specify processing characteristics such as process modularization, parallel and serial execution. Serial and parallel processing capabilities are provided by some ADLs, including *Rapide*.

It is desirable to allow capturing performance statistics, such as timing, states, and errors. These would be useful in system diagnostics and maintenance. It should be noted that *Rapide* does provide the capability to capture time-related data which could not be exercised in this study due to resource limitations.

For designing real-time systems, an ADL should define notions of duration in time of processes, mixed asynchronous and synchronous processing, spatial scope of a process or set of processes, algorithm and component complexity, and determinism in execution.

One of the benefits of rigorously specifying RCS designs is that it is possible to check the completeness and internal consistency of the reference model architecture before it is used as a basis for developing individual system designs. By providing a basis for formalized, or at least rigorous specification, most ADL products surveyed also provide a basis for development of automated analysis capabilities. In the case of *Rapide*, analysis is based on simulation of the execution of a system architecture and analysis of POSET traces. This proved to be valuable for visualizing, understanding, and verifying system behavior.

Other ADLs take different approaches using automated tools for analysis of specifications based on formal methods approaches. *SADL* [20] uses w-logic, a weak second-order logic, as a basis for proving the correctness of mappings between architectures at different levels of abstraction. *Wright* [6] uses First-Order Logic to specify constraints and a Communicating Sequential Processes (CSP) computational model to specify behavior of components and connections, providing a basis for a set of automated checks on specification consistency and completeness. Examples from *Wright* are checks that determine the existence of a deadlock condition within the specification of the behavior of an architecture and checks to determine compatibility between connections and components.

5.4 Transfer of ADL Concepts into Real-Time Development Tools

Presently, there are a number of public domain and commercially available software support tools for design and simulation of real-time software systems. These tools have well-developed facilities for designing and implementing individual software systems. However, they do not typically provide any guidance to users about how to structure their system or make other design decisions. ADLs introduce notions of software architecture that could potentially provide additional structure in order to improve the capabilities of these tools. Users or enterprises could set preferences in term of which architecture or architectural style is to be used in developing systems. The tools would then either guide designers as the system is being developed or could flag situations where the architecture or style are violated. Further effort is necessary to explore the potential of infusing ADL concepts into real-time development support tools. This avenue could provide the benefits of ADLs to end users while shielding them from having to learn a new language and concepts. The real-time development tools would guide users in constructing systems per rules for a prescribed architecture through their graphical user interfaces. The users would not be burdened with the underlying mechanics of the ADL specification. In addition to design, analysis and simulation capabilities from the ADLs could be incorporated into the tools. The tools could generate executable or source code. This would automatically assure traceability from the desired architecture through to the executable code. Eventually, tools using ADLs could support highly automated composition of real-time systems from existing or tailorable components.

5.5 ADLs and Component-Based Software Reuse

There is potentially a strong relationship between ADLs and component-based software reuse. ADLs go beyond providing just the signature specification for a component or subsystem. They allow developers to see the big picture, where their particular pieces fit in, and how the pieces are expected to behave or interact with the rest of the system. Simulation of components provides additional benefits not available in typical notations or descriptions of software components.

ADLs could be extended to support reuse with additions of specific language features based on reuse concepts from the literature on domain engineering [14] [24] [25], thus providing a basis for automation of software development. Domain engineering is the process of developing reusable software for a family of systems with similar requirements. An architecture specification may identify optional components, parameterizable components, or even entire subarchitectures that can be varied. Guidelines would be used by developers with the aid of support tools to select options and customize the specification for particular applications. This concept could be further extended by the use of software support tools that assist developers in selecting and modifying system designs and components. The resulting system specifications potentially could be automatically composed and generated using the support tools. An example of such a system for automated generation of system requirements is provided in [8].

6 Summary

This report has provided the results of an investigation into the use of architectural description languages to represent the RCS Reference Model Architecture and RCS software components. ADLs have the capabilities to represent RCS and to be useful tools for further developing RCS. However, several areas of research are suggested in order to make ADLs more effective tools for RCS software specifications. Transfer of ADL concepts into existing real-time software development tools is another important direction to pursue. It is the hope of the authors that this work provides a contribution towards both the development of ADLs as tools for software component technology and the formalization of the RCS Reference Model Architecture.

Acknowledgements The authors wish to thank John Kenney, David Luckham, and other members of the Stanford University *Rapide* Project for their generous assistance with the *Rapide* ADL and software support tools. Thanks is also provided to NIST staff members who reviewed this paper and the RCS prototype specification and provided critical commentary.

References

1. Albus, J.S., Lumia, R., Fiala, J., and Wavering, A. 1989. NASREM - The NASA/NBS Standard Reference Model for Telerobot Control System Architecture. Proc. of the 20th International Symposium on Industrial Robots, Tokyo, Japan.
2. Albus, J. S. 1991. "Outline for a Theory of Intelligence. IEEE Transactions on Systems, Man, and Cybernetics, Vol. 21, No. 3:473-509.
3. Albus, J.S., Lumia, R. 1994. The Enhanced Machine Controller (EMC): An Open Architecture Controller for Machine Tools. Journal of Manufacturing Review, Vol. 7, No. 3, pgs. 278-280.
4. Albus, J. S. 1995. The NIST Real-time Control System (RCS): An Application Survey. Proc. of the AAAI 1995 Spring Symposium Series, Stanford University, Menlo Park, CA.
5. Albus, J. S. 1997. 4-D/RCS: A Reference Model Architecture for Demo III. National Institute of Standards and Technology, Gaithersburg, MD, NISTIR 5994.
6. Allen, R. 1997. A Formal Approach to Software Architecture. PhD Thesis, Carnegie Mellon University, Pittsburgh, PA, Technical Report Number: CMU-CS-97-144.
7. Dabrowski, C., Huang, H., Messina, E., Horst, J., 1999. Formalizing the NIST 4-D/RCS Reference Model Architecture Using An Architectural Description Language. National Institute of Standards and Technology, Gaithersburg, MD, NISTIR 6443.
8. Dabrowski, C. and Watkins, C. 1994. A Domain Analysis of the Alarm Surveillance Domain. National Institute of Standards and Technology, Gaithersburg, MD, NISTIR 5494.
9. Garlan, D., and Perry, D. 1995. Introduction to the Special Issue on Software Architecture. IEEE Transactions on Software Engineering, Vol. 21, No. 4, pp. 269-274.
10. Garlan, D., and Shaw, M. 1994. Characteristics of Higher-Level Languages for Software Architecture. Software Engineering Institute, Carnegie-Mellon University, Pittsburgh, PA, CMU/SEI-94-TR-23.
11. Horst, J. A. 1993. Coal Extraction Using RCS. Proc. of the 8th IEEE International Symposium on Intelligent Control, Chicago, IL, pp. 207-212.
12. Horst, J. A., Messina, E., Kramer, T., Huang, H. M. 1997. Precise Definition of Software Component Specifications. Proc. of the 7th Symposium on Computer-Aided Control System Design (CACSD '97), Gent, Belgium, pp.145-150.

13. Huang, H. and Messina, E. 1996. NIST-RCS and Object-Oriented Methodologies of Software Engineering: A Conceptual Comparison. Proc. of the Intelligent Systems: A Semiotic Perspective Conference, Vol. 2: Applied Semiotics. Gaithersburg, MD, pp. 109-115.
14. Kang, K., Cohen S. , Hess J., Novak W., and Peterson S. 1990. Feature-Oriented Domain Analysis (FODA) Feasibility Study. Software Engineering Institute, Carnegie-Mellon University, Pittsburgh, PA, CMU/SEI-90-TR-21.
15. Luckham, D. 1996. Rapide: A Language and Toolset for Simulation of Distributed Systems by Partial Ordering of Events. Stanford University, Palo Alto, CA. CSL-TR-96-705.
16. Medvidovic, N. and Taylor R. 1999. Classification and Comparison Framework for Software Architecture Description Languages. Accepted for publication in IEEE Transactions on Software Engineering.
17. Messina, E., Horst, J., Kramer, T., Huang, H. Michaloski, J. 1999. Component Specifications for Robotics Integration. Autonomous Robots Journal, Volume 6, No. 3, pp. 247-264.
18. Messina, E., Horst, J., Kramer, T., Huang, H., Tsai, T., Amatucci, E. A Knowledge-Based Inspection Workstion. Proc. of the IEEE International Conference on Information, Intelligence, and Systems. Bethesda, MD. November, 1999.
19. Moriconi, M., Qian, X. and Riemenschneider, R. 1995. "Correct Architecture Refinement. IEEE Transactions on Software Engineering, Volume 21, Number 4, pp.356-372.
20. Moriconi, M and Riemenschneider, R. 1997. Introduction to SADL 1.0: A Language for Specifying Software Architecture Hierarchies. Stanford Research Institute, Palo Alto, CA, TR SRI-CSL-97-01.
21. OMG. 1999. RFP: UML Profile for Scheduling Performance, and Time Object Management Group Document ad/99-03-13. Object Management Group, Framingham, MA. `http://www.omg.org`.
22. Shaw, M. 1994. Comparing Architectural Design Styles. IEEE Software, November, 1994, pp. 27-41.
23. Shoemaker, C. M. and Bornstein, J. A. 1998. Overview of the Demo III UGV program. Proc. of the SPIE Robotic and Semi-Robotic Ground Vehicle Technology , Vol. 3366, pp.202-211.
24. SPC 1992. Domain Engineering Guidebook, Software Productivity Consortium. Herndon, VA. SPC-92019-CMC, Version 01.00.03.
25. STARS. 1993. Organizational Domain Modeling, Volume I - Conceptual Foundations, Process And Workproduct Description, Informal Technical Report for the Software Technology for Adaptable, Reliable Systems (STARS), Report Number STARS-UC-05156/024/00.
26. USPS. 1991. Stamp Distribution Network, Advanced Technology & Research Corporation, Burtonsville, MD. USPS Contract Number 104230-91-C-3127 Final Report.
27. Vera, J., Perrochon, L., Luckham, D. 1998. Event-Based Execution Architectures for Dynamic Software Systems. Proc. TC2 First Working IFIP Conference on Software Architecture (WICSA1). San Antonio, Texas, USA. Kluwer. pp. 303-317.
28. Vestal, S. 1993. A Cursory Overview and Comparison of Four Architecture Description Languages. Honeywell Technology Center, February 1993.

Conceptual Design, Functional Decomposition, Mathematical Modelling, and Perturbation Analysis

S. Dierneder and R. Scheidl

Department for Foundations of Machine Design
Institute for Mechanics and Machine Design
Johannes Kepler University of Linz
Altenbergerstrasse 69, A-4040 Linz, Austria
Phone: ++43 732 2468 9746, FAX: ++43 732 2468 9753
{dierneder, scheidl}@mechatronik.uni-linz.ac.at

Abstract. Conceptual Design methods place some intermediate solution steps between problem definition and final technical solution. Functional Decomposition is such a method which aims at finding solution concepts in terms of functions first and, after that, technical realisations for the individual functions. The role of mathematical models in conceptual design is to provide some coarse quantitative assessment of certain solution concepts, and furthermore, to clarify the relationship between the Functional Requirements and the Design Parameters. Later on, in the more detailed design steps refined mathematical models are used. What is their relation to the early models used in conceptual design? The concept of perturbation analysis seems to provide a framework for defining and understanding these relationships.

1 Conceptual Design and Functional Decomposition

Formally, design could be defined as a mapping from the space of required functionality (RF) on a certain system under consideration (SUC) to the space of feasible technical solutions (TSs), see Fig. 1. For even moderate complex design tasks, one never will know all TSs, even more, one cannot say something about the structure of this space.

Design is a bi-directional process (see Fig. 2). First - in a forward step - some seemingly possible solutions are picked up, which then - in a backward step - have to be analysed whether or not the given requirements are be fulfilled. Normally, a great number of forward (creative) steps and backward (analysis) steps must be run, until a satisfactory solution is obtained.

A direct way from the RF to the detailed TS mostly is impossible. Some intermediate planning levels - conceptual design stages - must be placed in between to find feasible solutions in reasonable time and at reasonable expense. So far, this is common practice in design, whereas the methods and principles applied in this conceptual design phase differ strongly. A vast number of conceptual design strategies based on various terms - often for the same idea - have been developed [4-7], there are different personal styles, and different needs of the diverse industrial branches.

F. Pichler, R. Moreno-Díaz, and P. Kopacek (Eds.): EUROCAST'99, LNCS 1798, pp. 38–45, 2000.

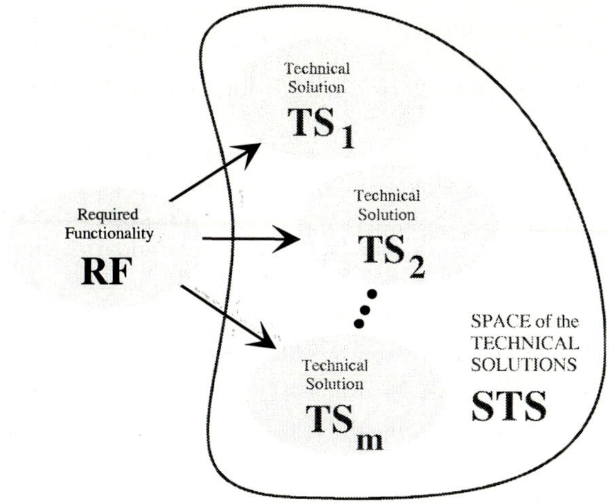

Fig. 1. Desing:= mapping Required Functionality to the Space of Technical Solutions

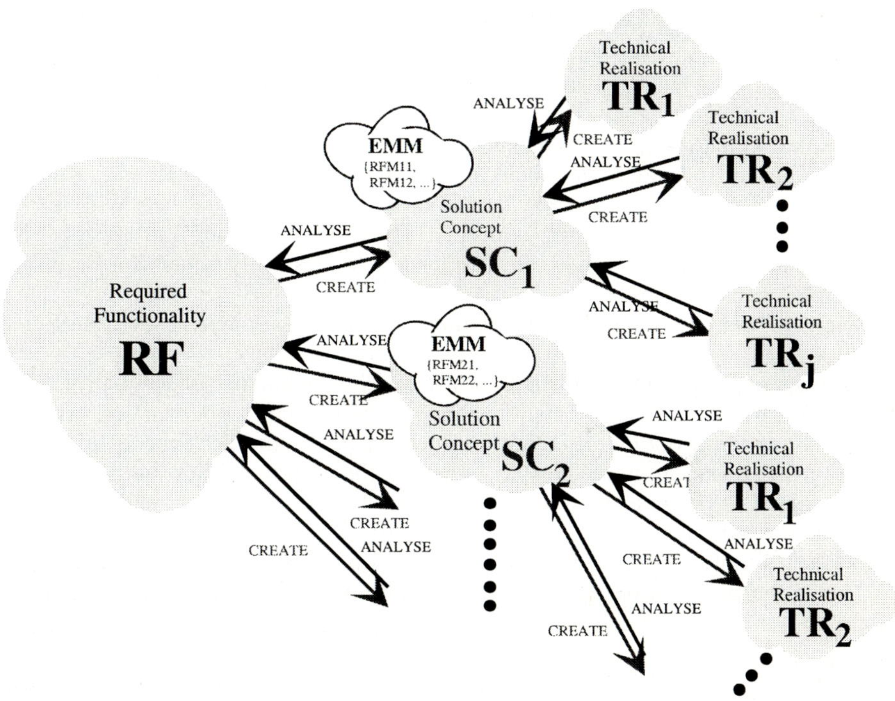

Fig. 2. Conceptual Design, forward creation and backward analysis

A direct way from the RF to the detailed TS mostly is impossible. Some intermediate planning levels - conceptual design stages - must be placed in between to find feasible solutions in reasonable time and at reasonable expense. So far, this is common practice in design, whereas the methods and principles applied in this conceptual design phase differ strongly. A vast number of conceptual design strategies based on various terms - often for the same idea - have been developed [4-7], there are different personal styles, and different needs of the diverse industrial branches.

A widely addressed principle in conceptual design is, to think in terms of functions and not in terms of technical solutions [4-6]. In a very clear and expressed way, this is proclaimed by Suh [7], who also introduces the terms *Functional Requirement* (FR) and *Functional Decomposition*. The clear definition of the design task in terms of a minimal number of independent FRs is a major issue to apply his design principles. This is a very challenging but decisive step for the final success. Of particular importance and difficulty is to work out the minimal number of FRs and to relate them to the needs of the customer.

In [1,2], the authors define a *Functional Solution Concept* (FSC) to be a class of technical realisations (TRs). In other words, an FSC constitutes a subsets of all the feasible TRs, which have some common properties. Since FSCs are defined in terms of physical functions, like, for instance, guiding or positioning of mechanical parts, or amplifying a voltage, they seem fully independent of the technical realisation. From a design viewpoint this is inadequate, because a concept for which not even one technical realisation exists, is useless. Even more, design engineers rarely think only in abstract terms. Very often, a new concept is initiated by specific technical solution which in a second step is elevated to the more abstract level of a functional solution concept.

The solution is generalised to a class of solutions by picking out some essentials. This supports to find other TSs, belonging to the same FSC as well as to seek for alternative FSCs. These diverse FSCs must be evaluated comparatively with respect to the RF to skip non promising concepts. This is much faster and cheaper then evaluating all the TSs. However, at abstract levels feasibility never can be judged completely, because obstacles can arise at a later, more detailed stage. Thus, fulfilment of the FR at the FSC-level only is a necessary but never a sufficient condition.

Design not only aims at fulfilment of requirements but also at optimality. Since in practice various criteria do exist, design is a multi-criteria optimisation problem, for which a strict optimum is not defined [10]. From the conceptual design viewpoint the question arises how to judge on the different FSCs with respect to their potential to become an "optimal" solution. To our knowledge, so far only Suh [7] has addressed this question - at least implicitly - by his design axioms. As a brief of his ideas, simple structured FSCs have a higher potential than complex ones.

From the practical design viewpoint a systematic conceptual design approach should be

- leading to a modular design and supporting parametric variant design - where advantageous
- acting as a guidance for an overall systematic design process (also for detail design)
- giving a sound strategy for the selection of special variant design parameters - where advantageous
- leading to a standardisation of the documentation of the design process
- giving good support for later detail design work

- giving a better understanding of the problem structure of very large and complex SUCs

A lot of the just described aims and purposes only can be realised for complex technical systems, if the design process, in particular also the conceptual phase, is computer assisted. For a broad practical acceptance, computer assistance should fulfil several common demands on modern software. It should be easy to handle, should have user friendly interfaces, should be based to a large extent on standardised and widespread software, and furthermore, it should be compatible to the other CAE-tools in use. Design and its documentation requires different software packages, ranging from common CAD tools, computer simulation environments and mathematical programs to simple editors and data bank systems for the description of the design process and knowledge data.

In [8] the authors have presented a software based on the data bank system MS Access to map the FD structure and all the relevant data on the computer.

2 Mathematical Modelling and Perturbation Analysis

A competent evaluation of the different FSCs also requires some qualitative evaluation of its structure and an assessment of its feasibility to meet the FRs also quantitatively. For this purpose mathematical models are useful whereas not always available. As a vague definition, if they reflect the essence of a certain FSC in the light of the given FRs, we call them Essential Mathematical Models (EMM).

The EMM should represent the relationship between the relevant physical balance and material laws, the FRs, and DPs. As indicated in the Dependency Graph (DG) in Fig. 3, the state X (or functionals of it) of the system and the FRs and DPs are involved. Clearly, the state X can be more complex than a vector, it is for instance an element of a function space if field problems are involved.

Suh's first of his two design axioms [7, page 47], the so called *independence axiom*, claims that a good solution concept is characterised by an uncoupled relation between the FRs and the DPs. This can be easily checked by the EMM.

If the state X is discrete, the relationship between the DPs and the state X may be expressed by the incidence matrix B:

$$X \cong B \cdot DP$$

with $X \in R^{N_X}$, $DP \in R^{N_{DP}}$ and $B \in \left(R^{N_X} \times R^{N_{DP}}\right)$; $\cong$ means depends on.

Furthermore, there is also a relationship between the FRs and the state X:

$$FR = \begin{Bmatrix} FR_\alpha \\ FR_\beta \end{Bmatrix}; \; FR_\alpha \cong C_\alpha \cdot X \quad ; \; FR_\beta \cong A_\beta \cdot DP$$

with $X \in R^{N_X}$, $FR_\alpha \in R^\alpha$, $C_\alpha \in \left(R^\alpha \times R^{N_X}\right)$, $FR_\beta \in R^{(N_{FR}-\alpha)}$, and $A_\beta \in \left(R^{(N_{FR}-\alpha)} \times R^{N_{DP}}\right)$

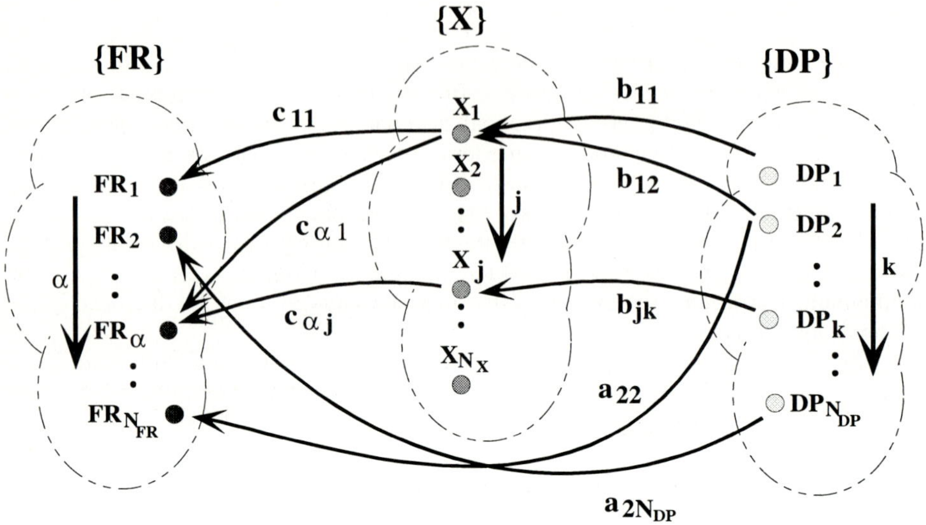

Fig. 3. Dependency Graph

Combining these dependency relations, we get the dependency between the FRs and the DPs.

$$FR_\alpha \cong C_\alpha \cdot X \cong C_\alpha \cdot B \cdot DP = T_\alpha \cdot DP;$$
$$FR_\beta \cong A_\beta \cdot DP = T_\beta \cdot DP$$

$$FR = \begin{Bmatrix} FR_\alpha \\ FR_\beta \end{Bmatrix} \cong \begin{Bmatrix} T_\alpha \\ T_\beta \end{Bmatrix} \cdot DP = T \cdot DP;\ DP \cong T^{-1} \cdot FR = V \cdot FR$$

A second purpose of mathematical models is to rule out whether the required functionality also can be met in quantitative terms, geometrically spoken, does the space of reachable FRs (see Fig. 4) include the required value of FRs. This, off course, needs some information on realistic values of the DPs, which is a technical statement. Quite often, from the general technical knowledge we have an idea of realistic values of certain parameters. For instance, strength parameters or density of metals, torque and power of certain classes of electrical drives, to name only a few.

EMMs preferably should be analytic models. Then, all the above manipulations can be done automatically in any modern symbolic manipulation program as outlined in more detail in [9].

The EMMs must incorporate the design parameters (DPs) but not always in technical terms, because the technical details appear only at later phases of the whole decomposition. Clearly, there must be some equivalence between the possibly abstract DPs in the EMM and the final technical DPs. Thus, the EMMs rule out the

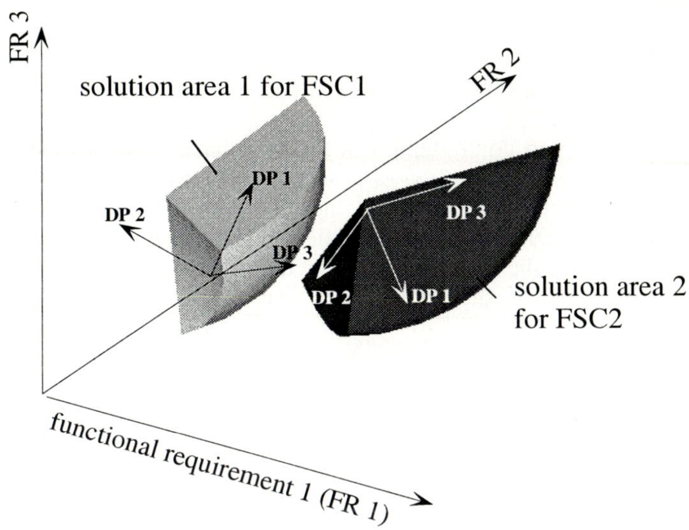

Fig.4. Quantitative Assessment

influence of the main parameters of a certain model. As an example we consider the bending stiffness of a beam in beam theory model and the corresponding cross sectional parameters of the technical beam.

At more detailed levels, further and more refined models (RFMs) are used. We expect them to better approximate reality of this detailed design than the EMMs. As the RFMs refer to a more detailed system they need more data, are more complex, need more time to be set up and to get results (run simulation). But, what is their relation to the EMMs?

As explained above, the results obtained by the EMMs should not be fully contradicted by the RFMs, otherwise the EMMs are of little use. But how to compare these models? At the RFM level more and possibly different DPs are involved, and the state space X is of higher dimension. What are the relevant criteria for neighbourhood of simple models, in particular EMMs, to reality or to RFMs from a design point of view? In more mathematical terms, is there an adequate "metric" to judge on the distance between different models, and, furthermore, somewhat like a universal rule, to assign the attribute "essential" to a model?

Following Nam P. Suh, see [7], the design is assessed by the fulfilment of the RF in terms of the FRs. In order to operate mathematically, these FRs must be mathematical expressions. They involve some functions of state of the system and map it to the "space of FRs".

These questions are related to the class property of a certain FSC. With the additional parameters of an RFM the system behaviour is modified. If we go to extremes of these parameters the basic functioning principles and the validity of the EMMs - quite often - can be destroyed. Consider again the beam model of beam theory and an actual beam with its real dimensions and its modelling as 3D elasticity problem. If slenderness of the beam is violated or loading conditions are beyond those of elementary beam theory, it is no more valid, even qualitatively.

Definition. The characteristic property of an EMM is, that for any related refined functional model (RFM), in principle a combination of all its parameters exists, which gives the same values of the FRs, and that the additional parameters of the RFMs - those not existing or having a pendant in the EMM - influence the FRs in a regular manner, in particular, as a regular perturbation.

Seemingly, this is in contradiction to the fact that a refined model with additional degrees of freedom quite often is a singular perturbation of the original, simpler model, for instance in dynamics, if the evolution of the states of both models is compared under general initial conditions. The discrepancy is resolved, because what counts is not the system behaviour under any general (e.g., initial) condition but the fulfilment of the FRs, which are functionals of state, under realistic conditions.

3 Conclusion

The Functional Decomposition method provides a hierarchically structured approach in the conceptual design of technical systems. It not only helps to find a good technical solution for a give problem but also reflects the "problem structure" as well as the design history, if it is properly documented. Its documentation even for moderately complex systems needs computer support. Data bank system are well suited for this purpose.

Mathematical models of different complexity should convene the design process. At the rather abstract level of functional solution concepts so called Essential Mathematical Models help to evaluate the diverse concepts. For the qualitative evaluation the application of Suh's independence axiom is recommended.

The refined mathematical models used at later, more detailed stages of design should be in reasonable neighbourhood of the results of the Essential Models. A concept for defining this neighbourhood is regular perturbation with respect to the Functional Requirements and the Design Parameters.

References

1. R. Scheidl, S. Dierneder, K. Mörwald, Computer Aided Conceptual Design by a Functional Decomposition Method for Process Oriented Heavy Machinery and its Relations to Mechatronization, Lancaster International Workshop on Engineering Design – CACD'98, 27th and 28th May 1998, Lancaster, Great Britain, Proc. CACD'98, Editors: Alan Bradshaw and John Counsell, Lancaster University Engineering Design Centre, 1998
2. S. Dierneder, R. Scheidl, K. Mörwald, J. Guttenbrunner, A Computer Aided Conceptual Design Method for Mechatronic Systems in Process Oriented Heavy Machinery, The 6th UK Mechatronics Forum International Conference – MECHATRONICS'98, 9th – 11th September 1998, Skövde, Sweden, Proc. MECHATRONICS'98, Editors: Josef Adolfsson and Jeanette Karlsén, Elsevier Science Ltd, 1998
3. J. Buur, Does mechatronics need a special design attitude?, Proc. Mechatronics Research Conf., 13th – 15th September 1990, St. Albans, England, I MECH E
4. N. Cross, Engineering Design Methods, 2nd edition, Wiley, Chichester, 1994

5. N. F. M. Roozenburg, J. Eekels, Product Design: Fundamentals and Methods, Wiley, Chichester, 1995
6. Pahl, Beitz, Konstruktionslehre, 3. neubearbeitete und erweiterte Auflage, Springer, Berlin – Heidelberg, 1993
7. Nam P. Suh, The Principles of Design, Oxford University Press, New York, Oxford, 1990
8. S. Dierneder, R. Scheidl, A Data Bank System for the Representation of the Functional Decomposition Method, Lancaster International Workshop on Engineering Design – CACD'99, 18th to 20th May 1999, Lancaster, Great Britain, Proc. CACD'99, Lancaster University Engineering Design Centre, 1999
9. R. Zurmühl, S. Falk, Matrizen 1 – Grundlagen, 6., vollständig neubearbeitete Auflage, Springer Verlag, Berlin – Heidelberg, 1992
10. H. Eschenauer, J. Koski, A. Osyczka, Multicriteria Design Optimization – Procedures and Applications, Springer Verlag, Berlin – Heidelberg, 1990

AV-Petri Systems:
How to Get Together Abstraction and Views for Petri Systems?

Gisbert Dittrich

Department of Computer Science
University of Dortmund
Otto-Hahn-Str. 16, 44221 Dortmund, Germany
dittrich@ls1.cs.uni-dortmund.de

Abstract. In order to model complex systems it is indispensable to structure the modelings. The mechanism of abstraction as well as the usage of views is very helpful for structuring. In this paper I will discuss these aspects in the context of Petri systems in order to get fundamental definitions. It turns out, that with respect to many different definitions of Petri systems only a "generic definition" is possible. To get final definitions for concrete types of Petri systems it requires much further work.

1 Introduction

It is very well known that Petri systems are well suited for the modeling of concurrent systems. In practice such modelings rapidly become very big. Thus this results in the task to structure modelings using Petri system approaches adequately. Therefore we have to answer the following question: What are appropriate structurings for Petri systems? Structuring especially means to show how to describe parts or aspects of systems in a comprehensible way as well as how to compose the total system from these partial descriptions.

Abstraction is a well known concept of structuring also applied in the world of Petri systems. Abstraction means to represent a whole subnet or subsystem by only one node in a coarsened presentation, or the other way around: a coarse node is refined into the associated subnet or subsystem. Usually Petri system descriptions using this presentation are called hierarchical. In the literature different variations of this concept are available (cf. e.g. [1], [2], [3]). I prefer the approach of Fehling ([1], [2]) because this one succeeds (in contrast to other definitions) in the modeling of the refinement of adjacent nodes (cf. [4], [5]). This is a very useful property.

A second form of structured modeling is available in tools for generating and editing graphics in the possibility to apply levels, transparencies or layers. With them a whole description of something can be generated by piling up transparencies where each of the transparencies presents the description of an aspect or - in other words - a view of the total system. This sort of structuring too has been demonstrated as meaningful in the context of Petri systems. The development of this second main idea

F. Pichler, R. Moreno-Díaz, and P. Kopacek (Eds.): EUROCAST'99, LNCS 1798, pp. 46–54, 2000.

for Petri nets as well as the problems emerging from the application for the modeling of systems can be found in [6].

Thus we have to solve the task to give a Petri systems definition including both possibilities of structuring simultanously. Its a pity that I am not able to give a comprehensive definition for that. But I‘ll suggest how to generate a collection of definitions to approach the aim mentioned above. This paper is a modified, partly enhanced and partly shortened version of [7].

2 Towards Defining AV-Systems

In order to apply structurings as mentioned above, it seems to be a good approach to „orthogonally“ enhance unstructured Petri net modelings by both possibilities of structuring. That means the enhancements have to be given independently from each other. I will give the answer in two steps: At first I will treat the problem only for the nets underlying the Petri systems. Then I will enhance the results to system descriptions.

2.1 Examples for Abstraction and Views

Let´s start with parts of examples for showing the phenomena of abstraction as well as views (already for systems). Here I am not interested in explaining the modeling of the concrete examples themselves.

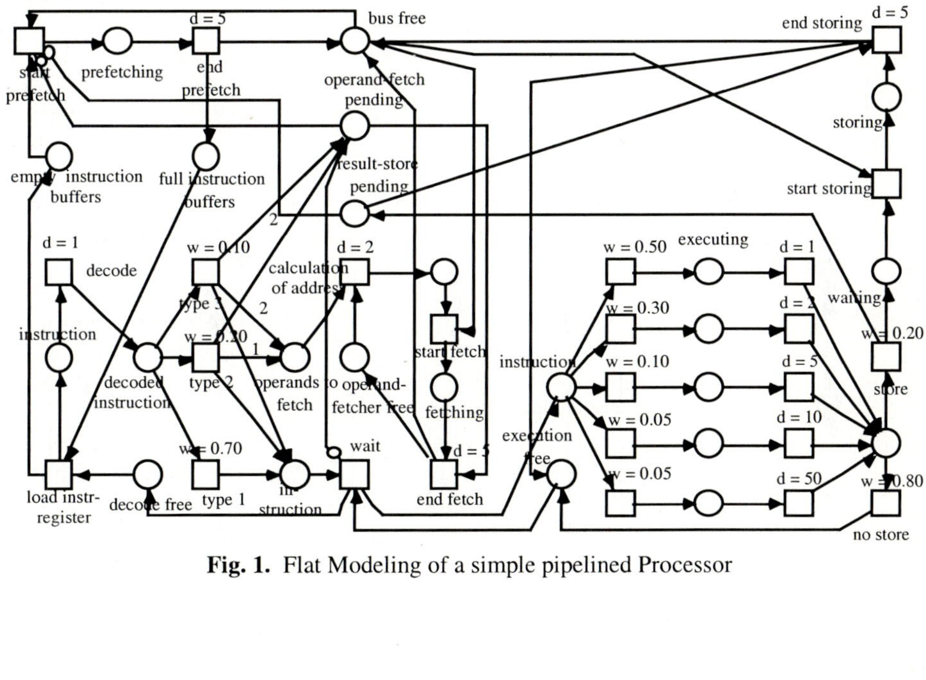

Fig. 1. Flat Modeling of a simple pipelined Processor

2.1.1 A Model of a Pipelined Processor

At first we will discuss abstraction by looking on an example. In Fig. 1 a flat modeling of a simple pipelined processor is given as a DSPN-modeling.

Even this small example of a complex modeling in this representation is not easy to explain. But the developer can easily decompose the whole modeling into different components as depicted in Fig.2.

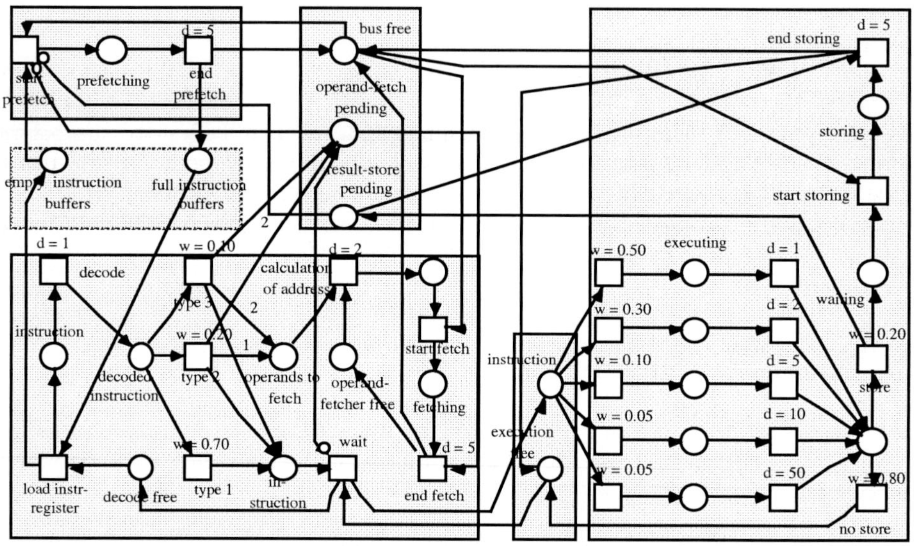

Fig. 2. Suggested components in the modeling of Fig. 1.

Thus it is possible to tell the whole story describing the total system by the smaller and thus comprehensible components and the interfaces between them. In addition a

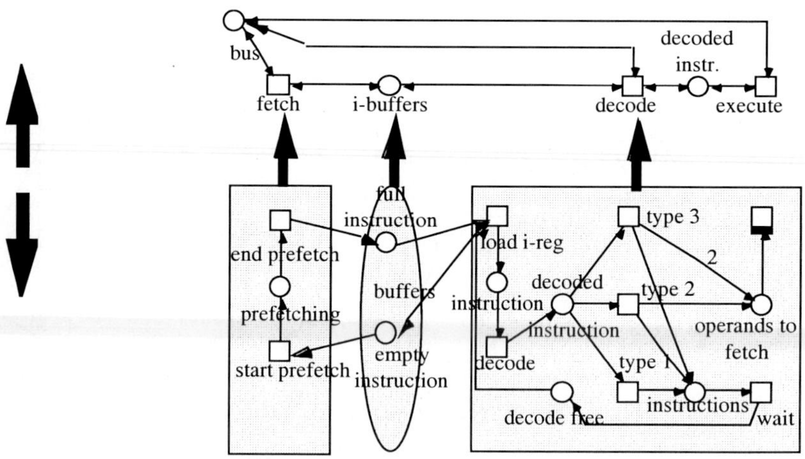

Fig. 3. Part of the morphism between the detailed and the coarsened description

rough description of the overall connections of the components would be very helpful for a good understanding.

Under some circumstances - which are fulfilled in the example above - the overview describing the connections of the components itself can be given by a Petri net, where the components are represented by single, coarsened nodes. The main condition is that the boundary of a component only contains elements of one sort: only places or only transitions. The coarsened node representing a component has to be from the same sort as the elements from the boundary. An edge between coarsened nodes will emerge when there exists an edge between the represented components. It turns out that this leads to a Petri net morphism from the detailed to the rough description. In Fig. 3. this relation is partly illustrated (only for 3 components).

This first example has discussed the aspects of abstraction using a bottom up-approach. This does not mean that abstraction is only useful in such cases. Abstraction is also useful in a top-down approach as well as in mixtured versions like “jojo”. (Cf. e.g. [8]). The discussion of abstraction using the example above can be found in more detail in [5].

2.1.2 A Model of Dialing

Now we will discuss the main idea of views by looking on another example.

In Fig. 4. an overall modeling of dialing between A and B as a condition/event system enhanced by inhibitior arcs is given.

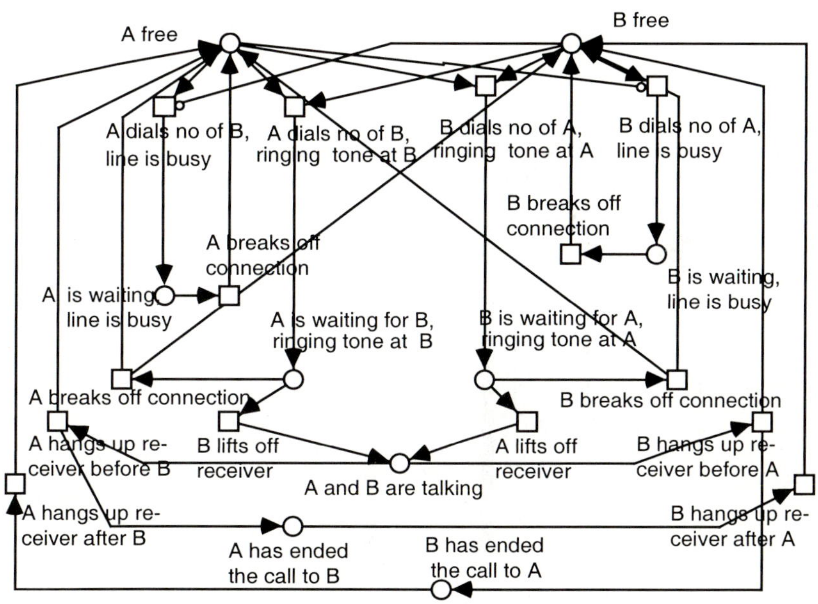

Fig. 4. Dialing between A and B, modeled by a condition/event system

In order to restrict to parts more easily to comprehend this description will be decomposed with respect to different aspects, here denoted as views described in layers.

A suggested division into different aspects will be to model that A dials the number of B. This will be done from the viewpoint of A giving the regular behaviour described in a layer 1 and the abnormal behaviour (exception handling) described in a layer 2. A full description of the whole story encompasses this behaviour from the viewpoint of B in a layer 3 and in addition all these aspects by interchanging A and B

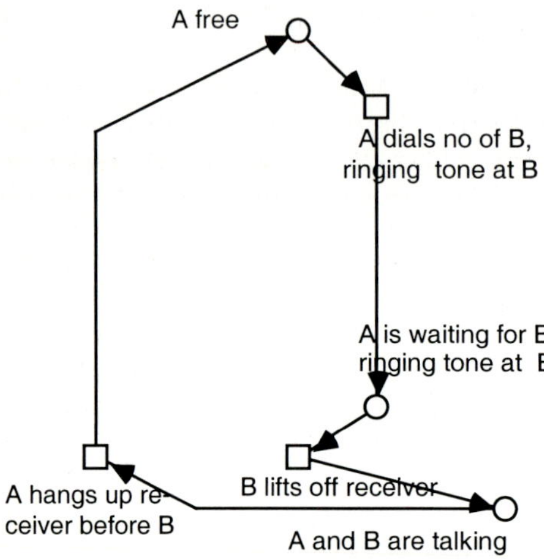

Fig. 5. Layer 1

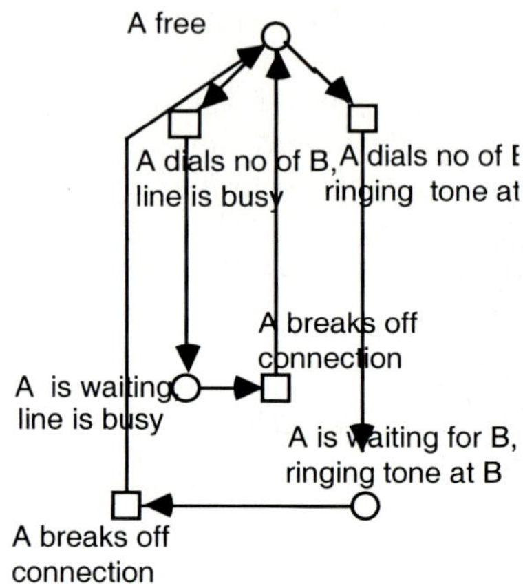

Fig. 6. Layer 2

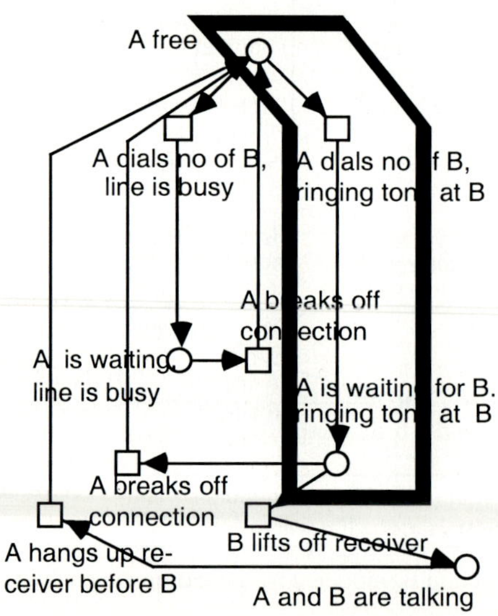

Fig. 7. Layer 1 + 2

(layers 4 - 6). Here only layer 1 and layer 2 will be shown as Fig. 5. and Fig. 6. In Fig. 7. the union of layer 1 and 2 is depicted. This points out that the intersection is nontrivial (in contrast to the case in Section 2.1.1).

To look at the full modeling please confer [6].

Thus the main idea behind views is to model the whole system as a "union" of (may be non disjoint) subsystems.

2.2 Abstraction and Views for Nets

In this section the approach will be discussed only regarding the underlying Petri nets of whole system descriptions.

2.2.1 In the approach of Fehling ([1], [2]) the modeling of abstraction of Petri nets is consequently elaborated. As suggested in the example of section 2.1.1 in this approach whole subnets fulfilling some conditions concerning their „borders" may be represented by nodes. The relations between the subnets again may be described by a Petri net, the overall net using the coarse nodes, which represent the subnets. Subnets represented by coarse nodes fulfill the relation „is part of" or „is disjoint".

The abstraction (coarse) nodes (in Fig. 8. depicted in black) represent the subnets „lying under that node". That means that subnet, that is spanned by the leaf nodes which are directly or indirectly associated to the abstraction node by the vertical, dashed depicted edges. These edges represent the „is in the refinement of" - function.

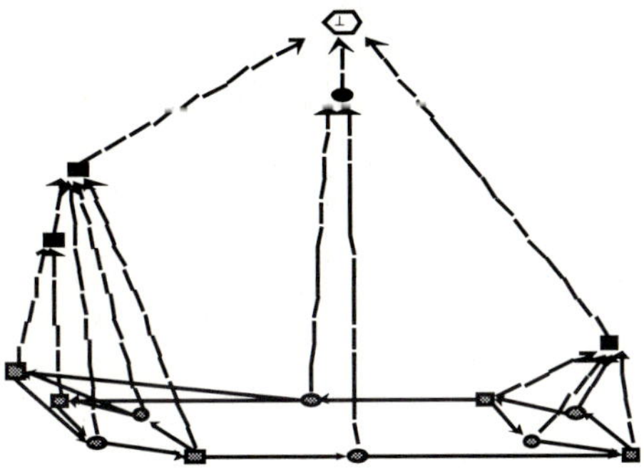

Fig. 8. Example of a „hierarchical" Petri net due to Fehling.

Fehling's definition of a hierarchical Petri net (cf. [1], p. 64) is essentially as follows (up to modifications in notation). We will denote the same concept as an A-net (because showing abstractions).

$HN = (P, T; F, (f, P_a, T_a, \perp))$ is an *A- net*

$:\Leftrightarrow$ 1. $N = (P \cup P_a, T \cup T_a; F)$ is a Petri net

(in general including many isolated nodes).

2. $f: X \rightarrow X \cup \{\perp\}$ with $X := P \cup P_a \cup T \cup T_a$ is a function.

3. Conditions (HN_i) for i = 1,2,3 are fulfilled.

The (here not explicitly given) conditions HN_i ensure: The abstractions are captured by the function f, representing a tree (in Fig. 8. depicted by the „vertical" edges). The nodes representing abstractions (elements from $P_a \cup T_a$) are the internal nodes of the tree. The actually modeled („flat") net (in Fig. 8. depicted by gray nodes and edges between them) is described by the nodes that are leaves under f (elements from $P \cup T$) and the edges F which only occur between leaf nodes. Edges from or to abstraction nodes are not explicitly described in this definition. But they can be induced from elements of F. Thus the flat net (P,T; F) is explicitely included in the definition of an A-net.

2.2.2 As suggested in the example of section 2.1.2 an other form of decomposition into subnets not obeying the above mentionend conditions has been approved as meaningful, namely to decompose into (in general non disjoint) subnets, such that the gluing of those subnets represents the whole net. In [6] this operation is introduced as the union of subnets. Obviously this leads to a second – from the description via abstractions independent – structured presentation of a whole net.

VN = (N, TN) is a *V-net*

:⇔ 1. N = (P, T; F) is a Petri net (in general with isolated nodes).
2. TN = $((P_i, T_i; F_i) | i = 1, ..., n)$ is a collection of n subnets of N with
$P = \cup_{i=1,..,n} P_i$, $T = \cup_{i=1,..,n} T_i$, $F = \cup_{i=1,..,n} F_i$.

Remark: The union of finitely many subnets again yields a subnet.

2.2.3 Thus we get as the basic definition of a Petri net with abstraction and views:

AVN = (N, f', TN) is an *AV-net*

:⇔ 1. (N, f') is an A-net with f' = $(f, P_a, T_a, \bot)$
2. (N, TN) is a V-net.

Obviously we got the required properties mentioned above, namely both structurings independently.

2.3 Abstraction and Views for Systems

Now I am interested in answering the following question: how to apply this basic idea developed to nets to models of systems?
Systems in addition to the description of the underlying net contain information to describe the dynamics on the net by informations attached to transitions, places and edges.

E. g. a place/transition system can be described as PTsys = (P, T, F; C, W, M) (e.g. cf. [9], [10]). There (P, T; F) denotes the (underlying) Petri net of PTsys, C is a function attaching capacities to nodes, W is a function attaching weights to the edges and M denotes a start marking. Additionally a firing rule is given. This firing rule in connection with C, W, and M are sufficient to derive the dynamics of the system description. In this sense we speak of a system and its underlying net.

The here suggested main approach is as follows: Only the flat net, the first component in the above given definition of structured nets, has to be enhanced to a system description.

But for that we have to fulfil the following condition: All sub*nets* induced by the structurings applying abstractions or views to a net indeed induce sub*systems*. This condition in general may be invalid. But this demand makes sense, because this ensures to decompose the whole system into (may be non disjoint) subsystems which can be analysed or validated in advance before doing so with the modeling of the total system.

In order to apply this idea independently from a system type, the approach will be formulated relatively to a given system type. Insofar the definition will be a parameterized one. Informally speaking a system type shall describe a class of Petri nets with a uniformly described dynamics on the nets. Thus let PST be a Petri system type. Examples of Petri system types are C/E systems, P/T systems, colored systems, ... (C.f. e.g. [11], [12], [13]).

Let C(PST) be the class of all system descriptions relative to the system type PST.
Let: Union(PST): C(PST) x C(PST) → C(PST)
(may be partial) operation (Union of system descriptions)
Intersec(PST): C(PST) x C(PST) → C(PST)
(may be partial) operation (Intersection of system descriptions).

AVS = (Sys, f', TN) is an *AV-system of type PST*

:⇔
1. Sys is element of C(PST).
2. (N, f', TN) is an AV-net, where N is the net underlying Sys.
3. Each sub*net* got by applying a structuring component induces a sub*system* of Sys.
4. The set of so generated subsystems is closed under Union(PST) and Intersec (PST).

Remarks:

- Union and intersection, respectively, of systems are in general not well defined in a natural manner. For further information to this point cf. e.g. [6], [7].
- Because of the given parametrization by PST the approach from above only yields a schema of definition. Thus one has to elaborate in each concrete case, what union and intersection does mean. Surely this is a disadvantage.
- On the other side it is an advantage to have a universal approach. Thus it is not neccessary to repeat the development of the main idea for each type separately.

3 Outlook

A lot of additional tasks remains to be handled.

- We have to elaborate some concrete AV-system definitions, that means we have to fix a Petri system type and define special operations union and intersection conform with the definition of AV-system.
- For sure it is important to work out some further basic concepts like „building block" in the context of this approach. The special case to model in an object-oriented manner using Petri systems seems to be a very interesting and important one.
- In order to apply the concepts developed above in reality it is indispensable to develop tools supporting those concepts.

4 References

1. Fehling, R.: Hierarchische Petrinetze, Verlag Dr. Kovac•, Hamburg (in german) (1992)
2. Fehling, R.: A Concept of Hierarchical Petri Nets with Building Blocks, in: Lecture Notes in Computer Science 674, Springer Verlag, pp. 148-168 (1993)
3. Jensen, K.: Coloured Petri Nets, EATCS Monographs on TCS, Springer Verlag Berlin (1992)
4. Dittrich, G.: Strukturierte Petrinetze, http://lrb.cs.uni-dortmund.de/Lehre/ Petri2_ SS98 (in german) (1998)
5. Dittrich, G.: Modeling of Complex Systems Using Hierarchically Represented Petri Nets, in: Procs. of IEEE SMC ´95, Vancouver, p.2694-2699 (1995)
6. Dittrich, G.: Layering as Means for Structuring Petri Nets, in: Procs. of IEEE SMC ´96, Bejing, p. 2294-2298 (1996)
7. Dittrich, G.: Towards a Generic Definition of Petri Systems Supporting Abstraction and Views, in: Procs. of IEEE SMC ´99, Tokyo, p. I-884 – I-887(1999)
8. Dittrich, G.: Tools for Modelling with Petri-Net like Nets, in Pichler,F., Moreno-Diaz, R. (Eds.): Computer Aided Systems Theory – EUROCAST '89, Springer Verlag, Berlin (1990)
9. Reisig, W.: Petrinetze – Eine Einführung, Springer Verlag, 2. Edition (in german) (1986)
10. Dittrich, G.: Petrinetze - Eine Einführung, http://lrb.cs.uni-dortmund.de/Lehre/Petri/ (in german) (1999)
11. Brauer, W., Reisig, W., Rozenberg, G. (Eds.): Petri Nets: Central Models and their Properties, Lecture Notes in Computer Science Vol. 254, Springer Verlag (1987)
12. Reisig, W., Rozenberg, G. (Eds.): Lectures on Petri Nets I: Basic Models, Lecture Notes in Computer Science 1491, Springer Verlag (1998)
13. Reisig, W., Rozenberg, G. (Eds.): Lectures on Petri Nets II: Applications, Lecture Notes in Computer Science 1492, Springer Verlag (1998)

Computer-Aided Analysis and Validation of Heterogeneous System Specifications

Giuseppe Del Castillo* and Uwe Glässer

Heinz Nixdorf Institut, Universität Paderborn,
Fürstenallee 11, D-33102 Paderborn, Germany
{giusp,glaesser}@uni-paderborn.de

Abstract In applications of formal methods, the problem of how to establish the correctness of the *initial* formalization step is an often underestimated aspect of the system design process. We propose a methodology based on the construction of a mathematical model which reflects the given system so closely that the correctness can be established by observation and experimentation (*ground model*). Complex technical systems are often heterogeneous, so that different system aspects are best modelled by means of different techniques. This leads to heterogeneous ground models. To achieve a consistent and coherent view of heterogeneous behavioural models, which is a prerequisite for any systematic analysis and validation, we introduce a common semantic framework (*meta-model*) based on the notion of Abstract State Machines. We exemplify our methodology by an industrial case study from automated manufacturing, the distributed control for a material flow system (MFS).

1 Introduction

Complex technical systems in virtually all application areas increasingly rely on *embedded* hardware/software components performing various control operations and supervision tasks. Typical examples are automotive control, industrial automation and extended telecommunication services. In general, one can observe a strong tendency towards *distributed* solutions running on *heterogeneous* system platforms that are interconnected through networks. As such they are characterized by their concurrent, reactive and object-based nature (performing operations that are often subject to external timing constraints). Complex embedded systems are frequently realized as loosely-coupled aggregations of disparate components performing specialized tasks rather than monolithic architectures with a regular structure [14,15]. Engineering of complex embedded systems usually involves several domain specific description techniques in order to deal with distinct facets of system behaviour at various abstraction levels. Heterogeneous modelling approaches offer additional expressiveness and flexibility (e.g. allow for more adequate and natural abstractions) resulting in more realistic and reliable descriptions. On the other hand, any systematic analysis and validation

* Partially supported by the DFG Schwerpunktprogramm "Softwarespezifikation".

F. Pichler, R. Moreno-Díaz, and P. Kopacek (Eds.): EUROCAST'99, LNCS 1798, pp. 55–79, 2000.

of heterogeneously specified system models requires a *coherent* and *consistent* view of the underlying behavioural models such that relevant system properties can be inspected. This situation calls for integrated specification and design approaches[1] allowing to combine different models of computation and data in such a way that the *interfaces* between the various system components and the dependency of *internal* system operations on *external* actions and events (e.g. as associated with the environment into which a system is embedded) become transparent.

Systems Theory faces the challenging task of finding the right *abstractions* to cope with complexity, diversity and the presence of "dirty" system features (which one usually encounters in real life systems). Of particular importance is the role of Systems Theory for *macro-architecting* providing means for a systematic construction, analysis and transformation of formal models as well as its ability to interpret different kinds of formal models with respect to domain-specific models [12]. In the work presented here, we concentrate on *discrete* behavioural models of distributed embedded systems with the objective to support high-level analysis and validation of dynamic system properties.

This paper is structured as follows. We begin, in Sect. 2, with a methodological discussion about the problem of mathematical modelling of non-mathematical reality (*initial formalization step*) and outline our approach (*ground model construction*). The problem of integrating heterogeneous system models (including ground models) is discussed in Sect. 3. The basic notions of Abstract State Machines, which are the foundation of our integration approach, are recalled in Sect. 4. Then, in Sect. 5, we present a case study from automated manufacturing to illustrate the proposed methodology (from both points of view, ground model construction and integration of heterogeneous models). Finally, in Sect. 6, we conclude with some remarks about lessons learned and future research.

2 Reliable Ground Models

Formal specification and verification methods and tools offer a variety of mathematical modelling and proof techniques supporting the entire system design process from abstract requirements specifications down to concrete realizations. At any given abstraction level, the correctness of a model that is obtained as result of some refinement step can in principle be proved. There is however no way of proving correctness of the *initial* formalization step, i.e. the relation between a formal (mathematical) model and the part of the real (physical) world this model is intended to describe. Consequently, one never gains absolute evidence on whether a formal model faithfully "implements" the user's (or customer's) intuitions about the expected system behaviour, nor whether the implications resulting from the stated requirements are completely and correctly understood so that any unexpected and undesirable behaviour is a priori excluded.

[1] Specification and design is meant here in a fairly general sense also including *reverse engineering* as well as *reusability aspects*.

In other words, a formal basis alone does not automatically lead to models that fulfill their requirements; even if a model is proved to have certain properties this does not necessarily mean that it is *correct* in the sense that the model really *fits* into a given (physical or system) environment, since assumptions on which such proofs are based may be incomplete or wrong. Practical experiences with formal methods indeed show that such "mismatchings" do frequently occur when real world phenomena are involved.

Now, the question then is: "How can we establish that a formal requirements specification of some computer-based system actually formalizes our intuitive understanding of system behaviour in the given context in which this behaviour is to be regarded?", or shorter, "*How can one establish that a model is faithful to reality?*". The approach documented here relies on the assumption that this can be done by constructing a mathematical model which reflects the given system so closely that the correctness can be established by observation and experimentation (cf. [7], p. 9). Such a model is called a *ground model*.

The quality of ground models considerably depends on the underlying abstractions for dealing with real-world phenomena, i.e. on the way in which basic objects and operations of a ground model are related to basic entities and actions as observed in the real world. This also influences the choice of the formalism: some formalisms are more appropriate than others to formalize particular kinds of systems in an intuitively transparent way, as their underlying abstractions are closer to the nature of the system under consideration.[2] Quite often, different formalisms are needed for modelling different parts or aspects of the same system, such that—in the end—they coexist in the same ground model (*heterogeneous modelling*). This leads to the *integration* problem and to the need for a unifying *meta-model*, as discussed in detail in the next section.

Regarding the construction of ground models (as part of the system analysis and design process), one can identify general principles for justifying their appropriateness and validity. In a discussion of methodological guidelines building on epistemological insights, Egon Börger convincingly argues that this justification process has three basic dimensions ([2], Sect. 2.2):

Conceptual justification aims at a direct comparison of the ground model with a given part of the real world (e.g. as described by an informal requirement specification).

Experimental justification can be provided for the ground model by accompanying it with clearly stated system test and verification conditions (system acceptance plan), i.e. falsifiability criteria in the Popperian sense [13] which lay the ground for objectively analyzable and repeatable experiments.

[2] In particular, the relation between the given system and its formal model should be made as evident as possible: in fact, the model should be readily understood by persons who are experts of the application domain, but not necessarily of the formalism, as their judgements are essential to establish whether the model is faithful to reality. In this respect, graphical formalisms with a precise semantics can be very useful.

Mathematical justification As opposed to the conceptual and experimental justification of a ground model, the mathematical justification aims at establishing its internal consistency and is essentially a problem of high–level reasoning and (where possible machine assisted) proof checking.

The reliability of ground models is an often underestimated aspect in applied systems engineering. Taking into account possible consequences of specification errors that are not discovered in early design phases, the validation of formal requirement specifications by means of conceptual and experimental means is certainly one of the most important steps in the entire system design and development process (albeit, the best we can achieve is a "*pragmatic foundation*" [2]).[3] Once we have established the adequacy and validity of a ground model with sufficient confidence and evidence, other models (also models expressed in other formalisms, e.g. as required for machine–supported analysis and proofs) may be derived by applying purely mathematical transformation techniques (where the correctness of each transformation step can in principle be proved in a strict mathematical sense).

3 Heterogeneous System Modelling

In this section, we first discuss the integration of heterogeneous system models from a general perspective. Then we focus on the integration of (heterogeneously specified) controller and device models.

3.1 Overview

Complex technical systems are, by their very nature, heterogeneous: they may consist of mechanical and electronic devices, sensors and actuators, measuring devices, communication media, computer programs, and sometimes even involve human beings (e.g., operators). In order to deal with all the distinct facets of such systems, many different modelling paradigms and languages have been developed over the years. By now, well established languages, methods and tools are available and used with success for the design of almost every single system aspect. For instance, even if we restrict attention to the IT-related parts of such systems, we find a large number of complementary modelling paradigms (such as synchronous, asynchronous, control-flow, data-flow, with or without real-time constraints) and corresponding formalisms (e.g., Statecharts, synchronous languages, SDL, VHDL, different Petri net variants, etc.). Not yet answered is, in general, the question, how to combine such heterogenous descriptions into one consistent and coherent system model which can undergo well-defined validation and analysis processes.

[3] Model-checking of system properties, the way we employed it in the work documented here, should also be considered as a means of *experimental* justification: in fact, it is (when applicable) a systematic and highly efficient way of conducting experiments.

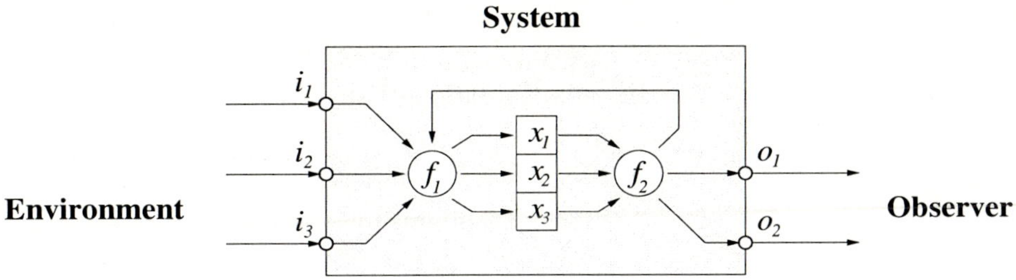

Fig. 1. Open System View

In order to model individual system components, it is convenient to follow an *open system view*, where the interaction between such a component and its environment is modelled by introducing well-defined *interfaces* (this also emphasizes the reactivity aspect). On the one hand, such a view is useful in order to describe the behaviour of individual components or subsystems without the need to model details of the surrounding environment. As a consequence, different components or subsystems can be specified using different (and most appropriate) formalisms. On the other hand, whenever one wants to validate a system model or verify particular properties of this model, one is usually interested in the *overall* system behaviour, and not in the behaviour of single components or subsystems (possibly under some boundary conditions, i.e., assumptions on the environment's behaviour).[4]

Thus, *composition mechanisms* are needed: they should allow to build—out of the component/subsystem specifications—a unified model of the whole system. Moreover, it is essential for these composition mechanisms to be able to deal with heterogeneous models, as they should act as *interfaces* between components or subsystems which are described in different languages/styles. Actually, there is nothing in the composition mechanisms themselves that hinders them from working in an heterogeneous setting (consider, for instance, mechanisms such as sharing of state variables, message passing, or connection of ports, which are very general). However, it is impossible to precisely understand and analyse the behaviour of the composed system in the absence of a common semantic framework (i.e., a *meta-model*) capturing both the behaviour of the system components as well as of the composition mechanisms in a uniform way.

In the remainder of this paper, we are not going to introduce new ways of composing systems or discuss issues of compositionality from a theoretical point of view. Instead, we propose the use of Abstract State Machines as a *meta-model* for the integration of heterogeneous descriptions. Starting from a concrete application (a distributed material flow system, or MFS, in automated manu-

[4] Note also that, in real-life situations, following an "academic" approach, where one tries to describe environment assumptions and deduce their implications as abstractly and generally as possible, may be not only difficult, but also unnecessary: for instance, it often happens that the environment of a component or subsystem is fixed (e.g, by contract) or subject to severe constraints (e.g., due to available technology).

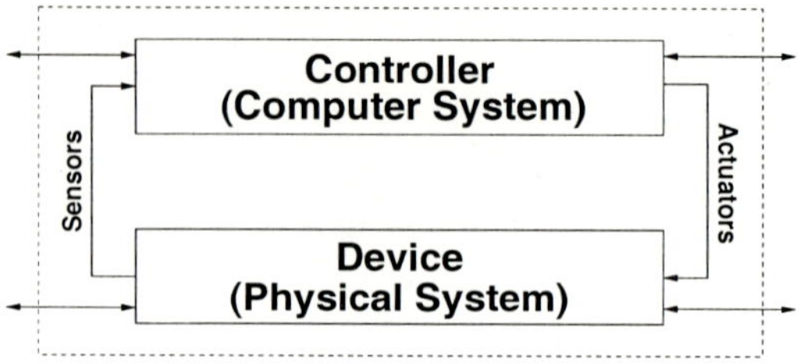

Fig. 2. Controller-Device Interaction in Embedded Control Systems

facturing), we identify description languages appropriate for describing parts of the system as well as composition mechanisms needed to put them together. To exemplify our approach, we consider a particular subsystem of the MFS (an "intelligent" switch module), show how different description languages can be used for modelling parts of the switch, and how the submodels can be integrated by embedding them into the meta-model. Finally, we validate the resulting system model, by means of simulation and symbolic model-checking.

3.2 Integration of Controller and Device Models

Design of embedded control systems requires to deal with two kinds of conceptually different units, the control units (*controllers*) and the controlled physical systems (*devices*), which are interfaced with each other by means of *sensors* and *actuators* (see Fig. 2). In addition to the connections corresponding to sensors and actuators, the picture also shows further connections to the external world. In fact, a complex technical system may consist of a network of such controller/device units (forming a *distributed* system, such as the material flow system from our case study).

A very important aspect in the specification of embedded control systems—especially in view of their validation and verification—is a clear separation between controller and device model, and the precise modelling of the device behaviour. Especially computer scientists often tend to overlook this aspect and only specify the controller behaviour (possibly very thoroughly and formally), but leave everything else (the "*environment*") out of the formal model.[5]

Note that the ultimate aim of the validation process is not to prove properties of the *control program* under some assumptions on the environment, but to show that the *controlled device* achieves the intended goals while ensuring some safety requirements (for instance, transporting something from one place to another, while avoiding collisions which can damage the system). Hence, we need formal models of both the controller and the device (where the device model, by itself, reflects the behaviour of the *uncontrolled* device): only on the basis of

[5] See, for instance, [9], where a production cell controller is formally specified using SDL, but no formal model of the device is given.

the combined controller/device model it is possible to precisely formulate and possibly prove properties of interest, which follow from the interaction of the control program with the physical system behaviour.

In general, the physical device is much more complex than the control program, as it involves a large number of parameters (including continuous ones). Due to this complexity, the development of a completely faithful and detailed device model is often not a feasible task. However, many details are not necessary in order to formulate and prove properties of interest. If appropriate *abstractions* are made, the task of modelling the physical device becomes affordable.

A typical example is *discretization*, where continuous values in the physical system are replaced by a finite set of "critical" values. Discretization is the basic abstraction tecnique employed in our MFS case study. For example, consider the "switch" device of the MFS, whose central part consists of a rotating plate: in a detailed physical model, the state of the switch plate is given by its rotation angle (a continuous value), but the essential information is whether it is in one of its stable states (left or right end position) or moving between them (clockwise or counterclockwise). A complete discretized model of the physical switch can be represented by means of a high-level Petri net, as shown in Sect. 5 (Fig. 5).

The methodological approach sketched above is very close to the one of [3], which deals with modelling and verification of the well-known case study "Production Cell" [10] by means of Statecharts. This paper also emphasizes the importance of a formal behavioural model of the physical device, as well as the need for clearly identified abstractions. Compared to [3], the originality of our approach is that it allows for *heterogeneous* models. The ability to employ different languages/formalisms within the system's ground model leads in many situations to descriptions which are more intuitive and thus easier to relate to the system being modelled (which is an essential requirement for a reliable ground model, as discussed in Sect. 2). Although this kind of improvement is clearly subjective and can not be quantified, we argue that, for instance, our Petri net model of the physical switch reflects the switch device much more directly than the Statechart model in [3] does for the Production Cell, where much encoding overhead is introduced[6] (in fact, using Statecharts for the device model there was mainly motivated by the verification tools requiring a Statechart model as input).

4 Abstract State Machines

Building on the *Abstract State Machine* (*ASM*) approach to mathematical modelling of discrete dynamic systems [7,2], we use here the special class of *multi-agent ASMs* as a formal basis for dealing with reactivity and concurrency. ASMs combine the abstract operational view of transition systems (for behavioural modelling) with declarative concepts of first-order logic (for data modelling).

[6] On the other hand, the Statechart formalism allows for a concise and intuitive specification of the finite state machine underlying the control unit.

They allow to formalize complex system behaviour in terms of executable models using familiar structures and notations from discrete mathematics and computer science in a direct and intuitive way. The scope of applications ranges from distributed control systems over discrete event simulation systems to formal semantics of system modelling languages (like VHDL and SDL [6]).[7] In this section, we recall some basic notions of ASM (see [7] for the complete definition). We first describe the computational model underlying ASMs, and then the syntax and semantics for a core subset of the ASM language.

4.1 Computational Model

Computations Abstract State Machines define a state-based computational model, where computations (*runs*) are finite or infinite sequences of states $\{S_i\}$, obtained from a given *initial state* S_0 by repeatedly executing *transitions*. Such runs can be intuitively visualized as

$$S_0 \xrightarrow{\delta_1} S_1 \xrightarrow{\delta_2} S_2 \ldots \xrightarrow{\delta_n} S_n \ldots$$

where the S_i are the states and the δ_i the transitions.

States The *states* are algebras over a given *signature* Σ (or *Σ-algebras* for short). A signature Σ consists of a set of *basic types* and a set of *function names*, each function name f coming with a fixed arity n and type $T_1 \ldots T_n \to T$, where the T_i and T are basic types (written $f : T_1 \ldots T_n \to T$, or simply $f : T$ if $n = 0$)[8]. A Σ-algebra (or state) S consists of: *(i)* a nonempty set $\mathcal{T}^S$ for each basic type T (the *carrier set* of T), and *(ii)* a function

$$\mathbf{f}_S : \mathcal{T}_1^S \times \ldots \times \mathcal{T}_n^S \to \mathcal{T}^S$$

for each function name $f : T_1 \ldots T_n \to T$ in Σ (the *interpretation* of the function name f in S). Function names in Σ can be declared as:

- *static*: static function names have the same (fixed) interpretation in each computation state;
- *dynamic*: the interpretation of dynamic function names can be altered by transitions fired in a computation step (see below);
- *external*: the interpretation of external function names is determined by the environment (thus, external functions may change during the computation as a result of environmental influences, but are not controlled by the system).

Any signature Σ must contain a basic type *BOOL*, static nullary function names (constants) *true* : *BOOL*, *false* : *BOOL*, the usual boolean operations ($\wedge$, $\vee$, etc.),

---

[7] For a comprehensive overview on ASM applications and other available material, see also the annotated ASM bibliography [1] as well as the following two URLs: `http://www.eecs.umich.edu/gasm/` and `http://www.uni-paderborn.de/cs/asm/`.

[8] Note that, while the definition given by Gurevich in [7] is untyped and uses classical first-order structures as states, we prefer to see states as multi-sorted algebras: from a conceptual point of view and for the purpose of this paper this makes no difference.

and the equality symbol $=$. Finally, there is a special constant *undef* $: T$ for any basic type T except *BOOL*. When no ambiguity arises we omit explicit mention of the state S (e.g., we write $\mathcal{T}$ instead of $\mathcal{T}^S$ for the carrier sets, and $\mathbf{f}$ instead of $\mathbf{f}_S$ for static functions, as they never change in the course of a computation).

Locations If $f : T_1 \dots T_n \to T$ is a dynamic or external function name, we call a pair $l = (f, \overline{x})$ with $\overline{x} \in \mathcal{T}_1 \times \dots \times \mathcal{T}_n$ a *location* (then, the *type* of l is T and the *value* of l in a state S is given by $\mathbf{f}_S(\overline{x})$). Note that two states S_1 and S_2 are equal iff the values of all locations in S_1 and S_2 are equal (i.e., they coincide iff they coincide on all locations).

Transitions Transitions transform a state S into its successor state S' by changing the interpretation of some dynamic function names on a finite number of points (i.e., by updating the values of a finite number of *locations*).

More precisely, the transition transforming S into S' results from firing a finite *update set* Δ at S, where the *updates* are of the form $((f, \overline{x}), y)$, where $(f, \overline{x})$ is the location to be updated and y the value. The state S' resulting from firing Δ at S is such that the carrier sets are unchanged and, for each function name f:

$$\mathbf{f}_{S'}(\overline{x}) = \begin{cases} y & \text{if } ((f, \overline{x}), y) \in \Delta \\ \mathbf{f}_S(\overline{x}) & \text{otherwise.} \end{cases}$$

The update set Δ—which depends on the state S—is determined by evaluating in S a distinguished *transition rule* P, called the *program*.[9] Note that the above definition is only applicable if Δ does not contain any two updates $((f, \overline{x}), y)$ and $((f, \overline{x}), y')$ with $y \neq y'$ (i.e., if Δ is *consistent*).

4.2 The ASM Language

Terms Terms are defined as in first-order logic: if $f : T_1 \dots T_n \to T$ is a function name in Σ, and t_i is a term of type T_i (for $i = 1, \dots, n$), then $f(t_1, \dots, t_n)$ is a term of type T (written $t : T$).[10] The meaning of a term $t : T$ in a state S is a value $S(t) \in \mathcal{T}$ defined by

$$S(f(t_1, \dots, t_n)) = \mathbf{f}_S(S(t_1), \dots, S(t_n)).$$

Transition rules While terms denote values, transition rules (*rules* for short) denote *update sets*, and are used to define the dynamic behaviour of an ASM: the meaning of a rule R in a state S is an update set $\Delta_S(R)$.

ASM runs starting in a given initial state S_0 are determined by the program P: each state S_{i+1} ($i \geq 0$) is obtained by firing the update set $\Delta_{S_i}(P)$ at S_i:

$$S_0 \overset{\Delta_{S_0}(P)}{\longrightarrow} S_1 \overset{\Delta_{S_1}(P)}{\longrightarrow} S_2 \dots \overset{\Delta_{S_{n-1}}(P)}{\longrightarrow} S_n \dots$$

The syntax and semantics of rules are as follows.

---

[9] In applications of ASM, the program consists usually of a set (block) of rules, describing system behaviour under different—usually mutually exclusive—conditions.

[10] If $n = 0$ the parentheses are omitted, i.e. we write f instead of $f()$.

Skip rule The simplest rule is the *skip* rule, which simply does nothing, i.e. its semantics is an empty update set: $\Delta_S(\texttt{skip}) = \{\,\}$.

Update rule The *update* rule has the syntax

$$R ::= f(t_1, \ldots, t_n) := t$$

where $f : T_1 \ldots T_n \to T$ is a dynamic function name in Σ, $t_i : T_i$ for $i = 1, \ldots, n$, and $t : T$. Such an update rule produces a single update:

$$\Delta_S(R) = \{((f, (S(t_1), \ldots, S(t_n))), S(t))\}.$$

Intuitively, the terms t_i and t are evaluated—in the state S—to values $x_i = S(t_i)$, $y = S(t)$; then, the interpretation of f on $(x_1, \ldots, x_n)$ is changed to y.

Block rule The *block* rule

$$R ::= R_1 \ldots R_n$$

combines the effects of more transition rules:

$$\Delta_S(R) = \bigcup_{i=1}^{n} \Delta_S(R_i).$$

Executing a block rule corresponds to *simultaneous* execution of its subrules.[11]

Conditional rule The *conditional* rule has the syntax

$$R ::= \texttt{if } G \texttt{ then } R_T \texttt{ else } R_F$$

where G is a boolean term. Its meaning is, obviously:

$$\Delta_S(R) = \begin{cases} \Delta_S(R_T) \text{ if } S(G) = \textbf{true} \\ \Delta_S(R_F) \text{ otherwise.} \end{cases}$$

The short form "`if` G `then` R" is also used instead of "`if` G `then` R `else skip`".

4.3 Multi-agent ASM

Concurrent systems can be modelled in ASM by the notion of multi-agent ASM (called *distributed ASM* in [7]). The basic idea is that the system consists of more *agents*: each agent $a \in AGENT$[12] executes its own program $prog(a)$ and can identify itself by means of a special nullary function $self : AGENT$, which is interpreted by each agent a as a.

In [7] several semantical models for multi-agent ASM are discussed, the most general being *partially ordered runs*. For our purposes, a simple interleaving semantics is sufficient and allows us to model concurrent systems in the basic ASM formalism as described in Sect. 4.2. In particular, we consider *self* as an external function, whose interpretation $\mathbf{self}_{S_i}$ determines the agent which fires at

[11] For example, a block rule `a` := `b`, `b` := `a` exchanges `a` and `b`. Note also that the use of block rules may lead to inconsistent update sets.

[12] Note that agents are identified with elements of the domain $AGENT$, which are actually a sort of "agent identifiers".

state S_i. We assume that there is one program P, shared by all agents, possibly performing different actions for different agents, e.g.:

$$\begin{array}{l} \texttt{if } \mathit{self} = a_1 \texttt{ then } \mathit{prog}(a_1) \\ \dots \\ \texttt{if } \mathit{self} = a_n \texttt{ then } \mathit{prog}(a_n) \end{array}$$

where $\{a_1, \dots, a_n\}$ are the agents and $\mathit{prog}(a_i)$ is the rule to be executed by agent a_i, i.e., the "program" of a_i.

4.4 The ASM-SL Language

The ASM language, including all constructs above, is supported by the "ASM Workbench" tool environment [4], which provides syntax- and type-checking of ASM specifications as well as their simulation and debugging. The source language for the ASM Workbench, called ASM-SL, includes some additional features which are necessary for practical modelling tasks: constructs for defining types, functions, and named transition rules ("macros"), as well as a set of predefined data types (booleans, integers, tuples, lists, finite sets, etc.): as the ASM-SL notation is quite close to usual mathematical notation, no further explanation of ASM-SL will be needed.

5 A Case Study from Automated Manufacturing

In this section, we present an industrial case study, a distributed material flow system (MFS), to illustrate the methodological approach sketched in the previous section. First, we give an overview of the case study and the related problems, then we concentrate on a subsystem (the switch module) and use it as a running example to discuss the proposed modelling and analysis techniques.

5.1 Overview

The subject of our case study is the distributed control for a modular *material flow system*. The case study has been provided by the Computer Integrated Manufacturing (CIM) group of our institute, with which we cooperate in the research project *ISILEIT*.[13]

Our MFS is based on a modern modular material flow technology that allows to build complex transportation topologies by composing standard modules (essentially: straight and curved *tracks*, *switches*, and *halting points*, see Fig. 3). Special vehicles called *shuttles* are employed to transport pieces over the railway between halting points (corresponding, for instance, to machines or stores).

[13] *ISILEIT* is a project funded by the DFG (the German Research Foundation) within the program "Integration of Software Specification Techniques for Engineering Applications" and is a cooperation between our group, the software engineering group, and the CIM group of the Heinz Nixdorf Institut.

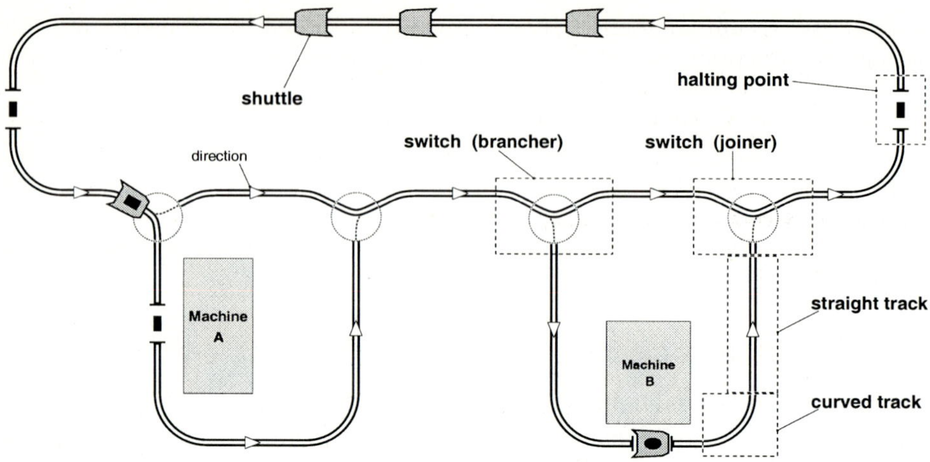

Fig. 3. Example of Modular MFS

While operating, tracks are permanently supplied with current, such that shuttles keep moving over the railway in a fixed direction, unless they are arrested by means of *stopping cams* (placed at certain points along the tracks). Finally, each shuttle has a built-in device that avoids collisions by enforcing a minimum distance from the foregoing shuttle.

Traditionally, the control of a MFS is centralized: a central control unit drives the physical MFS according to predefined transport plans and routes, which implement the material flow for a given manufacturing process. However, the modular structure of the MFS sketched above suggests an alternative solution, based on *distributed control*. Instead of being controlled by a central unit, each of the modules of the MFS (e.g., shuttle, switch, or halting point) is controlled by a corresponding *local* control unit (or *control node*), cooperating with other control nodes in order to achieve the global goals of the MFS, i.e., to execute the given transportation tasks.[14] Thus, the control of the MFS is implemented by a network of concurrently operating and asynchronously communicating controllers. The only non-distributed part of the system is an host PC connected to the network for the purpose of user interaction (such as definition, transmission, and visualization of the transportation tasks), which however plays no role in the actual MFS control.

Such a distributed and modular architecture has obvious advantages over a centralized one in terms of scalability, reconfigurability, and fault tolerance,

[14] The control nodes are not necessarily dedicated processing units *physically bound* to the devices being controlled. Instead, each control node is a process *logically associated* to the controlled device. In our MFS, for instance, the control nodes for the switches are processing units physically located by the switch, while the control nodes for the shuttles—for technical reasons, such as cabling—are not located on the shuttles, but implemented on computers placed outside the physical MFS.

but its software is more difficult to design, implement, and validate. In fact, the implementation does not immediately reflect the system behaviour to be realized. Instead, the overall system behaviour results from the interplay of quite a large number of processes, each one executing its own protocol, having only quite limited knowledge of the system's state. How the proposed methodology can help in the design and validation of this kind of systems is what we are going to illustrate in the sequel of this section. In order to do this, we consider a subsystem of the MFS, namely the *switch* module.

5.2 The Switch Module

As shown in Fig. 3, there are two kinds of switch modules, called *brancher* and *joiner*, respectively. As the names suggest, a brancher directs shuttles to the one or the other destination, whereas a joiner reunites two paths into one. The brancher and the joiner differ slightly, both in their physical composition and in their control. The following discussion refers to the brancher, which is slightly more complex and interesting than the joiner.[15]

The switch module, depicted in Fig. 4, is built around a *switch drive* (*SD*), which can modify the state of connection between MFS tracks by inducing a rotation of the switch plate around its center. A few additional components are needed to ensure the correct and safe operation of the switch, namely:

- an *identification unit* (*ID*), which detects the passage of a shuttle and, at the same time, ascertains its identity (given by a conventional *shuttle id*);
- a *stopping cam* (*SC*), which may stop a shuttle or let it pass (recall that shuttles keep moving, if not hindered from doing so);
- *passing sensors* (*PS*), which detect the passage of a shuttle (without identifying it).

Note that the length of the track segment between ID and SC is such that at most one shuttle can occupy this segment. This is ensured by the hardware enforcement of the mininum distance, mentioned in Sect. 5.1 (see Fig. 4).

The interface between the physical switch and the corresponding control node is represented by an appropriate set of messages, as shown in the picture. The communication takes place over a bidirectional channel. Messages received by the controller are notifications from the sensors, messages sent by the controller are commands for the actuators. In particular, we distinguish the following message types:

- *identification*(*sh*) is sent by the identification unit whenever the shuttle with id *sh* passes over it;
- *stop*(*b*) controls the stopping cam: if $b = true$ it switches to the stop position, if $b = false$ it is released;

[15] In particular, the brancher needs to distinguish between the individual shuttles in order to direct them to the correct destination, while the joiner must simply let any incoming shuttle pass.

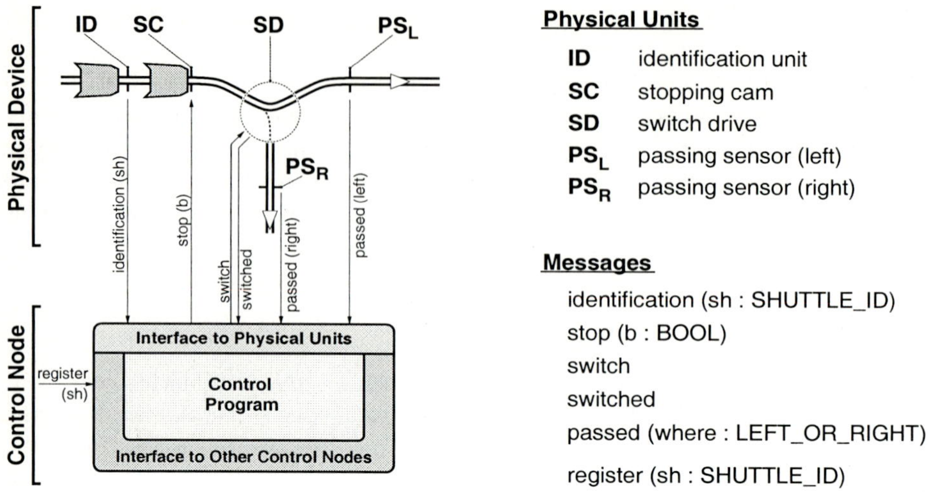

Fig. 4. The Switch Module

- *switch* is sent to the switch drive in order to let it start its rotation: when the rotation is completed, the switch notifies that the end position is reached by sending back the acknowledgement message *switched*;
- *passed*(*left*) and *passed*(*right*) are sent by the left and right passing sensors, respectively, whenever a shuttle passes over them.

We call the protocol regulating the communication between a control node and the corresponding physical device (which involves, in the case of the switch, the messages listed above) *low-level protocol*. A low-level protocol is responsible for ensuring local properties of the MFS modules, e.g., safety properties such as "*a shuttle does not move onto the switch plate while the plate is moving*".

The *high-level protocols*, instead, are run between neighbouring control nodes in order to direct the overall shuttle traffic in the proper way and thus cooperatively achieve the high-level goals of the MFS (i.e., executing the requested transportation tasks). In the case of a brancher, the high-level protocol is fairly simple, as its task is just to send the incoming shuttles in the "correct" direction. The information about the correct direction is provided by the halting point where a shuttle stopped before coming to the switch under consideration. As soon as the shuttle leaves the halting point, the control node of the latter "registers" the shuttle by the next (brancher) switch, to inform it about the direction that the shuttle has to take.[16]

[16] Actually, as shown in Fig. 4, the *register* message does not mention the direction. In fact, our model follows an existing implementation that, in order to reduce message traffic, distinguishes between a "standard" and a "non-standard" direction: shuttles going in the non-standard direction are registered, while non-registered shuttles are supposed to go in the standard direction (the one expected to be taken most often).

5.3 Physical Switch: A Petri Net Model

The physical switch can be formally described by means of a high-level Petri net, shown in Fig. 5. This model reflects both the topology and the behaviour of the physical device. In particular, different track segments within the switch module are represented by places **S1**, **S2**, ..., each of which contains a list of shuttle ids as its only token.[17] The movement of shuttles between these places is represented by corresponding transitions t1, t2, ..., which update the lists appropriately (the condition on t1 reflects the assumption about the length of the track segment **S2** between ID and SC, stated informally in Sect. 5.2). The other places represent other aspects of the physical switch state, namely:

- the position of the stopping cam (which can be either in stop position or released, as reflected by the place *passing_enabled*), and
- the state of the switch plate (which can either stay in one of the end positions *left* or *right*, or be rotating from left to right or from right to left, as represented by the places *moving_lr* and *moving_rl*, respectively).

A peculiarity of the net model shown in Fig. 5 is that some transitions are annotated by an input or by an output symbol (whose graphical representation is borrowed from SDL). In this way, we model the interaction of the physical switch with the environment according to an open system view. Intuitively, the meaning of an input annotation is that the given transition is fired depending on a given input signal (i.e., the transition is triggered by an external event). Similarly, an output annotation means that a given output signal is emitted when the corresponding transition occurs. Note that the occurrence of such a transition is "spontaneous", in the sense that the transition happens as a possible consequence of the internal workings of the system, and not as a necessary reaction to an external stimulus.[18]

The semantics of the input and output annotations will be made precise in Sect. 5.5, where it will also be shown how the net model of the physical switch can be composed with the SDL model of its controller. By going from an open system view over to a closed system view, meaningful statements about the system behaviour can be formulated.

---

[17] We use this representation in order to reflect the sequence of shuttles present on a track segment in a given state. This is important, as shuttles leave a track segment in the same order as they entered it, i.e., they can not overtake each other.

[18] For instance, as shuttles always keep moving (unless explicitly stopped), a given shuttle will eventually leave the track segment it occupies and enter the next one. This event is reflected by the occurrence of a transition in our net model. These transitions are typical examples of "spontaneous" or "internal" transitions. Optionally, such a transition can emit an output to notify the event to the environment, making the event visible to an external observer (in Fig. 5, *t1*, *t4l* and *t4r* are of this kind).

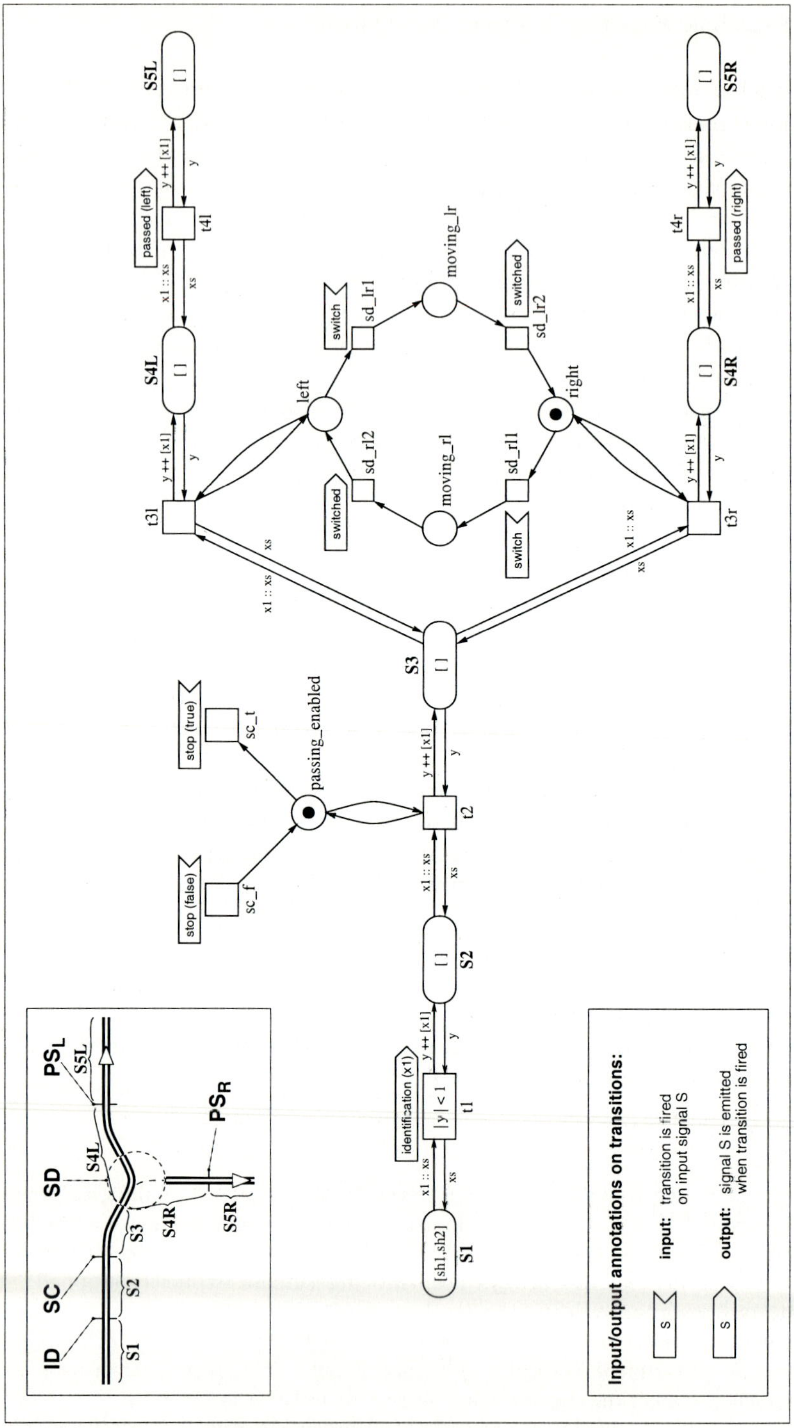

Fig. 5. Physical Model of the Switch

5.4 Switch Controller: An SDL Model

The switch controller is specified by means of SDL *process diagrams*, which represent a kind of extended finite state machines. The process diagrams for its three control states (*stable*, *waiting*, and *switching*) are shown in Fig. 6. Within the diagrams the following (local) state variables are used (the corresponding SDL declaration and initialization parts are not shown here for reasons of space):

- *direction* : { *left*, *right* }, initialized as the standard direction[19] *std*, corresponds to the controller's knowledge about the switch position;
- *registered* : *SHUTTLE_ID-set*, initialized as an empty set, keeps track of the registered shuttles;
- *passing* : *INT*, initialized as 0, is a counter keeping track of the number of shuttles occupying the critical area of the switch (i.e., the switch plate and its immediate surroundings): this information is important because it is safe to activate the switch drive only when this area is free, i.e., when *passing* $= 0$.

Note that the state variables of the controller must be consistent with the physical system state: as this must hold, in particular, in the initial state, it implies that, when the MFS is started, its (physical) state must correspond to the controller state, e.g., all the switches must be in the standard direction.

We do not go into further details here, as Fig. 6 already contains the complete specification of the switch controller. However, we try to explain the basic idea of how the controller works in a typical case. When a shuttle arrives to the switch, it passes over the identification unit, which informs the controller. The controller, depending on the identity of the incoming shuttle, checks if the switch is already in the right position. If so, it lets the shuttle pass; otherwise it puts the stopping cam in the stop position and: *(a)* if the critical area is free, it activates the switch drive and goes into "switching" mode, where it waits for the acknowledgement message *switched* from the switch drive; *(b)* if the critical area is not free, it goes into "waiting" mode, where it waits for the critical area to become free before it can begin switching.

While executing these steps, care must be taken to keep the physical switch state and the controller state consistent. If inconsistencies arise, the controller may end in the "ERROR!" state (for which no behaviour is defined), or may behave unpredictably. Note however that, if the protocol is correctly implemented, this can only happen in case of hardware failures (e.g., if the identification unit detects passage of shuttles when there is none). Thus, a result to be expected from the analysis of the integrated controller/device model is that the controller never reaches the "ERROR!" state, as the device model (the Petri net) reflects only the failure-free behaviour of the physical switch.

---

[19] We make use of a constant *std* standing for the standard switch direction (either *left* or *right*, depending on the MFS configuration).

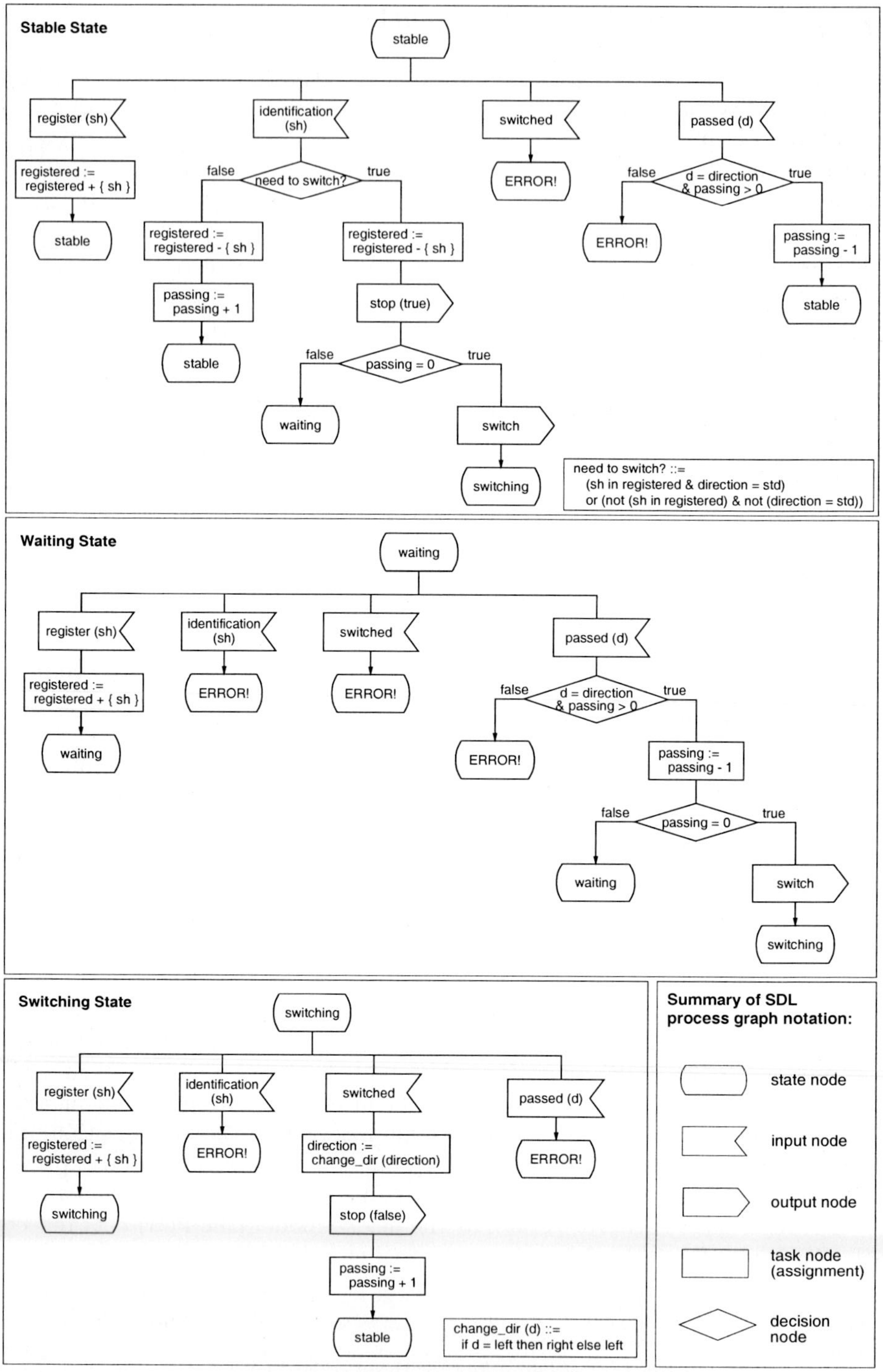

Fig. 6. Switch Controller

5.5 Integrating Device and Controller Models

As a prerequisite for discussing details of the integration technique, we have to make precise the communication model and the semantics of the input and output annotations on net transitions in Fig. 5. We follow SDL in assuming an asynchronous communication model with buffered channels. Consequently, the meaning of input and output symbols in the net model can be formalized by adding to the net, for each (input or output) message queue, a place containing a list of messages and, for each input or output symbol on a transition, the corresponding edges to check and update the corresponding message queue, as shown in Fig. 7.

An aspect which is not explicitly specified, neither in the net model nor in the SDL process graphs, is the association between individual input/output commands (or individual messages) and the channels (message queues) over which they are to be sent/received. This association can in general be described by means of SDL *block diagrams*. Here, we omit the block diagram describing the connections between device and controller, as it is quite trivial: we assume that the communication between controller and device takes place over a bidirectional channel, such that two message queues (one for each channel direction) are needed. We call these message queues *sensQ* and *actQ*, standing for "sensor queue" and "actuator queue", respectively.[20]

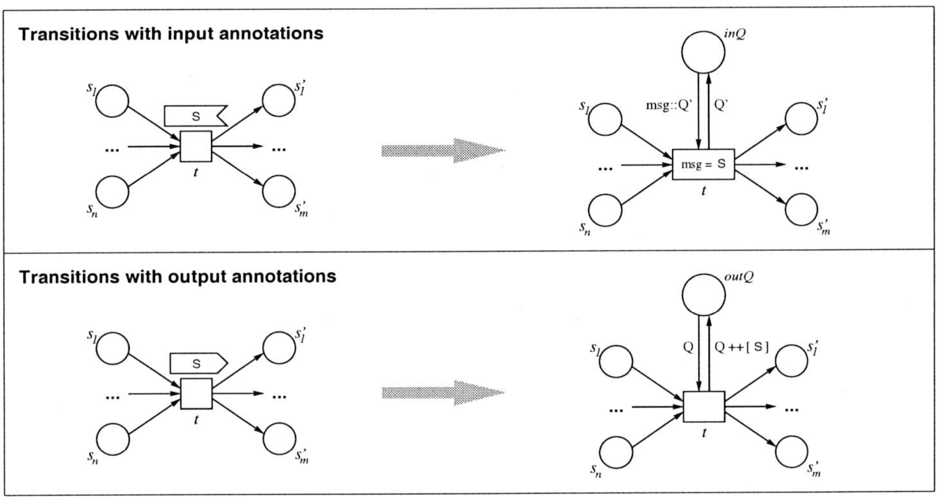

Fig. 7. Input/Output Annotations on Net Transitions

[20] Related to the high-level protocol, there is an additional message queue *registrQ* corresponding to the input channel on which the preceding halting point(s) send the *register* message to the switch, in order to communicate the arrival of a given shuttle. However, there will be no further mention of this in the rest of this paper, as we focus on the device/controller integration.

The integration of the two switch models, the device model and the controller model, into a uniform behavioural model amenable to computer-aided analysis (simulation, model-checking) is then achieved as follows:

1. both the net model of the physical device and the SDL model of the controller are mapped to equivalent ASM models
2. interaction between the resulting models of the two subsystems takes place by *sharing* message queues[21]
3. an additional ASM rule specifies the coordination between activities of concurrent subsystems of the switch module and thereby complements their behavioural specifications.

The mapping from high-level Petri nets to ASMs is realised by mapping each place to a dynamic function and by expressing the behaviour of each transition of the net by means of a boolean function and a transition rule in the ASM model. The boolean function corresponds to the *enabling condition* of the transition, while the transition rule specifies the *state change* produced when the transition occurs. We do not go into formal details of the transformation, which is quite straightforward, and illustrate it by the example of the switch's transition *t1* instead, which is shown in Fig. 8.[22]

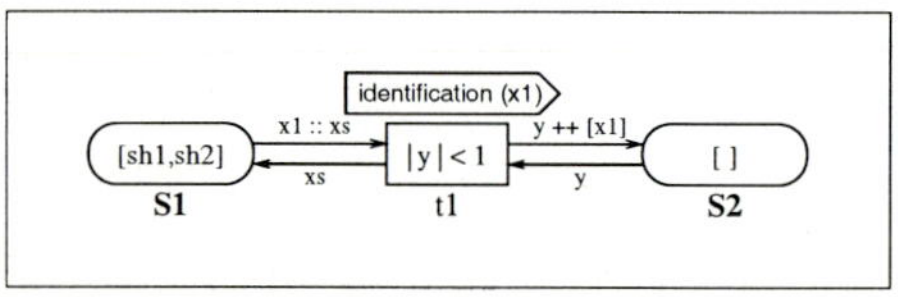


```
derived function t1_enabled ==
case (S1, S2, sensQ) of
  (x1::xs, y, Q) : length (y) < 1;
  otherwise false
endcase
```

```
transition t1_action ==
case (S1, S2, sensQ) of
  (x1 :: xs, y, Q) :
    if length (y) < 1
    then S1 := xs
         S2 := y @ [ x1 ]
         sensQ := Q @
           [ identification (x1) ]
    endif
endcase
```

Fig. 8. Mapping from High-Level Petri Nets to ASMs (transition: t1)

The mapping from SDL process graphs to ASMs is based on the SDL semantics defined in [6,16]. However, the corresponding ASM rules are simplified with respect to those which can be obtained from the SDL semantics of [6]. In particular, we assume that an SDL transition is executed in one ASM step, whereas the single actions of a transition should actually be executed in more

[21] Note, in fact, that the inputs for the controller are the outputs of the device (and vice versa). The unifying ASM model of the controller-device system makes the interaction explicit by showing how the controller and the device actually read/write those messages from/into the same message queue.

[22] See also [5] which deals with the integration of Pr/T nets and ASMs.

steps, sequentially. This simplication is important for the subsequent analysis of the model (especially by model-checking) in order to reduce the size of the state space. Of course, care should be taken in order to preserve the SDL transition semantics. Fig. 9 shows the ASM rule corresponding to the process graph for the switching state.

```
transition Switching ==
  case sensQ of
    msg :: restQ :
      case msg of
        identification (sh) : control_state := ERROR ;
        passed (d)          : control_state := ERROR ;
        switched            : direction   := change_dir (direction)
                              actQ := actQ @ [ stop (false) ]
                              passing := passing + 1
                              control_state := stable

      endcase
  endcase
```

Fig. 9. Mapping from SDL Process Graphs to ASMs (state: switching)

The mechanism which allows the interaction of the controller and device models consists in sharing the message queues *actQ* and *sensQ*, which are declared in the ASM model as

```
dynamic function actQ  :LIST(MESSAGE)  initially [ ]
dynamic function sensQ :LIST(MESSAGE)  initially [ ].
```

Note, for instance, that the transition t1 of the device model in Fig. 8 writes to `sensQ`, while the process graph for the switching state of the controller model in Fig. 9 reads `sensQ`, triggering an SDL transition if an appropriate message is present at the head of `sensQ` (which is the input queue for the SDL process).

Finally, to complete the model integration, *coordination* rules have to be specified to define a discipline by which all concurrently operating parts of the system (represented, in our model, by the single net transitions and the control process) cooperate to achieve the overall system goals.[23] We adopt a simple interleaving model of concurrency and define a main coordination rule (the ASM *program*) which non-deterministically chooses whether the controller or the physical switch makes a move (`phys_switch_step` and `controller_step` are the subrules corresponding to a move of the physical switch or of the controller, respectively):

[23] Coordination rules are either derived from the actual physical realization of the system (thus stating assumptions on the environment) or express requirements regarding the control software (thus stating guidelines for its implementation).

```
external function choose_round
with choose_round in { phys_switch, controller }

transition Main ==
  case choose_round of
    phys_switch : phys_switch_step ;
    controller  : controller_step
  endcase
```

Moreover, we have to model the implicit assumption that, while the transitions corresponding to shuttles passing track segment boundaries ("shuttle transitions") happen after some time, the transitions corresponding to *reactions* of both the controller and the physical switch to incoming messages ("reaction transitions", triggered by sensors/actors) happen "immediately".[24] A simple way to do this is to prioritise the transitions of the net model, such that shuttle transitions always have lower priority than reaction transitions. This can be achieved by an appropriate definition of `phys_switch_step`, the coordination rule for the physical switch (net model).

5.6 Analysis and Validation

Analysis and validation essentially are performed by means of simulation and model-checking. In particular, the combination of both is very effective in "debugging" the high-level system specification. The model-checker can be used to find counterexamples, i.e., runs which contradict the expected behaviour (specified by a set of properties expressed in temporal logic). Each counterexample can then be fed into the simulator in order to find out the origin of the wrong behaviour. Debugging features, which are very helpful in this regard, include possibility of executing single steps forward and backward, a sequence of steps until a given condition is satisfied, inspection of the system state, etc. After the specification is fixed, the whole process is reiterated, until all properties are satisfied.

The properties to be checked are specified in CTL (Computation Tree Logic), the temporal logic supported by the model checker SMV [11], which we employ for the verification task. We consider both *safety* and *liveness* properties, which altogether build an abstract *requirements specification* for the switch module:[25]

[24] "Immediately" does not mean here that these actions take no time, but that the time taken is neglectible compared to the time needed for other actions, in particular for a shuttle to traverse a track segment (in fact, the order of magnitude is of seconds for the latter, of milliseconds for the former). Note that we want to avoid explicit mention of time constraints here, as this would be a kind of over-specification at this level of abstraction.

[25] Intuitively, the *safety* properties correspond to requirements of the kind "*something bad never happens*", the *liveness* properties to requirements of the kind "*something good eventually happens*" (the system should not only operate safely, but also accomplish its task).

1. Safety (controller). "*The controller never reaches the* ERROR! *state*" (in fact, the ERROR! state indicates that some hardware failure happened):

$$\neg \mathsf{EF}\ (\text{control\_state} = \text{ERROR!})$$

2. Safety (device). "*A shuttle does not move onto the switch plate while the plate is moving*". This is ensured by requiring that the critical area of the switch, consisting of the switch plate and its immediate surrounding (places S3, S4L, S4R in the model), is always free whenever the switch plate is moving:

$$\mathsf{AG}\ (\text{moving\_lr} \vee \text{moving\_rl}\ \Rightarrow\ \text{S3} = [\,]\ \wedge\ \text{S4L} = [\,]\ \wedge\ \text{S4R} = [\,])$$

3. Liveness. "*All shuttles entering the switch will eventually leave it: unregistered shuttles leave it in the standard direction, registered ones in the non-standard direction.*": for each shuttle id "sh"

$$\mathsf{AG}\ (\mathit{contains}(\text{S1},\text{sh})\ \wedge\ \neg(\text{sh} \in \mathit{registered})\ \Rightarrow\ \mathsf{AF}\ \mathit{contains}(\text{S5L},\text{sh}))$$
$$\mathsf{AG}\ (\mathit{contains}(\text{S1},\text{sh})\ \wedge\ \text{sh} \in \mathit{registered}\ \Rightarrow\ \mathsf{AF}\ \mathit{contains}(\text{S5R},\text{sh}))$$

   where *contains* is a static predicate which tests if a given list contains a given element (for simplicity, we show the formulae for the special case $std = left$; for the case $std = right$, simply exchange S5L and S5R).

Note that, thanks to our integration approach, we can freely mix state variables of the controller model (which are actually *program variables*) and state variables of the device model (which are an abstraction of the *physical state* of the switch) within our abstract requirements specification. The best example of this is given by the two properties in (3.), where variables of both models occur within the same formula, in order to formalize the causal relation between the controller state at a given moment and the resulting physical system behaviour.

We could verify the properties above for instances of the problem with one, two, and three incoming shuttles. During the validation process (carried out according to the methodology sketched above, based on iteration of the model-checking/counterexample-simulation cycle), we detected a bug in the controller specification (in the control state *switching*, we forgot to increase the *passing* counter on releasing the stopping cam), and found out that the assumptions about immediate reaction of some transitions—mentioned at the end of Sect. 5.5—are essential for the correct working of the system.

6 Conclusions

In this paper, we presented an approach to heterogeneous system modelling and specification in a very early design phase (ground model construction). Heterogeneous modelling is particularly convenient in this phase, as it often allows for descriptions which are more intuitive and thus easier to relate to the corresponding real-world phenomena (if the most appropriate description formalisms for each part or aspect of the system are chosen). We have shown how Abstract

State Machines can be employed as meta-model for the mathematically well-founded integration of heterogeneous behavioural models, such that they also become amenable to computer-aided analysis and validation. In this context, the combination of model-checking and simulation provides a considerable help for the validation of ground models.

The MFS case study demonstrated the principal feasibility of our approach. Obviously, the switch module discussed in this paper is a very simple example (still, we have reported that, even in such a simple scenario, the validation process uncovered problems and mistakes). Difficulties are to be expected for larger and more complex systems (such as a complete MFS, as resulting from the interaction of several switches, shuttles, halting points, etc.), in particular because of the well-known problem of state space explosion in model-checking. Thus, future research should deal with the question, how state-of-the-art verification techniques (e.g., abstraction, compositional model-checking) can be integrated in the proposed methodology, in order to make it applicable to large systems.

References

1. E. Börger and J. Huggins. Abstract State Machines 1988-1998: Commented ASM Bibliography. *Bulletin of EATCS*, 64, February 1998.
2. E. Börger. High level system design and analysis using Abstract State Machines. In D. Hutter, W. Stephan, P. Traverso, and M. Ullmann, editors, *Current Trends in Applied Formal Methods (FM-Trends 98)*, volume 1641 of *LNCS*, pages 1–43. Springer, 1999.
3. W. Damm, H. Hungar, P. Kelb, and R. Schlör. Using graphical specification languages and symbolic model checking in the verification of a production cell. In *[10]*.
4. G. Del Castillo. Towards comprehensive tool support for Abstract State Machines: The ASM Workbench tool environment and architecture. In D. Hutter, W. Stephan, P. Traverso, and M. Ullmann, editors, *Current Trends in Applied Formal Methods (FM-Trends 98)*, volume 1641 of *LNCS*, pages 311–325. Springer, 1999.
5. U. Glässer. Modelling of concurrent and embedded systems. In F. Pichler and R. Moreno-Díaz, editors, *Computer Aided Systems Theory–EUROCAST'97 (Proc. of the 6th International Workshop on Computer Aided Systems Theory, Las Palmas de Gran Canaria, Spain, Feb. 1997)*, volume 1333 of *LNCS*, pages 108–122. Springer, 1997.
6. U. Glässer, R. Gotzhein and A. Prinz. Towards a new formal SDL semantics based on Abstract State Machines. In G. v. Bochmann, R. Dssouli and Y. Lahav, editors, *9th SDL Forum Proceedings*, pages 171-190. Elsevier Science B.V., 1999.
7. Y. Gurevich. Evolving Algebras 1993: Lipari Guide. In E. Börger, editor, *Specification and Validation Methods*, pages 9–36. Oxford University Press, 1995.
8. Y. Gurevich. The sequential ASM thesis. *Bulletin of the EATCS*, February 1999.
9. S. Heinkel and T. Lindner. The Specification and Description Language applied with the SDT support tool. In *[10]*.
10. C. Lewerentz and T. Lindner (eds.). *Formal Development of Reactive Systems – Case Study Production Cell*, volume 891 of *LNCS*. Springer, 1995.
11. K. McMillan. *Symbolic Model Checking.* Kluwer Academic Publishers, 1993.

12. F. Pichler. Systems Theory for Macro-Architecting in the Computer- and Information Sciences. In *Cybernetics and Systems '98*, ISBN 3-85206-139-3, Austrian Society for Cybernetic Studies, R. Trappl (ed.), pages 50-53, Vienna, 1998.
13. K. Popper. *Logik der Forschung.* 1935.
14. W. H. Wolf. Hardware-software co-design of embedded systems. *Proceedings of the IEEE*, 82(7):967–989, 1994.
15. W. H. Wolf. *Hardware-software co-design of distributed embedded systems.* Kluwer Academic Publishers, 1996.
16. *The Formal Semantics of SDL.* Technical report, Beijing University of Posts and Telecommunication. (See `http://tseg.bupt.edu.cn/`).

Patterns for Embedded Systems Design

M. Švéda

Department of Computer Science and Engineering
Technical University of Brno
Bo et•chova 2, 612 66 Brno, Czech Republic
sveda@dcse.fee.vutbr.cz

Abstract. This paper deals with embedded systems architecture components called as application patterns, and with their employment for design reuse. The first part of this contribution introduces the concepts of application patterns and relates them to the well-known object-oriented design abstractions. Employing application patterns that demonstrate the concrete examples of reusability, the kernel of this contribution presents two case studies, which are based on real design projects: petrol pumping station dispenser controller and multiple lift control system. To reuse an architectural component whose implementation usually consists both of software and hardware, it means to reuse its formal specification. The paper deals with behavioral specifications employing state or timed-state sequences and with their closed-form descriptions by finite-state or timed automata. The contribution focuses on identification, creation, and initial classification of reusable application patterns while retrieval, adaptation, and storage reuse tasks with case-based reasoning support are treated briefly at the conclusion as an introductory information about launching research.

1 Introduction

Methods and approaches in systems engineering are often based on the results of empirical observations or on individual success stories (Robillard, 1999). Every real-world embedded system design stems from decisions based on an application domain knowledge that includes facts about some previous design practice. Evidently, such decisions relate to systems architecture components, called in this paper as application patterns, that determine not only a required system behavior but also some presupposed implementation principles. Application patterns should respect those particular solutions that were successful in previous relevant design cases. While focused on the system architecture range that covers more than software components, the application patterns look in many features like object-oriented design concepts such as reusable patterns, see Coad and Yourdon, 1990, design patterns, see Gamma et al., 1995, and frameworks, see Johnson, 1997. Of course, there are also other related concepts such as use cases, see Jacobson, 1992, architectural styles, see Shaw and Garlan, 1996, or templates, see

F. Pichler, R. Moreno-Díaz, and P. Kopacek (Eds.): EUROCAST'99, LNCS 1798, pp. 80–89, 2000.

Turner, 1997, that could be utilized for the purpose of this paper instead of introducing a novel notion. Nevertheless, the system architectures of embedded applications are dealt preferably with no implicit reference to object-oriented modeling or to other special features contemplated in specifications of the above mentioned notions.

The following section of this paper introduces the principles of design reuse applied by way of application patterns. Then, employing application patterns fitting a class of real-time embedded systems, the kernel of this contribution presents two design projects: petrol pumping station dispenser controller and multiple lift control system. Via identification of the identical or similar application patterns in both design cases, this contribution proves the possibility to reuse substantial parts of formal specifications in a relevant subdomain of embedded systems. The last part of the paper deals with intended knowledge-based support for this reuse process applying case-based reasoning paradigm.

2 Design Reuse

To reuse an application pattern, whose implementation usually consists both of software and hardware components, it means to reuse its formal specification, development of which is very expensive and, consequently, worthwhile for reuse. This paper is aimed at behavioral specifications employing state or timed-state sequences, which correspond to the Kripke style semantics of linear discrete time temporal or real-time logics, and at their closed-form descriptions by finite-state or timed automata (Alur and Henzinger, 1992).

Comparing to systems design reuse, software design reuse is currently highly published topic, see e.g. Arora and Kulkarni, 1998, Sutcliffe and Maiden, 1998, Mili, Mili, and Mittermeir, 1997, Holtzblatt et al., 1997, Henninger, 1997. Namely the *state-dependent specification-based approach* discussed by Zaremski and Wing, 1997, and by van Lamsweerde and Willemet, 1998, inspired the application patterns handling presented in the current paper. To relate application patterns to the previously introduced software oriented concepts more definitely, the inherited characteristics of the archetypal terminology, omitting namely their exclusive software and object-orientation, can be restated as follows. A *pattern* describes a problem to be solved, a solution, and the context in which that solution works. Patterns are supposed to describe recurring solutions that have stood the test of time. Design patterns are the micro-architectural elements of frameworks. A *framework* -- which represents a generic application that allows the creation of different applications from an application (sub)domain -- is an integrated set of patterns that can be reused. While each pattern describes a decision point in the development of an application, a *pattern language* is the organized collection of patterns for a particular application domain, and becomes an auxiliary method that guides the development process (see the pioneer work Alexander, 1977, which belongs surprisingly to the field of buildings architecture).

Application patterns correspond not only to design patterns but also to frameworks while respecting multi-layer hierarchical structures. Embodying domain knowledge, application patterns deal both with requirement and implementation specifications. In fact, a precise characterization of the way, in which implementation specifications and requirements differ, depends on the precise location of the interface between an embedded system, which is to be implemented, and its environment, which generates requirements on system's services. However, there are no strict boundaries in between: both implementation specifications and requirements rely on designer's view, i.e. also on application patterns employed.

A design reuse process involves several necessary reuse tasks that can be grouped into two categories: supply-side and demand-side reuse, see Sen, 1997. Supply-side reuse tasks include identification, creation, and classification of reusable artifacts. Demand-side reuse tasks include namely retrieval, adaptation, and storage of reusable artifacts. For the purpose of this paper, the reusable artifacts are represented by application patterns.

The following two sections of this contribution describe two case studies based on implemented design projects, using application patterns that enable to discuss the concrete examples of application patterns reusability.

3 Petrol Dispenser Control System

The first case study pertains to a petrol pumping station dispenser with a distributed, multiple microcomputer counter/controller, see Švéda, 1996. A dispenser controller is inter-connected with its environment through an interface with volume meter (input), pump motor (output), main and by-pass valves (outputs) that enable full or throttled flow, release signal (input) generated by cashier, unhooked nozzle detection (input), product's unit price (input), and volume and price displays (outputs).

3.1 Two-Level Structure

The first employed application pattern is the *two-level structure* proposed by Xinyao et al., 1994: the higher level behaves as an event-driven component, and the lower level behaves as a set of real-time interconnected components. The behavior of the higher level component can be described by the following state sequences of a finite-state automaton with states "blockcd-idlc," "rcady," "full fuel," "throttled" and "closed," and with inputs "release," (nozzle) "hung on/off," "close" (the preset or maximal displayable volume achieved), "throttle" (to slow down the flow to enable exact dosage) and "error":

blocked-idle $\xrightarrow{\text{release}}$ ready $\xrightarrow{\text{hung off}}$ full_fuel $\xrightarrow{\text{hung on}}$ blocked-idle

blocked-idle $\xrightarrow{\text{release}}$ ready $\xrightarrow{\text{hung off}}$ full_fuel $\xrightarrow{\text{throttle}}$ throttled $\xrightarrow{\text{hung on}}$ blocked-idle

blocked-idle $\xrightarrow{\text{release}}$ ready $\xrightarrow{\text{hung off}}$ full_fuel $\xrightarrow{\text{throttle}}$ throttled $\xrightarrow{\text{close}}$ closed $\xrightarrow{\text{hung on}}$ blocked-idle

blocked-idle $\xrightarrow{\text{error}}$ blocked-error

blocked-idle $\xrightarrow{\text{release}}$ ready $\xrightarrow{\text{error}}$ blocked-error

$$\text{blocked-idle} \xrightarrow{\text{release}} \text{ready} \xrightarrow{\text{hung off}} \text{full\_fuel} \xrightarrow{\text{error}} \text{blocked-error}$$

$$\text{blocked-idle} \xrightarrow{\text{release}} \text{ready} \xrightarrow{\text{hung off}} \text{full\_fuel} \xrightarrow{\text{throttle}} \text{throttled} \xrightarrow{\text{error}} \text{blocked-error}$$

The states "full_fuel" and "throttled" appear to be hazardous from the viewpoint of unchecked flow because the motor is on and the liquid is under pressure -- the only nozzle valve controls an issue in this case. Also, the state "ready" tends to be hazardous: when the nozzle is unhooked, the system transfers to the state "full_fuel" with flow enabled. Hence, the accepted fail-stop conception necessitates the detected error management in the form of transfers to the state "blocked-error." To initiate such transfers for issue blocking, the error detection in the hazardous states are necessary. On the other hand, the state "blocked-idle" is safe because the input signal "release" can be masked out by the system that, when some failure is detected, performs the internal transition from "blocked-idle" to "blocked-error."

3.2 Incremental Measurement

The volume measurement and flow control represent the main functions of the hazardous states. The next applied application pattern, *incremental measurement*, means the recognition and counting of elementary volumes represented by rectangular impulses, which are generated by a photoelectric pulse generator. The maximal frequency of impulses and a pattern for their recognition depend on electro-magnetic interference characteristics. The lower-level application patterns are in this case a *noise-tolerant impulse detector* and a *checking reversible counter*. The first one represents a clock-timed impulse-recognition automaton that implements the periodic sampling of its input with values 0 and 1. This automaton with n states recognizes an impulse after n/2 (n>=4) samples with the value 1 followed by n/2 samples with the value 0, possibly interleaved by induced error values, see the following timed-state sequence:

$$(0, q_1) \xrightarrow{\text{inp}=0} \dots \xrightarrow{\text{inp}=0} (i, q_1) \xrightarrow{\text{inp}=1} (i+1, q_2) \xrightarrow{\text{inp}=0} \dots \xrightarrow{\text{inp}=0} (j, q_2) \dots \xrightarrow{\text{inp}=1} (k, q_{n/2+1}) \xrightarrow{\text{inp}=1} \dots$$
$$\dots \xrightarrow{\text{inp}=1} (m, q_{n-1}) \xrightarrow{\text{inp}=0} (m+1, q_n) \xrightarrow{\text{inp}=1} \dots \xrightarrow{\text{inp}=1} (n, q_n) \xrightarrow{\text{inp}=0/\text{IMP}} (n+1, q_1)$$

i, j, k, m are integers: $0 < i < j < k < m < n$

For the sake of fault-detection requirements, the incremental detector and transfer path are doubled. Consequently, the second, identical noise-tolerant impulse detector appears necessary.

The subsequent lower-level application pattern is the *checking reversible counter*, which starts with the value (h + l)/2 and increments or decrements that value according to the "impulse detected" outputs from the first or the second recognition automaton. Overflow or underflow of the pre-set values of h or l indicate an error. Another counter that counts the recognized impulses from one of the recognition automata maintains the whole measured volume. The output of the letter automaton refines to two displays with local memories not only for the reason of robustness (they can be compared) but also for functional requirements (double-face stand). To guarantee the overall fault detection capability of the device, it is necessary also to consider checking the counter. This task can be maintained by an *I/O watchdog* application pattern that can compare input impulses from the photoelectric pulse generator and the changes of the total value; the appropriate automaton performs again reversible counting.

3.3 Fault Management

To prevent unregistered flow, the fail-stop conception used appraises as more acceptable the forced blocking of the dispenser with frozen actual data on displays instead of an untrustworthy issue. The application patterns, so far introduced stepwise, cooperate so that they accomplish the consequent application pattern, *fault management based on fail-stop behavior approximation*, in the form of (a) hazardous state reachability control and (b) hazardous state maintenance. In all safe states ("blocked-idle," "closed," and "blocked-error"), any fuel flow is disabled by power hardware construction; in the same time, the contents of all displays are protected against any change required by possibly erroneous control system. The system is allowed to reach hazardous states ("ready," "full_fuel," and "throttled") when the installed processors successfully have passed start-up checks and interprocessor communication initiation. The hazardous state maintenance includes doubled input path check for detected product impulses and I/O watchdog check. The danger of explosion in the case of uncontrolled petrol flow is eliminated by hard kernel items such as the nozzle with hydraulic shut-off and mechanical blocking the hooked nozzle.

4 Multiple Lift Control System

The second case study deals with the multiple lift control system based on a dedicated multiprocessor architecture, see Švéda, 1997. An incremental measurement device for position evaluation, and position and speed control of a lift cabin in a lift shaft can demonstrate reusability. In fact, that device is contained in the lift control system via multiple instances: one for each lift shaft in a multiple lift system.

The applied application pattern, *incremental measurement*, means in this case the recognition and counting of rectangular impulses that are generated by an electromagnetic or photoelectric sensor/impulse generator, which is fixed on the bottom of the lift cabin and which passes equidistant position marks while moving along the shaft. That device communicates with its environment through interfaces with impulse generator and drive controller. So, the first input, I, contains the values 0 or 1 that are altered with frequency equivalent to the cabin speed. The second input, D, contains the values "up," "down," or "idle." The output, P, resembles to the actual absolute position of the cabin in the shaft.

4.1 Two-Level Structure

The next employed application pattern is the *two-level structure*: the higher level behaves as an event-driven component:

initialization $\rightarrow$ position_indication $\rightarrow$ fault_indication

and the lower level behaves as a set of real-time interconnected components. The specification of the lower level can be developed by refining the higher level state "position_indication" into three communicating lower level automata: two noise-tolerant impulse detectors and checking reversible counter.

4.2 Incremental Measurement

The first automaton models the *noise-tolerant impulse detector*, see the following timed-state sequence:

$$(0, q_1) \xrightarrow{inp=0} \ldots \xrightarrow{inp=0} (i, q_1) \xrightarrow{inp=1} (i+1, q_2) \xrightarrow{inp=0} \ldots \xrightarrow{inp=0} (j, q_2) \ldots \xrightarrow{inp=1} (k, q_{n/2+1}) \xrightarrow{inp=1} \ldots$$
$$\ldots \xrightarrow{inp=1} (m, q_{n-1}) \xrightarrow{inp=0} (m+1, q_n) \xrightarrow{inp=1} \ldots \xrightarrow{inp=1} (n, q_n) \xrightarrow{inp=0/IMP} (n+1, q_1)$$

i, j, k, m are integers: $0 < i < j < k < m < n$

The information about a detected impulse is sent to the counting automaton that can also access the indication of the cabin movement direction through the input D. For the sake of fault-detection requirements, the impulse generator and the impulse transfer path are doubled. Consequently, a second, identical noise-tolerant impulse detector appears necessary. The subsequent application pattern is the *checking reversible counter*, which starts with the value (h + l)/2 and increments or decrements the value according to the "impulse detected" outputs from the first or second recognition automaton. Overflow or underflow of the preset values of h or l indicate an error. This detection process sends a message about a detected impulse and the current direction to the counting automaton, which maintains the actual position in the shaft. To check the counter, an *I/O watchdog* application pattern employs again reversible counting that can compare the impulses from the sensor/impulse generator and the changes of the total value.

4.3 Fault Management

The approach used accomplishes a consequent application pattern, *fault management based on fail-stop behavior approximation*, in the form of (a) hazardous state reachability control and (b) hazardous state maintenance. In safe states, the lift cabins are fixed at any floors. The system is allowed to reach any hazardous state when all relevant processors successfully passed the start-up checks of inputs and monitored outputs and of appropriate communication status. The hazardous state maintenance includes operational checks and consistency checking for execution processors. To comply with safety-critical conception, all critical inputs and monitored outputs are doubled and compared. When the relevant signals differ, the respective lift is either forced (with the help of a substitute drive if the shaft controller is disconnected) to reach the nearest floor and to stay blocked, or (in the case of maintenance or fire brigade support) its services are partially restricted. The basic safety hard core includes mechanical, emergency brakes.

5 Application Patterns Reuse

The two case studies presented above demonstrate the possibility to reuse effectively substantial parts of the design dealing with a petrol pumping station technology for a lift control technology project. While both cases belong to embedded control systems, both the application domains and the technology principles differ: volume measurement and dosage control seems not too close to position measurement and control. Evidently, the similarity is observable by employment of application patterns.

The reused upper-layer application patterns presented include the automata-based descriptions of incremental measurement, two-level (event-driven/real-time) structure, and fault management stemming from fail-stop behavior approximations. The reused lower-layer application patterns are exemplified by the automata-based descriptions of noise-tolerant impulse detector, checking reversible counter, and I/O watchdog.

Clearly, while all introduced application patterns correspond to design patterns in the above explained interpretation, the upper-layer application patterns can be related also to frameworks. Moreover, the presented collection of application patterns creates a base for a pattern language supporting reuse-oriented design process for industrial real-time embedded systems.

6 Knowledge-Based Support

Industrial scale reusability requires a knowledge-based support, e.g. by case-based reasoning, which was successfully used by the author and his colleagues previously in another application (Švéda, Babka and Freeburn, 1997). *Case-based reasoning*, see e.g. Kolodner, 1993, differs from other rather traditional methods of Artificial Intelligence relying on case history. For a new problem, the case-based reasoning strives for a similar old solution. This old solution is chosen according to the correspondence of a new problem to some old problem that was successfully solved by this approach. Hence, previous significant cases are gathered and saved in a case library. Case-based reasoning stems from remembering a similar situation that worked in past. For software reuse, case-based reasoning utilization is currently studied from several viewpoints, see e.g. Henninger, 1998, and Soundarajan, 1998.

6.1 Case-Based Reasoning

The case-based reasoning method contains elicitation, which means collecting those cases, and implementation, which represents identification of important features for the case description consisting of values of those features. A case-based reasoning system can only be as good as its case library (Kolodner, 1993): only successful and sensibly selected old cases should be stored in the case library. The description of a case should comprise the corresponding problem, solution of the problem, and any other information describing the context for which the solution can be reused. A feature-oriented approach is usually used for the case description.

Case library serves as a knowledge base of the case-based reasoning system. The system acquires knowledge from old cases while learning can be achieved accumulating new cases. Solving a new case, the most similar old case is retrieved from the case library. The suggested solution of the new case is generated in conformity with this retrieved old case. Search for the similar old case from the case library represents important operation of case-based reasoning paradigm. The problem to be solved arises how to measure the similarity of state-based specifications for retrieval. Retrieval schemes proposed in the literature can be classified based upon the technique used to index cases during the search process (Atkinson, 1998): (i) classification-based schemes, which include keyword or feature-based controlled vocabularies; (ii) structural schemes, which include signature or structural characteristics matching; and (iii) behavioral schemes; which seek relevant cases by comparing input and output spaces of components.

6.2 Case-Based Reasoning Application Concepts

The primary approach to the current application includes some equivalents of abstract data type signatures, belonging to structural schemes, and keywords, belonging to classification schemes. While the first alternative means for this purpose to quantify the similarity by the topological characteristics of associated finite automata state-transition graphs, such as number and placement of loops, the second one is based on a properly selected set of keywords with subsets identifying individual patterns.

7 Conclusions

The above presented case studies, which demonstrate the possibility to reuse concrete application patterns, have been in fact excerpted from two realized design cases. The application patterns, originally introduced as "configurations" in the project of petrol pumping station technology, see Švéda, 1996, were effectively -- but without any dedicated development support -- reused for the project of lift control technology, see Švéda, 1997.

This contribution informs about case studies and tools available at the beginning of research aiming at knowledge-based support for industrial embedded systems design.

Case-based reasoning was successfully utilized by the author and his colleagues previously for another industrial application (Švéda, Babka and Freeburn, 1997). Hopefully, also this meta-knowledge can be successfully reused.

Acknowledgment. The author gratefully acknowledges the contributions to the presented work by Ota Babka from the University of Macau, Macau. This research has been partly funded by the Czech Ministry of Education in frame of the Research intention No. CEZ: J22/98: 262200012 - Research in information and control systems.

References

Alexander, C. (1977). A Pattern Language: Towns / Buildings / Construction. Oxford University Press.

Alur, R., Henzinger, T.A. (1992). Logics and Models of Real Time: A Survey. In: (de Bakker, J.W., et al.). Real-Time: Theory in Practice. Springer-Verlag, LNCS 600, 74-106.

Arora, A., Kulkarni, S.S. (1998). Component Based Design of Multitolerant Systems. IEEE Transactions on Software Engineering, 24(1), 63-78.

Atkinson, S. (1998). Modelling Formal Integrated Component Retrieval. Proceedings of the Fifth International Conference on Software Reuse, IEEE Computer Society, Los Alamitos, California, 337-346.

Coad, P., Yourdon, E.E. (1990) Object-Oriented Analysis. Yourdon Press, New York.

Gamma, E., Helm, R., Johnson, R., Vlissides, J. (1995). Design Patterns -- Elements of Reusable Object-Oriented Software. Addison-Wesley.

Henninger, S. (1997). An Evolutionary Approach to Constructing Effective Software Reuse Repositories. Transactions on Software Engineering and Methodology, 6(2), 111-140.

Henninger, S.. (1998). An Environment for Reusing Software Processes. Proceedings of the Fifth International Conference on Software Reuse, IEEE Computer Society, Los Alamitos, California, 103-112.

Holtzblatt, L.J., Piazza, R.L., Reubenstein, H.B., Roberts, S.N., Harris, D.R. (1997). Design Recovery for Distributed Systems. IEEE Transactions on Software Engineering, 23(7), 461-472.

Jacobson, L. (1992). Object-Oriented Software Engineering: A User Case-Driven Approach. ACM Press.

Johnson, R.E. (1997). Frameworks = (Components + Patterns). Communications of the ACM, 40(10), 39-42.

Kolodner, J, *Case-based Reasoning,* Morgan Kaufmann, San Mateo, CA, USA, 1993.

Mili, R., Mili, A., Mittermeir, R.T. (1997). Storing and Retrieving Software Components: A Refinement Based System. IEEE Transactions on Software Engineering, 23(7), 445-460.

Robillard, P.N. (1999). The Role of Knowledge in Software Development. Communications of the ACM, 42(1), 87-92.

Shaw, M., Garlan, D.(1996). Software Architecture. Prentice Hall.

Sen, A. (1997). The Role of Opportunity in the Software Reuse Process. IEEE Transactions on Software Engineering, 23(7), 418-436.

Soundarajan, N., Fridella, S. (1998). Inheritance: From Code Reuse to Reasoning Reuse. Proceedings of the Fifth International Conference on Software Reuse, IEEE Computer Society, Los Alamitos, California, 206-215.

Sutcliffe, A., Maiden, N. (1998). The Domain Theory for Requirements Engineering. IEEE Transactions on Software Engineering, 24(3), 174-196.

Švéda, M., Babka, O., Freeburn J. (1997). Knowledge Preserving Development: A Case Study. Proceedings of the Engineering of Computer-Based Systems. IEEE Computer Society, Los Alamitos, California, pp. 347-352.

Švéda, M. (1996) Embedded System Design: A Case Study. Proceedings of the Engineering of Computer-Based Systems. IEEE Computer Society, Los Alamitos, California, 260-267.

Švéda, M. (1997) An Approach to Safety-Critical Systems Design. In: (Pichler, F., Moreno-Diaz, R.). Computer Aided Systems Theory. Springer-Verlag, LNCS 1333, 34-49.

Turner, K.J. (1997). Relating Architecture and Specification. Computer Networks and ISDN Systems, 29(4), 437-456.

van Lamsweerde, A., Willemet, L. (1998). Inferring Declarative Requirements Specifications from Operational Scenarios. IEEE Transactions on Software Engineering, 24(12), 1089-1114.

Xinyao, Y., Ji, W., Chaochen, Z., Pandya, P.K.. (1994) Formal Design of Hybrid Systems. In: (Langmaack, H., de Roever, W. P., Vytopil, J.) Formal Techniques in Real-Time and Fault-Tolerant Systems. Springer-Verlag, LNCS 863, 738-755.

Zaremski, A.M., Wing, J.M. (1997). Specification Matching of Software Components. ACM Trans. on Software Engineering and Methodology, 6(4), 333-369.

Towards Verifying Distributed Systems Using Object-Oriented Petri Nets

Milan Češka, Vladimír Janoušek, and Tomáš Vojnar

Department of Computer Science and Engineering,
Brno University of Technology
Božetěchova 2, CZ–612 66 Brno, Czech Republic
{ceska,janousek,vojnar}@dcse.fee.vutbr.cz

Abstract. The article discusses the notion of state spaces of object-oriented Petri nets (OOPNs) associated to the tool called PNtalk and the role of identifiers of dynamically appearing and disappearing instances within these state spaces. Methods of working with identifiers based on sophisticated naming rules and mechanisms for abstracting names are described and compared. Some optimizations of state space generating algorithms for the context of OOPNs are mentioned, as well. Finally, some possibilities of specifying properties of systems to be checked over the state spaces of their OOPN-based models are discussed.

1 Introduction

Current complex distributed applications require dealing with dynamically arising and disappearing objects which can communicate, synchronize their actions, and migrate among particular nodes of the distributed environment they are running in. Particularly, distributed operating systems, groupware allowing for a concurrent work of several people on the same project, or applications exploiting the technology of agents or mobile agents can be listed as examples of the above-mentioned applications.

A language called PNtalk based on object-oriented Petri nets (OOPNs) [Jan98] has been developed at the DCSE, TU Brno in order to support modelling, investigating, and prototyping complex distributed object-oriented software systems. PNtalk supports intuitive modelling all the key features of these systems, such as object-orientedness, message sending, parallelism, and synchronization. This is achieved through working with active objects encapsulating sets of processes described by Petri nets. Processes inside the objects communicate via a shared memory, while objects themselves communicate by message passing.

Simulation is one of the ways of examining systems modelled by OOPNs and it is already supported by a prototype version of a tool called PNtalk [ČJV97]. Moreover, models created in PNtalk can be used as a basis of prototypes of the modelled systems. In such a case, some objects are likely to be implemented in Smalltalk exploiting the fact that PNtalk allows Smalltalk-based and OOPN-based objects to transparently communicate in both directions.

Although we have started with simulation, this article considers more the first steps made towards exploiting formal analysis and verification methods in the

F. Pichler, R. Moreno-Díaz, and P. Kopacek (Eds.): EUROCAST'99, LNCS 1798, pp. 90–104, 2000.

context of OOPNs. This approach can be considered an alternative to simulation because although we are not always able to fully verify or analyse the behaviour of a system, even partial analysis or verification can reveal some errors which tend to be different from the ones found by simulation [Val98]. We believe that object-orientation should allow for a relatively easy extraction of the subsystems to be verified together with a suitable abstraction of their surroundings.

Among the different attitudes to performing formal analysis or verification, using state spaces appears to be the most straightforward way for the case of OOPNs. Methods based on state spaces are quite universal, can be almost fully automated, and allow for relatively easy implementation. There have been proposed many different ways of alleviating their main deficiency — the state explosion problem [Val98]. Now these methods should be adapted and optimized for the context of OOPNs. Apart from that, it is necessary to solve some new problems accompanying state spaces of OOPNs as a formalism with dynamic instantiation, such as the problem how to efficiently treat the identifiers of dynamically appearing and disappearing instances.

At the beginning of building a theory of state space analysis over OOPNs (or more generally over any formalism with dynamic instantiation of some kind of components) it is necessary to pay careful attention to treating identifiers of objects (or in general to identifiers of some other kind of dynamically appearing and disappearing instances). Otherwise, many unnecessary states can be generated and the state space can even unnecessarily grow to infinity. This naming problem can be solved either by introducing some sophisticated rules for assigning identifiers to instances or by not considering concrete names of instances to be important when checking states to be equal.

The work with instance identifiers influences not only generating state spaces of OOPNs, but also analyzing them. This is because we need to be able to describe expected properties of the systems being examined without referring to the concrete names of the instances involved in states and events of their state spaces. These names are semantically not important and, what is more, modellers can hardly work out what identifiers will be used in different states.

In the article, we first present the main ideas behind the OOPN formalism. Then we briefly discuss the naming problem arising in state spaces of OOPNs, together with some further optimizations to be used when generating them. Finally, we suggest a method of querying over states spaces of OOPNs.

2 Key Concepts of OOPNs

The OOPN formalism [ČJV97] is characterized by a Smalltalk-based object-orientation enriched with concurrency and polymorphic transition execution, which allow for message sending, waiting for and accepting responses, creating new objects, and performing primitive computations. An example demonstrating the notation of OOPNs is shown in figure 1.

This section rephrases the basic ideas of the definition of OOPNs, however, due to space limitations, without making the description formal and complete. We explain the necessary notions only. A bit deeper introduction to the OOPN formalism can be found in [ČJV97] and the entire definition of OOPNs in [Jan98].

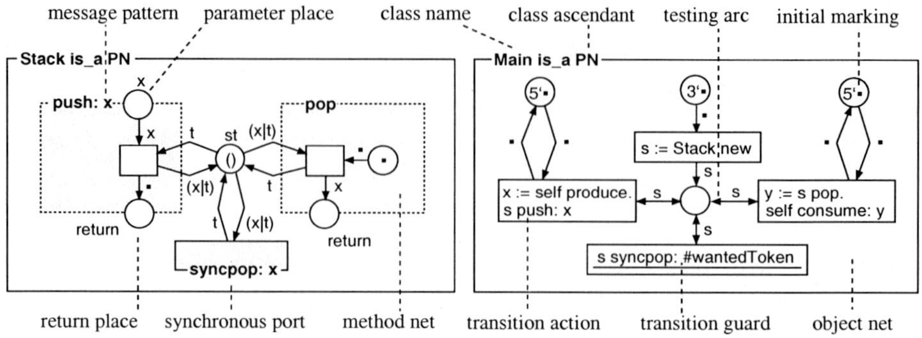

Fig. 1. An OOPN example (Main's methods produce and consume are not shown).

2.1 The OOPN Structure

An *object-oriented Petri net* is a triple (Σ, c_0, oid_0) where Σ is a system of classes, c_0 an initial class, and oid_0 the name of an initial object from c_0.

Σ contains sets of OOPN elements which constitute classes. It comprises constants $CONST$, variables VAR, net elements (such as places P and transitions T), class elements (such as object nets $ONET$, method nets $MNET$, synchronous ports $SYNC$, and message selectors MSG), classes $CLASS$, object identifiers OID, and method net instance identifiers MID. We denote $NET = ONET \cup MNET$ and $ID = OID \cup MID$. The universe U of an OOPN contains (nested) tuples of constants, classes, and object identifiers. Let $BIND = \{b \mid b : VAR \longrightarrow U\}$ be the set of all bindings of variables.

Object nets consist of places and transitions. Every place has some initial marking. Every transition has conditions (i.e. inscribed testing arcs), preconditions (i.e. inscribed input arcs), a guard, an action, and postconditions (i.e. inscribed output arcs). *Method nets* are similar to object nets but, in addition, each of them has a set of parameter places and a return place. Method nets can access places of the appropriate object nets in order to allow running methods to modify states of objects which they are running in.

Synchronous ports are special transitions which cannot fire alone but only dynamically fused to some other transitions which "activate" them from their guards via message sending. Every synchronous port embodies a set of conditions, preconditions, and postconditions over places of the appropriate object net, and further a guard, and a set of parameters. Parameters of an activated port s can be bound to constants or unified with variables defined on the level of the transition or port that activated s.

A *class* is specified by an object net (an element of $ONET$), a set of method nets (a subset of $MNET$), a set of synchronous ports (a subset of $SYNC$), and a set of message selectors (a subset of MSG) corresponding to its method nets and ports. Object nets describe possible independent activities of particular objects, method nets reactions of objects to messages sent to them from outside, and ports allow for remotely testing and changing states of objects in an atomic way. The inheritance mechanism of OOPNs allows for an incremental specification of classes. Inherited methods and synchronous ports can be redefined and new methods and synchronous ports can be added. A similar mechanism applies for object net places and transitions.

2.2 The Dynamic Behaviour of OOPNs

The dynamic behaviour of OOPNs corresponds to the evolution of a system of objects. An *object* is a system of net instances which contains exactly one instance of the appropriate object net and a set of currently running instances of method nets. Every *net instance* entails its identifier $id \in ID$ and a marking of its places and transitions. A *marking of a place* is a multiset of elements of the universe U. A *transition marking* is a set of invocations. Every *invocation* contains an identifier $id \in MID$ of the invoked net instance and a stored binding $b \in BIND$ of the input variables of the appropriate transition.

A state of a running OOPN has the form of a *marking*. To allow for the classical Petri net-way of manipulating markings, they are represented as multisets of token elements. In the case of a transition marking, the identifier of the invoked method net instance is stored within the appropriate binding in a special (user-invisible) variable `mid`. Thus a formal compatibility of place and transition markings is achieved and it is possible to define a token element as a triple consisting of the identifier of the net instance it belongs to, the appropriate place or transition, and an element of the universe or a binding. Then we can say for a marking M that:

$$M \in [(ID \times P \times U) \cup (ID \times T \times BIND)]^{MS}.$$

A step from a marking of an OOPN into another marking can be described as the so-called *event*. Such an event is a 4-tuple

$$E = (e, id, t, b)$$

including (1) its type e, (2) the identifier $id \in ID$ of the net instance it takes place in, (3) the transition $t \in T$ it is statically represented by, and (4) the binding tree b containing the bindings used on the level of the invoked transition as well as within all the synchronous ports (possibly indirectly) activated from that transition. There are four kinds of events according to the way of evaluating the action of the appropriate transition: `A` – an atomic action involving trivial computations only, `N` – a new object instantiation via the message `new`, `F` – an instantiation of a Petri-net described method, and `J` – terminating a method net instance. If an enabled event E occurs in a marking M and changes it into a marking M', we call this a *step* and denote it by

$$M[E\rangle M'.$$

For a given OOPN, its *initial marking* M_0 corresponds to a single, initially marked object net instance from the initial class c_0 identified as oid_0. The *set of all markings* of an OOPN is denoted as MA and the *set of all events* as EV.

Finally, let us introduce the following notation. Given $id \in ID$, $net(id)$ denotes the net $n \in NET$ such that id identifies an instance of n, and $oid(id)$ denotes $oid \in OID$ such that id identifies a net instance belonging to the object identified by oid. Note that an object is identified by the identifier of its object net instance.

3 The Notion of State Spaces of OOPNs

This section discusses the notion of state spaces of OOPNs as a representative of formalisms with dynamic instantiation of some kind of components. Two methods for dealing with names of instances of the components being instantiated are suggested and discussed here trying to minimize the impact of the presence of names in states upon the state space explosion problem.

State spaces can generally be defined [Val98] as 4-tuples consisting of a set of states, a set of structural transitions, a set of semantic transitions (i.e. links between states and structural transitions), and an initial state. This concept can be used when dealing with state spaces of OOPNs (or any other formalism with dynamic instantiation), as well. However, in the context of such formalisms, it is necessary to pay careful attention to efficiently handling the naming information present in states in order not to worsen the state space explosion problem. Let us denote this phenomenon as the *naming problem*.

The naming information present in states of dynamically structured formalisms is used for uniquely identifying the just existing instances which allows for separating their local states and expressing references among them. Working with instance identifiers, e.g. in the form of addresses of objects, is common when running object-oriented programs or simulating object-oriented models. However, in the context of state spaces, the naming information can significantly enlarge the state space explosion problem. This is due to the possibility of unnecessarily generating many states differing only in the names of the involved instances even if the names cannot influence the future behaviour of the system being examined in any way (up to renaming). What is worse, sometimes the naming information can make state spaces of evidently finite-state systems grow to infinity — it suffices to keep creating and destroying an instance identifying it using still new identifiers.

There are at least two methods for solving the naming problem — using sophisticated naming rules for assigning identifiers to newly arising instances and the so-called name abstraction as a specialization of the symmetry method for reducing state spaces [Jen94]. The latter method is based on not considering concrete names of instances to be important when checking states to be equal, which leads to working with renaming equivalence classes of states rather than with the individual states. Both of these methods together with their pros and cons are discussed in the following in the context of OOPNs. However, first of all we define full state spaces of OOPNs in order to obtain a basis to be reduced using one of the mentioned methods.

Still before discussing the two mentioned principles in more detail we should note that the problem they should solve cannot be avoided by simply presenting an algorithm for transforming the formalism with dynamic instantiation under question into some kind of low-level formalism which should serve as a basis for formal analysis. When we for example try to transform object-oriented Petri nets into some kind of "plain" high-level nets, as for example in [SB94], there must appear a construction generating identifiers which then become a distinguishing part of tuples representing tokens of originally different net instances folded together. Thus the problem of naming is carried into the domain of non-object nets and must be solved within their analysis process.

3.1 Full State Spaces of OOPNs

Full state spaces of OOPNs can be defined using the general concept of state spaces mentioned above. For a given OOPN, states will correspond to reachable markings and structural transitions to applicable events. Semantic transitions will be defined in accordance to the firing rules of OOPNs. Finally, the initial state will be the initial marking, of course.

Definition 1 (Full State Spaces of OOPNs).
Let an object-oriented Petri net $OOPN$ with its set of markings MA, its initial marking M_0, and its set of events EV be given. We define the (full) state space *of $OOPN$ to be the 4-tuple $StSp = (S, T, \Delta, M_0)$ such that:*

1. $S = [M_0\rangle$.
2. $T = \{(M_1, E, M_2) \in S \times EV \times S \mid M_1[E\rangle M_2\}$.
3. $\forall M_1, M_2 \in MA\ \forall E \in EV[(M_1, E, M_2) \in T \Leftrightarrow (M_1, (M_1, E, M_2), M_2) \in \Delta]$.

A consequence of the definition of full state spaces of OOPNs ignoring the naming problem is that when we try to create the first instance of some net whose domain of its instance identifiers is infinite we immediately obtain infinitely many possible target markings. Moreover, requiring sets of possible identifiers of nets to be finite will not solve the problem because (1) it can change the semantics of the model by artificially restricting the number of concurrently existing instances and (2) there can still be generated unnecessarily many target markings.

3.2 Using Sophisticated Naming Rules

Sophisticated rules for assigning identifiers to newly arising instances attempt to decrease the degree of nondeterminism potentially present in the management of names of dynamically arising and disappearing instances and thus to decrease the number of reachable states. Such rules can be made a part of the semantics of the given modelling language. However, they can also be applied without being integrated into the modelling language. Then their application can be viewed as a tool for reducing state spaces and we have to show that using them we do not lose any important information wrt. the definition of the appropriate formalism. The proof can be based on showing that the application of such rules is in fact a "safely-redundant" implementation of the principle of (complete) name-abstracted state spaces discussed later.

The simplest nontrivial rule for naming instances is assigning identifiers according to some ordering over them. A deficiency of this attitude is that when we are cyclically creating and destroying some instance we will again obtain an infinite state space. This can be solved by recycling identifiers, i.e. by identifying newly emerging instances by the lowest and currently not used identifiers. However, even when we use recycling, there can still be generated many different states which are obviously semantically equal. Such a situation arises when some configuration of instances characterized by the number of the involved instances, their types, their trivial marking, and their mutual relations can be reached via several state space paths in which the instances are created in different orders and using various auxiliary instances with differently overlapped lifetimes. Then there can be generated several states containing the given configuration of instances and differing only in the names of some of the involved uninterchangeable instances (distinguished by their contents or by the way they are referred to).

The problem of generating unnecessarily many states, which can hardly be avoided even under elaborated naming schemes, can be alleviated to some degree when using partial order reduction techniques [Val98]. This is because these techniques reduce numbers of paths leading to particular states and thus also possibilities to obtain different permutations of identifiers of the involved characteristic instances. Nevertheless, the problem is not fully solved this way as it is not always possible to choose only one interleaving out of a set of the possible ones. Partial order techniques can ignore different orders of actions only in the case they are invisible and do not collide. Furthermore, finding optimal stubborn (persistent, ample) sets can be too time-consuming and so an approximation is often taken (especially in the case of high-level formalisms).

3.3 Abstracting Away the Naming Information

With respect to the previous discussion, we now suggest another possible method for solving the naming problem based on not considering concrete values of names of instances to be important when checking states to be equal. In other words, we are going to define two markings to be equal if there exists a suitable permutation over the set of all identifiers whose application makes the states identical. As a consequence, we will replace working with particular states by working with renaming equivalence classes of them. In the following, we will try to describe the method at least partially in a formal way — a fully formal description, together with proofs of the propositions, can be found in [Voj00].

It should be noted here that the concept of name abstraction is a specialization of the general notion of symmetries [Jen94] applied for reducing state explosion caused by the presence of concrete names of instances in states. Unlike general symmetries, renaming symmetries are universal in the domain of OOPNs, i.e. they can be used for all OOPN-based models. Furthermore, since renaming symmetries are highly specialised, they allow (1) for formulating more exact propositions over the state spaces based on them and (2) for using more effective methods of treating them within generating state spaces.

The idea of abstracting away the naming information can only be applied due to the fact that the behaviour of OOPN-based models does not depend on concrete values of identifiers. From this point of view, it is crucial that the definition of OOPNs does not allow for using instance identifiers in expressions and that there cannot be performed trivial computations depending on concrete values of instance names. Therefore it can be proved that starting from some state concrete names of instances do not influence the future evolution of the appropriate OOPN in any way (up to renaming). Thus it is not necessary to distinguish states equal up to renaming because of the future behaviour they can lead to. Furthermore, it is not necessary to be afraid of losing the identity of an object within some marking stemming from its history and to differentiate markings equal up to renaming only because their histories cannot be made equal even when applying renaming. If we are interested in one particular history of some object within a marking, we can always concentrate on that history additionally and ignore all other possible histories.

We have said that we want two markings to be equal if there exists a suitable permutation over the set of all the identifiers allowed in the appropriate OOPN whose application makes the states identical. However, we do not accept all permutations. An acceptable permutation must preserve the information about (1) to which object a given instance belongs to, (2) to which net the instance belongs, and (3) it cannot change the identifier of the initial object, which is important for the garbage collecting mechanism. Permutations conform to the just described requirements will be called *renaming permutations* in the following.

Definition 2 (Renaming Permutations).
Suppose we have an object-oriented Petri net $OOPN$ with its set of instance identifiers ID and its initial object identifier oid_0. We define renaming permutations over $OOPN$ to be the bijections $\pi : ID \leftrightarrow ID$ such that:

1. $\pi(oid_0) = oid_0$.
2. $\forall id \in ID \; [net(id) = net(\pi(id))]$.
3. $\forall id \in ID \; [\pi(oid(id)) = oid(\pi(id))]$.

The concept of renaming permutations provides a basis for defining the so-called *renaming symmetries*, i.e. bijections on sets of markings and sets of events. The formal definition of renaming symmetries can be obtained by a simple but a little longer extension of bijections working over identifiers to bijections over markings and events — we will skip the definition here. We denote the renaming symmetry induced by a renaming permutation π as ϱ^{π}. Now we can define two markings M_1, M_2 to be *equal up to renaming* iff there exists a renaming permutation π such that $\varrho^{\pi}(M_1) = M_2$. The same can be done for events. In the following, we will denote the renaming equivalence relation by $\sim$. Members of its equivalence classes will be referred to using the black board alphabet, i.e. $\mathbb{M}$ or $\mathbb{E}$, or via their representatives, i.e. $[M]$ or $[E]$. Finally, quotient sets wrt. $\sim$ will be denoted using $\sim$ as a subscription, as e.g. $MA_{\sim}$.

The notion of renaming symmetries allows us to formalize the already mentioned proposition that concrete names of instances cannot influence anything else than again names of instances present in the future behaviour of an OOPN-described system starting from a given state. Such a property is crucial in the theory of symmetrically reduced state spaces.

Proposition 1. *Let us have an object-oriented Petri net OOPN with its set of markings MA, its set of events EV, and the corresponding set of all renaming permutations Π. Then the following holds for every $M_1, M_2 \in MA$, $E \in EV$, and $\pi \in \Pi$: $M_1[E\rangle M_2 \Leftrightarrow \varrho^\pi(M_1)[\varrho^\pi(E)\rangle\varrho^\pi(M_2)$.*

Renaming symmetries allow us to propose the expected notion of *name-abstracted state spaces (NA state spaces)* of OOPNs. When generating a name-abstracted state space, concrete identifiers of instances will not be taken into account and two states or events will be considered equal if they are equal up to renaming. The definition of name-abstracted state spaces will be based again on the general concept of state spaces. However, this time states will correspond to reachable *name-abstracted markings*, i.e. equivalence classes of MA wrt. $\sim$, and structural transitions to useful *name-abstracted events*, i.e. equivalence classes of EV wrt. $\sim$. Semantic transitions will be defined in accordance to the firing rules of OOPNs and to the semantics of renaming. The initial state will be equal to the equivalence class comprising the initial marking and only the initial marking.

Definition 3 (Name-Abstracted State Spaces).
Let an object-oriented Petri net OOPN with its set of markings MA, its initial marking M_0, its set of events EV, and its renaming equivalence relation $\sim$ be given. We define the name-abstracted state space (NA state space) *of OOPN to be the 4-tuple $NAStSp = (S_N, T_N, \Delta_N, [M_0])$ such that:*

1. $S_N = \{[M] \in MA_\sim \mid M \in [M_0\rangle\}$.
2. $T_N = \{([M_1], [E], [M_2]) \in S_N \times EV_\sim \times S_N \mid M_1[E\rangle M_2\}$.
3. $\forall \mathbb{M}_1, \mathbb{M}_2 \in MA_\sim\ \forall \mathbb{E} \in EV_\sim\ [(\mathbb{M}_1, \mathbb{E}, \mathbb{M}_2) \in T_N \Leftrightarrow (\mathbb{M}_1, (\mathbb{M}_1, \mathbb{E}, \mathbb{M}_2), \mathbb{M}_2) \in \Delta_N]$.

The above proposed name-abstracted state spaces are based on projecting away the naming information present in particular states and structural transitions of the classical state spaces. However, this does not say much about what kind of information they preserve or whether they even contain exactly the same information as full state spaces. More precisely, we could say that NA state spaces preserve all information present in full state spaces if it was possible to reconstruct the appropriate full state spaces from them. Unfortunately, this is not the case. NA state spaces preserve information about reachable states and events but their interconnection is preserved only partially. This fact is formalized in the proposition 2 from whose power we can guess that NA state spaces do not contain information about which particular instances are manipulated by events when going from one state into another.

Proposition 2. *Let us have an OOPN with its state space StSp and the corresponding name-abstracted state space NAStSp. Then the following holds:* $\forall n \geq 1\ \forall \mathbb{M}_1, \ldots, \mathbb{M}_n \in MA_\sim\ \forall \mathbb{E}_1, \ldots, \mathbb{E}_{n-1} \in EV_\sim\ \forall i \in \{1, \ldots, n\}\ \forall M_i \in \mathbb{M}_i$ [$\langle \mathbb{M}_1, \mathbb{E}_1, \ldots, \mathbb{M}_i, \ldots, \mathbb{E}_{n-1}, \mathbb{M}_n \rangle$ *is a path in NAStSp if and only if* $\exists M_1 \in \mathbb{M}_1, \ldots, M_{i-1} \in \mathbb{M}_{i-1}, M_{i+1} \in \mathbb{M}_{i+1}, \ldots, M_n \in \mathbb{M}_n\ \exists E_1 \in \mathbb{E}_1, \ldots, E_{n-1} \in \mathbb{E}_{n-1}$ *such that* $\langle M_1, E_1, \ldots, M_i, \ldots, E_{n-1}, M_n \rangle$ *is a path in StSp*].

The information we are losing in NA state spaces is obviously not important when we are dealing with isolated states only. On the other hand, if we need to be able to explore sequences of states and events, this information might become necessary even if we do not consider concrete names of instances to be important. This is because it can be useful to know how a particular instance given by its identifier within some arbitrarily chosen representative of the appropriate NA state behaves within the events surrounding the state being examined. However, if we need the information, which is lost in NA state spaces, it is not difficult to preserve it. For this reason, we define the so-called *complete name-abstracted state spaces (CNA state spaces)*.

We define CNA state spaces as labelled NA state spaces. Every NA state will be labelled by a tuple consisting of a representative marking belonging to the equivalence class represented by the NA state and a set of self-renaming permutations (i.e. permutations which map the representative marking to itself). Every structural transition will be labelled by a set of tuples consisting of a representative event and a renaming permutation. We require that every representative event must be firable from the appropriate representative source marking leading to the appropriate representative target marking *after* applying the given renaming. Furthermore, every event firable from the representative source marking and leading to a marking equal up to renaming to the representative target marking must be derivable (up to the name of an eventually newly arising instance) from some of the representative events via a permutation from the set of source self-renaming permutations. Self-renaming permutations can decrease the number of target markings we have to process [Jen94].

Note that there can exist multiple CNA state spaces for a single NA state space. This is because the choice of representatives is not deterministic. However, all such CNA state spaces are equal because (as we will mention later) it is possible to build the appropriate full state space from all of them.

In the definition of CNA state spaces, we will use a predicate *eideq* which is fulfilled when applied to two "existing identifier equal" events. Such events can differ only in the identifier of the newly created object in the case of `N` events and in the identifier of the newly started method net instance in the case of `F` events.

Definition 4 (Complete Name-Abstracted State Spaces).
Suppose we have an $OOPN$ with its set of markings MA, its set of events EV, and its set of renaming permutations Π. We define the complete name-abstracted state space (CNA state space) *of $OOPN$ to be the triple $CNAStSp = (NAStSp, m, e)$ such that:*

1. $NAStSp = (S_N, T_N, \Delta_N, [M_0])$ *is the NA state space of $OOPN$.*
2. $m : S_N \to MA \times 2^{\Pi}$ *such that* $\forall \mathbb{M} \in S_N\ [m(\mathbb{M}) = (M, \Phi) \Rightarrow (M \in \mathbb{M}\ \wedge \forall \varphi \in \Phi\ [\varrho^{\varphi}(M) = M])]$.
3. $e : T_N \to 2^{EV \times \Pi}$ *such that for all* $(\mathbb{M}_1, \mathbb{E}, \mathbb{M}_2) \in T_N$ *with* $m(\mathbb{M}_1) = (M_1, \Phi_1)$ *and* $m(\mathbb{M}_2) = (M_2, \Phi_2)$, $e((\mathbb{M}_1, \mathbb{E}, \mathbb{M}_2))$ *is the smallest set such that* $\forall E' \in \mathbb{E}\ \forall M_2' \in \mathbb{M}_2\ [M_1[E'\rangle M_2' \Rightarrow \exists (E, \pi) \in e((\mathbb{M}_1, \mathbb{E}, \mathbb{M}_2))\ [M_1[E\rangle \varrho^{\pi}(M_2) \wedge \exists \varphi \in \Phi_1\ [eideq(E', \varrho^{\varphi}(E))]]]$.

The time complexity of generating CNA state spaces is almost the same as in the case of NA state spaces. This is because state representatives and renaming symmetries must be computed even when generating NA state spaces. The only difference is that, in the case of NA state spaces, the computed renaming symmetries are thrown away after determining the target nodes of particular semantic transitions. On the other hand, the information about them is stored within CNA state spaces which leads to slightly increased memory requirements.

Now it is the right time to express the fact that CNA state spaces contain exactly the same information as the corresponding full state spaces. In other words, it is possible to obtain a full state space from its CNA-variant. These are the contents of the below proposition. Its proof (skipped here) would be based on showing how the full state space can be derived from its CNA-version.

Proposition 3. *Let us have an object-oriented Petri net $OOPN$ with its CNA state space $CNAStSp$ and its set of renaming permutations Π. Then it is possible to reconstruct the full state space of $OOPN$ from $CNAStSp$ and Π.*

NA state spaces remove all the redundancy associated to names of dynamically appearing and disappearing instances and thus can save more memory than the attitudes based on sophisticated naming rules. This is a consequence of always ignoring all the different possibilities of identifying uninterchangeable instances within otherwise identical states without any respect to the way how they were created. At the same time we know that this is not the case of using sophisticated naming rules where the order in which the instances characteristic by their contents or by the way they are referred to from the rest of the system is significant. This implies that an exponential number of classical states can be folded onto one name-abstracted state (or its representative).

The source of the above reductions is less significant when using partial order reduction techniques. However, even in this case, it is not guaranteed that all the redundancies stemming from the naming problem are fully removed. Limitations of partial order techniques were already mentioned at the end of subsection 3.2. The idea that renaming symmetries can remove some redundancies preserved by sophisticated naming rules combined with partial order techniques can also be supported by the fact that it is generally advantageous to combine partial order reduction methods and methods based on symmetries [Val98]. This is because they fight against different sources of redundancies in state spaces.

So, it seems that renaming can save more memory than sophisticated naming rules, even in the case of using partial order reduction. On the other hand, we might have to pay for using renaming quite a lot in terms of the time complexity because testing systems of objects to be equal up to renaming can multiply the overall time of generating state spaces by $O(n!)$ where n is the maximal number of concurrently existing instances. Fortunately, this is the worst case scenario only and we can usually decrease the time complexity using intelligent construction of models and some heuristics briefly mentioned in the next subsection. These heuristics are based on renaming insensitive hashing techniques decreasing numbers of states to be compared and on exploiting the structure of states for selecting instances whose identifiers it is sensible to permute.

As a conclusion, we can say that more studies are still needed to answer the question whether and in which cases it is better to allow bigger memory consumption and when to use name abstraction.

3.4 Generating State Spaces of OOPNs

In the context of OOPNs, the efficiency of the classical algorithm of generating full state spaces [Jen94] as well as its plain or partial order reduced depth first variants typically used for formal verification [Pel96] can be improved in several ways supposed in [Voj00] and briefly mentioned below.

Reducing Numbers of States to be Compared. We can save a lot of time by decreasing numbers of states to be compared. This can be achieved by representing state spaces as hash tables indexed via hash functions working over states. The employed hashing procedure must be insensitive to the mechanism of name abstraction which can be fulfilled when we replace instance identifiers by the appropriate typing information plus eventually some associative identification of the instances based on their trivial marking.

Improving the Efficiency of Testing the Renaming Equivalence. The worst case complexity of testing the renaming equivalence cannot be decreased, but the average one can be improved by exploiting the structure of states instead of blindly testing all permutations of identifiers. The basic principle of this is first trying to match the identifiers which are present in somehow unique tokens.

More Efficient Garbage Collecting. The definition of OOPNs makes garbage collecting a part of every event. However, this computation is not necessary in every step because not every step makes some instance obsolete. Events which can cause a loss of some instance can be detected according to the way they work with transition and port variables.

Computing Enabled Events in an Incremental Way. The set of events enabled in a state can be computed in an incremental way starting with the set of events enabled in a predecessor state of the given state and just adding or removing some events according to re-checking the firability of some transitions. More precisely, we have to examine all the transitions which are connected to at least one input place whose marking was changed. Furthermore, we have to check transitions whose guards can use objects whose state was changed in a visible way. An object is changed in a visible way if there is a port in the class of the object which can read the contents of an object net place whose marking within the object was changed or which contains a visibly changed object.

4 Specifying Properties of Systems To Be Evaluated

In this section, we discuss different ways of specifying properties to be evaluated over state spaces of systems being examined using OOPNs.

Most of the common ways of specifying properties to be checked over state spaces of models based on different modelling languages [Val98] can be used in the context of OOPN-based models, too. We can think of using the following attitudes:

- evaluating *state space statistics* such as numbers of states, numbers of strongly connected components, bounds of places, Petri net live transitions, etc.,
- proposing a *versatile state space query language* allowing for user-controlled traversing through state spaces examining the encountered states,
- *instrumenting models* by property labels such as end-state labels, progress labels, assertions, etc. or by property automata,
- using a *high level specification language* such as some temporal logic.

Most of the above listed attitudes have to be slightly accommodated for the context of OOPNs and their state spaces. For example, bounds of places of OOPNs should be computed separately for particular instances and then a maximum should be chosen, particular property labels should be joint in a suitable way either with places or transitions of OOPNs, etc. However, there arises one more general problem here which influences almost all of the mentioned attitudes (more precisely all of them up to state space statistics). This problem is querying particular states and events of OOPNs.

The main problem to be solved when querying states and events of OOPNs stems from the dynamism of OOPNs. We have to prepare tools for exploring properties of sets of states and events in a way which respects the fact that the sets of existing instances, their names and relations can be different in every encountered state and cannot be fully predicted. Therefore it is not possible to use as simple queries as e.g. in Design/CPN, such as "take the marking of some place p from the static net instance unambiguously identified by *id*".

Within a prototype of an OOPN state space tool, we have suggested a solution of the above problem based on two groups of functions. First of all, we use the so-called *instance querying functions.* They allow us to begin with the unique initial object net instance or with sets of the just existing instances of certain nets given statically by their types. Subsequently, they allow for recursively deriving sets of the net instances or constants straight or transitively referenced from the already known instances via the marking of some of their places or transitions. There also exists a function returning the set of method net instances just running over some given objects. (Objects are represented by the corresponding object net instances here.)

Instance querying functions are intended to be combined with the so-called *set iterating functions* in order to obtain the appropriate characteristics of states. Set iterating functions allow for searching somehow interesting instances or constants in the sets of them returned by the instance querying functions. We can for example take all the just existing instances of some net, select the ones which contain some constant in some place, and then go on by exploring some other instances referenced from the selected ones.

So far we have been speaking about functions for querying OOPN states only. However, examining events seems to be a little easier. It is enough to have tools for accessing the particular items of events, i.e. their type, the transition they are bound to, the instance they are firing in, and the appropriate binding.

The functions for querying states and events can be straight used as a part of a versatile OOPN state space query language for examining the encountered states and events. Moreover, they can be used for describing terms embedded in temporal logic formulae specifying properties of systems to be verified over their OOPN-based models. Finally, they can also be applied when specifying legal termination states, progress events, or system invariants.

We will now a little more describe some of the instance querying functions. We describe them in the form of Prolog predicates as they are declared in the prototype tool using them. They all take the current state to be implicit and return the result via their last parameter. The predicate `init(Is)` returns the set with the initial object net instance. The predicate `inst(Cs,Ns,Is)` returns the set of the just existing instances belonging to the nets from the set `Ns` and running over objects belonging to the classes from `Cs`. The predicate `token(Is,Ps,Cs,Ms)` returns the set of tokens belonging to the classes from `Cs` and stored in the places from `Ps` within the instances from `Is`. The predicate `invoc(Is,Ts,Cs,Ns,Bs)` returns the set of invocations of the transitions from `Ts` within the instances from `Is`. The invocations are represented by the appropriate bindings and only the ones are selected which launch nets from `Ns` over objects of the classes from `Cs`. Finally, the predicate `over(Is1,Ns,Is2)` collects all the instances of the nets in `Ns` which run over the objects in `Is1`.

Out of the group of the set iterating functions, we can mention for example the following ones. The predicate `sforall(S,X,P,Y)` returns `true` in `Y` iff the predicate `P` over `X` is fulfilled over every element of the non-empty set `S` whose elements are one-by-one bound to `X`. Otherwise, a counter-example is found and bound to `Y`. The predicate `select(S1,X,P,S2)` selects all the elements `X` from `S1` which fulfill the predicate `P` over `X`.

Let us now present a very simple example of examining states of OOPNs. Below we define a predicate `ex_depth(N)` allowing for finding out whether some of the stack instances in the model from figure 1 can grow up to a given depth. A check whether an arbitrary stack can become deeper than $N-1$ can than be implemented by a state space query which evaluates the predicate `ex_depth(N)` over every state and collects the states where it holds. A more abstract approach would be checking the validity of the CTL formula EF `ex_depth(N)`.

```
ex_depth(N) :-
  inst([stack],[[stack,object]], S),
  select(S,Si,(token([Si],[st],all,[L]),length(L,N)),SN),
  empty(SN,false).
```

5 Conclusions

We have briefly described the notion of object-oriented Petri nets and some of the problems accompanying generating their full state spaces. We have especially mentioned the phenomenon of worsening the state space explosion problem due to working with identifiers of dynamically arising and disappearing net instances. Two possible approaches of dealing with the identifiers, namely sophisticated naming rules and name abstraction, have been described and compared.

We have also discussed a method allowing for asking analysis or verification questions over OOPN state spaces. This method avoids referring to uninteresting and unknown concrete names of instances.

The notions included in the article are supposed to be exploited within formal analysis and verification on suitably reduced OOPN state spaces which is one of the goals of our future research. We further intend to do more research on using OOPNs for modelling distributed systems, and especially the software ones.

Acknowledgment. This work was done within the research intention No. CEZ:J22/98: 262200012 - "Research in Information and Control Systems" and it was also supported by the Grant Agency of the Czech Republic under the contract 102/98/0552 "Research and Applications of Heterogeneous Models".

References

[ČJV97] M. Češka, V. Janoušek, and T. Vojnar. PNtalk – A Computerized Tool for Object-Oriented Petri Nets Modelling. In F. Pichler and R. Moreno-Díaz, editors, *Proceedings of the Computer Aided Systems Theory and Technology – EUROCAST'97*, volume 1333 of *Lecture Notes in Computer Science*, pages 591–610, Las Palmas de Gran Canaria, Spain, February 1997. Springer-Verlag.

[Jan98] V. Janoušek. *Modelling Objects by Petri Nets.* PhD thesis, Department of Computer Science and Engineering, Technical University of Brno, Czech Republic, 1998. (In Czech).

[Jen94] K. Jensen. *Coloured Petri Nets: Basic Concepts, Analysis Methods and Practical Use, Vol. 2: Analysis Methods.* EATCS Monographs on Theoretical Computer Science. Springer-Verlag, 1994.

[Pel96] D. Peled. Combining Partial Order Reductions with On-the-fly Model-Checking. *Journal of Formal Methods in Systems Design*, 8 (1):39–64, 1996. Also appeared in 6th International Conference on Computer Aided Verification 1994, Stanford CA, USA, LNCS 818, Springer-Verlag, 377-390.

[SB94] C. Sibertin-Blanc. Cooperative Nets. In R. Valette, editor, *Proceedings of the 15th International Conference on Application and Theory of Petri Nets*, volume 815 of *Lecture Notes in Computer Science*, pages 471–490, Zaragoza, Spain, June 1994. Springer-Verlag.

[Val98] A. Valmari. The State Explosion Problem. In W. Reisig and G. Rozenberg, editors, *Lectures on Petri Nets I: Basic Models*, volume 1491 of *Lecture Notes in Computer Science*, pages 429–528. Springer-Verlag, 1998.

[Voj00] T. Vojnar. *State Spaces of Object-Oriented Petri Nets.* PhD thesis, Department of Computer Science and Engineering, Brno University of Technology, Czech Republic, to be published in 2000.

Representing Petri Nets in an Action Based Formalism

Ramón P. Otero and José M. Rodríguez

Computer Science Department, University of Corunna,
Campus de Elvina, 15071 Corunna, Galicia, Spain.
Tel: +34-981-167000 ext 1276 Fax: +34-981-167160
{otero,josema}@dc.fi.udc.es

Abstract. This paper addresses the possibility of representing and simulating *Petri Nets* using an Artificial Intelligence formalism for Reasoning about Actions and Change called *Generalized Magnitudes*. As a result, we obtain an alternative logical-based formalization for Petri Nets, allowing to apply common techniques in Reasoning about Actions and Change in the analysis of Petri Nets models. Futhermore, we provide the alternative of representing concurrency and synchronization in a Generalized Magnitudes system using a Petri Net.

1 Introduction

The formal study of evolving systems has been tackled under many different perspectives and inside independent (and frequently too disconnected) research areas. Despite of this lack of interconnection, it is interesting to note how different approaches, and specially, different representational formalisms, share many features thanks to a common underlying domain: dynamic systems. Two of the areas for which some interrelation results have been established are Temporal Reasoning in Artificial Intelligence (AI) and Systems Modeling and Simulation. For instance, in previous works [1, 5, 6] a strong relationship between the Generalized Magnitudes (GMs) AI formalism and *Discrete Events Systems* representations (under DEVS formalism) was studied, leading to the introduction of new AI features into Computer Aided Systems Theory.

In this paper we continue these interconnection studies analysing the relation between the already mentioned framework for temporal reasoning (GMs) and a well-known formalism for dealing with concurrency and synchronization: *Petri Nets* [8, 9]. More concretely, we study the representation of low-level Petri nets in the GMs formalism.

As a result of this study, we obtain benefits for both formalisms. In the GMs side, we extend the range of applicability (temporal expert systems, DEVS, etc), making it feasible to build hybrid systems where the concurrency and synchronization tasks are modeled as Petri Nets. In the Petri Nets side, the representation in GMs provides an alternative theoretical reformulation for Petri Nets thanks to the logical formalization of GMs called L^2 [7], and used for Reasoning

F. Pichler, R. Moreno-Díaz, and P. Kopacek (Eds.): EUROCAST'99, LNCS 1798, pp. 105–115, 2000.

about Actions and Change. The translation of a Petri Net specification into an action theory, allows applying to Petri Nets common techniques in Reasoning about Actions and Change (*explanation*, *planning*,...). Another additional advantage, which will be analysed in future work, is the improvement in efficiency aspects for simulation of Petri Nets due to the efficiency features [2] for execution of GMs systems that have been incorporated to the practical shell *Medtool* [4].

This paper is organized as follows. In section 2 and 3, we recall the basic definitions and foundations of the Petri Nets and GMs formalisms. Section 4 gives a complete description of how to represent Petri Nets in the GMs formalism and how the simulation can be performed. We outline some ideas about representing Timed Petri Nets and Coloured Petri Nets in section 5. In section 6, we show how AI techniques can be applied to a Petri Net represented in the GMs formalism. Finally, in section 7, we present some conclusions and address future directions of work.

2 Petri Nets

Although we assume the reader is familiar with basic Petri Nets concepts, we will recall some basic definitions [8, 9] in order to make this paper self-contained.

A classical Petri Net is a bipartite directed graph which consists of two node types called *places* (represented as circles) and *transitions* (represented as bars). Places can only be connected to transitions and vice versa. A place p is called an *input place* of a transition t if there exists a directed arc from p to t, whereas p is called an *output place* of t if there exists a directed arc from t to p. Arcs can be labeled with a positive integer number called the arc *weight*.

Definition 1 (Petri Net) A Petri Net is a 4-tuple $\mathcal{N} = (P, T, I, O)$ where P and T are two finite and non-empty disjoints sets of *places* and *transitions*, and I and O are the *input* and *output* incidence functions from $P \times T$ to the set of positive integer numbers. □

The dynamic behaviour of the modeled system is represented by tokens flowing through the net. A *token* is graphically represented by a dot inside a place. Each place may contain zero or more tokens, and its number may change during the execution of the net. A *marking* of a Petri Net is the distribution of tokens at a given time.

Definition 2 (Marking) Let $\mathcal{N} = (P, T, I, O)$ be a Petri Net. A marking M from P to positive integers, $\mathcal{M} : P \longrightarrow \mathbf{N}^+$ is an assignment of tokens to the places of $\mathcal{N}$. We write $(\mathcal{N}, \mathcal{M})$ for a Petri Net $\mathcal{N}$ with marking $\mathcal{M}$ □

A transition t is *enabled* if each of its input places contains at least as many tokens as the weight of the arc connecting it with t. Formally:

Definition 3 (Enabled transition) Let $\mathcal{N} = (P, T, I, O)$ be a Petri Net with marking $\mathcal{M}$, $t \in T$ a transition and $\mathcal{M}(p)$ the number of tokens contained in some $p \in P$. The transition t is enabled in $(\mathcal{N}, \mathcal{M})$ if and only if $\forall p \in P$: $\mathcal{M}(p) \geq I(p, t)$ □

A transition t may be *fired* whenever it is enabled. In most of cases, it is required the arrival of an event to fire an enabled transition. Firing a transition means removing from its input places as many tokens as the weight of the corresponding input and, simultaneosly, adding as many tokens as the weight of the arcs to each output place.

Definition 4 (Firing a transition) Let $\mathcal{N} = (P, T, I, O)$ be a Petri Net with marking $\mathcal{M}$, $\mathcal{M}(p)$ the number of tokens contained in $p \in P$ and $t \in T$ an enabled transition. Firing the transition t in $(\mathcal{N}, \mathcal{M})$ results in a new marking $\mathcal{M}'$ given by $\mathcal{M}'(p_i) = \mathcal{M}(p_i) - I(p_i, t) + O(p_i, t)$. □

Sometimes, Petri Nets contain a special type of arcs called *Inhibitor Arcs* [8, 9] which allow incorporating negative preconditions to transitions. A place p is called an *Inhibitor Place* of a transition t iff there exists an inhibitor arc from p to t. A Petri Net with inhibitor arcs is called an *Inhibitor Petri Net*. In these nets, a transition t is enabled by a marking $\mathcal{M}$ iff not only every input place p_i of t contains at least $I(p, t)$ tokens, but also every inhibitor place contains zero tokens.

3 GMs Formalism

The GMs formalism is intended for temporal representation and deals with causality in reasoning about actions and change. Briefly, it consists in a modular representation in which the basic units are called *Generalized Magnitudes* (GMs). These units are used for characterizing the properties and the relationships of the domain. A GM is defined by a *name* and the set of *values* it can take in different situations (only one of them at each moment). Knowledge is represented by attaching a *Knowledge Expression* (KE) to GMs. A KE is an expression containing the usual relational, arithmetic and logical operators, plus a conditional constructor *if*; all of them applied to other GMs or to their previous value (using the operator `previous`). Each KE is always directed towards a single GM, being the unique and complete way of obtaining its value. The value of a GM can be set directly too, providing *input facts*, which play the role of actions.

An implicit causal relation is defined between a GM B and the GMs A_i ocurring in its KE. We say that A_i may cause B, or more appropriately, B *depends* on A_i. This causal relation allows to establish an ordered evaluation of the KEs and to identify the relevant part of the knowledge base. Causality is used both for the non monotonic assumption of inertia and for computing ramifications in a directional way, in a similar way as proposed in [10]. If a GM depends directly or indirectly on an input fact, it is said to be *pertinent*. Otherwise, it is said to be *persistent* - the *inertia* principle is applied to it and so, it maintains its previous value. The expression `pert(A)` can be used in a KE for checking whetter a given GM `A` is pertinent at the current evaluation.

The evaluation of the KEs occurs at discrete time instants located in a continuos time basis. A special event (`now`) is defined to represent the time point of the current evaluation (`timeof(now)`). The new GMs values obtained at the evaluation (i.e. the pertinent values) will have this temporal instant as an associated time. An operator, `timeof(A)`, allows referring to the associated time of a given GM `A`, maintaining at every moment, the instant in which `A` took its current value.

4 Petri Nets Modeling and Simulation in the GM Scheme

In this section we introduce the translation process for representing a Petri Net inside the GMs formalism. First we propose an equivalence between the basic concepts of both formalisms.

4.1 Representation of Places and Transitions

Places are represented with numeric GMs. A single GM is created for each place in the network. The value assigned to this GM represents the number of tokens contained in the represented place. The KEs of this kind of GM are in charge of the computation of the next distribution of tokens after transitions are fired.

Transitions are represented with a pair of boolean GMs. The first GM determines the transition state, i.e. whether it is fired or not. The other one is an input fact and it represents the arrival of the associated external event that eventually fires the corresponding transition.

Therefore, given a transition t with input places $p_{in}(i)$ and output places $p_{out}(j)$, we define the GMs `event_t` and `fire_t`. The KE of this last GM will be:

$$\texttt{fire\_t: event\_t and } \bigwedge_{i=1}^{n} \texttt{previous(}p_{in}(i)\texttt{)} \geq \texttt{I(}p_{in}(i)\texttt{,}t\texttt{);}$$

where $p_{in}(i)$ represents the GM defined for each input place of t. We obtain the previous value of the GM applying the `previous` operator.

The KEs of the GMs defined for representing places must compute the number of tokens after transitions are fired. Given a place p being an output place of $t_{in}(i)$ transitions and an input place of $t_{out}(j)$ transitions, a GM `p` with the bellow KE is defined:

$$\begin{aligned}&\texttt{p: previous(p) +}\\ &\quad + \textstyle\sum_{i=1}^{n} \texttt{O(}p\texttt{,}t_{in}(i)\texttt{)} * \texttt{(pert(}fire\_t_{in}(i)\texttt{)) and } fire\_t_{in}(i)\texttt{)}\\ &\quad - \textstyle\sum_{i=1}^{m} \texttt{I(}p\texttt{,}t_{out}(i)\texttt{)} * \texttt{(pert(}fire\_t_{out}(i)\texttt{)) and } fire\_t_{out}(i)\texttt{);}\end{aligned}$$

Inhibitor Petri Nets can be represented easily in the GMs formalism changing properly the KE of transitions and places connected by inhibitor arcs. In GMs representing the firing conditions we must check that inhibitor input places are empty. That is,

$$\begin{aligned}\texttt{fire\_t: event\_t and } &\bigwedge_{i=1}^{m} \texttt{previous(}p_{in}(i)\texttt{)} \geq \texttt{I(}p_{in}(i)\texttt{,}t\texttt{) and}\\ &\bigwedge_{i=m+1}^{n} \texttt{previous(}p_{in}(i)\texttt{)} = \texttt{0;}\end{aligned}$$

The KE expression of the inhibitor place must guarantee that the number of tokens remains to zero, and therefore, we need to add the next line to the KE representing the inhibitor place connected to the transition $t_{out}(j)$.

`0 if pert(`$fire\_t_{out}(j)$`) and `$fire\_t_{out}(j)$`;`

4.2 An example

We show the translation of a Petri Net to the GMs formalism with the next example (from [9]). Let us suppose two wagons C_1 and C_2 moving along the segments $A - B$ and $C - D$ respectively. Initial positions are A and B. When a button M is pushed both wagons begin moving to points C and D perhaps at different velocities. The return to the initial positions is only started when both wagons have already reached points C and D.

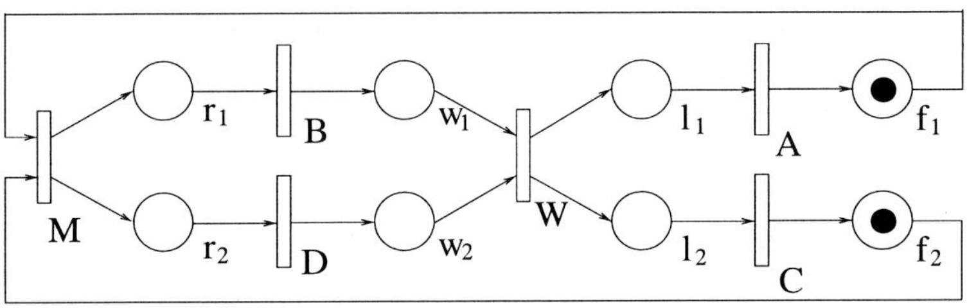

Fig. 1. A Petri Net for the wagons example.

Figure 1 shows a Petri Net modeling the proposed example. The transition M represents the button pushing, whereas A, B, C and D represent the arrival of the wagons to each corresponding point. The transition W represents that the two wagons are in points B and D. The places r_1, r_2, l_1 and l_2 represent that the corresponding wagons are moving to the right or to the left. The waiting of the wagons until the buttom M is pushed is modeled with places f_1 and f_2. Finally, places w_1 and w_2 are used to guarantee that both wagons are on the right side before they return to their initial positions.

The corresponding representation of this Petri Net in the GMs formalism is shown in figure 2. Note that we want the transition W to be fired automatically when the two wagons are on the right side, without need of any external event. Therefore, we need to make a new system evaluation without inputs facts. In this evaluation the transition W is fired, removing one token from places w_1 and w_2 and adding another token in places l_1 and l_2. Note that the KE of the GM representing the transition W includes the expression `timeof(now)` instead of the test for the arrival of an external event. It gives pertinence to GM `fire_W` even if there is not any input fact and allows checking whether W can be fired.

```
gm event_A { }
gm event_B { }
gm event_C { }
gm event_D { }
gm event_M { }

gm fire_A { event_A and previous(l1) >= 1; }
gm fire_B { event_B and previous(r1) >= 1; }
gm fire_C { event_C and previous(l2) >= 1; }
gm fire_D { event_D and previous(r2) >= 1; }
gm fire_M { event_M and previous(f1) >= 1 and previous(f2) >= 1; }
gm fire_W { timeof(now) and previous(w1) >= 1)) and previous(w2) >= 1)); }


gm r1 { previous(r1) + 1 * (pert(fire_M) and fire_M) - 1 * (pert(fire_B) and fire_B); }
gm r2 { previous(r2) + 1 * (pert(fire_M) and fire_M) - 1 * (pert(fire_D) and fire_D); }
gm f1 { previous(f1) + 1 * (pert(fire_A) and fire_A) - 1 * (pert(fire_M) and fire_M); }
gm f2 { previous(f2) + 1 * (pert(fire_C) and fire_C) - 1 * (pert(fire_B) and fire_B); }
gm i1 { previous(l1) + 1 * (pert(fire_W) and fire_W) - 1 * (pert(fire_A) and fire_A); }
gm i2 { previous(l2) + 1 * (pert(fire_W) and fire_W) - 1 * (pert(fire_C) and fire_C); }

gm w1 { previous(w1) + 1 * (pert(fire_B) and fire_B) - 1 * (pert(fire_W) and fire_W); }
gm w2 { previous(w2) + 1 * (pert(fire_D) and fire_D) - 1 * (pert(fire_W) and fire_W); }
```

Fig. 2. A GMs specification for the wagons example.

4.3 Simulation

The execution of the Petri Nets is controled by the number and distribution of the tokens and the events that eventually fire the transitions. The GMs representing the arrival of events determine the steps in the simulation. For each set of external events (that become system input facts) a new evaluation is done. Internal transitions can also force a new evaluation without input facts.

When an evaluation occurs, the inference engine tries to apply the relevant KEs in order to obtain new values for the pertinent GMs. When representing Petri Nets, the pertinent GMs will be those ones representing the transition associated to the event, and those ones representing the input and output places connected to that transition. As a consequence, the simulation process is locally restricted and therefore is very efficient.

Pertinence is widely used in the KEs of the GMs for places and transitions, and is used to establish the relevant part of the net that must be analysed. In this way, the amount of computations is drastically reduced, since it is usual that large nets imply in practice a small deal of changes after each set of events.

Note that the events correspond to actions in our formalism, and the number of tokens in each place along with the transitions states correspond to fluents.

5 Representing other Petri Nets Formalisms

The classical Petri Net model has been used in many application areas. Many authors have developed new extensions and classes of Petri Nets. In this section we outline several guidelines to represent them in the GMs formalism.

5.1 Timed Petri Nets

Petri Nets theory was one of the first concurrent formalisms for dealing with real-time by using an extension known as *timed* Petri Nets. An important property of the GMs formalism is that it allows reasoning along time, making inferences about the evolution of the modeled domain. We can take advantage of this property to represent Timed Petri Nets.

Several concepts of Petri Nets with *time* have been proposed assigning for instance firing times to the transitions and/or places [11]. In *t-timed* nets, deterministic firing times are assigned to transitions. Each transition takes a time to execute its firing. When a transition t is enabled, a firing can be initiated by removing tokens from input places. After the firing time, tokens are added to the output places.

The representation of this kind of Petri Nets in the GMs formalism requires the use of the *timeof* operator. In order to add tokens to an output place p of transition $t_{in}(i)$ after a time $t_t(i)$ we define the GM for p as:

$$
\begin{aligned}
&\texttt{p: previous(p)}\\
&\quad + \textstyle\sum_{i=1}^{n} \texttt{O}(p, t_{in}(i))\\
&\qquad * \ (fire\_t_{in}(i)\ \texttt{and (timeof(now) = timeof(}fire\_t_{in}(i)\texttt{)} + t_t(i)\texttt{))}\\
&\quad - \textstyle\sum_{i=1}^{m} \texttt{I}(p, t_{out}(i)) \ * \ \texttt{(pert(}fire\_t_{out}(i)\texttt{) and } fire\_t_{out}(i)\texttt{);}
\end{aligned}
$$

At the end of a given evaluation, the system is able to compute which new value of `timeof(now)` would cause some condition to change its truth value, and automatically advance the simulation until such moment. [1]

In our example, once `fire_t` becomes pertinent, the system would conclude that the next activation time should be `timeof(fire_t)` + t_t and would fire a new transition at that moment. The output places would be modified at his delayed evaluation.

There exist, however, some limitations when using this mechanism which are still under study. The main drawback to be solved is that the autonomous activation only allows a single next activation and not a set of them. This disables the possibility, present in timed Petri Nets, of firing a delayed transition when the previous firing has not been executed yet (this would imply storing the two delayed evaluations).

5.2 Coloured Petri Nets

The classical Petri Net definition is inadequate for modeling most of the current real systems, which are usually very complex and extremely large. To solve this problem, higher level nets as *Coloured Petri Nets* (CPNs) [3] have been introduced.

CPNs are a combination of Petri Nets, (for the description of the synchronization of concurrent process), and programming languages, that provide the

[1] For a more detailed study of this feature, called *autonomous activation*, see [5].

flexibility of the definition of data types and the mechanism for manipulating the data values.

As opposed to classical Petri Nets, in CPNs each token carries a data value. Places have associated types, which can be arbitrarily complex, and that characterize the class of tokens they contain. Transitions and arcs may have also an associated expression that is applied to the tokens. In CPNs, a transition is enabled when in its input places there are as many tokens as determined by the arc expressions and, its associated expression, if defined, is satisfied. When an enabled transition is fired, the determined tokens are removed from the input places and the tokens indicated by the arc expressions are added to the output places.

The representation of CPNs into the GMs formalism can be very complex and large. Whereas in classical Petri Nets it is only needed to represent the number of tokens in each place (and therefore we can use a single GM), in CPN we must store also the value associated to each token. As only numeric or text GMs can be defined, we would need a different GM must be defined for each item of a complex data type. In CPNs we must represent implicitly each token. The main problem representing CPN arises when the number of tokens in a place is unbounded. This would require to define new GMs dynamically during the evaluation, and this feature is not allowed in the GMs formalism.

6 Planning in Petri Nets

As explained before, the availability of a logical formalization for GMs formalism, allows applying common AI techniques to Petri Nets when represented in GMs. As an example we show in this section how *planning* techniques can be applied. Given an initial and a goal situation, a planning problem tries to provide a *plan* which specifies the sequence of inputs that, starting at the initial situation, causes the system to pass through several intermediate situations until reaching the goal one. A single planning problem may be solved by differents plans.

In the GMs formalism, the planning process will rely on a goal-driven algorithm that should deal with a dynamic domain and temporal sequences of situations. Facts contained in the goal state are seen as wanted effects. Thus, for each wanted effect, its associated knowledge expression is analysed backwards, obtaining the possible causes that can make such fact true. Then, these causes become new goals to be satisfied, repeating this process iteratively, building a tree, until no new assumptions can be done. At that point, the reasoning for the current state has finished. The planner examines now if the leaf solutions include a reference to the `previous` operator. If this is so, the process goes on at a precedent state. The goal for the precedent state is constructed with all the facts affected by the `previous` operator in the current state. When a solution is consistent with respect to the initial state, the branch is not expanded any more and we obtain a valid plan.

Figure 3 shows the initial marking of a single Petri Net which has been represented in the GMs formalism as indicated in section 4 (see figure 5). However,

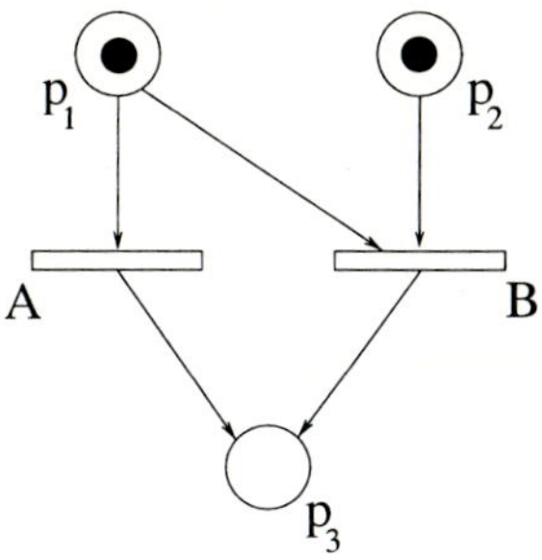

Fig. 3. Initial marking of the Petri Net.

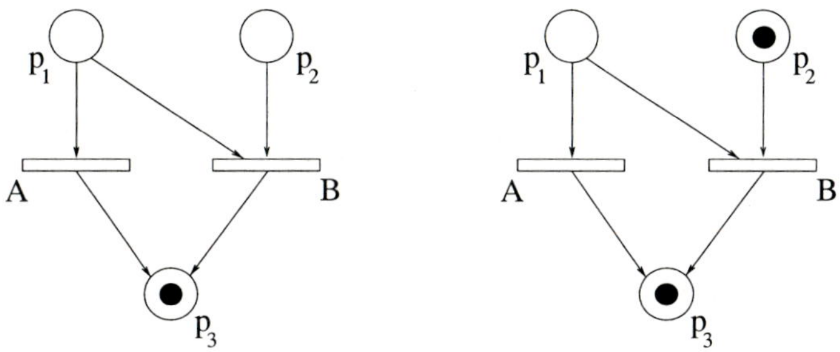

Fig. 4. Two final markings.

and in order to make more understandable the planning process, we assume that places can have at most one token, and therefore we can make a propositional description of the Petri Net. We want to know what could have happened to reach another marking where we only know that place p_3 has one token. Figure 4 shows two possible reachable states from the initial one depending on which transition `A` or `B` is fired.

The planning process of the GMs formalism, given the above data, obtains (see figure 6) two valid ways of reaching the goal: (1) the arrival of an event that fires transition `A`; and (2) the arrival of an event that fires `B`.

```
gm event_A { }
gm event_B { }

gm fire_A { event_A and previous(p1); }
gm fire_B { event_B and previous(p1) and previous(p2); }

gm p1 { false if (pert(fire_A) and fire_A) or (pert(fire_B) and fire_B); }
gm p2 { false if (pert(fire_B) and fire_B); }
gm p3 { true if (pert(fire_A) and fire_A) or (pert(fire_B) and fire_B); }
```

Fig. 5. The GMs specification for the planning example.

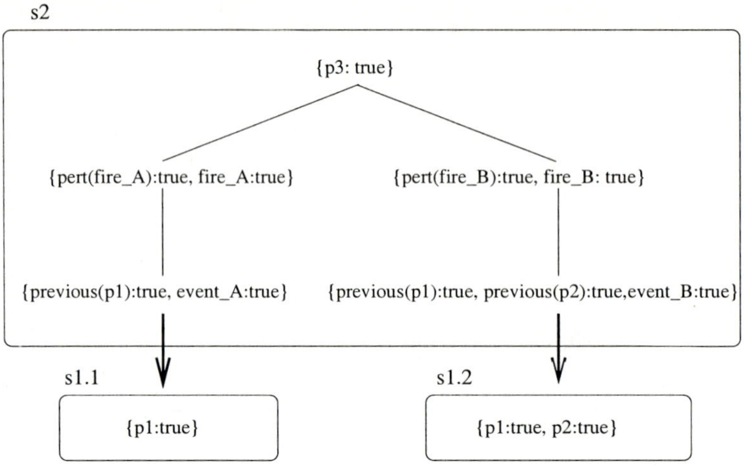

Fig. 6. Solutions tree for the planning example.

7 Conclusions and Future Work

In this work a study about the integration of two formalisms from two different areas has been presented. We have shown that both formalisms can take advantage from this integration. Specifically, an alternative theoretical reformulation for Petri Nets, based on a logic for Reasoning about Actions and Change was presented. This allows to apply common techniques on Artificial Intelligence (such as explanation, planning, ...) to Petri Nets represented in the GMs formalism. Moreover, simulation with Petri Nets can benefit from efficient features of GMs. We have seen that the use of *pertinence* can optimize the simulation process, restricting the part of the knowledge base to be considered in each evaluation. Finally, we can directly apply developed techniques on concurrent evaluation of GMs. For the GMs formalism, we have introduced an hybrid system where tasks of concurrency and synchronization can be modeled as Petri Nets, represented in the GMs formalism.

Given that low level Petri Nets are not up to the task of modeling complex systems, we will try to represent in the GMs formalism *Coloured Petri Nets*. Therefore, our future work is focused on solving the main problems when representing CPNs. In this way, we will study how dynamic creation and deletion of GMs can be added to the formalism. We must establish some important points about when GMs are created or deleted (during the evaluation or after it), when they can be evaluated for first time, etc...

Besides, when trying to emulate *Timed Petri Nets* some limitations for representing them properly appear in the GMs formalism. A future line of research will be the study on how to modify the *autonomous activation* mechanism in order to allow several delayed activations.

Acknowledgements

This work was supported in part by project XUGA10501B98 from the Government of Galicia, and in part by project PB97-0228 from the Government of Spain.

References

[1] Cabalar P., Otero R. P., Cabarcos M., Barreiro A.: Introducing Planning in Discrete Event Systems. Computer Aided Systems Theory, Lecture Notes on Computer Science. Vol 1333, (1997) pp 149-159. Springer, Berlin.

[2] Cabarcos M., Otero M., Cabalar P., Otero R. P.: Efficient concurrrent execution of Medtool expert systems. Conference on Artificial Intelligence Applications (EXPERSYS'96), Paris.

[3] Jensen K.: Coloured Petri Nets. Basic Concepts, Analysis Methods and Practical Use. Volume 1: Basic Concepts. Volume 2, Analysis Methods. Monographs in Theoretical Computer Science, (1997) Springer-Verlag.

[4] Otero R.: Medtool, una herramienta para el desarrollo de sistemas expertos. PhD thesis, 1991, Universidade de Santiago.

[5] Otero R. P., Barreiro A., Cabalar P., Lorenzo D.: Discrete event simulation in an environment for temporal expert systems. Lecture Notes in Computer Science (1996) 1030:271-282.

[6] Otero R. P., Barreiro A., Praehofer H., Pichler F., Mira J.: Stims-Medtool: Integration of expert systems with systems modelling and simulation. Lecture Notes in Computer Science (1994) 763:347-356.

[7] Otero R. P., Cabalar P.: Pertinence and Causality. Proceedings of the 3rd Workshop on Nonmonotonic Reasoning, Action and Change (NRAC), IJCAI'99, Stockholm, Sweden.

[8] Peterson J. L.: Petri net theory and the modeling of systems. Prentice-Hall. Englewood Cliffs (1991) New Jersey.

[9] Silva, M.: Las Redes de Petri en la Automática y la Informática. Ed. AC. Madrid, (1985) Spain.

[10] Thielscher M.: Ramification and Causality. Artificial Intelligence Journal (1997).

[11] Zuberek W. M.: Timed Petri Nets: Definitions, Properties, and Applications. Microelectronics and Reliability, vol.31, no.4, (1991) pp.627-644.

Simplification of Proof Procedures Based on the Path Condition Concepts

Mireille Larnac, Janine Magnier, and Vincent Chapurlat

LGI2P
Ecole des Mines d'Alès - Site EERIE
Parc Scientifique Georges Besse
30035 - NIMES cedex 1
France
Phone: +33 (0)466387026 - Fax: +33 (0)466387074
Mireille.Larnac@site-eerie.ema.fr

Abstract. The formal proof of properties of a system first requires the expression of the behavior of the system into a formal language.
The scope of this paper is the simplification of the proof procedure of properties of systems which are represented by discrete time models (Finite State Machines or extensions) The formulas which are manipulated are decomposed and the global proof is reduced to the study of a small subset of elementary proofs. This method was obtained by re-using some work developed in the framework of the management of the Path Condition in Symbolic Simulation.

1 Introduction

The goal of this paper is to lower the complexity of proof procedures which are involved in the verification process of discrete time models (Finite State Machines and extensions) which are based on the formal manipulation of temporal logic formulas. Some simplification using previous work developed within the framework of symbolic simulation are proposed.

The Path Condition concept for Symbolic Simulation will first be presented. Then, a method which uses temporal logics for proving properties on discrete time models will be shown. Finally, the Path Condition concepts will be applied on the former proof process in order to simplify it.

2 Symbolic Simulation and the Path Condition

2.1 Symbolic Simulation vs. Numerical Simulation

Symbolic simulation consists in simulation in which the data which are manipulated (e.g. inputs) remain symbolic and do not necessarily have a numerical value.

The main advantage of this approach (with respect to classical simulation) is that a symbolic simulation run gathers a (possibly infinite) set of numerical simulation runs.

F. Pichler, R. Moreno-Díaz, and P. Kopacek (Eds.): EUROCAST'99, LNCS 1798, pp. 116–126, 2000.

This technique obviously increases the power of simulation, but a problem for managing conditions arises. Indeed, a symbolic simulation run has to execute conditional actions (i.e. choose an execution branch of the execution tree). If the encountered condition cannot be directly evaluated (because the variables which are involved have a symbolic value), the user must decide the truth value of this condition and the execution can then continue.

But it may happen that the choices which have already been made on the execution path contain the value of the current condition. In this case, the user should not intervene. This constitutes the purpose of the Path Condition management [LAR92,LAR93].

2.2 The Path Condition Management

The Path Condition (PC) gathers the truth values of conditions which have already been met during the current simulation run. The value of PC is obviously $True$. PC is the logical AND between all the conditions which define the execution path. These conditions can involve any types of data (Boolean, integer, real, etc.).

Let C be a condition which is met by the symbolic simulator. In order to maintain the consistency of the execution, it is necessary to study if PC contains the truth value of C or not. In order to do this, the **Choice operation** and the **Free Choice variable** have been defined.

The value of the Choice operation $\triangleright$ is Boolean. The value of $PC \triangleright C$ is:

- $False$ if the truth value of C is included in PC ($PC \supset C \equiv True$ or $PC \supset \neg C \equiv True$),
- $True$ if the choice of the value of C is free.

The Free Choice variable $FC_{PC,C}$ is defined as follows: $FC_{PC,C} = PC \triangleright C$.

The evaluation of FC is decomposed in two steps:

- if the definition sets of PC and C are disjoint, $FC_{PC,C}$ is $True$
- if the definition sets of PC and C are not disjoint, the solution sets of PC and C must be examined.

2.3 Properties of the Choice Operation

In order to evaluate the Free Choice variable associated with complex expressions,the following properties have been established [LAR92]:

1. $FC_{PC,C_1 \wedge C_2} \equiv FC_{PC,C_1} \vee FC_{PC,C_2}$
2. $FC_{PC,C_1 \vee C_2} \equiv FC_{PC,C_1} \wedge FC_{PC,C_2}$
3. $FC_{PC_1 \wedge PC_2,C} \equiv FC_{PC_1,C} \wedge FC_{PC_2,C}$
4. $FC_{PC_1 \vee PC_2,C} \equiv FC_{PC_1,C} \vee FC_{PC_2,C}$
5. $FC_{PC_1 \wedge PC_2,C_1 \wedge C_2} \equiv (FC_{PC_1,C_1} \wedge FC_{PC_2,C_1}) \vee (FC_{PC_1,C_2} \wedge FC_{PC_2,C_2})$
6. $FC_{PC_1 \wedge PC_2,C_1 \vee C_2} \equiv FC_{PC_1,C_1} \wedge FC_{PC_1,C_2} \wedge FC_{PC_2,C_1} \wedge FC_{PC_2,C_2}$
7. $FC_{PC_1 \vee PC_2,C_1 \wedge C_2} \equiv FC_{PC_1,C_1} \vee FC_{PC_1,C_2} \vee FC_{PC_2,C_1} \vee FC_{PC_2,C_2}$
8. $FC_{PC_1 \vee PC_2,C_1 \vee C_2} \equiv (FC_{PC_1,C_1} \wedge FC_{PC_1,C_2}) \vee (FC_{PC_2,C_1} \wedge FC_{PC_2,C_2})$

Properties 5 to 8 can immediately be generalized to expressions made up of more than two terms.

It follows that the evaluation of the Free Choice variable can be split into very simple subcases.

3 The Proof of Properties of Discrete Time Systems

3.1 Formal Representation

A FSM model is defined by a 5-tuple:

$$FSM = < \mathcal{S}, \mathcal{I}, \mathcal{O}, \delta, \lambda > \text{ where:}$$

- $\mathcal{S}$ is a finite, non-empty set of states
- $\mathcal{I}$ is a finite, non-empty set of inputs
- $\mathcal{O}$ is a finite set of outputs
- δ is the transition (next state) function: $\delta : \mathcal{I} \times \mathcal{S} \to \mathcal{S}$
- λ is the output function: $\lambda : \mathcal{I} \times \mathcal{S} \to \mathcal{O}$

#S, #I and #O are the cardinalities of S, I and O respectively.

Only deterministic machines which obey the following rules are considered:

- each state has one and only one following state for each relevant input
- no two distinct inputs can be applied simultaneously
- no two distinct outputs can appear simultaneously

Moreover, the machines which are studied are completely specified. This means that for all states, the next state and output are specified for all inputs [1].

3.2 Expression in Temporal Logic

The DUX System: DUX is a Linear Time Temporal Logic (LTTL) [GAB80]; it is well-known and has been widely used for the verification of programs [AUD90]. It constitutes a temporal interpretation of the modal logics defined by Manna and Pnueli [MAN82]. The suitability of this tool lies in its expressiveness power, and in its properties (completeness and decidability). DUX is defined by:

- a set of propositional variables: $V_p = \{p, q, r, \ldots\}$
- the classical logical operators: $\neg$ (not), $\wedge$ (and), $\vee$ (or), $\supset$ (implication)
- temporal operators:
  - ⋆ unary: $\bigcirc$(next), $\Box$(always), $\Diamond$(sometimes)
  - ⋆ binary: $\mathcal{U}$(until)
- True, False

[1] Note that it is possible to deal with incompletely specified machines within the framework of this model by defining a new type of variable for representing the unspecified inputs or outputs

The building rules of formulas are:

(i) Every propositional variable of Vp, True, False are formulas,
(ii) If A and B are formulas, then
  - $\star$ $\neg$A, A$\wedge$B, A$\vee$B, A$\supset$B
  - $\star$ $\bigcirc$A, $\Box$A, $\Diamond$A
  - $\star$ A $\mathcal{U}$B

  are formulas,
(iii) Any formula is obtained by application of rules (i) and (ii).

Interpretation:

- $\bigcirc$A (next A) means that "A will be true in the next (1-future) instant"
- $\Box$A (always A) means that "A is true for all future instants (including the present one)"
- $\Diamond$A (sometimes A) means that "A will be true for some future instant (possibly the present one)"
- A$\mathcal{U}$B (A until B) means that "there is a future instant where B holds, and such that until that instant, A continuously holds"

Remark: We denote by $\bigcirc^n$ the n-future instant.

Expression of the Behavior of a *FSM*: Let us consider a machine M. The sets **S**, **X** and **Z** are defined as follows [MAG90,MAG94]:

- **S** is the set of state type propositions: $\mathbf{s_i} \in \mathbf{S}, \forall s_i \in S, i = 0, \ldots, \#S - 1, \mathbf{s_i}$ is *True* when the state of M is s_i
- **X** is the set of input type propositions: $\mathbf{x_j} \in \mathbf{X}, \forall x_j \in X, j = 0, \ldots, \#I - 1, \mathbf{x_j}$ is *True* when the present input of M is i_j
- **Z** is the set of output type propositions: $\mathbf{z_k} \in \mathbf{Z}, \forall z_k \in Z, k = 0, \ldots, \#O - 1, \mathbf{z_k}$ is *True* when the present output of M is o_k

These definitions allow us to describe the temporal evolution of M (by expressing the behavior of the transitions of M) into the DUX temporal logic, called **Elementary Valid Formula** (EVF).

Let us suppose that we have $\delta(s_i, i_j) = s_k$ and $\lambda(s_i, i_j) = o_l$; it follows

$$EVF ::= \Box(\mathbf{s_i} \wedge \mathbf{x_j} \supset \bigcirc \mathbf{s_k} \wedge \mathbf{z_l})$$

whose interpretation is: "it is always true ($\Box$ operator) that if s_i is the current state (and therefore $\mathbf{s_i}$ is *True*) and i_j is the current input ($\mathbf{x_j}$ is *True*), then the next state ($\bigcirc$ operator) will be s_k ($\mathbf{s_k}$ will be *True*) and the current output is o_l ($\mathbf{z_l}$ becomes *True*)".

It follows that the set of all the EVF's (each of which expresses the existence of a transition of the FSM) provides an equivalent representation of the behaviour of the FSM model. This statement is true if we also take into account the set of formulae which represent the determinism constraints. The first set contains the state determinism concept, which says that at a given time step, there is one and only one current state. This determinism formula can be written, using the DUX formalism:

$$DF1 ::= \Box[\mathbf{s_i} \supset \neg \mathbf{s_j}] \forall \mathbf{j} \neq \mathbf{i}, \mathbf{i}, \mathbf{j} \in \{\mathbf{0}, \ldots, \#\mathbf{S} - \mathbf{1}\}$$

Similarly, $DF2$ and $DF3$ express that at a given time step, the machine cannot have two different inputs, and cannot produce two different outputs:

$$DF1 ::= \Box[\mathbf{x_i} \supset \neg \mathbf{x_j}] \forall \mathbf{j} \neq \mathbf{i}, \mathbf{i}, \mathbf{j} \in \{\mathbf{0}, \ldots, \#\mathbf{X} - \mathbf{1}\}$$

$$DF1 ::= \Box[\mathbf{z_i} \supset \neg \mathbf{z_j}] \forall \mathbf{j} \neq \mathbf{i}, \mathbf{i}, \mathbf{j} \in \{\mathbf{0}, \ldots, \#\mathbf{Z} - \mathbf{1}\}$$

Within the framework of a verification process, it is often necessary to consider time intervals. This leads to the definition of state, input and output sequences, which are noted, respectively:

$$\mathbf{s_i^n} ::= \mathbf{s_{i_1}} \wedge \bigcirc \mathbf{s_{i_2}} \wedge \bigcirc^2 \mathbf{s_{i_3}} \wedge \ldots \wedge \bigcirc^{n-1} \mathbf{s_{i_n}}$$

$$\mathbf{x_j^n} ::= \mathbf{x_{j_1}} \wedge \bigcirc \mathbf{x_{j_2}} \wedge \bigcirc^2 \mathbf{x_{j_3}} \wedge \ldots \wedge \bigcirc^{n-1} \mathbf{x_{j_n}}$$

$$\mathbf{z_k^n} ::= \mathbf{z_{k_1}} \wedge \bigcirc \mathbf{z_{k_2}} \wedge \bigcirc^2 \mathbf{z_{k_3}} \wedge \ldots \wedge \bigcirc^{n-1} \mathbf{z_{k_n}}$$

Then, in order to provide the user with a more global view of the system evolution, first the concept of temporal event (E_t) which represents the possible effects of the machine functioning has been defined, and then all the conditions which lead to obtaining a given temporal event are gathered into one single formula, called **Unified Valid Formula** (UVF). A temporal event will be a future state ($E_t = \bigcirc \mathbf{s_i}$), a future state within n time steps ($E_t = \bigcirc^n \mathbf{s_i}$), a state sequence ($E_t = \mathbf{s_i^n}$), a present output ($E_t = \mathbf{z_k}$), a n-future output ($E_t = \bigcirc \mathbf{z_k}$), or an output sequence ($E_t = \mathbf{z_k^n}$). The Unified Valid Formula associated with a temporal event E_t is thus:

$$UVF(E_t) = \bigvee_{(p,q):\mathbf{s_p} \wedge \mathbf{x_q^n} \supset E_t} \mathbf{s_p} \wedge \mathbf{x_q^n}$$

Obviously, an UVF is obtained by formal reasoning on the set of EVF's. More precisely, if the Temporal Event E_t is a next state or an output variable then the subset of the EVF's which contain E_t in the right hand side of the formula (after the implication operator) is constructed; $UVF(E_t)$ is then the logical OR between the left part of the EVF's of this subset.

Furthermore, if the temporal event is a n-future state or output, then the former process is performed on the associated 1-future state or present output (called E_t'). $UVF(E_t')$ then contains all the possibilities to reach E_t' within one time step; each of them is constituted by the requirement to be into a given state and then to apply a given input. In order to obtain $UVF(E_t)$, it is then necessary to iterate the process on the states which appear in $UVF(E_t')$.

Similarly, $UVF(E_t)$ can be iteratively built for E_t being a sequence of next states or outputs.

Note that another method for construction $UVF(E_t)$ for any kind of temporal event has been elaborated using Graph Theory results. It is then possible to establish the formula which corresponds to $UVF(E_t)$ for E_t being a n-future event, the value of n not being fixed but being "generic" [CHE97,LAR97].

Property of a Fixed Time Step UVF: It is very interesting to note that for a "classical" UVF (which does not concern generic future but a fixed time step), thanks to the determinism properties of the FSM, it is possible to express the UVF formula as an exclusive-or between the conditions (this is stronger than the or operation which has been shown before):

$$UVF(E_t) = \bigoplus_{(p,q):\mathbf{s_p}\wedge\mathbf{x_q^n}\supset E_t} \mathbf{s_p} \wedge \mathbf{x_q^n}$$

3.3 Verification of Properties

Let us come back to the goal of this work; it addresses the formal verification of properties of systems which are represented thanks to a FSM model. The first question to answer is: "what kind of properties can be proved?". To start with, the structure of the FSM can be exploited, and properties of some states can be of great interest. For instance, it is worth knowing if two states are equivalent, or if a state is a source or a sink. More sophisticated properties consist in establishing the conditions (on input sequences) to make a state being a "functional" sink, even though it is not a structural one, or to generate input sequences to synchronize the machine into a given state.

In most cases, the state evolution of the machine is not available, and the only means for the user to get some information on the machine evolution is to examine the outputs. So the analysis process of output sequences is very important, and a tool for generating input sequences in order to obtain outputs, or to distinguish internal states must be provided.

Last, it seems very important to be able to formally establish the influence of a current factor (input or state) on the future evolution. This relates to the "sensitivity" of the future with respect to a present situation or decision.

The verification method is based on two approaches:

- In some cases (for some properties), it is sufficient to analyze the EVF's or UVF's (either search if a given formula exists, or what its form is). For example, if s_p is a sink state, it means that all the transitions which leave s_p go back to this state. It follows that all the EVF's which contain s_p in their left part must be of the form:
  $EVF ::= \Box(\mathbf{s_i} \wedge \mathbf{x_j} \supset \bigcirc\mathbf{s_p} \wedge \mathbf{z_l})$, for all $\mathbf{x_j}$. Similarly, if s_p is a source state, it means that if there exists some transitions whose destination is s_p, they come from s_p. The consequence is that either $UVF(\bigcirc\mathbf{s_p}$ is empty, or that it has the following form:
  $UVF(\bigcirc\mathbf{s_p} = \bigvee(\mathbf{s_p} \wedge \mathbf{x_j})$.

– Unfortunately, this approach of verification based on the study of EVF's and UVF's is not sufficient for analyzing the influence of the present on the future. This is the reason why a formal tool for analyzing the sensitivity has been defined: the **Temporal Boolean Difference**.

Temporal Boolean Difference: The Temporal Boolean Difference (TBD) is the extension of the classical Boolean Difference [KOH78] defined on propositional logic, to temporal logic, especially the DUX system [MAG90,MAG94].

Definition 1. *The Temporal Boolean Difference of a function f with respect to a variable v_q is defined as the exclusive or between the restriction of the function with v_q set at $True$ and the restriction of f with the variable v_q set at $False$. Then the representation of the influence of the variable v_q on a function $f(v_1, \ldots, v_q, \ldots, v_n)$ is:*

$$\frac{\partial f}{\partial v_q} = f(v_1, \ldots, False, \ldots, v_n) \oplus f(v_1, \ldots, True, \ldots, v_n)$$

The result is a formula which contains the conditions for v_q to make f change value when itself changes value. It is called the *sensitivity* of f regarding v_q.

For example, he representation of the influence of having $\mathbf{s_j}$ as current state in order to be in $\mathbf{s_h}$ in n time units is:

$$\frac{\partial UVF(\bigcirc^n \mathbf{s_h})}{\partial \mathbf{s_j}} = C_t(\mathbf{s_j}) \oplus C_t(\neg \mathbf{s_j})$$

$$C_t(\mathbf{s_j}) = \bigvee_{(l):\mathbf{s_j} \wedge \mathbf{x_l}^n \supset \bigcirc^n \mathbf{s_h}} [\mathbf{s_j} \wedge \mathbf{x_l}^n]$$

$$C_t(\neg \mathbf{s_j}) = \bigvee_{(k,m):k \neq j, \mathbf{s_k} \wedge \mathbf{x_m}^n \supset \bigcirc^n \mathbf{s_h}} [\mathbf{s_k} \wedge \mathbf{x_m}^n]$$

$C_t(\mathbf{s_j})$ is the set of all the sequences of $(\mathbf{x_i})$ with the origin state $\mathbf{s_j}$ which permit to obtain $\bigcirc^n \mathbf{s_h}$. $C_t(\neg \mathbf{s_j})$ is the set of all the sequence of $(\mathbf{x_i})$ with the origin state $\mathbf{s_k}$ where $\mathbf{s_k} \neq \mathbf{s_j}$ ($\mathbf{s_k} \in \mathbf{S}$) and which also permit to obtain $\bigcirc^n \mathbf{s_h}$.

We note $DVF(E_t, v)$ the **Derived Valid Formula** of E_t with respect to v. So $DVF(E_t, q)$ is:

$$DVF(E_t, q) = \frac{\partial UVF(E_t)}{\partial q}$$

The result of the calculation of $DVF(E_t, q)$ can be:

– $False$. This means that $UVF(E_t)$ is independent of q; in other words, the fact that q changes value has no influence on the fact that E_t will occur or not
– not $False$. In this case, we obtain a Temporal Logic formula which expresses the sensitivity of $UVF(E_t)$ to changes in q, i.e. the conditions for $UVF(E_t)$ to pass from $True$ to $False$ (or conversely) when q changes value.

For example, the result of $DVF(\bigcirc^4 \mathbf{s_x}, \mathbf{s_i})$ is a temporal logic formula. If it equals $False$, it means that $UVF(\bigcirc^4 \mathbf{s_x})$ is totally independent on $\mathbf{s_i}$. The interpretation is that even though $\mathbf{s_i}$ turns from $True$ to $False$ or from $False$ to $True$, the fact that within 4 time steps the state will be or not be $\mathbf{s_x}$ does not change. On the other hand, it the result is not $False$, the formula contains the conditions which make the value of $\mathbf{s_i}$ influence the fact that $\mathbf{s_x}$ will be the 4-future state.

It is important to note that the result of this calculation is double: first, it indicates if a present factor can influence a future temporal event or not; second, if it appears that this influence exists, the formula contains all the sensitivity conditions. This can then be used either for generating simulation input sequences, or this information can be exploited by the user to modify the system so that the sensitivity cases cannot occur, ...

The important point is that there is an strong analogy with the Boolean Difference of Boolean functions, but the fundamental differences are that the formulae are expressed in Temporal Logic, and that the variables which are manipulated are typed (states, inputs, outputs) and non independent (because of the determinism properties).

Properties of the TBD: Similarly to the classical Boolean Difference, the Temporal Boolean Difference has got some distribution properties. The most interesting here concerns the TBD of the exclusive-or of two functions: it is equivalent to the exclusive-or of the TBD's of each function. It immediately follows that $DVF(E_{t_1} \oplus E_{t_2}, q) = DVF(E_{t_1}, q) \oplus DVF(E_{t_2}, q)$

Extension to Generic Future: Furthermore, the calculation of the TBD can also be performed on an UVF which manipulates Generic Future [CHE97]. The applications of TBD are various. It permits to generate input sequences for resynchronizing the machine into a given "initial" state, or for distinguishing the inner states (which are unknown) through the generation of distinct output sequences. Moreover, a very wide field of application is the study of the impact of a decision (or a current event) on the future evolution of the system. Further, even though the user has got no possibility to change the present, he knows all the conditions which, when made $True$, make the system evolve into a given way. It is then up to him to choose his strategy for modifying some parameters and then determine what he wants to get into the future. In conclusion, we have defined a formal method for providing the user with an equivalent symbolic representation of the behaviour of the Finite State Machine model, and then with a tool which support proof of properties and formal analysis. In order to do this, it has been necessary to define the concept of Temporal Boolean Difference for evaluating the sensitivity of the evolution of a system with respect to some variable change. The details of the modeling and verification approach, as well as the demonstrations of all the theorems can be found in [MAG90]. The limitations of this approach are linked to the weak expressiveness of the FSM model, which only handles Boolean data, and which needs to express any data influencing the system through inputs or states. It follows that the number of states or transitions tend to increase exponentially as soon as new data have to be taken into account. This is the reason why we have defined an extension of the FSM, called the **Interpreted Sequential Machine** model (ISM). All the details on this model and its verification can be found in [VAN95,LAR97]

4 The Choice Operator for the Simplification of the Verification Process

The FSM model and its verification principles have been presented. The simplification of the verification process is now studied.

It has been shown that the proof of properties of a system modeled by a FSM is based on the study of EVF's, UVF's of DVF's. The calculation of a DVF is based on the evaluation of the TBD of an UVF.

We have seen that an UVF for a fixed next time step is an exclusive-or of conditions (which are themselves constituted a conjunction of an initial state proposition and an input sequence). In this case, thanks to the exclusive-or distribution property, the evaluation can be split into subcases. For each of them, the TBD is $False$ if the condition does not depend on the variable with respect to which the FVD is calculated. This means that this comes down to calculating the Free Choice variable of each of the conditions and of the variable.

Let us illustrate this approach with an example.

Let us consider the FSM shown in Figure 1. Each of the transitions which appear on this FSM can be expressed as an EVF (with the associated propositional variables):

- $EVF1 ::= \Box(\mathbf{s_1} \wedge \mathbf{x_1} \supset \bigcirc \mathbf{s_2} \wedge \mathbf{z_1})$
- $EVF2 ::= \Box(\mathbf{s_1} \wedge \mathbf{x_2} \supset \bigcirc \mathbf{s_3} \wedge \mathbf{z_1})$

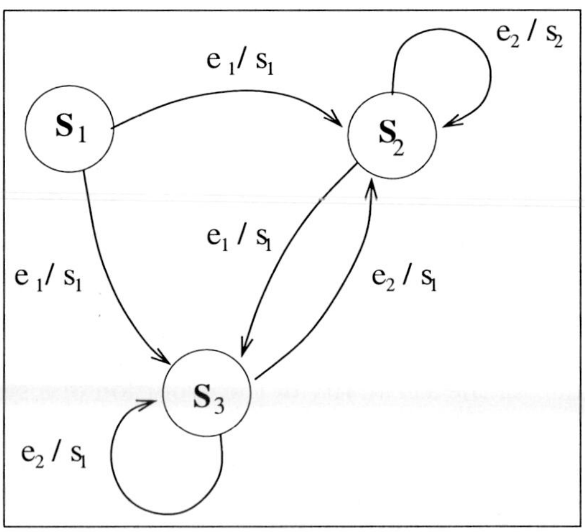

Fig. 1. Example of a FSM

- $EVF3 ::= \Box(\mathbf{s_2} \wedge \mathbf{x_1} \supset \bigcirc \mathbf{s_3} \wedge \mathbf{z_1})$
- $EVF4 ::= \Box(\mathbf{s_2} \wedge \mathbf{x_2} \supset \bigcirc \mathbf{s_2} \wedge \mathbf{z_2})$
- $EVF5 ::= \Box(\mathbf{s_3} \wedge \mathbf{x_1} \supset \bigcirc \mathbf{s_3} \wedge \mathbf{z_1})$
- $EVF1 ::= \Box(\mathbf{s_3} \wedge \mathbf{x_2} \supset \bigcirc \mathbf{s_2} \wedge \mathbf{z_1})$

Let then study the way to reach state S_3 in two time steps. This requires to establish the following UVF:
$UVF\ (\bigcirc^2 \mathbf{s_3}) ::= (\mathbf{s_1} \wedge \mathbf{x_1} \wedge \bigcirc \mathbf{x_1}) \oplus (\mathbf{s_1} \wedge \mathbf{x_2} \wedge \bigcirc \mathbf{x_1}) \oplus (\mathbf{s_2} \wedge \mathbf{x_2} \wedge \bigcirc \mathbf{x_1}) \oplus (\mathbf{s_2} \wedge \mathbf{x_1} \wedge \bigcirc \mathbf{x_1}) \oplus (\mathbf{s_3} \wedge \mathbf{x_1} \wedge \bigcirc \mathbf{x_1}) \oplus (\mathbf{s_3} \wedge \mathbf{x_2} \wedge \bigcirc \mathbf{x_2})$
Then, the study of $FVD(\bigcirc^2 \mathbf{s_3}, \mathbf{s_1})$ first requires to calculate:

1. $(\mathbf{s_1} \wedge \mathbf{x_1} \wedge \bigcirc \mathbf{x_1}) \rhd \mathbf{s_1}$
2. $(\mathbf{s_1} \wedge \mathbf{x_2} \wedge \bigcirc \mathbf{x_1}) \rhd \mathbf{s_1}$
3. $(\mathbf{s_2} \wedge \mathbf{x_2} \wedge \bigcirc \mathbf{x_1}) \rhd \mathbf{s_1}$
4. $(\mathbf{s_2} \wedge \mathbf{x_1} \wedge \bigcirc \mathbf{x_1}) \rhd \mathbf{s_1}$
5. $(\mathbf{s_3} \wedge \mathbf{x_1} \wedge \bigcirc \mathbf{x_1}) \rhd \mathbf{s_1}$
6. $(\mathbf{s_3} \wedge \mathbf{x_2} \wedge \bigcirc \mathbf{x_2}) \rhd \mathbf{s_1}$

Obviously, these Free Choice variables are respectively:

1. *False*
2. *False*
3. *True*
4. *True*
5. *True*
6. *True*

It is necessary to evaluate the FVD only for the subformulas whose Free Choice variables are *False*. This means that:
$FVD(\bigcirc^2 \mathbf{s_3}, \mathbf{s_1}) = \mathrm{FVD}(\mathbf{s_1} \wedge \mathbf{x_1} \wedge \bigcirc \mathbf{x_1}, \mathbf{s_1}) \oplus \mathrm{FVD}(\mathbf{s_1} \wedge \mathbf{x_2} \wedge \bigcirc \mathbf{x_1}, \mathbf{s_1})$.
Finally, by applying the definition of the Temporal Boolean Difference, it follows:
$FVD(\bigcirc^2 \mathbf{s_3}, \mathbf{s_1}) = (\mathbf{x_1} \wedge \bigcirc \mathbf{x_1}) \oplus (\mathbf{x_2} \wedge \bigcirc \mathbf{x_1})$

5 Conclusion

A formal verification method for systems which are modeled by Finite State Machines or extensions have been defined. The proof is based on the study of temporal logic formulas which express the behavior of the system. In order to be able to verify interesting properties (like the sensitivity of the future behavior with respect to present actions or decisions), a formal tool called the Temporal Boolean Difference has been defined.

This verification method involves the manipulation of sometimes large formulas which can be considered independently. This is the reason why the Choice operator (which had been defined for a very different purpose: the management of complex conditions in symbolic simulation) can lower the complexity of the proof processing by first evaluating which subformulas will have no influence on the result.

The next step of this study can consist in evaluating if the method which was developed for managing the Path Condition in symbolic simulation can help the proof process of a FSM by permitting to eliminate some temporal operators in the formulas.

References

AUD90. Audureau, E., Enjalbert, P., Farinas del Cerro, L.: Logique Temporelle - Sémantique et validation de programmes parallèles. Masson, Paris (1990)

CHE97. Chenot, B., Larnac, M.: Utilization of graph theory notions in the Interpreted Sequential Machine. SOCO-IIA'97, to appear (1997)

GAB80. Gabbay, D., Pnueli, A., Shelah, S., Stavi, J.: On the temporal analysis of fairness. 7th ACM Symposium on Principles of Programming Languages (1980)

HAR66. Hartmanis, J., Stearns, R.E.: Algebraic Structure Theory of Sequential Machines. Prentice hall, Englewood Cliffs, N.J. (1966)

KOH78. Kohavi, Z.: Switching and Finite Automata Theory. Tata McGraw Hill, Computer Science Series (1978)

LAR92. Larnac, M.: Gestion de la Condition de Chemin dans la Simulation Symbolique. PhD Thesis, University of Montpellier II (1992) (in French)

LAR93. Larnac, M., Giambiasi, N., Magnier, J., Rogacki, R.: Verification of VHDL Behavioral Descriptions by Symbolic Simulation. Proceedings of VHDL Forum for CAD in Europe, Innsbruck, Austria (1993)

LAR95. Larnac, M., Magnier, J., Vandermeulen, E., Dray, G., Chapurlat, V.: Temporal and Functional Verification of a Symbolic Representation of Complex Systems. EUROCAST'95, Lecture Notes in Computer Science, volume 1030, Springer Verlag (1995)

LAR97. Larnac, M., Magnier, J., Chapurlat, V., Chenot, B.: Formal Representation and Proof of the Interpreted Sequential Machine Model. EUROCAST'97, Lecture Notes in Computer Science, volume 1333, Springer Verlag (1997)

MAG90. Magnier, J.: Représentation symbolique et vérification formelle de machines séquentielles. PhD Thesis, University of Montpellier II, France (July 1990)

MAG94. Magnier, J., Pearson, D., Giambiasi, N.: The Temporal Boolean Derivative Applied to Verification of Sequential Machines. European Simulation Symposium, Istanbul, Turkey (1994)

MAN82. Manna, Z., Pnueli, A.: How to cook a temporal proof system for your pet language. Report No STAN-CS-82-954, Department of Computer Science, Stanford University (1982)

VAN95. Vandermeulen, E., Donegan, H.A., Larnac, M., Magnier, J.: The Temporal Boolean Derivative Applied to Verification of Extended Finite State Machines. Computers and Mathematics with Applications, Vol.30, N. 2 (January 1995)

Parallel Processor Array for Tomographic Reconstruction Algorithms*

Thomas Schmitt, Dirk Fimmel, Mathias Kortke, and Renate Merker

Dresden University of Technology,
Institute of Circuits and Systems,
Mommsenstraße 13,
D-01062 Dresden, Germany
{schmitt,fimmel,kortke,merker}@iee1.et.tu-dresden.de,
http://www.iee.et.tu-dresden.de/iee/st/

Abstract. In this paper we derive exemplarily a parallel processor array for algorithms of commonly used tomographic reconstruction methods by using the tools of the design system DESA. The algorithms represent a group of computationally intensive image processing algorithms requiring high throughput and real-time processing.
The design process is characterized by the consideration of hardware constraints and performance criteria. In particular, we determine one common parallel processor array for two different reconstruction techniques. Finally, the array is adapted to hardware constraints given by the target architecture which can be an application specific integrated circuit or a system of parallel digital signal processors.

1 Introduction

Parallel architectures such as parallel processor arrays enable the potential for producing scalable and efficient designs for computationally intensive applications in signal processing especially with real-time requests. A parallel processor array has a piecewise homogeneous structure with respect to the processor functions and the interconnection network between processors. This structure causes high parallelism, intensive pipelining and distributed memories. Algorithms which can be described as systems of affine recurrence equations (SARE) [11] are well suited for mapping onto processor arrays since they match the parallel computation structure and the piecewise regular interconnection scheme. In order to support the automatic design of processor arrays a wide range of methods has been developed e.g.[1,24,21,7,14]. The main parts of the processor array design process represent the transformations *allocation* which specifies the processors for the evaluation of the operations of the algorithm, and *scheduling* specifying the evaluation time of these operations. Several methods for integrating hardware constraints in the design process have been developed [25,4,5].

* The research was supported by the "Deutsche Forschungsgemeinschaft", in the project A1/SFB 358.

F. Pichler, R. Moreno-Díaz, and P. Kopacek (Eds.): EUROCAST'99, LNCS 1798, pp. 127–141, 2000.

Tomographic reconstruction techniques represent a kind of algorithms which are characterized by a high computational intensity and a structure well suited for parallelization in software or hardware. The commonly used reconstruction algorithms are the *filtered back projection* (FBP) [10] and the *algebraic reconstruction technique* (ART) [6]. Some work on parallelization of reconstruction methods has been published including both software and hardware solutions. In [18] the implementation of the FBP algorithm on parallel general purpose computers using parallel software for the Fast Fourier Transform (FFT) was presented. A custom processor for tomographic reconstruction methods which is superior to general-purpose processors was presented in [9]. In this CMOS VLSI chip for FBP the possible parallelism in processing the independent projection data is exploited.

In this paper we propose one common processor array for FBP and ART which enables running either the one or the other algorithm at the same hardware. As target architectures we consider application specific integrated circuits. Furthermore a parallel software solution running on a parallel system of digital signal processors (DSPs) is presented. The used design system DESA includes the basic design methods and new techniques which lead to an efficient adaptation of the processor array to hardware constraints such as number of processors, I/O capacities and to performance criteria such as minimum chip area and minimum computation time (latency). The parameters of the DSP system, such as computation and communication time, size and access time of the memory, are included likewise into the software design method [12].

In section 2 the basic methods of processor array design including a presentation of the additional design tools implemented in DESA [15,4] are introduced. Section 3 specifies the considered reconstruction algorithms, and in section 4 the parallel processor array for FBP and ART is derived. Section 5 gives results of the parallel software implementation of the reconstruction algorithms.

2 Array Design Methods

2.1 Basic Methods

Generally, algorithms in form of *systems of affine recurrence equations* (SARE) can be mapped onto processor arrays. These systems of affine recurrence equations can be transformed into *systems of uniform recurrence equations* (SURE) [11] by a localization / uniformization [20,3] In this paper we assume that the initial algorithm is described as a SURE.

Definition 1 (System of uniform recurrence equations). *A system of uniform recurrence equations is a set of equations of the following form:*

$$y_i[\mathbf{i}] = F_i(..., y_j[\mathbf{i} - \mathbf{d}_{ji}^k], ...) \quad \mathbf{i} \in \mathcal{I}_i \quad 1 \leq i, j \leq m, \tag{1}$$

where the equations are defined in index spaces $\mathcal{I}_i$ *being polytopes:*

$$\mathcal{I}_i = \{\mathbf{i} \mid \mathbf{A}_i \mathbf{i} \geq \mathbf{a}_{0i}\} \tag{2}$$

with $\mathbf{i} \in \mathcal{I}_i$ *are index points,* $\mathbf{i} \in Z^n, \mathbf{A}_i \in Q^{m_i \times n}, \mathbf{a}_{0i} \in Q^{m_i}$.

In (1) the *dependence vectors* $\mathbf{d}_{ji}^k \in Z^n, 1 \leq k \leq m_{ij}$ are constant vectors, called dependence vectors, and F_i, $1 \leq i \leq l$, are arbitrary functions. We suppose that the SURE has a single assignment form (every instance of a variable y_i is defined only once in the algorithm) and that there exists a partial order of the instances of the equations that satisfies the data dependencies. Variables which appear only on the right-hand side of equation 1 are denoted as independent variables. Variables which appear on the left-hand side of equation 1 are denoted as dependent variables. Due to exploitation of degrees of freedom during the design process we assume that in equation 1 the dependent data are localized, i.e. their indices are described by $\mathbf{i} - \mathbf{d}_{ji}^k$, and the indices of independent data can still be affine functions of the index point i. This version of the SURE is called single assignment code (SAC) and will be used in the following as initial algorithm description for the design process.
Next we introduce a graph representation of the data dependencies of the SURE.

Definition 2 (Reduced dependence graph (RDG)). *The equations of the SURE build the m nodes $v_i \in \mathcal{V}$ of the reduced dependence graph $\langle \mathcal{V}, \mathcal{E} \rangle$. The directed edges $(v_i, v_j) \in \mathcal{E}$ are the data dependencies weighted by the dependence vectors $\mathbf{d}_{ij}^k$.*

The weight of an edge $e \in \mathcal{E}$ is called $\mathbf{d}(e)$, the source of this edge $\sigma(e)$ and the sink $\delta(e)$.

Generally, the basic transformations of the MPPA design are the uniform affine scheduling and the uniform affine allocation. They lead to *full size arrays* which means that the size of the array depends on the size of the initial algorithm, and that they keep the regularity of the algorithm in the resulting processor array [19]. For each index space $\mathcal{I}_i$ of the SURE these transformations can be described as follows:

Definition 3 (Uniform affine scheduling). *An* uniform affine scheduling *assigns an evaluation time to each instance of the i-th equation of the SURE with:*

$$\tau_i : Z^n \to Z : \quad \tau_i(\mathbf{i}) = \boldsymbol{\tau}^T \mathbf{i} + t_i, \quad 1 \leq i \leq m, \tag{3}$$

where $\boldsymbol{\tau} \in Z^n, t_i \in Z$.

The index points of an index space lying on the same hyperplane defined by the *scheduling vector $\boldsymbol{\tau}$* are evaluated at the same time.

Definition 4 (Uniform affine allocation). *An* uniform affine allocation *assigns an evaluation processor to each instance of the i-th equation of the SURE with:*

$$\pi_i : Z^n \to Z^{n-1} : \quad \pi_i(\mathbf{i}) = \mathbf{S}\mathbf{i} + \mathbf{p}_i, \quad 1 \leq i \leq m, \tag{4}$$

where $\mathbf{S} \in Z^{n-1 \times n}$ is of full row rank, $\mathbf{p}_i \in Z^{n-1}$. Since $\mathbf{S}$ is of full row rank, the vector $u \in Z^n$, which is coprime and satisfies $\mathbf{S}u = \mathbf{0}$ as well as $u \neq \mathbf{0}$, is uniquely defined except to the sign and called projection vector.

The index points of an index space lying on a line spanned by the projection vector $\mathbf{u}$ are mapped onto the same processor.

Allocation and scheduling have to satisfy the constraints $\forall e \in \mathcal{E} : \boldsymbol{\tau}^T \mathbf{d}(e) > 0$ as well as $\boldsymbol{\tau}\mathbf{u} \neq 0$. The application of the allocation and the scheduling to the original SURE results in a full size array. The *interconnections* $\mathbf{v}(e)$ of the array are determined by the allocation with $\mathbf{v}(e) = \mathbf{Sd}(e) + \mathbf{p}_{\delta(e)} - \mathbf{p}_{\sigma(e)}, \forall e \in \mathcal{E}$ and the *time delays* w associated to the interconnections $\mathbf{v}$ by the scheduling with $w(e) = \boldsymbol{\tau}^T \mathbf{d}(e) + t_{\delta(e)} - t_{\sigma(e)}, \forall e \in \mathcal{E}$.

With the uniform affine scheduling the *latency* of the MPPA, i.e. the time to compute the SURE on the MPPA , is given by

$$L = \max_{1 \le i \le m} \max_{\mathbf{i} \in \mathcal{I}_i} \tau_i(\mathbf{i}) - \min_{1 \le i \le m} \min_{\mathbf{i} \in \mathcal{I}_i} \tau_i(\mathbf{i}). \tag{5}$$

2.2 Inclusion of High Level Synthesis Techniques

In general, a parallel processor array consists of a set of processors and an interconnection network. The processors have to evaluate the operations, store intermediate results and control the communication with other processors.

For the further array design process we include methods of the high level synthesis and consider modules which are responsible for evaluating certain operations of a processor. The number and the kind of modules realized in one processor is called processor functionality. Instead of assuming a fixed processor functionality and determining a resource constrained scheduling [24] we want to specify the scheduling and the processor functionality concurrently since both influence each other. The aim is to determine a scheduling function which minimizes the latency under consideration of hardware constraints. First, we consider some parameters describing the hardware constraints.

We assume a given set of modules $\mathcal{M}$, where each module $m_l \in \mathcal{M}$ is able to execute one or several operations needed to implement the SURE.

- To every module $m_l \in \mathcal{M}$ we assign a delay d_l in clock cycles needed to execute the operation of the module m_l, a necessary chip area c_l needed to implement the module in silicon and the number n_l of instances of that module which are implemented in one processor.
- If a module $m_l \in \mathcal{M}$ has a pipeline architecture we assign a time offset o_l to that module which determines the time delay after that the next computation can be started on this module, otherwise $o_l = d_l$.
- Some modules are able to compute different operations. We assign to such modules different delays $d_{l,i}$ and offsets $o_{l,i}$ depending on the operations F_i.

The assignment of a module $m_l \in \mathcal{M}$ to an operation F_i is denoted as $m(i)$.

Using the introduced hardware description we are able to define two constraints for the design process. In order to ensure a valid partial order preserving the data dependencies, the scheduling function has to satisfy the causality constraint:

$$\boldsymbol{\tau}^T \mathbf{d}(e) + t_{\delta(e)} - t_{\sigma(e)} \ge d_{m(\sigma(e)),\sigma(e)}, \quad \forall e \in \mathcal{E}. \tag{6}$$

The above inequation has to guarantee the existence of all data needed to evaluate an equation of the SURE at the evaluation time of that equation.

The second constraint (resource constraint) is responsible for the prevention of access conflicts to the modules. For more details we refer to [5].

Considering these constraints optimum scheduling and processor functionality can be determined using standard packages for solving the optimization problem. The optimization criterion is the minimum product of latency L of the processor array and the chip area c_P of a single processor.

3 Tomographic Reconstruction Methods

3.1 Principle of Computed Tomography

Computed tomography is known from applications in medical diagnostics as X-ray CT [2,8], MRI, and SPECT or in technical diagnostics e.g. non-destructive testing. It consists in the reconstruction of an n-dimensional image space from its (n-1)-dimensional projections. The principle is illustrated in figure 1. From a given set of projections $p(k,m)$ a slice image $b(i,j)$ can be calculated. The variables k and m address the k-th ray of the m-th projection(angle), i and j indicate the column and row indices of the slice image.
The set of projections obtained from e.g. transmission measurements of the object is modeled by line integrals. The integral transformation was formulated first by J. Radon in 1917 [17] and is called the Radon transform.

3.2 Filtered Back Projection Algorithm

Filtered back projection is the most usual method in computed tomography. From the system-theoretical view the back projection is the inverse Radon transform [10]. The projection value $p(k,m)$ is assigned to each pixel (i,j) of the trace

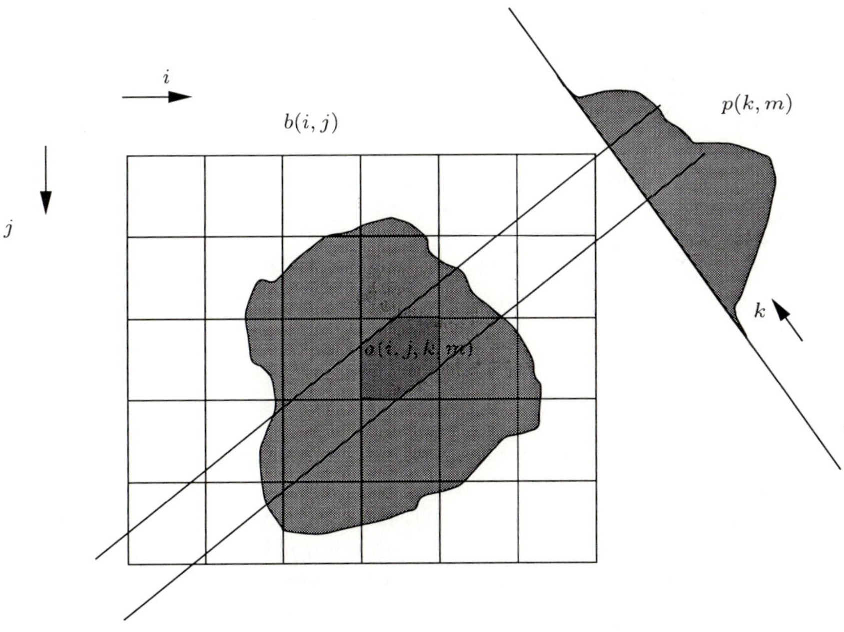

Fig. 1. Principle of computed tomography

path back from the detector to the source. The back-traced values are finally added from each projection to built up the slice image intensities. Since the back projection of the unfiltered Radon transform $p(k,m)$ results in a low-pass filtered version of the original image (caused by integration in the forward Radon transform and accumulation in back projection) in the FBP algorithm the projections $p(k,m)$ are prefiltered with a high-pass system which can be done in the space domain using a convolution kernel h or in the space frequency domain using the Fast Fourier Transform (FFT). Computer implementations of the FBP algorithm are often based on the equations 7 and 8. The projection values are thereby multiplied by an individual weighting factor $a(i,j,k,m)$ indicating whether and to which degree the projection contributes to the pixel. Geometrically, the factor $a(i,j,k,m)$ represents the overlapping area of a projection ray (k,m) with a pixel (i,j) (see figure 1)

$$g(k,m) = \sum_{n=-H/2}^{H/2} p(k+n,m) \cdot h(n) \qquad 0 \le k < K, 0 \le m < M \tag{7}$$

$$b(i,j) = \sum_{k=0}^{K-1} \sum_{m=0}^{M-1} a(i,j,k,m) \cdot g(k,m) \qquad 0 \le i,j < N \tag{8}$$

The equations are implemented in the following nested loop program.

```
for ( k = 0; k < K; k++ )
  for ( m = 0; m < M; m++ )
    for ( n = -H_2; n <= H_2; n++)
      g[k,m] = g[k,m] + p[k+n,m]*h[n];
for ( i = 0; i < N; i++ )
  for ( j = 0; j < N; j++ )
    for ( k = 0; k < K; k++ )
      for ( m = 0; m < M; m++ )
        b[i,j] = b[i,j] + a[i,j,k,m]*g[k,m];
```

Prog. 1. Nested loop program for FBP

3.3 Algebraic Reconstruction Techniques

Algebraic reconstruction techniques (ART) for CT have been introduced by Gordon in [6]. These methods are characterized by an iterative calculation of the slice image intensities $b(i,j)$. For each iteration step s a mapping of the slice image, called the forward projection $f(k,m)$, is determined according to equation 9. From a comparison to the measured projection $p(k,m)$ a correction term for updating the slice image is derived. There exist several modifications of ART differing from the updating procedure . The relationships for the simple additive ART are given with equation 10. The value $r(k,m)$ describes the number of pixels contributing to $f(k,m)$.

$$f(k,m)^{(s)} = \sum_{i=0}^{N-1}\sum_{j=0}^{N-1} a(i,j,k,m) \cdot b(i,j)^{(s)} \quad 0 \leq k < K, 0 \leq m < M \quad (9)$$

$$b(i,j)^{(s+1)} = max[0, b(i,j)^{(s)} + \frac{p(k,m) - f(k,m)^{(s)}}{r(k,m)}]$$
$$a(i,j,k,m) > 0, 0 \leq i,j < N, 0 \leq k < K, 0 \leq m < M \quad (10)$$

A computer implementation of the additive ART is given with the following nested loop program.

```
for ( s = 0; s < S; s++){
  for ( m = 0; m < M; m++){
    for { k = 0; k < K; k++){
      for ( j = 0; j < N; j++ )
        for ( i = 0; i < N; i++ )
          f[k,m] = f[k,m] + a[i,j,k,m]*b[i,j];
      c[k,m] = (p[k,m] - f[k,m])/r[k,m];
      for ( i = 0; i < N; i++ )
        for ( j = 0; j < N; j++ )
          if( b[i,j] + c[k,m] > 0 ){
            if ( a[i,j,k,m]!=0 ) b[i,j] = b[i,j] + c[k,m];
          else b[i,j] = 0; }     }}}
```

Prog. 2. Nested loop program for (additive) ART

4 Array Design for Reconstruction Algorithms

4.1 Generation of the Single Assignment Code

According to the design flow of DESA which exploits Lamport's hyperplane method [13] both algorithms have to be described by an SAC. As shown in the following, the original dimensions of both algorithms can easily be reduced so that they are embedded in a common three-dimensional index space $\mathcal{I} = (k\ m\ n)^T, 0 \leq k < K, 0 \leq m < M, -H/2 \leq n < 2N^2$.

FBP:

Since the variables i and j addressing the pixels in the reconstructed slice image generally occur as a tuple they can be concentrated in a single variable n. Hence the output image matrix becomes a 1D data stream of length N^2. Furthermore this variable can be concentrated with the variable n of the convolution kernel, finally the two initial nested loops are concatenated as shown in program 3.

ART:

The variables i and j are concentrated in one variable n again. Since the computation of the forward projection $f(k,m)$ has to be finished before the updating of the image matrix can start the loops are concatenated so that the variable n runs from 0 to $2N^2$. The iteration index s describes the step in the iteration. If we assume that a resulting processor array can be used with small modifications in all iteration steps only one step s ($s = 0$) is considered and the iteration index can be eliminated. The modified program is shown in program 4.

```
for ( k = 0; k < K; k++ ){
  for ( m = 0; m < M; m++ ){
    for ( n = -H_2; n < N * N + H_2; n++){
      if ( n <= H_2 )
        if (( k + n >= 0 )&&( k + n < N))
          g[k,m] = g[k,m] + p[k+n,m]*h[n];
      if ( n > H_2 )
        b[n-H_2] = b[n] + a[k,m,n-H_2]*g[k,m]; }}}
```

Prog. 3. Modified C program for FBP

```
for ( k = 0; k < K; k++){
  for ( m = 0; m < M; m++ ){
    for ( n = 0; n < 2 * N * N; n++ ){
      if ((n < N * N))
        f[k,m] = f[k,m] + a[k,m,n]*b_in[n];
      if ( n >= N * N )
        if (a[k,m,n] > 0)
          b_out[n-N*N] = b_in[n] + (p[k,m] - f[k,m])/r[k,m]); }}}
```

Prog. 4. Modified program for (additive) ART

The required single assignment code (SAC) can be directly derived from the nested loop programs 3 and 4. In contrast to nested loop programs, the iteration sequence of the SAC is not fixed but only limited by data dependencies. For the SAC conversion a reindexing and renaming of some variables of the nested loop programs is required. To observe the successive updates of the variable b in the FBP algorithm the new variables bb_1 and bb_2 are introduced.
Analogous the variables bf_1 and bf_2 have to be introduced in the ART code. The output variable bf_{out} is used as input bf_{in} in the next iteration cycle. Note that the calculation of $max[0, b_{in}[n] + (p[k,m] - f[k,m])/r[k,m])]$ according to equation 10 is performed outside of the processor array. The SAC with the recurrence equations of both algorithms is shown in program 5.

The SAC can be represented in a more comprehensive way by a dependence graph (DG) (see figure 2).

For reasons of clarity we show one separate DG for each algorithm. It should be noted that the DG for the ART algorithm was slanted by $-H/2$ as well as the DG of the FBP algorithm to obtain identical scheduling functions.

As it can be seen both DGs show a very similar shape. They essentially differ from the limits of the n direction and from the operations which have to be evaluated. Note, that only the dependent variables are indicated. The FBP algorithm is started with the convolution in plane $k + n = 0$, the convolution is finished in plane $k + n = H/2$, subsequently the back projection starts.

$$g(k,m,n) = p(k+n,m) * h(n) \qquad \forall (k\ m\ n)^T \in \mathcal{I} \wedge k+n=0$$
$$g(k,m,n) = g(k,m,n-1) + p(k+n,m) * h(n) \qquad \forall (k\ m\ n)^T \in \mathcal{I} \wedge 1 \le k+n \le \tfrac{H}{2}$$
$$g(k,m,n) = g(k,m,n-1) \qquad \forall (k\ m\ n)^T \in \mathcal{I} \wedge \tfrac{H}{2}+1 \le k+n \le \tfrac{H}{2}+N^2$$
$$bb_1(k,m,n) = g(k,m,n) * a(k,m,n+k-\tfrac{H}{2}) \quad \forall (k\ m\ n)^T \in \mathcal{I} \wedge m=0 \wedge \tfrac{H}{2}+1 \le k+n \le \tfrac{H}{2}+N^2$$
$$bb_1(k,m,n) = bb_1(k,m-1,n) + g(k,m,n) * a(k,m,n+k-\tfrac{H}{2})$$
$$\forall (k\ m\ n)^T \in \mathcal{I} \wedge 1 \le m \le M-1 \wedge \tfrac{H}{2}+1 \le k+n \le \tfrac{H}{2}+N^2$$
$$bb_2(k,m,n) = bb_1(k,m,n) \qquad \forall (k\ m\ n)^T \in \mathcal{I} \wedge k=K-1 \wedge m=M-1 \wedge \tfrac{H}{2}+1 \le k+n \le \tfrac{H}{2}+N^2$$
$$bb_2(k,m,n) = bb_2(k+1,m,n-1) + bb_1(k,m,n)$$
$$\forall (k\ m\ n)^T \in \mathcal{I} \wedge 0 \le k \le K-2 \wedge m=M-1 \wedge \tfrac{H}{2}+1 \le k+n \le \tfrac{H}{2}+N^2$$
$$bb_{out}(n) = bb_2(k,m,n) \qquad \forall (k\ m\ n)^T \in \mathcal{I} \wedge k=0 \wedge m=M-1 \wedge \tfrac{H}{2}+1 \le k+n \le \tfrac{H}{2}+N^2$$
$$f(k,m,n) = a(k,m,n) * bf_{in}(n) \qquad \forall (k\ m\ n)^T \in \mathcal{I} \wedge k+n=0$$
$$f(k,m,n) = f(k,m,n-1) + a(k,m,n) * bf_{in}(n) \qquad \forall (k\ m\ n)^T \in \mathcal{I} \wedge 1 \le k+n \le N^2$$
$$f(k,m,n) = f(k,m,n-1) \qquad \forall (k\ m\ n)^T \in \mathcal{I} \wedge N^2+1 \le k+n \le 2*N^2$$
$$bf_1(k,m,n) = sign(a(k,m,n)) * (p(k,m) - f(k,m,n))/r(k,m)$$
$$\forall (k\ m\ n)^T \in \mathcal{I} \wedge m=0 \wedge N^2+1 \le k+n \le 2*N^2$$
$$bf_1(k,m,n) = bf_1(k,m-1,n) + sign(a(k,m,n)) * (p(k,m) - f(k,m,n))/r(k,m)$$
$$\forall (k\ m\ n)^T \in \mathcal{I} \wedge 1 \le m \le M-1 \wedge N^2+1 \le k+n \le 2*N^2$$
$$bf_2(k,m,n) = bf_1(k,m,n) \qquad \forall (k\ m\ n)^T \in \mathcal{I} \wedge k=K-1 \wedge m=M-1 \wedge N^2+1 \le k+n \le 2*N^2$$
$$bf_2(k,m,n) = bf_2(k+1,m,n-1) + bf_1(k,m,n)$$
$$\forall (k\ m\ n)^T \in \mathcal{I} \wedge 0 \le k \le K-2 \wedge m=M-1 \wedge N^2+1 \le k+n \le 2*N^2$$
$$bf_{out}(n) = bf_{in}(n) + bf_2(k+1,m,n-1) + bf_1(k,m,n)$$
$$\forall (k\ m\ n)^T \in \mathcal{I} \wedge k=0 \wedge m=M-1 \wedge N^2+1 \le k+n \le 2*N^2$$

Prog. 5. Combination of the algorithms in a common single assignment code

Analogous the ART algorithm starts with the calculation of the forward projections which are available in plane $k+n=N^2$, then the determination of the correction term and the updating of b is performed. Finally, the b values are successively available in the index points $(0, M-1, n)$.

4.2 Determination of Allocation Functions

The common SAC of both algorithms is the starting point of the automatic design process with DESA. The embedding in one dependence graph ensures that allocation and scheduling valid for both algorithms can be determined.

In the first step possible affine uniform allocations due to equation 4 are calculated. In table 1 several allocations described by the projection vector $\boldsymbol{u}$ and the appropriate number of processors are given.

Obviously, the allocation in direction $\boldsymbol{u} = \begin{pmatrix} 0 & 0 & 1 \end{pmatrix}$ should be chosen, thus a two-dimensional processor array with the minimum number of $K*M$ processors adapted to both algorithms arises. Each other allocation leads to processor arrays

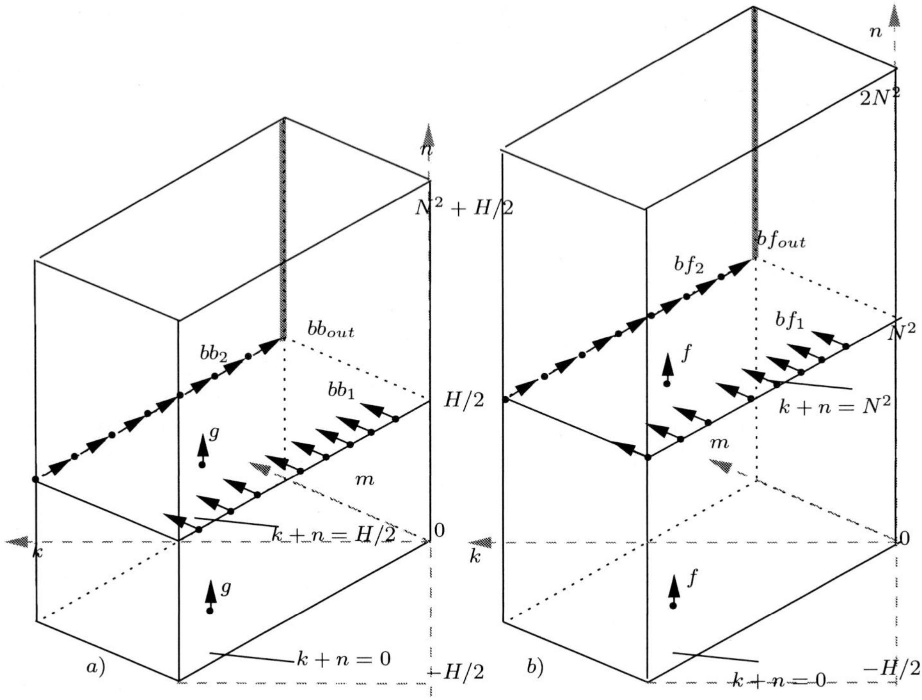

Fig. 2. Dependence graphs a) for FBP and b) for ART

Table 1. Processor allocation using different projection vectors

Projection vector	Number of processors
$\boldsymbol{u}_1 = (0\ 0\ 1)^T$	$K * M$
$\boldsymbol{u}_2 = (1\ 0\ -1)^T$	$(2N^2 + 1) * M$
$\boldsymbol{u}_3 = (0\ 1\ 0)^T$	$(2N^2 + 1) * K$
$\boldsymbol{u}_4 = (0\ 1\ 1)^T$	$(2N^2 + K - 1) * K$

with a higher number of processors only matching the requirements of the ART algorithm which means that a number of processors will not be used if the FBP algorithm runs on the array. Furthermore the chosen allocation has the advantage that the size of the processor array is only specified by the size of the acquisition system (K detectors, M acquisition angles), it does not depend on the size of the slice image which has to be calculated.

In figure 3 the structure of the arising processor array is given. The enlarged areas show one single processor with FBP and ART functionality. As it can be seen several modules can be used by both FBP and ART. Note that the boundary processors have slightly different functionalities.

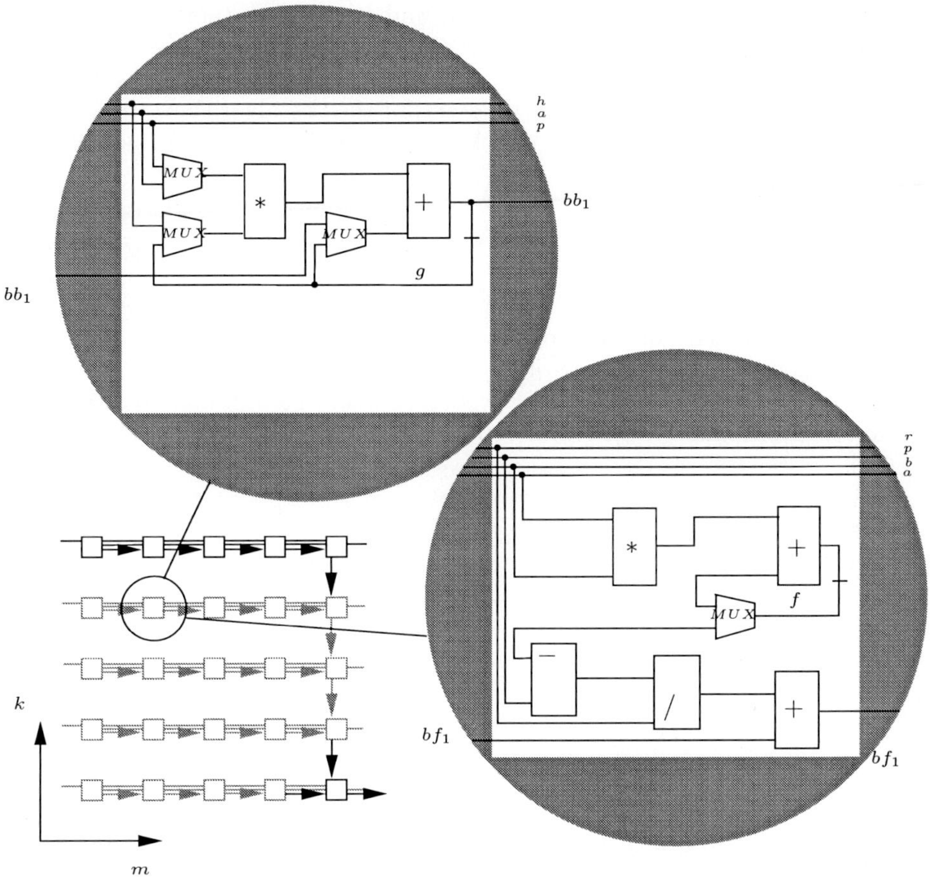

Fig. 3. Processor array with FBP functionality (top) and ART functionality (right)

4.3 Determination of Optimum Scheduling and Hardware Components

For the given allocation using the projection vector $\boldsymbol{u}_1$ the optimization of the scheduling function and the selection of the hardware components for the processor functions according to section 3 is performed. Table 2 gives an overview

Table 2. Hardware components: parameters and selection

Module	Function (o) $(o = d)$	a
m0	mac(3),add(1),sub(1),div(10)	8
m1	div(6)	12
m2	mac(1)	10
m3	add(1),sub(1)	3

Modules	c_P	$c_P \cdot L_{ART}$	$c_P \cdot L_{FBP}$
m0,m1	25	61660	38060
m1,m2,m3	20	77025	13600
2*m0	16	82128	17408

on available modules with their functions and parameters (normalized values) and shows the results of the hardware components selection.
The module selection was optimized for both algorithms separately. Although the optimum module selections are different for the two algorithms the selection m0,m1 should be preferred since the ART algorithm is more expensive caused by its iterative application.

5 Parallel Program Design for Reconstruction Algorithms

In the previous sections methods for the design of parallel hardware were presented. This section contains the application of analogous methods for the design of parallel software for a system of digital signal processors (DSPs).

The architecture of the used DSP system consists of a host and up to 16 DSPs of the types TMS320C40 and TMS320C44 [23,26]. The host computer includes

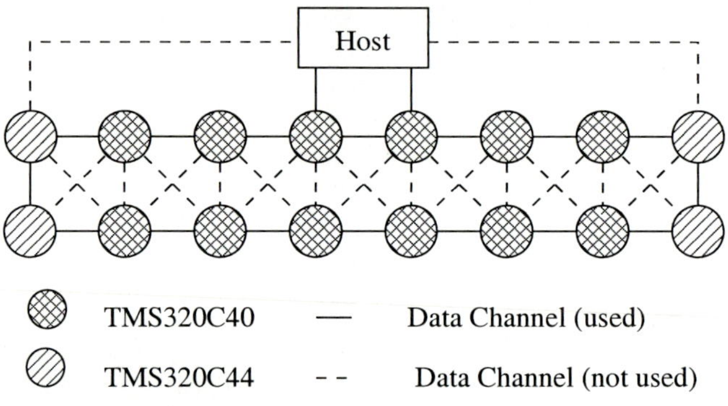

Fig. 4. Used topology of the DSP system

a SPARC 5 processor, the mass storage and I/O-facilities. Each DSP of the system contains an on-chip memory of 2×4 kByte and an external SRAM of at least 2×512 kByte. For the bidirectional communication with other processors, a TMS320C40 has 6 and a TMS320C44 has 4 communication ports (comports). The use of flexible cables for the communication channels allows variable topologies. Figure 4 shows the configuration used for the examples below.

The starting point for the parallel program design is also the single assignment code (see section 4.1). By applying a *processor allocation* and a *scheduling function* to each index point, a processing element and a processing time is assigned [16,19]. The result is a problem size dependent processor array. The adaptation of such a processor array to the number of available processors (in that case: DSPs) is realized by *partitioning* methods [3,22]. The partitioned processor array is interpreted as a parallel program for the given processor system.

This design process is illustrated exemplarily with two recurrence equations

$$g(k,m,n) = g(k,m,n-1) + p(k+n,m) * h(n) \qquad \forall (k\ m\ n)^T \in \mathcal{I} \wedge 1 \le k+n \le \tfrac{H}{2}$$

$$bb_1(k,m,n) = bb_1(k,m-1,n) + g(k,m,n) * a(k,m,n)$$

$$\forall (k\ m\ n)^T \in \mathcal{I} \wedge 1 \le m < M \wedge \tfrac{H}{2} < k+n \le \tfrac{H}{2} + N^2$$

of the FBP reconstruction algorithm. The transformation with scheduling vector $\boldsymbol{\tau} = (0\ 1\ 1)^T$ and projection vector $\boldsymbol{u} = (0\ 0\ 1)^T$ leads to the equations:

$$g(t, p1, p2) = g(t-1, p1, p2) + p(t+p1, p2) * h(t) \qquad \forall (t\ p1\ p2)^T \in \widetilde{\mathcal{I}} \wedge 1 \le t+p1 \le \tfrac{H}{2}$$
$$bb_1(t, p1, p2) = bb_1(t, p1, p2-1) + g(t, p1, p2) * a(t, p1, p2)$$
$$\forall (t\ p1\ p2)^T \in \mathcal{I} \wedge 1 \le p2 < M \wedge \tfrac{H}{2} < t+p1 \le \tfrac{H}{2} + N^2$$

in the transformed domain $\widetilde{\mathcal{I}} = \{(t\ p1\ p2)^T \mid 0 \le p1 < K \wedge 0 \le p2 < M \wedge 0 \le t+p1 < 2N^2 + K\}$. The size of the problem size dependent processor array is $p1 \times p2 = K \times M$. By partitioning, this processor array is adapted to the number of $M = 16$ available DSPs. All operations of the processors $p1$ are scheduled sequentially on a processor $p2$ by introducing the time $t1 = p1$.

$$g(t, t1, p2) = g(t-1, t1, p2) + p(t+t1, p2) * h(t) \qquad \forall (t\ t1\ p2)^T \in \widetilde{\mathcal{I}} \wedge 1 \le t+t1 \le \tfrac{H}{2}$$
$$bb_1(t, t1, p2) = bb_1(t, t1, p2-1) + g(t, t1, p2) * a(t, t1, p2)$$
$$\forall (t\ t1\ p2)^T \in \mathcal{I} \wedge 1 \le p2 < M \wedge \tfrac{H}{2} < t+t1 \le \tfrac{H}{2} + N^2$$

The resulting topology is a one dimensional line of 16 DSPs.

The recurrence equations above can be interpreted as follows as a part of the parallel program for the DSP $p2$ presented in the language C:

```
for( t = -H_2; t <= H_2; ++t )
  for( t1 = 0; t1 < K; ++t1 )
    if( t+t1 >= 1 && t+t1 <= H_2 )
      g[t][t1] = g[t-1][t1] + p[t+t1] * h[t+H_2];

for( t = -H_2; t <= H_2 + N * N; ++t )
  for( t1 = 0; t1 < K; ++t1 )
    if( t+t1 > H_2 && t+t1 <= H_2 + N * N ){
      bb_1[t][t1] = comport_read(InChannel)+g[t][t1]*a[t][t1];
      comport_write(OutChannel, bb_1[t][t1]); }
```

The transfer of the variables $bb_1(t, t1, p2-1)$ of the DSP $p2-1$ to the DSP $p2$ is realized by the function comport_read(InChannel), where InChannel determines the number of the comport to the DSP $p2-1$. The transport of the variables $bb_1(t, t1, p2)$ to the next DSP $p2+1$ is performed by the function comport_write(OutChannel, bb_1[t][t1]).

Finally, the collection of all program parts obtained by all recurrence equations results in a parallel program, which still fulfills the single assignment condition.

The time for the loading of the projection data to the DSPs, the computation time and the time for transferring the image data to the host are measured for different implementations of the FBP algorithm (see table 3).

Our parallelization strategy leads to parallel single assignment programs, which can be optimized "by hand" to get better results. The speedup for this method of 15.8 is calculated by comparing the parallel (t_{par}) with the one DSP single assignment implementation (t_{sing}). A more practical speedup of 4.9 is given by the ratio of the overall sequential t_{seq} and parallel computation time t_{par}.

Table 3. Implementation results of the FBP algorithm

Action	**Host** t_{host} in μs	**1 DSP** sequential t_{seq} in μs	**1 DSP** single assign. t_{sing} in μs	**16 DSPs** t_{par} in μs	**Speedup** 1 $\frac{t_{sing}}{t_{par}}$	**Speedup** 2 $\frac{t_{seq}}{t_{par}} \approx \frac{t_{host}}{t_{par}}$
Load	-	30	31	34		
Computation	40566	41153	133344	7955		
Store	-	153	154	153		
Overall	40566	41800	134124	8464	15.8	4.9

6 Conclusion

Tomographic reconstruction algorithms are suitable for a parallelization in hardware or software. Using the tools of the design system DESA we derived a common processor array for two different reconstruction algorithms. It is intended to implement this processor array in silicon.

The used design methods are also applicable to the generation of parallel software for multiprocessor systems. Currently, the work is continued with the treatment of larger image matrices and the inclusion of new partitioning methods [3].

References

1. D. Baltus and J. Allen. Efficient exploration of nonuniform space-time transformations for optimal systolic array aynthesis. In L. Dadda and B. Wah, editors, *Application Specific Array Processors*, pages 428–441, 1993.
2. A. M. Cormack. Representation of a function by its line integrals, with some radiological applications. *J. Appl. Phys.*, 34:2722–2733, 1963.
3. U. Eckhardt and R. Merker. Hierarchical algorithm partitioning at system level for an improved utilization of memory structures. *IEEE Transactions on CAD*, 18(1):14–24, January 1999.
4. D. Fimmel and R. Merker. Design of processor arrays for real-time applications. In *Proc. Int. Conf. Euro-Par '98*, pages 1018–1028, Southampton, 1998. Lecture Notes in Computer Science, Springer.
5. D. Fimmel and R. Merker. Determination of processor allocation in the design of processor arrays. *Microprocessors and Microsystems*, 22(3-4):149 – 155, August 1998.
6. R. Gordon. A tutorial on ART (algebraic reconstruction techniques). *IEEE Trans. on Nucl.Sc.*, NS-21:78–93, 1974.
7. P. Held, P. Dewilde, E.F. Deprettere, and P. Willage. Hifi: From parallel algorithm to fixed-size vlsi prozessor array. In F. Cathoor and L. Svensson, editors, *Application-Driven Architecture Synthesis*, pages 71–94. Kluwer Academic Publishers, 1993.
8. G. Hounsfield. Computerized transverse axial scanning (tomography): part 1. description of system. *Br. J. Radiol.*, 46:1016–1022, 1973.

9. K. Wayne I. Agi, P.J. Hurst. An image processing IC for backprojection and spatial histogramming in a pipelined array. *IEEE J. of Solid-State Circuits*, 28:14–23, 1993.
10. A.C. Kak and M. Slaney. *Principles of Computerized Tomographic Imaging.* IEEE Press, 1988.
11. R.M. Karp, R.E. Miller, and S. Winograd. The organisation of computations for uniform recurrence equations. *JACM*, 14(3):563–590, July 1967.
12. M. Kortke, D. Fimmel, and R. Merker. Parallelization of algorithms for a system of digital signal processors. In *EUROMICRO Workshop on Digital System Design 1999*, pages 46–50, Milan, 1999. .
13. L. Lamport. Parallel execution of do loops. *Communication of ACM*, 2(17), 1974.
14. H. Leverge, Ch. Mauras, and P. Quinton. A language-oriented approach to the design of systolic chips. In E.F. Deprettere and A.-J. van der Veen, editors, *Algorithms and Parallel VLSI-Architectures*, volume A, pages 309–327. Elsevier Science Publishers B.V., 1991.
15. R. Merker, D. Fimmel, U. Eckhardt, and H. Schreiber. A system for designing parallel processor arrays. In F. Pichler and R. Moreno-Diaz, editors, *Computer Aided Systems Theory - EUROCAST 97*, pages 3–12. Springer, lecture notes in computer science edition, 1997.
16. P. Quinton. The systematic design of systolic arrays. In *Automata Networks in Computer Science*, pages 229–260. Manchester University Press, 1987.
17. J. Radon. Über die Bestimmung von Funktionen durch ihre Integralwerte längs gewisser Mannigfaltigkeiten. *Bericht der Sächsischen Akademie der Wissenschaften*, 69:262–277, 1917.
18. P. V. R. Raman, R.D. Kriz (Chair), and A. L. Abbott. Parallel implementation of the filtered back projection for tomographic imaging. Master's thesis, Dept. Electrical Engineering, Virginia Polytechnic Institute and State University, Blacksburg, Virginia, January 1995.
19. S.K. Rao. *Regular Iterative Algorithms and their Implementations on Processor Arrays.* PhD thesis, Stanford U., 1985.
20. V. Roychowdhury, L. Thiele, S.K. Rao, and T. Kailath. On the Localization of Algorithms for VLSI Processor Arrays. In *VLSI Signal Processing III*, pages 459–470, New York, 1989.
21. J. Teich. *A Compiler for Application-Specific Processor Arrays.* PhD thesis, University of Saarland, Verlag Shaker, Aachen, 1993.
22. J. Teich and L. Thiele. Partitioning of processor arrays: a piecewise regular approach. *INTEGRATION, the VLSI Journal*, 14(2):297–332, 1993.
23. Texas Instruments. *TMS320C4x User's Guide*, 1991.
24. L. Thiele. Compiler techniques for massive parallel architectures. In P. Dewilde and J. Vandewalle, editors, *Computer Systems and Software Engenieering*, pages 101–149. Kluwer Academic Publishers, 1992.
25. L. Thiele. Resource constraint scheduling of uniform algorithms. In L.Dadda and B.Wah, editors, *Proc. Application-Specific Array Processors 1993*, pages 29–40, 1993.
26. Transtech Parallel Systems Ltd. *VME TIM-40 Motherboard Manual, TDM407 User Guide, TDM442 Manual*, 1996.

A Formalized Description Approach to Continuos Time Systems

Elena Jharko

Institute of Control Sciences,
Profsoyuznaya 65, Moscow 117806, Russia
zharko@ipu.rssi.ru

Abstract. The paper presents an approach to formalize a continuous-time system description and a suitable approach to elaborating intelligent computer aided design system. Corresponding principles and a structure scheme are proposed to implement the approach. The approach enables one to use both design rules and design algorithms by an effective manner in dependence on task requirements. The organization of calculations is considered under modeling continuous-time systems on the basis of using flexible simulating complex. Also, an application of the flexible simulating complex to design a universal model of a nuclear power plant is considered.

1 Introduction

Recent time researches are characterized by increasing interest to use of hierarchical and through design technique within design of complex continuos-time control systems. In turn, solving such a problem is faced with a need of a unified description of the model components under automatic design a complex model at any hierarchical level. Lack of a unified model description considerably influences on task description language and makes designers to use the apparatus of formal parameters as well as complicates generation of the model program code [1]. The present paper considers an approach to formalize a continuous-time system description.

2 Basic Description

Let us consider a complex system S, which consists of elements S_i, $i = \overline{1,N}$, e.g. fig.1. Each of the elements S_i is described by the equations of the form:

$$\left.\begin{aligned} \dot{\vec{x}}_i &= \vec{f}_i(\vec{x}_i, \vec{u}_i, t) \\ \vec{y}_i &= \vec{\varphi}_i(\vec{x}_i, \vec{u}_i, t) \end{aligned}\right\}, \tag{1}$$

F. Pichler, R. Moreno-Díaz, and P. Kopacek (Eds.): EUROCAST'99, LNCS 1798, pp. 142–153, 2000.

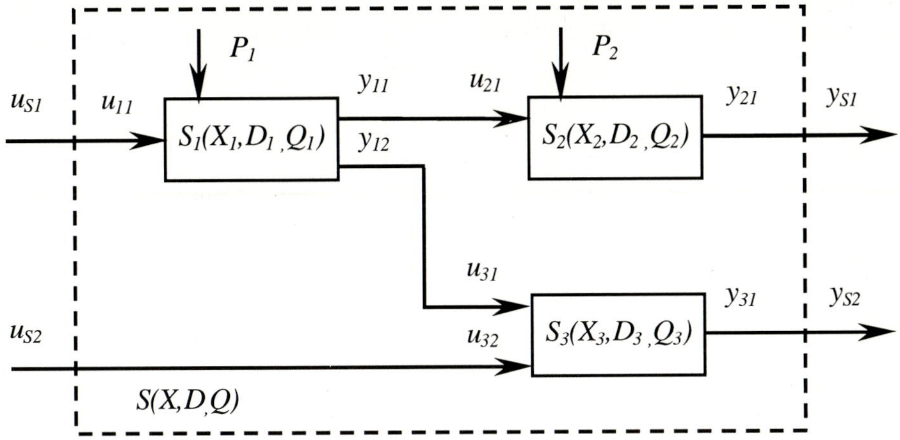

Fig. 1. Complex system S

with $\vec{x}_i$ being the state variable vector, $\vec{u}_i$, input variable vector, $\vec{y}_i$, output variable vector. In turn, $\vec{f}_i$ and $\vec{\varphi}_i$ are some nonlinear vector-valued functions.

Aggregation of the elements S_i into the system S is given by a one-to-one conjugation operator:

$$\left[y_{ij}\right]^* = R\left(\left[u_{lk}\right]^*\right) \tag{2}$$

which handles link of input variable of the l-th subsystem with output variable of the i-th subsystem. In (2), $\left[y_{ij}\right]^*$ is a subset of the output set of the system elements, $\bigcup_{i=1}^{N} y_i$, and $\left[u_{lk}\right]^*$ is a subset of the input set of the system, $\bigcup_{l=1}^{N} u_l$. Elements of the sets $\left[y_{ij}\right]^*$ and $\left[u_{lk}\right]^*$ are referred as common, or input/output, variables.

3 System Formalization

For the system S, the aggregated vector $\vec{x}_s(t) = (\vec{x}_1(t), \vec{x}_2(t), \ldots, \vec{x}_N(t))$, is the model S state at the time moment t. Here $\vec{x}_i(t)$ is the state vector of the functional element S_i at time moment t, $i = \overline{1, N}$. Again, the system S may be described by the input/output/state equations (3) and by the link equation $\left[y_{ij}\right]^* = R\left(\left[u_{lk}\right]^*\right)$

$$\left.\begin{aligned}
\dot{\vec{x}}_1 &= \vec{f}_1(\vec{x}_1, \vec{u}_1, t) \\
\vec{y}_1 &= \vec{\varphi}_1(\vec{x}_1, \vec{u}_1, t) \\
\dot{\vec{x}}_2 &= \vec{f}_2(\vec{x}_2, \vec{u}_2, t) \\
\vec{y}_2 &= \vec{\varphi}_2(\vec{x}_2, \vec{u}_2, t) \\
&\cdots\cdots \\
\dot{\vec{x}}_N &= \vec{f}_N(\vec{x}_N, \vec{u}_N, t) \\
\vec{y}_N &= \vec{\varphi}_N(\vec{x}_N, \vec{u}_N, t)
\end{aligned}\right\} \tag{3}$$

Under aggregation of the elements S_i into the system S, the set of the input/output variables will be denoted as Q', i.e. $Q' = \left(\bigcup_{i=1}^{N} u_i\right) \cap \left(\bigcup_{i=1}^{N} y_i\right)$, where $u_i = \{u_{ij}\}$, $(j = 1,\ldots,m)$ are sets of the system S_i input variables, and $y_i = \{y_{ik}\}$, $(k = 1,\ldots,n)$. Then, the input variables of the system S will include only those input variables of the functional elements S_i which are not the input/output variables, i.e. $u_S = \left(\bigcup_{i=1}^{N} u_i\right) \setminus Q'$.

For the total model, all the elements of the set $\bigcup_{i=1}^{N} y_i$ are the input variables. Consideration of complex systems leads to gathering of excessive information. So, instead of total models, macromodels are frequently used. Then, the input variables of the system are elements of a subset y_s^* meeting the condition $y_s^* \subset \bigcup_{i=1}^{N} y_i$.

Each element S_i may be characterized by a set of parameters, which describe some characteristics of the element, and functions, which are the coefficients in equations (1). Due to the system S is described by equations (3), the system S parameter set is the union of parameter sets of the system S components. In general, the system S may involve both stationary and non-stationary elements. So, the system parameter set will involve both constant and time-varying parameters. Let the set of constant parameters of the i-th element be denoted as P_i, and the set of time-varying parameters as Q_i''. Then, the following relationships hold $P_s = \bigcup_{i=1}^{N} P_i$, $Q_s'' = \bigcup_{i=1}^{N} Q_i''$. Here P_s stands for the set of the system S constant parameters, Q_s'', for the time-varying parameters.

Frequently, especially when describing complex system models, equations (3) are useful to be rewritten in the form

$$\dot{\vec{x}} = \vec{f}(\vec{x}, \vec{u}, \vec{Q}'', t), \tag{4}$$

$$\vec{y} = \vec{\varphi}(\vec{x}, \vec{u}, \vec{Q}''', t), \tag{5}$$

$$\vec{Q}''' = \vec{\psi}(\vec{x}, \vec{u}, t), \tag{6}$$

where $\vec{Q}'''$ is the vector of intermediate variables. Introducing into consideration the set Q''' is motivated by the following reasons. For a number of cases, the used component of equations (4) and (5) is comfortable to calculate in advance (equation (6)), with, in sequel, the calculation result to be used in equations (4) and (5). The set Q''' may be formed also when involving equations of correction of state variables or derivatives of the state variables. Under state variable correction, when integrating vector $\vec{x}$, the vector of state variables without errors, $\vec{x}_{\delta k}$, is formed. After improving the vector $\vec{x}_{\delta k}$ in accordance with equations of correction, the vector of corrected state variables, $\vec{x}$, is formed. For the case, elements of the set $\{x_{\delta k}\}$ also may be considered as intermediate state variables.

The set of intermediate variables corresponding to the correction of vector of state variable derivatives is formed in the same manner. Within this, the intermediate variables are those of calculated in accordance with equation (4), with the corrected variables forming the vector of state variable derivatives.

Let us introduce the set Q to be referred as the set of internal variables as follows: $Q = Q' \cup Q'' \cup Q'''$, i.e. the set unites input/output variables, time-varying parameters and intermediate variables. Then, any continuos-time system described by equations (1) may be characterized by the following sets: X, D, U, Y, Q, P, where X is the set of state variables, D is the set of state variable derivatives, U is the set of inputs, Y is the set of outputs, Q is the set of internal variables, and P is the set of parameters. The complete continuos-time system description is also required a set of equations R. Then, the continuos-time system model will be described by the following manner:

$$M_s = \langle X, D, U, Y, Q, P, R \rangle. \tag{7}$$

Benefit of such a representation is motivated by the reason that it enables one to describe uniformly complex systems as well as theirs components. Also, the complex model description design process based on the model components may be implemented in completely automatic manner.

Consider aggregating the elements S_i, $i = \overline{1, N}$ into the system S. A model m_i of each S_i is described by the equation

$$m_i = \langle X_i, D_i, U_i, Y_i, Q_i, P_i, R_i \rangle. \tag{8}$$

The system S model may be described by the equation of type (7), and in accordance with above consideration the following relationships between the sets X, D, U, Y, Q, P, R and $X_i, D_i, U_i, Y_i, Q_i, P_i, R_i$ hold

$$X = \bigcup_{i=1}^{N} X_i,; \quad D = \bigcup_{i=1}^{N} D_i; \quad P = \bigcup_{i=1}^{N} P_i; \quad Y = \bigcup_{i=1}^{N} Y_i; \tag{9}$$

$$U = \left(\bigcup_{i=1}^{N} u_i \right) \setminus \left[\left(\bigcup_{i=1}^{N} u_i \right) \cap \left(\bigcup_{i=1}^{N} y_i \right) \right]; \tag{10}$$

$$Q = \left(\bigcup_{i=1}^{N} Q_i \right) \cap \left[\left(\bigcup_{i=1}^{N} u_i \right) \cap \left(\bigcup_{i=1}^{N} y_i \right) \right]; \tag{11}$$

$$R = \bigcup_{i=1}^{N} R_i . \tag{12}$$

Involving common variables into the set Q enables one to eliminate the equations of linking. Thus, using expression (8) to describe each element models and relationships (9)-(12) permits to design a description of complex system model (7), with the formalized description of a functional element model being designed in automatic manner by use the element equations.

4 An Approach to Elaborating Intelligent Computer Aided Continuos Time System Design

At present, the following two methodologies of computer aided design may be noted. The first one is based on representation of a designer knowledge as a set of rules, with the rules being used by expert systems. Within the second methodology, knowledge may be represented as design algorithms. The present paper considers a suitable approach to elaborating an intelligent computer aided design system, which combines both the above methodologies. Corresponding principles and a structure scheme are proposed (fig. 2) to implement the approach. The approach enables one to use both design rules and design algorithms by an effective manner in dependence on task requirements [2].

When elaborating an intelligent computer aided design system, the following principles are proposed to be involved:

- providing a correct transformation of general system description into a structure scheme;
- parallel design of processing part and control part of the structure scheme of the elaborated system, what enables one global improving of the total structure;
- design process organization by a manner which provides the solution obtained at each step to reduce the design space with simultaneous possibility of investigating alternative variants at next steps, what enables one to implement problem oriented solution search in the design space;

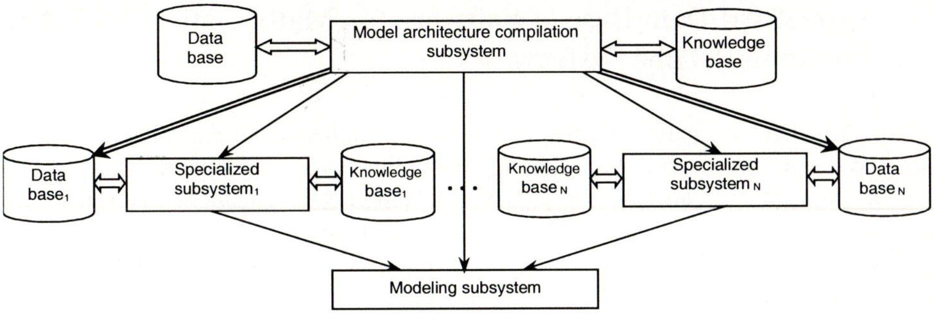

Fig. 2. Structure of an intelligent computer aided design system

- representing knowledge both as rules and algorithms, what enables one to increase effectiveness of the design structure of the elaborated system;
- using hierarchical structure of the computer aided design system and module nature of components of the system;
- providing the computer aided design system software to be extendable and easy to update.

The system software has the following two-level organization:

- the upper level is the model architecture compilation subsystem;
- the lower level corresponds to specialized subsystems.

The model architecture compilation subsystem has the following functions:

- analysis of the initial description and decomposition of the elaborated system onto separate functionally completed subsystems, selecting isochronous areas;
- selecting the time scheme;
- modeling the structure elaborated, analysis of the results obtained and decision making.

Basic functions of the specialized subsystems involve:

- analysis of the initial description and decomposition of the subsystem onto separate functionally completed subsystems, selecting isochronous areas;
- selecting time scheme;
- forming structural description of the functional subsystem;
- modeling the structure elaborated, analysis of the results obtained and decision making.

Each subsystem has an access to the corresponding data base which contain basic structures of the functional elements and subsystems. Besides that, an upper level subsystem has an access to databases of the specialized subsystems.

The all subsystems have access to the corresponding knowledge bases which contain design rules and design algorithms. Comprehensive modeling of the functional subsystems and the total system is implemented by the total modeling subsystem.

5 Specialized Data Base of Software for Mathematical Modeling Continuous Time Systems

Using specialized databases (SDB) of software for mathematical modeling continuous time systems enables one to simplify modeling task description language, to implement semantic analysis of a task, to introduce supplements into the task automatically. A SDB contains information about plants subject to investigation as well as about applied program codes, which are used under modeling. Also, a SDB has a user interface.

A database elaborating is concerned with constructing a conceptual description of enterprise, deriving database scheme, and its software realization. The conceptual description is used for abstract enterprise representation. This enables one to reflect corresponding enterprise features, objects and theirs properties, which are of the most importance, from a user point of view, as well as connections between the objects, which will be described in the SDB.

The following two types of objects may be noted to provide the process of modeling complex continuous time systems. These are information on models of the investigated systems and information on tools supporting the modeling process, i.e. the applied software.

A class of the investigated objects forms a hierarchical structure having additional connections and, in general, may be represented as a network whose vertexes are systems and elements of the class *S*, with the elements are the terminal vertexes. Arcs of the networks represent the connections between the objects. The class of the modeled objects will be characterized by notions corresponding to really existing elements and systems, which are interconnected by relationships of the form “the part” and “the whole”. A graph obtained by the manner will be referred as the graph of notions. Terminal vertexes of such a graph are the notions characterizing the functional elements, which are not disjointed in sequel under consideration of the systems of the given class.

The all vertexes of the graph are separated onto subgroups, of the form “element” and “system”. The first subgroup involves notions characterizing the functional elements of the null level of hierarchy. The second subgroup involves notions describing the functional elements at the others levels of hierarchy.

Each vertex of the graph will be characterized by a list of attributes and theirs values. Since in the graph of notions the all attributes of objects relating to elements are inherited by objects relating to systems, what is achieved by introducing relationship “part-whole’, the lists of attributes for elements and systems will be partially different ones. If elements are characterized by variables and operators then systems are characterized, in the first turn, by the list of elements involved into the systems.

Another important part of the conceptual description of the mathematical modeling process is the information on software tools used under modeling. Within the problem of mathematical modeling of continuos time systems, the following types of applied programs may be noted: calculating input actions, integration, processing intermediate results, forming delivery documents, delivering documents in accordance with results of investigations, processing results of modeling.

Each type of the applied programs may be described, in general, by attributes which are appropriate for all programs of the class, with each program being characterized by proper attributes and theirs values.

In accordance with the conceptual description, in the SDB, two parts are accentuated. The parts are oriented on storing information on investigated subjects and conditions of modeling.

Model of data used within the SDB has some advantages in comparison to such known models as the hierarchical one, network one, relational one. In contrast to the hierarchical model, the proposed model enables one to operate without preliminary fixing of the structure embedding level; enables one to realize references of an object onto itself, what is not allowed within the network model; has a control system which is more effective in time than that of relational model. The scheme of database is presented at fig. 3 and 4.

Fig. 3 demonstrates a part of the SDB used to represent investigated objects of some class. In the scheme, the following objects are accentuated: ELEMENT, OPERATOR, VARIABLE, SYSTEM, SYSTEM-INPUTS, SYSTEM-OUTPUTS, CONNECTIONS. Each objects is characterized by certain proper attributes. Object ELEMENT has the following attributes: ELEMENT-NAME, PRM-NAME indicating a connection with corresponding program code which realizes the model. Object OPERATOR is characterized by attributes NUMBER, APPLICATION-VECTOR, OUTPUT-VECTOR. Object VARIABLE has the following attributes: IDENTIFIER, ARRAY, INDEX, TITLE, DIMENSION-UNIT, DEFAULT-VALUE. Each element may have more than one OPERATOR's and more than one VARIABLE's. Objects SYSTEM is characterized by attributes SYSTEM-NAME, PRM-NAME. Objects SYSTEM-INPUTS and SYSTEM-OUTPUT are characterized by attributes OLD-NAME, NEW-NAME. OLD-NAME of the input or output is written to designate inputs and outputs of the functional elements at lower hierarchy levels. NEW-NAME determines the inputs and outputs of the investigated system. For a number of cases, the new and old names may coincide. To assign object CONNECTION one should to assign INPUT-NANE and OUTPUT-NAME which determine both input and output identifiers and identifiers of the corresponding functional elements. Each system may have several SYSTEM-INPUT's, SYSTEM-OUTPUT's, and CONNECTION's, may consist both of ELEMENT's and SYSTEM's.

Another part of the SDB describes modeling process support tools, i.e. applied programs. The following objects are accentuated: APPLIED-PROGRAM, INTEGRATION-METHOD, RESULT-PROCESSING-PROGRAM, DOCUMENT-FORMING-PROGRAM, DOCUMENT-DELIVERY-PROGRAM, PAPAMETER. In turn, each object is characterized by a set of proper attributes. Corresponding scheme of the SDP part considered is presented at fig. 4.

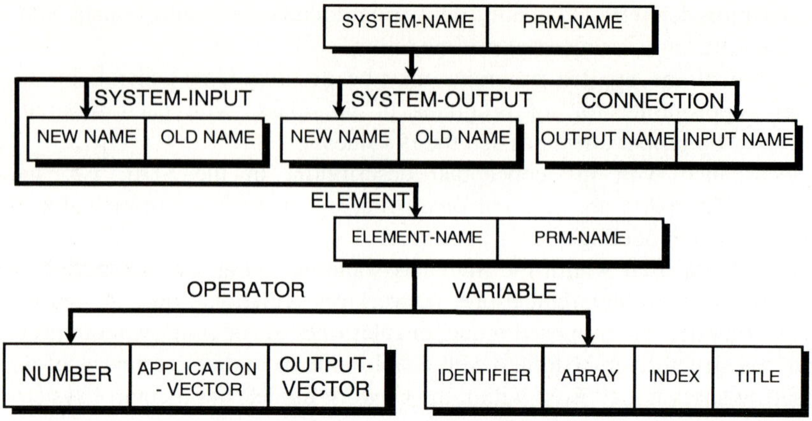

Fig. 3. A part of the SDB scheme used for presentation of investigated subjects of some class

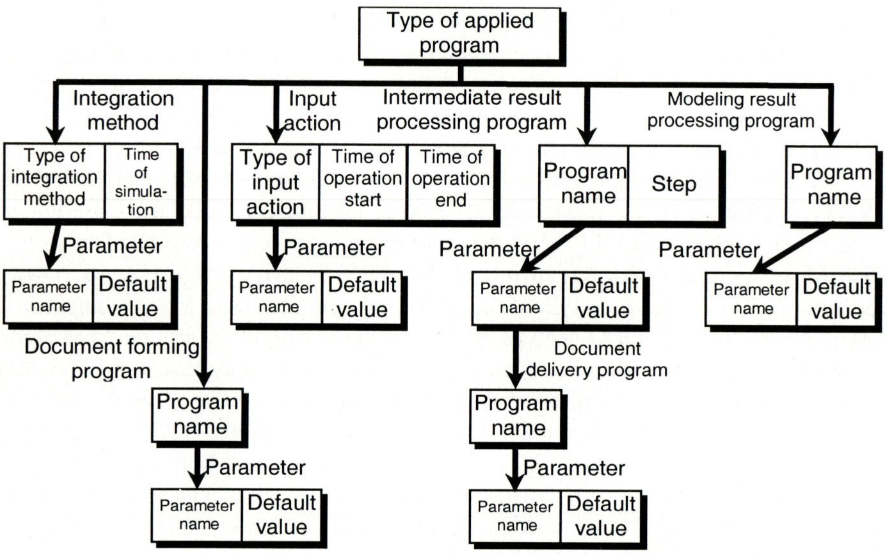

Fig. 4. A part of the SDB scheme used for presentation of applied program

6 Computation Organization for Modeling Continuous Time Systems

Within the approach presented, modeling continuous time system process is implemented by dialogue program complex which is based on the following main principles:

- separate description of model and modeling conditions;
- separate description of structure and model parameters;
- implementation of computing input actions, computing derivatives, integration, and result delivery within separate specific program modules;
- dynamic loading of models;
- use of interpreting approach to call the above modules.

Consider the organization scheme in details. Modeling task written in a special problem-oriented language is transformed into a set of tables containing information on the model, input actions, intermediate and final modeling results. Based on the information, corresponding computation scheme is composed, in which names of called programs are written.

Under modeling, corresponding control is implemented by use a modeling monitor which is partitioned onto four regions. These are: initialization, dynamics, changing, and termination. The region of initialization serves to load programs, to set, in a dialogue mode, initial conditions, model parameters and modeling conditions. The dialogue is initiated by computer. The user is provided to look through default parameter values and change those of them which will not be found suitable. Such an approach considerably simplify entering data and theirs modification in sequel.

The region of dynamics is used for cyclic call of programs of computing input actions, right-hand-side parts of dynamic equations, delivering intermediate results, integration. Call of programs is based on interpretation procedure in accordance with accepted scheme of computations. Using such a computation scheme provides user interference into the computation process. Under a request on interruption, modeling is terminated, and control is transferred to the region of changes. In the region, in a dialogue mode, it is found out which concrete changes are to be involved into the program, and corresponding actions are made. To assign another input actions or another method of integration it is sufficient to insert corresponding changes into the control tables and load new programs instead of those used before. To change model parameters or input action parameters it is sufficient to change parameters of the corresponding programs. At the same time, modification of the model structure requires new forming program of computation of right-hand-side parts of dynamic equations.

The region of termination serves for final processing of modeling results.

The approach described provides comparatively fast modeling program modification, preserving simultaneously its high efficiency.

7 Flexible Simulating Complex for Design Problems

Conventionally, a design process, concerned with both large scale models and industrial plants, requires large expenditures of design time and a number of other resources. In contrast, a computer modeling which uses a flexible simulating complex (FSC) is an effective alternative way, in costs and various engineering decision making criteria, of solving the problem [3]. The entity of such a flexible modeling is multiple using various program modules as unified and standard components of different complexes of programs.

Within the flexible modeling approach, a universal model of a nuclear power plant (NPP) unit has been elaborated. Since there exist no absolutely equal NPP units, there exist a need to elaborate a FSC which could be tuned in operative mode to any existing or designed NPP unit, and which could also be tuned to any change in the NPP unit equipment or in the NPP unit control system. Thus elaborated, the FSC involves models having different complexity and detailed elaboration, what enables one to implement prediction computations for NPP unit industrial processes without essential loss of accuracy. A structure scheme of the FSC performance corresponding to a NPP unit parameter modeling is presented at fig. 5.

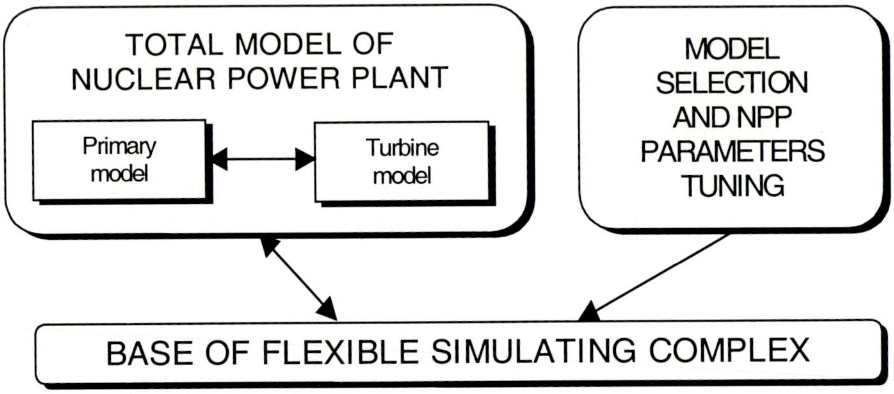

Fig. 5. A structure scheme of FSC performance corresponding to a NPP unit parameter modeling

The FSC elaborated is built by use of module principle. Within it, the all industrial equipment is partitioned onto groups, with each group being described by a separate functional module (block). Most of these functional modules have several variants of computer realization. These variants differ one to another in details and completeness of description of corresponding parts of equipment and technological processes, and, consequently, computational characteristics (CPU time, on-line storage, etc.).

Design and modeling tools composed within the FMC have the functional structure presented at fig. 6.

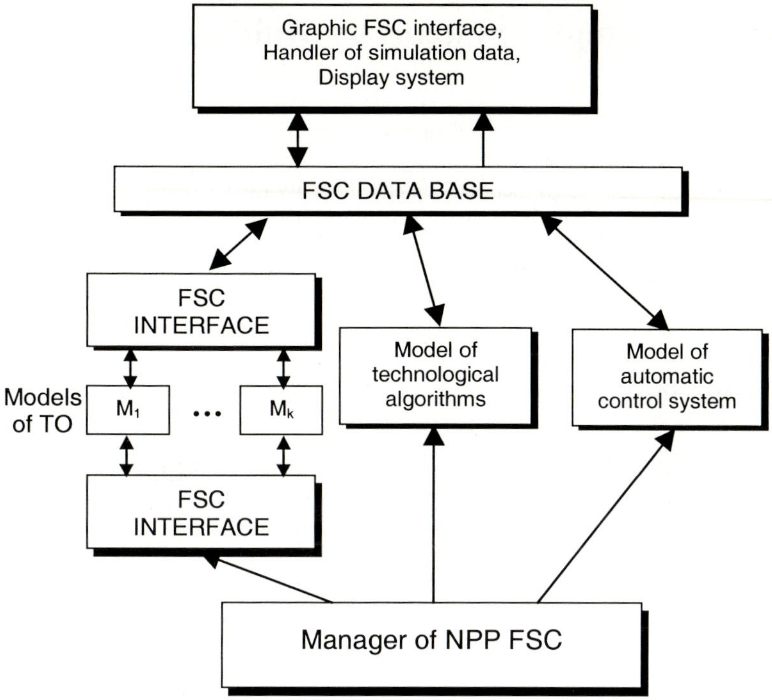

Fig. 6. Scheme of functional module of the NPP FSC

References

1. Kalinichenko, A.A.: Integration of heterogeneous databases: methods and tools. Nauka Publ., Moscow (1983) (in Russian)
2. Jharko, E.Ph.: Design experience of flexible simulating complexes. In: Proceedings of International Conference on Theory of Active Systems. Institute of Control Sciences, Moscow (1999) 108-109 (in Russian)
3. Jharko, E.Ph., Motulevich, A.V.: A flexible simulating complex for design of energetically and ecologically optimal industrial devices and systems. In: Modeling and Control of High Risk Industrial Plants. Institute of Control Sciences, Moscow (1997) 122-135 (in Russian)

Modeling Complex Systems by Multi-agent Holarchies

Franz Pichler

Johannes Kepler University Linz
Institute of System Sciences
Systems Theory and Information Engineering
Altenbergerstraße 69, A-4040 Linz
pichler@cast.uni-linz.ac.at

1 Introduction

Arthur Koestler (1967, 1969, 1978) introduced into science the concepts of a "holon" and of a "holarchy". A holarchy is according to A. Koestler a tree-like hierarchy where the nodes of the tree - the components of the hierarchy - are autonomous intelligent acting I/0 systems.

Koestler calls such components "holons". Holarchies provide according to Koestler the appropriate conceptual framework for modeling and simulation of self organizing open hierarchical systems as they can be found in biology, medicine, sociology and management science. Especially he was interested to use holarchies for modeling the function of the human brain. In the past years artificial intelligence developed the topic of agent systems. Agents are intelligent acting systems which are able to perform certain tasks in an autonomous way. For the user agents are usually invisible. In case that agents are information processing systems their realization can be done in software (software agents). The elaboration of agent methodology led to the concept of a "multi-agent", a network of cooperating agents (e.g. J. Ferber 1999). Agents have a conceptual similarity to the holons of A. Koestler. It seems to be natural to try to make use of the available agent technology to construct a network of systems which constitute a holarchy in the sense of A. Koestler. We call such system a Multi-Agent Holarchy.

In this paper we try to make a first step towards such an investigation. Besides of the work of Koestler and the existing models for multi-agent systems we base our paper here on earlier work of this author (Pichler 1995, 1999).

2 Holons and Holarchies

A holon, according to Koestler is a model-component with a "Janus face" - one side looking "down" and acting as an autonomous system giving directions to "lower" components and the other side looking "up" and serving as a part of a "higher" holon.

Holons, in the sense of Koestler, are essential in hierarchical systems with intelligent performance. They allow the modeling of complex phenomena in a non-reductionistic way. In a multi-strata hierarchy, in the sense of M. Mesarovic (1970) (a hierarchical ordered system where every level is a domain specific abstract

F. Pichler, R. Moreno-Díaz, and P. Kopacek (Eds.): EUROCAST'99, LNCS 1798, pp. 154–168, 2000.

version of the overall complex real system under consideration) holons are the components for modeling parts of the system at different levels. They emerge in this case from the dependent holons in the model of the next lower level.

In the case of multi-strata hierarchies the mathematical concept of structural "morphisms" – used to relate models of different levels onto each other - plays an important role. Using this concept there is a good chance that a rigorous mathematical approach to construct such models does exist and strong mathematical oriented means for their analysis will be available.

A different situation is given by hierarchies which are multi-layer systems (in the sense of M. Mesarovic). These are hierarchies which model the overall system, where the components receive "orders" from components above and transmit "orders" to components on the next lower layer of the model. When Arthur Koestler introduced his concept of a "Selforganizing Open Hierarchical Order" (SOHO) he had a multi-layer hierarchy in mind with holons as the components.

Holons in the sense of Koestler are important modeling means for components of any hierarchical model of a real system with complex behavior.

Systems Theory, the scientific discipline which provides formal models for solving complex problems in science and engineering, has the task to elaborate the concept of holon and related holonic models and to provide methods and computerized tools for its application.

In the following we will try to explain the approach of Arthur Koestler.

First, we mention again the concept of the "Janus face" nature of a holon, which is emphasized by Koestler. This concept assures compactness of a SOHO structure with respect to the vertical coupling of its layers as well as the independence of holons (which are the components) at each layer. Immediately one is confronted here with the question, how the "Janus face" of holons laying in the most upper and the most lower layer of the "holarchy" is realized. We believe that such holons have to rely for their function in a great deal on "art". The highest level upper holons need, for their role in giving guidance and orientation to the holons below, a reference systems – mostly available to them by intuition – on which to base their decisions. The "Janus face" of those holons, with respect to its looking "up", can usually not be achieved by learning alone but needs a certain talent to build the proper reference system.

On the other hand, holons which are situated on the lowest layer need for the "down" looking "Janus face" a skill comparable to handcraft, to realize and integrate the processes which define the lower boundary of the SOHO-structure. This skill can again be considered as a kind of art which can only be acquired by practical experience. From this point of view, the holons on the upper and lower boundaries in a SOHO-structure play an important role and deserve special attention. It is mainly their proper functioning which defines the quality of performance of a SOHO-structure. Although such considerations might be evaluated as being trivial, it might be interesting to the reader to construct his own example. In the organizational structure of a company, the people at the highest management level and the workers on the lowest level are in that sense critical holons, which realize the input/output processes on the interfaces of a SOHO-structure which is embedded in an environment consisting of the market.

Fig. 1. Communicating a reference system to the most upper holons of a holarchy (from a book by Robert Fludd)

By our above arguments the holons of the layers between are classified as a type of administrative helpers. The realization of their "Janus face" does not need the kind of "art" or "skill" necessary for the holons on the upper and lower bounds of a holarchy (SOHO-structure). However, this should not make them less important. To avoid "bureaucracy" these holons have to perform their Janus faced function in an intelligent manner and have to possess a certain autonomy. Furthermore the holons on intermediate levels are strongly responsible for the selforganizing property of a SOHO-structure. Using constraints they detect the need for restructuring the hierarchical order, including the cancellation of holons and their depending parts or the addition of new holons in order to adapt to requirement changes and improve performance. Although flat hierarchies with a small number of intermediate levels are often desirable, the complexity which exists or is demanded of a real system very often requires a certain number of intermediate levels.

The choice of words which we use suggests to the reader the organizational structure of a country, a company, or a governmental administrative division as a valid example of a SOHO-structure. However, Koestler's conceptual framework has a much wider domain of application, for example any living organism, a forest but also biological cells or a fully automated manufacturing system, are real systems to which this framework can be applied.

Hierarchies are already models in a decomposed form. The different control- and communication channels between the holons constitute the coupling system of the decomposition. A hierarchy with holons as its components, a holarchy, constitutes generally a very desirable decomposition of the overall system.

In the case of a "multi-strata holarchy" (in the sense of Mesarovic-Koestler) the overall model is decomposed into different levels, where each level models the real system in discussion from a certain domain-specific and abstract point of view. Within a domain-specific view of modeling, the model represented by a multi-strata hierarchy at a certain level is a refinement of the models of the levels above. Very often this refinement is realized by an accompanying decomposition, such that to a component of a level-model several components are assigned in the refined version of this component at the next lower level. In this case components of a certain level in a multi-strata model have a "Janus face" in the sense of Arthur Koestler. The question is, whether the necessary additional features of components of this kind can be found, such that they can be considered as holons. The answer can be positive if we assume, that the performance of the model has also a stratified structure such that every level-model has to meet certain performance criterion as determined for that level. Then "intelligent" behavior of the level components are required and self-organisation of some kind might be necessary to meet the performance requirements.

Examples are given by "design hierarchies" as used in the design of microelectronic circuits or generally in the use of the methodology of systems engineering. Other examples are given by "evolution hierarchies" as represented by models of the evolution of living systems or also in some respect by the evolution of machines such as transportation devices (e.g., cars, railways, airplanes, ships) or machining tools (e.g., lathes, tool making machines, robots).

Another type of decomposition of a model as discussed above is given by a multi-layer representation of the model. There the individual components are ordered hierarchically and depending on the level in the hierarchy they have to fulfill specific functions. Examples of typical applications of multi-layer hierarchies are given by organization charts of a company which determines the responsibility in decision making, supervision, and workload distribution. While for multi-strata decomposition of formal models many systemstheoretical methods do exist, not so many methods for multi-layer decomposition are known. Their existence depends very often on an evolutionary process over a rather longer period of time. As mentioned already in the introductory part, multi-layer hierarchies are however well suited as the framework of a holarchy. The performance of such a "multi-layer holarchy" depends strongly on the degree of autonomy of the individual holons. The extreme case, that the holons depend in their functioning completely on the leading holons of the most upper layer represents dictatorship with a central organization enforcing bureaucracy. The other extreme case that the individual holons of a multi-layer holarchy are completely independent determines an uncoordinated system which most likely performs in a chaotic manner. To find a balance between these extreme cases is an important goal in the design of complex systems. However it seems, that a mathematical approach to support such a design is not currently available. This results in practice, for example, in permanently changing architectures of socio-economic systems depending on the political orientation of the decision makers. In the case that the upper holons of the holarchy emphasize the free market they trust that selforganization will eventually bring the system into the wanted balance. In many practical cases this might not happen during the envisaged time horizon and the goal is not reached. On the other hand a completely planned and controlled market has always the danger that certain holons or clusters of holons of the hierarchy will not function as planned and will not contribute towards the wanted balance.

Arthur Koestler defines for the SOHO-structure explicitly what he means by the "Janus face property" of a holon. When looking "down", a holon represents a quasi-autonomous whole (self-assertion tendency) such that the depending holons of the next level have for performing their main function no need in coupling their input- and output channels to other holons.

On the other hand, looking "up", a holon integrates its functions into a existing or developing whole (integration ability).

In the case of living systems Koestler points out that in adult holons the self-assertion tendency is realized by the emphasis on rituals caused by instincts and by stereotypical thinking caused by past experience. The ability to integrate is supported by the creativity of a holon to adapt to new needs of the associated whole.

According to the hierarchical functioning of a holarchy Koestler distinguishes between input-hierarchies and output-hierarchies. Input hierarchies in the sense of Koestler operate to achieve from the signals and states associated with holons on lower levels an abstraction or generalization represented by the signals and states of holons on upper levels. Input hierarchies have therefore the main function to compute the emergent properties in a holarchy.

Output-hierarchies, on the other hand are defined by Koestler as holarchies which operate in the opposite direction of an input hierarchy. They take signals and states from holons of upper levels and transform them to specific concrete signals and states suitable for the proper operation in holons of the lower levels of a holarchy.

Further properties which are introduced by Arthur Koestler to specify a SOHO-structure concern the degree of arborisation of a holarchy and the degree of reticulation of such a structure. Further he discusses the importance of regulation channels, which take care that in a holarchy signals are transmitted only one step at a time, up or down. The holons of a SOHO-structure have to be balanced between being "mechanized" and having a certain degree of "freedom". Holons on higher levels have usually more freedom for their operation while holons at lower levels will usually have to follow more mechanized patterns in their operation. Another important property of a SOHO-structure concerns its degree of performance between dynamical equilibrium and complete disorder. Dynamical equilibrium is achieved if the self assertion tendency and the integration tendency of the holons counterbalance each other. Disorder appears if those tendencies dominate each other. In this context it might be interesting to mention that Koestler states that the rules of a social holon are not reducible to the rules which conduct its members.

The final statements in Koestler's definition of the properties of a SOHO-structure are devoted to regeneration. Critical challenges caused by the environment of a holarchy result in changes of rules for operating holons such that an adaption to new circumstances is realized by a reached new state of equilibrium.

3 Agents and Multi-agent Systems

In the definition of "agents" and "multi agents" we follow directly J. Ferber (1999). An agent is a physical or virtual entity which

(a) is capable of acting in an environment,
(b) can communicate directly with other agents,
(c) is driven by a set of tendencies (in the form of individual objectives or of a satisfaction/survival function which it tries to optimize),
(d) possesses resources of its own,
(e) is capable of perceiving its environment (but to a limited extent),
(f) has only a partial representation of this environment (and perhaps non at all),
(g) possesses skills and can offer services,
(h) may be able to reproduce itself,
(i) whose behavior tends towards satisfying its objectives, taking account of the resources and skills available to it and depending on its perception, its representations and the communications it receives.

Having the properties (a) - (i) an agent can be considered as an "intelligent system". An agent can also be considered as an I/0 Systems serving an user as part of the environment.

A multi-agent system (or MAS) is according to J. Ferber a system which is defined by the following elements:

(1) An environment, E, that is, a space which generally has a volume.
(2) A set of objects, O. These objects are situated, that is to say, it is possible at a given moment to associate any object with a position in E. Some objects are passive, that is, they can be perceived, created, destroyed and modified by active objects.
(3) An assembly of agents A, which are specific objects ($A \subseteq O$), representing the active objects of the system.
(4) An assembly of relations, R, which link objects (and thus agents) to each other.
(5) An assembly of operations, Op, making it possible for the agents of A to perceive, produce, consume, transform and manipulate objects from O.

(6) Operators with the task of representing the application of these operations and the reaction of the world to this attempt at modification, which we shall call the laws of the universe.

From a systems point of view a MAS can be viewed as an open system consisting of a network A of agents and a set of O objects on which they act. Both sets are coupled to the environment E which includes also the model of the user. In case that there are no objects in a MAS ($O=\emptyset$) the MAS is called a purely communicating multi-agent system (CMAS).

In case that the set A of agents and the set O of objects of a MAS together with channels between them have a tree-like hierarchical order we are conceptually close to the definition of a holarchical system of A. Koestler. The individual agents of the tree-structured MAS can be considered as holons. The objects which are situated on the same level of an agent (holon) can be considered as the (physical) work-pieces associated with it. the environment E models the outer world of the holarchical system which is defined by the MAS.

In MAS research and development two different schools of thought are appearent. The "cognitive school" assumes that each agent acts as an "intelligent" system.

A MAS can then be considered as a topic of "distributed artificial intelligence" (DAI). By contrast the second school, the "reactive school" assumes that the individual agents of a MAS have not necessary to proof a certain degree of intelligent behavior in order that the MAS as a whole demonstrates intelligent behavior. The intelligence of MAS is, in this case, considered to emerge as a system property generated by subsystems.

A holarchical system of A. Koestler does not distinguish sharply between these two viewpoints. Holons have to be intelligent systems, however it is assumed also that the whole holarchy has in reaction to the intelligent behavior of the individual holons an overal intelligent behavior which emerges as a new system property.

4 Multi-agent Holarchies: Conceptual Foundation

After our reviewing of the concept of a "holon" and of "holarchy" as introduced by A. Koestler and the discussion of the definition of an "agent" and of a "multi-agent system" as given by C. Ferber we want to compare these two approaches in modeling complex distributed intelligent systems. Our goal is to point out that holons can be considered as agents and that a holarchy (a model which has a SOHO-structure as defined by A. Koestler) is a specific multi-agent system. If we are able to show this, the existing multi-agent methodology together with the associated multi-agent software technology can be used to model and simulate holarchical systems.

Let us start in identifying Koestler's holons as agents. Each holon of a SOHO-structure has the ability to interact with the overal environment such that (a) of the definition of an agent is fulfilled. Furthermore for each holon there exist communication channels to its "master holon" above and to all holons which are subholons of it on the next layer. This means that (b) is valid. Koestler assumes from each holon of a SOHO-structure that it performs according to a overal performance criterion given for the whole holarchy. Therefor a holon fulfills (c). Koestler does not assume expressis verbis the properties (d), (e), (f) for a holon. However these properties of an agent do not contradict with Koestler's holons and might be assumed. The property (g) of an agent is for Koestler of special importance. Each holon of a SOHO-structure is assumed to have special skills to offer services to its associated "master holon". The property (h) of an agent, the ability of self-reproduction is a property which is specifically emphasized in Koestler's definition of a holon. The same applies to the property (i) of an agent. A holon tries always to meet a given performance criterion under the given constraints in resources, skills and available information. We conclude that holons can be considered as agents.

Next we compare the properties of a holarchical system HS (a model with a SOHO-structure) with the properties (1) to (6) as assumed by C. Ferber (1999) in the definition of a multi-agent system MAS.

In (1) it is stated that the environment of a MAS should have the property of a space with nonempty volume (measure). Since a HS being an open system interacts with an environment assumption (1) is acceptable also for a HS. A MAS has a set O of objects which are positioned in E. Some of the objects are according to (2) passive and can be perceived, created, destroyed and modified by other objects, the active one, which are according to (3) agents. In a HS holons can be considered as objects of a MAS, which are because of their janus-face active (acting as master-holons) and passive (acting as sub-holons) at the same time. With this interpretation of holons (2) and (3) as required for a MAS are by a HS fulfilled. A HS fulfills furthermore the properties (4) to (6) of a MAS however with the restrictions which are given by the tree-hierarchy of a HS. A holon will apply operations from Op to perceive, produce, consume, transform or manipulate only to the associated sub-holons.

This concludes our discussion to compare HS and MAS. We have pointed out that the conceptual framework of a SOHO-structure as introduced by Koestler is fully compatible with the elaborated modern concept of a multi-agent system. Since MAS methodology and the associated MAS software technology (regarding modeling and simulation) has reached a high standard, we are optimistic that they can provide the proper means to model and simulate the kind of complex biological or socio-economic systems which Koestler had in mind. Besides it should also be possible to apply MAS technology to the field of "Holonic Manufacturing Systems" (Kawamura (1997)), a field which has found international interest in Robotics and Manufacturing (Jacak (1999)).

To approach the concept of a "Multi-Agent Holarchy" we introduce the concept of an organisation in a MAS. After Morin 1977 (as sited in Ferber 1999) an organisation is defied as

"an arrangement of relationships between components or individuals which produces a unit, or system, endowed with qualities not apprehended at the level of the components or individuals. The organisation links, in an interrelational manner, diverse elements or events or individuals, which thence forth become the components of a whole. It ensures a relatively high degree of interdependence and reliability, thus providing the system with the possibility of lasting for a certain length of time, despite chance disruptions".

Therefor if we assume for a MAS to constitute a organisation we get a hierarchically ordered multilevel system.

Ferber 1999 comments on this part (p. 90):
"The concept of levels of organisation makes it possible to consider the embedding of one level into another. In the same way that, in biology, a cell is considered as being an organisation of macromolecules and at the same time an individual being for the multicellular organism of which it forms a part, we can similarly consider that an organisation is an aggregation of elements of a lower level and a component in organisations of a higher level)."

Furthermore this saying is supported by the following Figure 2.

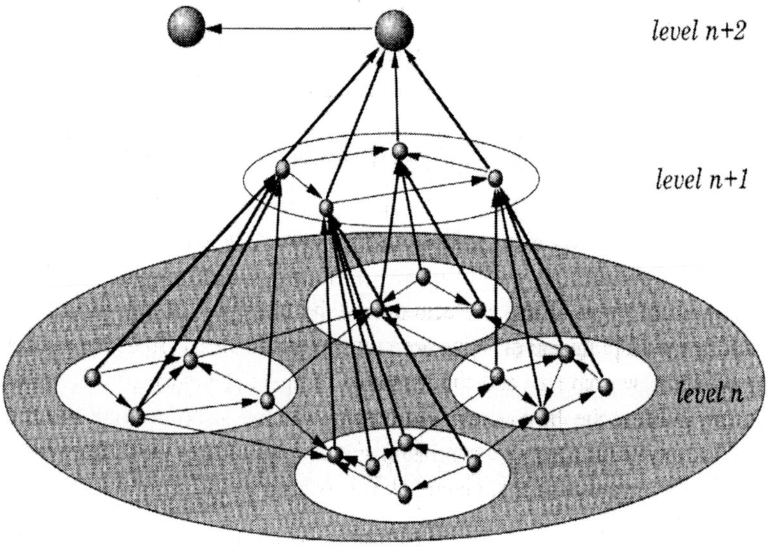

Fig. 2. Agents at level n are grouped into organisations which can be considered at level $n+1$ as individual entities. Inversely, individual entities at level $n+1$ can be seen at level n as organisations. The process can be repeated on any number of levels (from C. Ferber (1999))

We are now at the point, where we are able to compare A. Koestler's concept of a holarchy with the concept of a multi agent system which is an organisation (organisational multi agent system OMAS).

Both are tree-like hierarchical structures with autonomous intelligent acting components being the nodes of the tree-structure. The architecture and the kind of components seems to be of the same kind. Although A. Koestler had with his "holons" models from the biological fields in mind he did not specify in detail their functional behavior or their structural internal properties. Here the conceptualisation is in case of agents on a advanced state of development and can be used for the construction of special types of holarchies. This applies also to concepts for a formal specification of holons where different formalism for agents are known. A further progress in methodological matters of "agent technology" with respect to Koestler's framework its operationality with respect to software implementation. This allows the effective construction of simulators for multi agent systems and can naturally also be used to simulate holarchies. At the time of the appearance of Koestler's work (1967) the science and art of model based simulation was still in an early stage. Furthermore Koestler himself had, as we guess, no close relations to computers and programming and did therefor probably not explore any application of simulation.

Agents, following the terminology of Ferber (1999) can realize different organisational function such as being a supplier (servicing customers), a mediator(managing execution requests), a planner (determining actions to be taken), a coordinator (distribution of actions and execution requests), a decision maker (taking the choice between different possible actions) or a executive (realizing actions).

Such functions can also be observed for holons. A difference is, however, that the holons of Koestler are according to their role in fulfilling an organisational function hierarchically ordered but the agents in an OMAS are seemingly all on the same level and they define as group-collections only virtually agents of conceptual abstract kind on levels above.

In the next chapter we will see that this difference is only artificial and we are able to remove this by assigning to holons the proper organisational function.

5 On Construction of Multi-agent Holarchies

In the following we try to construct a formal model for the architecture of a holarchy such that its components are possible candidates to perform organisational functions as agents of a OMAS. In consequence the existing multi-agent system methodology can be used to realize specific agents in line with Koestler's concept of a holon and to assure properties which are requested for a holarchy. If this is successfully done, we call such a holonic organisational multi agent system (HOMAS). Although in this first attempt we will not be able to show the construction of a HOMAS in detail. However, thanks to the developed state of multi agent systems methodology we can claim that this is possible for many cases of specific SOHO structures. In our proposed construction we follow Pichler (1999).

We start with the principal task of describing the systems architecture of a holarchy with the usual systems-theoretical means. A holarchy in the sense of Koestler can be classified as a multi-level system in a kind of a multi-layer system (in the sense of Mesarovic-Takahara (1970)) with – as seen top-down – a graph-theoretical tree-structure. Its components M(i,j) (= the j'th component on the i'th level) interact with the network and with the environment by the following couplings:

(1) each component M(i,j), (i=0,1,2,...,n-2) is coupled to its associated "parts" which consist of components M(i+1,k), k=l(1),l(2),...,l(i,j), of level i+1

(2) from each component M(i,j) (i=1,2,...,n-1) we have couplings to exactly one associated component M(i-1, k(i,j)) of the level i-1 above, the associated

"whole". Level 0 contains only one component M(0,1), the "root" of the network. M(0,1) has, according to our definition, no associated "whole" above it.

(3) each component M(i,j) of a holarchical network is coupled to the environment E of the network. In this sense all components M(i,j) can be considered as "open".

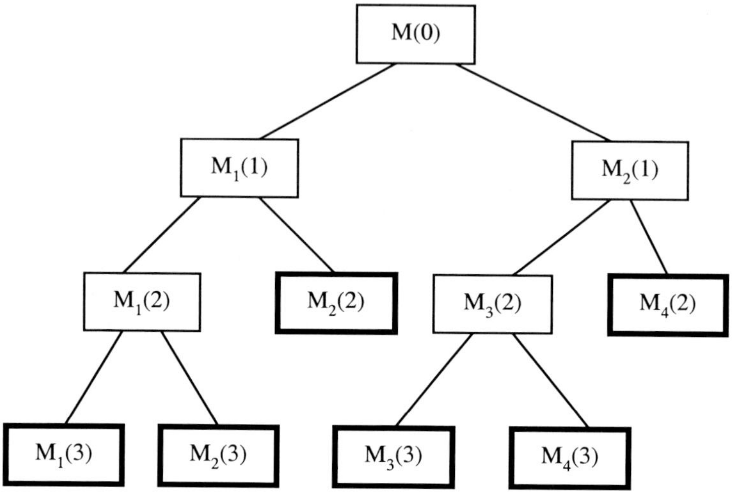

Fig. 3. Example of the tree-like architecture of a holarchy

To model the couplings of a holarchy different concepts can be used. In the simplest case coupling can be defined by joining variables, dynamic or static, of the individual components. In more complex cases functions or relations (static or dynamically generated by some kind of "machines" such as, for example, marked petri-nets) can be used to define the couplings. In general, since in practice the different levels have a different semantic meaning, coupling operations have to be transducers to establish the proper connection between the different levels of language use.

In the simplest case a component M of a holarchy can be modeled by an I/0 relation $M \subset X \times Y$, where X is the set of inputs and Y the set of outputs. A component of this kind is also called a "general system" (in the sense of M. Mesarovic (1975)) or a "black box" (if nothing is stated about how the I/0 relation M is constructed). However, as seen from the systems-theoretical point of view, the most desirable formal model for a component of a holarchy is given by a dynamical system with input and output DS. Most often a DS will be defined by an associated "generator" in the form of equations which describe locally the rate of state-changes. Examples are here given by differential- or difference equations, by marked petri-nets, by formal grammars, by the rule-bases of expert systems or by finite state machines.

A hierarchical decomposition of a complex system, as requested by A. Koestler for building a holarchy, results in a division of the workload among the different components. While components M(i,j) of "higher" levels (i=1,2,...), to fulfill the functional requirements as posted by the environment E, contribute mostly to the planning part, the components M(i,j), (i=n-1,n-2,...) of the "lower" levels have to put an emphasis on the "physical" realization of the "tasks" as required by the environment. Besides observing the fulfillment of functional requirements, as they are stated via the associated "whole" M(i-1,k) above, they have to contribute to satisfying non-functional requirements which are typical for the associated level i. As a consequence the formal models to be used in the different levels of a holarchy have to differ in their granularity. While on higher levels symbolic operations which reflect qualitative features will be dominant, the models of the lower levels will have to perform numerical and quantitative operations.

To make the mathematical analysis of a holarchy feasible, it is advisable to choose the model-types for the different components M(i,j) as homogeneously as possible. To give an example, if we use for all components formal models of a dynamical system with input and output, then we are able to apply the concept of "dynamorphism" as introduced by M. Arbib (Pichler 1983) to describe the couplings between the different levels. Furthermore, in case of the existence of an effective method to decompose a specific type of dynamical system, we are able to build the network in a top-down manner starting from the dynamical system given on top.

The most important feature of an holarchical network, as defined by Arthur Koestler, concerns its self-regulating property. It is assumed that such a network adapts automatically to new requirements (as given by the environment) and automatically corrects internal faults by establishing a new configuration of its architecture.

Self-regulation in that sense is only possible if every component possesses a certain degree of autonomy to be able to respond in the appropriate intelligent manner. The model-types of classical systems theory, which are by their tradition rather "mechanical", have to be extended towards "biological" features such as learning, adaptation and others. In the following we try to sketch principal approaches to construct such extensions.

As we know, each component M(i,j) of an holarchical N(M) represents, for its associated "parts" $T_1,T_2,\ldots,T_l$ which "live" on the level i+1 below, a "whole". There exists no direct coupling relation between the parts; the necessary "communication" has to be established via the whole M(i,j). If we answer the question, how for a given component M(i,j) (i=0,1,...,n-1) the associated parts $T_1,T_2,\ldots,T_l$ can be constructed, we have also the means to design by iteration "top-down" the whole network N(M) starting on level 0 with a initial given "machine" M. By E we denote the set of all functional and non-functional requirements which M has to fulfill. In this case we write M~E.

The principal idea for our approach consists of the first step in the construction of a decomposition $K(M_0,M_1,\ldots,M_l)$ of M(i,j) into components M_i(i=0,1,...,l) which fulfill individually associated requirements E_i (i=0,1,...,l) such that their "union" covers E; $E_0 \cup E_1 \cup \ldots \cup E_l \supset E$. In this case we write $K(M_0,M_1,\ldots,M_l)\ \sigma$ M(i,j). In a second step we distribute the workload (= the tasks to be solved) of M(i,j). We determine that the component M_0 of the decomposition $K(M_0,M_1,\ldots,M_l)$ will be executed by M(i,j) on level i. However, the components $M_1,M_2,\ldots,M_l$, which are left, are only planned by M(i,j) on level i. Their execution is delegated to the associated parts $T_1,T_2,\ldots,T_l$ on level i+1 of the network. This requires that each part T_i

(i=1,2,...,l) simulates the associated component M_i in such a way, that the set E_i of requirements which are assigned to M_i are fulfilled. A simulation of M_i by T_i is denoted by $T_i \sigma_i M_i$. Figure 1 shows, using a block-diagram the relation between the "whole" M, the decomposition $K(M_0,M_1,...,M_l)$ and the associated "parts" $T_1,T_2,...,T_l$.

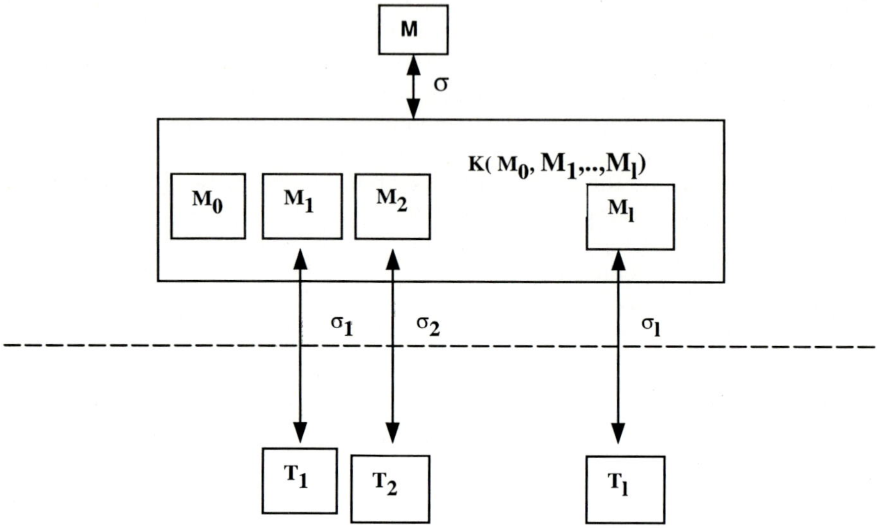

Fig. 4. Requirement delegation from M to the components $T_1,T_2,...,T_l$

For the construction of a decomposition $K(M_0,M_1,...,M_l)$ of M two different types of methods can be distinguished: for a given pair (E,M) one type is given by methods which establish a proper selection of components $M_0,M_1,...,M_l$ (known to the designer by past experience) together with an applicable coupling relation K so that $K(M_0,M_1,...,M_l)$ is a decomposition of M. Methods of this kind are here called "inductive". The second type of methods, which we call "deductive methods", compute by mathematical-logical reasoning, starting from a given M, an associated decomposition $K(M_0,M_1,...,M_l)$. In this case the components M_i(i=0,1,2,...,l) of the decomposition are specific homorphic images of M which reflect certain structural and behavioral properties. The coupling relation K ensures it provable that $K(M_0,M_1,...,M_l)$ simulates M such that M~E is valid.

The design of the overal holarchical network is established by a top-down procedure, which starts with the set E(0) of requirements to be fulfilled by the model M(0) at top-level 0. The decomposition $K(M_0(0),M_1(0),...,M_l(0))$ derived for M(0) determines the partition of the workload between level 0 and level 1. $M_0(0)$ describes the execution part of systems activity at level 0, $M_1(0),...,M_l(0)$ determines the planning part of M(0) at level 0. The task of execution of this planning is delegated to the associated parts $T_1(0),T_2(0),...,T_l(0)$ at the next level 1 below. Each part $T_i(0)$, i=1,2,...,l(0) is now considered as a formal model $M_i(1)$ on level 1 which is the basis for a continuation of the procedure that is: decomposition into an execution part and a planning part, followed by the delegation of the planning part to the next level. After k steps the procedure stops at formal models $M_i(k)$ which can be fully executed and do not require the delegation of a planning part to parts at a next level below.

So far we did not mention any relation of the constructed abstract architecture to organisational multi-agent systems (OMAS). However, the reader which is familiar with the concept of OMAS has most likely already seen this relation. A component M at a level i (see Figure 3) can be seen as model which represent the emerging collective behavior of the models $M_0,T_1,T_2,...,T_l$. M is mainly a planning agent which determines by $M_1,...,M_l$ the kind and distribution of actions for the agents $T_1,T_2,...,T_l$ of level i+1 and by M_0 the action for (own) execution on level i. The simulation assignments $\sigma_1,\sigma_2,...,\sigma_l$ realize a "membrane" (Ferber (1999)) between level i and level i+1. The simulation assignment σ assures the decomposition of M into $K(M_0,M_1,...,M_l)$.

We see that the proposed interpretation of model-decomposition leads on the one hand to a hierarchical order in the sense of A. Koestler which is needed for a holarchy and on the other hand also to the concept of a organisational multi-agent system, where the different levels model the emerging properties of a collective. The used "trick" is, to have on level i an agent which reflects by M the emerging collective properties of $T_1,T_2,...,T_l$.of the agents on level i+1 and which determines by M_0 its own execution function and determines by $M_1,M_2,...,M_l$ the tasks for $T_1,T_2,...,T_l$. of level i+1.

6 Conclusion

The paper considers in an informal manner the concepts of "holon" and "holarchy" as introduced 30 years ago by A. Koestler to model complex systems in the biological and socio-economic fields. In addition it discusses the modern concept of "agent" and "multi-agent system" which are currently in use in modeling and implementing distributed software-systems which assist intelligently the user in solving certain tasks. The comparison of these two approaches in modeling shows that multi-agent systems are appropriate conceptual means to model Koestler's complex systems with SOHO-structure. Furthermore they provide the computational means for the simulation of such systems. This gives us for the future some hope that the work of Koestler in modeling complex systems such as the dynamics of the evolution of species or brain functions has a chance to be continued.

On the other hand by following such research goals of A. Koestler might also be fruitful to the further elaboration and extension of multi-agent systems methodology and the associated tools.

References

Koestler A. (1967) The Ghost in the Machine.Hutchinson & Co. Ltd. London

Koestler A. (1969) Beyond reductionism. New perspectives in the life Sciences. The Alpbach Symposium 1968. Hutchinson & Co Ltd. London

Koestler A. (1978) Janus. A Summing Up. Hutchinson & Co Ltd. London

Kawamura K. (1997) Holonic Manufacturing Systems: An Overview and Key Technical Issues. 4th IFAC Workshop on Intelligent Manufacturing Systems: IMS'97, Seoul, Korea, pp. 33-36

Mathews J. (1996): Holonic Organisational Architectures Human Systems Management, Vol 15, No 1, pp. 27-54, 1996

Mesarovic M., Macko D., Takahara Y. (1970) Theory of Hierarchical, Multi-Level Systems. Academic Press, New York 19

Pichler F. (1997) Auf der Suche nach Arthur Koestlers Holonen - eine systemtheoretische Sicht.
In: Peter Weibel: Jenseits von Kunst, Passagen Verlag, ISBN 3-8165-254-1, pp. 452-455

Pichler F. (1999) Holarchical Multidisciplinary Modelling and Simulation.
In: Complex Systems Thinking Revisited, Publikation des Schweizerischen Wissenschaftsrats (ed. Thomas Bernold), pp. 28-32 (in print)

Ferber J. (1999) Multi-Agent Systems: An Introduction to Distributed Artificial Intelligence. Addison Wesley Longman

Jacak W. (1999) Intelligent Robotic System: Control, Planning and Design. Kluwer Verlag Plenum, New York

Partition of Systems by General System Logical Theory (GSLT)

Germano Resconi

Dipartimento di Matematica,
Università cattolica via Trieste 17
25100 Brescia, Italy

Abstract. In this paper we study the partition of a systems by a new theory denoted General System Logical Theory GSLT). The partition of a system in subsystems creates a conflict situation. In fact when we move from the system to the subsystems, new variables grow up. These variables are the elements that represent the interaction of one subsystem with the others. The interaction is computed by recursive equations that compute the interaction terms. So any subsystem can be isolated from the total system but we must join external variables or sources or inputs that represent the action of the other subsystems. The GSLT language is used to detect the conflict between the system and the subsystems and solve the conflict by the computation of the internal action among the subsystems. We can associate with any subsystem a global variable that changes in time only for the action of the other subsystems.

1 Introduction to the General System Logical Theory (GSLT)

The General System Logical Theory (GSLT) used in this paper is based on the input-output paradigm at different levels and was first proposed by Resconi and Jessel (Resconi and Jessel, 1986). It has since been applied to deriving the basic rules of tensor analysis in non-Euclidean space (Mignani et al., 1993), the study of robot (Resconi and Tzvetkova, 1991; Petrov et al., 1994), extensions of Markov chains (Kazakov and Resconi, 1994), General Relativity (Mignani et al., 1993), the action control of sound of waves (Resconi Jessel 1986) and the intelligent machines (Fatmi et al., 1990). Using the language of GSLT different contexts are connected at different levels. In fact at the first level inside a given contexts we define operators that connect variables in input and in output. At the second level an agent with information in the first context determines the wanted input and output value in a second context. The agent tries also to build operators in the second context coherent with the first as we can see in fig. 1

Or

$$\begin{bmatrix} O_0 & O_2 \\ O_1 & \{O_{12}, O_{21}\} \end{bmatrix} = \begin{bmatrix} I & M_0^0 \\ M_1^0 & \{M_{12}^1 M_1^0, M_{21}^2 M_2^0\} \end{bmatrix} O_0 \tag{1}$$

For example in fig. 1 the operator M_1^0 can be a transformation of the position, M_2^0 and M_{12}^1 the derivative as to time and M_{21}^2 the transformation in the context

F. Pichler, R. Moreno-Díaz, and P. Kopacek (Eds.): EUROCAST'99, LNCS 1798, pp. 169–181, 2000.

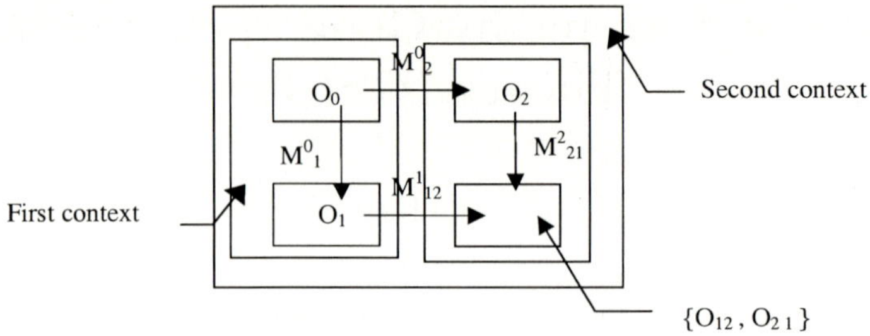

Fig. 1. Elementary Logical System at the Second level or ELS(2)

of the velocity. The couple $\{O_{12}, O_{21}\}$ represent the conflict situation between the context of the position and the context of the velocity. The agent can make some errors in the definition of the operator in the second context errors that can be *compensated.* At the third level as we can see in fig. 2

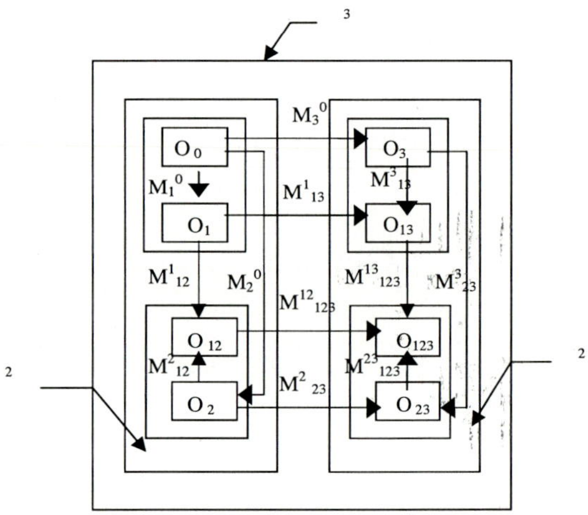

Fig. 2. Elementary Logical System at the third level or ELS(3)

Or

$$\left\{ \begin{array}{cc} O_0 & O_3 \\ O_1 & \{O_{13}, O_{31}\} \\ \{O_{12}, O_{21}\} & \{O_{123}, O_{213}, O_{132}, O_{321}, O_{231}, O_{312}\} \\ O_2 & \{O_{23}, O_{32}\} \end{array} \right\} = ELS(3) \qquad (2)$$

another agent, with the information in the two contexts generates the input and output wanted values for other two contexts. In this way he controls the transformation between two new contexts. The same agent creates connection between the new contexts in a coherent way as to the previous one Other agents can generate higher levels. We are led to particular n-dimensional diagrams or Elementary Logic Systems ELS(n) generated by a simple recursive rule.

The role of GSLT in problem development and solution hinges compensation notions. Compensation gives an important connection with the suited notion of compensation prevalent in physical systems. Compensation creates coherent relations between levels of control.

2 Space Partition of the Wave Propagation System

When the model of a system is represented by differential equations in many cases we cannot use this model for the difficulty to obtain the solution. One way to simplify a complex model of a system, is to divide the system in subsystems. A set of equations is associated with any system that we represent in this symbolic way

$$\mathbf{P\,n} = \mathbf{S}. \tag{3}$$

Where $\mathbf{n}$ is a vector of variables, $\mathbf{P}$ a differential operator and $\mathbf{S}$ a vector of input values or sources. For the wave equation

$$\left(\frac{\partial^2}{\partial x^2} - \frac{1}{c(x)^2}\frac{\partial^2}{\partial t^2}\right) n(x,t) = S \tag{4}$$

and

$$P = \left(\frac{\partial^2}{\partial x^2} - \frac{1}{c(x)^2}\frac{\partial^2}{\partial t^2}\right). \tag{5}$$

With the transformations μ_{j} with $\mathrm{j} = 1, 2, \ldots \mathrm{p}$ we can generate from the set of variables $\mathbf{n}$ the variables $\mathbf{n_j} = \mu_{\mathrm{j}}\mathbf{n}$. In a matrix form

$$\mathbf{n_j} = \begin{bmatrix} n_{11} \; n_{12} \ldots n_{1q} \\ n_{21} \; n_{22} \ldots n_{2q} \\ \ldots \;\ldots \ldots n_{3q} \\ n_{p1} \; n_{p2} \ldots n_{pq} \end{bmatrix} = \begin{bmatrix} \mu_{11} \; \mu_{12} \ldots \mu_{1q} \\ \mu_{21} \; \mu_{22} \ldots \mu_{2q} \\ \ldots \;\ldots \ldots \ldots \\ \mu_{p1} \; \mu_{p2} \ldots \mu_{pq} \end{bmatrix} \begin{bmatrix} n_1 \\ n_2 \\ \ldots \\ n_q \end{bmatrix}$$

where

$$\sum_j \mu_{i,j} = 1\,, \quad \sum_{\mathrm{j}} \mathrm{n_j} = n. \tag{6}$$

Given the prototype operators $\mathbf{P}_{\mathrm{j}}$ for any row of the previous matrix the associate sources are given by the formula

$$\mathbf{P}_{\mathrm{j}}\mathrm{n_j} = \begin{bmatrix} \mathbf{P}_1(n_{11}, n_{12}, \ldots, n_{1q} \\ \mathbf{P}_2(n_{21}, n_{22}, \ldots, n_{2q} \\ \ldots \\ \mathbf{P}_{\mathrm{p}}(n_{21}, n_{22}, \ldots, n_{2q} \end{bmatrix} = \begin{bmatrix} \mu_{11} \; \mu_{12} \ldots \mu_{1q} \\ \mu_{21} \; \mu_{22} \ldots \mu_{2q} \\ \ldots \;\ldots \ldots \ldots \\ \mu_{p1} \; \mu_{p2} \ldots \mu_{pq} \end{bmatrix} \begin{bmatrix} S_1 \\ S_2 \\ \ldots \\ S_q \end{bmatrix} + \begin{bmatrix} \mathbf{S}_1^{ex} \\ \mathbf{S}_2^{ex} \\ \ldots \\ \mathbf{S}_{\mathrm{q}}^{ex} \end{bmatrix}$$

where

$$\begin{bmatrix} \mathbf{S}_1 \\ \mathbf{S}_2 \\ \ldots \\ \mathbf{S}_{\mathrm{q}} \end{bmatrix} = \begin{bmatrix} \mu_{11} \; \mu_{12} \ldots \mu_{1q} \\ \mu_{21} \; \mu_{22} \ldots \mu_{2q} \\ \ldots \;\ldots \ldots \ldots \\ \mu_{p1} \; \mu_{p2} \ldots \mu_{pq} \end{bmatrix} \begin{bmatrix} S_1 \\ S_2 \\ \ldots \\ S_q \end{bmatrix}$$

and

$$\begin{bmatrix} \mathbf{S}_1^{ex} \\ \mathbf{S}_2^{ex} \\ \dots \\ \mathbf{S}_q^{ex} \end{bmatrix} = \begin{bmatrix} \Gamma_{11}\,\Gamma_{12}\dots\Gamma_{1q} \\ \Gamma_{21}\,\Gamma_{22}\dots\Gamma_{2q} \\ \dots\ \dots\ \dots\ \dots \\ \Gamma_{p1}\,\Gamma_{p2}\dots\Gamma_{pq} \end{bmatrix} \begin{bmatrix} S_1 \\ S_2 \\ \dots \\ S_q \end{bmatrix}$$

so we have the formula

$$\mathbf{P}_j \mathbf{n}_j = \mathbf{S}_j + \mathbf{S}_j^{ex} = (\mu_j + \Gamma_j)\mathrm{S} \tag{7}$$

so

$$\mathbf{n} = \sum_j \mathbf{P}_j^{-1} \left(\mathbf{S}_j + \mathbf{S}_j^{ex}\right). \tag{8}$$

The connection between the original system in (3) and all the different subsystems with the prototype operator $\mathbf{P}_j$ is given by the relation

$$\sum_j \mathbf{P}_j \mathbf{n}_j = \sum_j \mathbf{S}_j + \mathbf{S}_j^{ex} = \mathbf{P}\,\mathbf{n} = \mathbf{S}$$

where

$$\sum_j \mathbf{S}_j^{ex} = 0. \tag{9}$$

The exchange sources $\mathbf{S}_j^{ex}$ give the difference among the change of the variables $\mathbf{n}$ and the associate variables $\mathbf{S}$ in (3). The exchange sources $\mathbf{S}_j^{ex}$ have sum equal to zero.

The variable $\mathbf{n}(t)$ and the projection operator μ_j form the context of the states of the system (see fig. 3). When we change the context and from $\mathbf{n}(t)$ we move to the source or input S, the operator associated with the previous operator μ_j is formally defined in this way $(\mu_j + \Gamma_j)$. The unknown operator Γ_j is the compensate term that we introduce when we move from the context of the states to the context of the sources We show the two contexts in one picture in fig. 3.

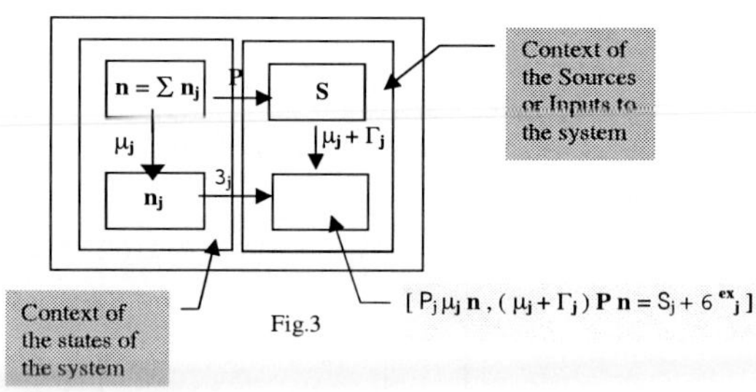

Fig. 3.

The operator μ_j changes in $(\mu_j + \Gamma_j)$ when we move from one context to the other. The two operators are operators of the first type. The operators $\mathbf{P}$ and

$\mathbf{P}_j$ that change the contexts are operators of the second type. As we can see in fig. 3 in the sources context we have the couple of results

$$[\mathsf{P}_j \mu_j \, \mathbf{n}, (\mu_j + \Gamma_j) \mathbf{P} \, \mathbf{n}] \tag{10}$$

where

$$\mu_j \, \mathbf{P} \, \mathbf{n} + \Gamma_j \, \mathbf{P} \, \mathbf{n} = \mathsf{S}_j + \mathbf{S}_j^{ex}. \tag{11}$$

For the coherence between the function n(t) and the sources S, we must have the identity condition

$$\mathsf{P}_j \mu_j \mathbf{n} = (\mu_j + \Gamma_j) \mathbf{P} \, \mathbf{n} \tag{12}$$

so we have

$$(\mathsf{P}_j \mu_j - \mu_j \mathbf{P}) \, \mathbf{n} = \Gamma_j \mathrm{S} = \mathbf{S}_j^{ex} \tag{13}$$

but with P_j^{-1} from the sources context we can come back to the solutions n(t) by the relation (13) see fig. 4.

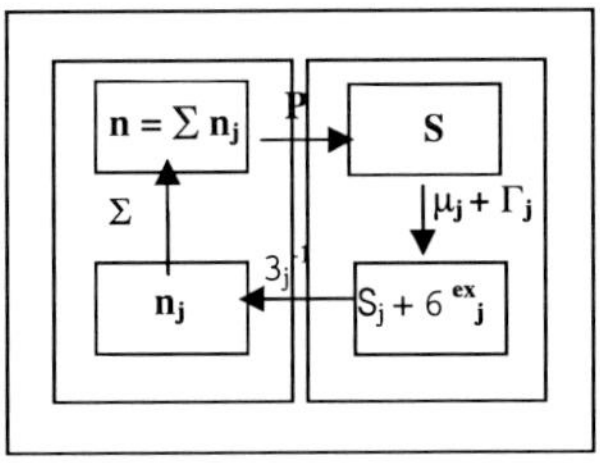

Fig. 4. feedback form by ELS(2)

In conclusion we have the feedback

$$(\mathsf{P}_j \mu_j - \mu_j \mathbf{P}) \, \Sigma \mathsf{P}^{-1} j \, (\mathsf{S}_j + \mathbf{S}_j^{ex}) = \mathbf{S}_j^{ex}. \tag{14}$$

Remark
With the operator $\mathbf{P}^{-1}$ we have the diagram

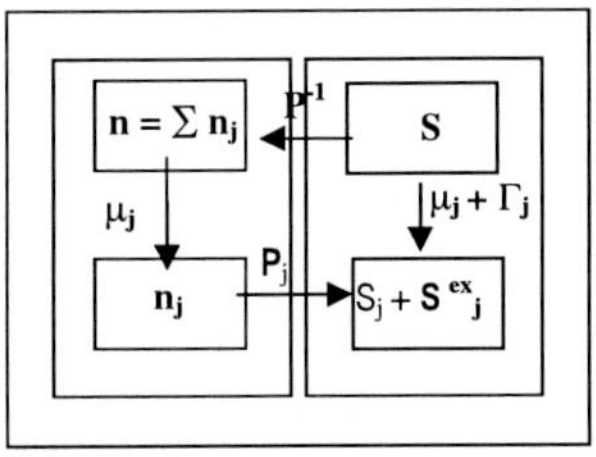

Fig. 5. inverse operator of P

where

$$(\mathsf{P}_j\mu_j - \mu_j\mathbf{P})\,\mathbf{P}^{-1}\mathbf{S} = \mathbf{S}_j^{ex}.$$

But in general we cannot solve the equation (3) so we have not the operator P^{-1}. In this case we use another path to obtain $\mathbf{n}$ as we can see in fig. 5. When we know the $\mathbf{S}_j^{ex}$ with the feedback, we can know the solution of the system equation (3). In fact with the equation (8) we can obtain $\mathbf{n(t)}$ from $\mathbf{S}_j^{ex}$. In fig. 6 we collect in one picture all the previous considerations.

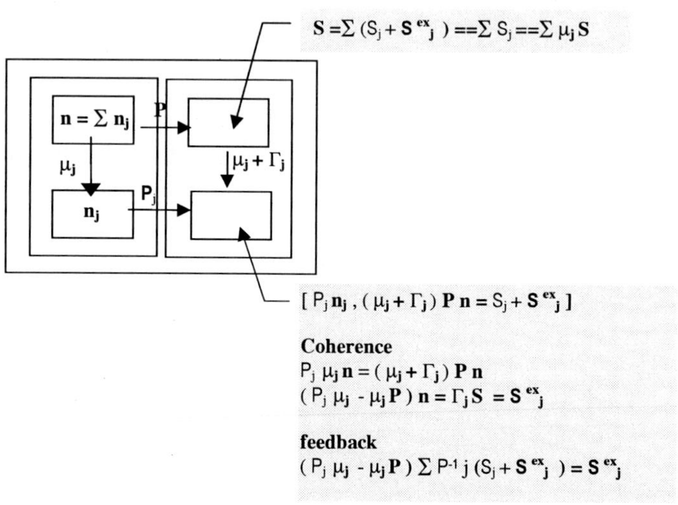

Fig. 6. feedback and ELS(2)

To obtain a feedback that converges to the fixed point in any case, we change the previous feedback in another by the Theorem 1.

Theorem 1. *When we write the feedback in this way*

$$\mathbf{S}_j^{ex}(\mathrm{n}+1) = (\mathsf{P}_j\mu_j - \mu_j\mathbf{P})\,\Sigma\mathsf{P}^{-1}\mathrm{j}\,(\mathsf{S}_j + \mathbf{S}_j^{ex}(\mathrm{n})) - \mathbf{S}_j^{ex}(\mathrm{n}) + \mathbf{S}_j^{ex}(\mathrm{n}) \tag{15}$$

for

$$(\mathsf{P}_j\mu_j - \mu_j\mathbf{P})\,\Sigma\mathsf{P}^{-1}\mathrm{j}\,(\mathsf{S}_j + \mathbf{S}_j^{ex}(\mathrm{n})) - \mathbf{S}_j^{ex}(\mathrm{n}) = \mathrm{F}(\mathbf{S}_j^{ex}(\mathrm{n}). \tag{16}$$

The feedback becomes

$$\mathsf{S}_j^{ex}(\mathrm{n}-1 = \frac{-\mathrm{k}^2 F(\mathsf{S}_j^{ex},\mathrm{n})}{\left[1+\mathrm{k}^2 F(\mathsf{S}_j^{ex},\mathrm{n})^2\right]^2}\,\frac{\mathrm{d}F(\mathsf{S}_j^{ex},\mathrm{n})}{\mathrm{d}\mathsf{S}_j^{ex}} + \mathsf{S}_j^{ex}(\mathrm{n}). \tag{17}$$

When there is a fixed point the feedback in the last form converges to this fixed point and to a point with derivative equal to zero. The parameter k is the step of the recursive process.

Proof. Because we have

$$\frac{-k^2 F(S_j^{ex},n)}{\left[1+k^2 F(S_j^{ex},n)^2\right]}\frac{dF(S_j^{ex},n)}{dS_j^{ex}}=\frac{-k^2 F(S_j^{ex},n)}{\sqrt{\left[1+k^2 F(S_j^{ex},n)^2\right]}}\frac{d}{dS_j^{ex}}\frac{F(S_j^{ex},n)}{\sqrt{\left[1+k^2 F(S_j^{ex},n)^2\right]}}$$

for

$$R(F)=\frac{k^2 F(S_j^{ex},n}{\sqrt{\left[1+k^2 F(S_j^{ex},n)^2\right]}}$$

we have

$$H(F)=\frac{-k^2 F(S_j^{ex},n)}{\left[1+k^2 F(S_j^{ex},n)^2\right]^2}\frac{dF(S_j^{ex},n)}{dS_j^{ex}}=-R(F)\frac{dR(F)}{dS_j^{ex}}=-\frac{1}{2}\frac{dR(F)^2}{dS_j^{ex}}.$$

So the recursive process can be writen in this form

$$S_j^{ex}(n+1)=-\frac{1}{2}\frac{dR(F)^2}{dS_j^{ex}}+S_j^{ex}(n)$$

where we have that when

$$F(S_j^{ex},n)\to 0 \quad \text{also} \quad \frac{dR(F)^2}{dS_j^{ex}}\to 0 \quad \text{and} \quad R(F)\to 0$$

as we know $R(F)^2 \geq 0$ so $R(F)^2 = 0$ only when $F = 0$

Because when $F\left((\mathbf{S}_j^{ex})\right) = 0$ also $R\left((\mathbf{S}_j^{ex})\right)^2 = 0$ with $R\left((\mathbf{S}_j^{ex})\right)^2 > 0$

$$\frac{dR(F)^2}{dS_j^{ex}}=0.$$

Given an $\mathbf{S}_j^{ex} > (\mathbf{S}_j^{ex})^*$ where $F\left((\mathbf{S}_j^{ex})^*\right) = 0$, we have

$$\frac{dR(F)^2}{dS_j^{ex}}>0$$

and

$$S_j^{ex}(n+1)<S_j^{ex}(n).$$

Given an $\mathbf{S}_j^{ex} < (\mathbf{S}_j^{ex})^*$ we have

$$\frac{dR(F)^2}{dS_j^{ex}}<0$$

we have

$$S_j^{ex}(n+1)>S_j^{ex}(n)$$

so in any case we have that

$$\mathbf{S}_j^{ex}\to(\mathbf{S}_j^{ex})^* \quad \text{where} \quad F\left((\mathbf{S}_j^{ex})^*\right)=0.$$

Example 1. When we compare the previous iteration method with the Newtonian formula

$$\mathsf{S}_j^{ex}(n+1) = \frac{-F(\mathsf{S}_j^{ex}, n)}{\frac{dF(\mathsf{S}_j^{ex}, n)}{d\mathsf{S}_j^{ex}}} + \mathsf{S}_j^{ex}(n).$$

We remark Sidney Yakowitz (1989) that for the function

$$F(p) = 1 - 2\exp(-|p|).$$

Given by the graph

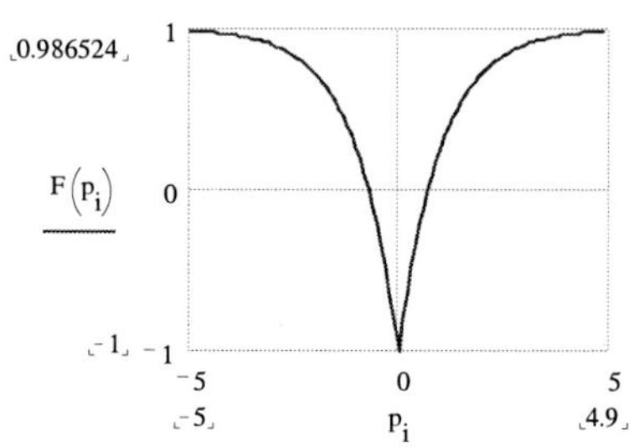

Fig. 7. the Newton method cannot find the zero values for the given function

The Newton method always diverges. With the new method for k = 0.5, and 100 steps we have p = 0.69316 with F(0.69316) = 1.341 10^{-5}. The behaviour to the convergence is

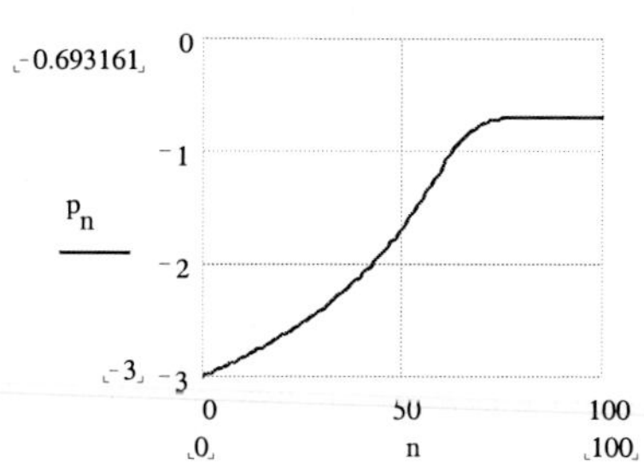

Fig. 8. Behaviour of the recursive process in this paper for the function in fig. 7

With the recursive process

$$p_{n+1} := -k^2 - \frac{F(p_n)}{\left(1 + k^2 \cdot F(p_n)^2\right)^2} \cdot D(p_n) + p_n$$

where D(pn) is the derivative of F(p).

3 Physical Example

One example is the physical system given in fig. 9

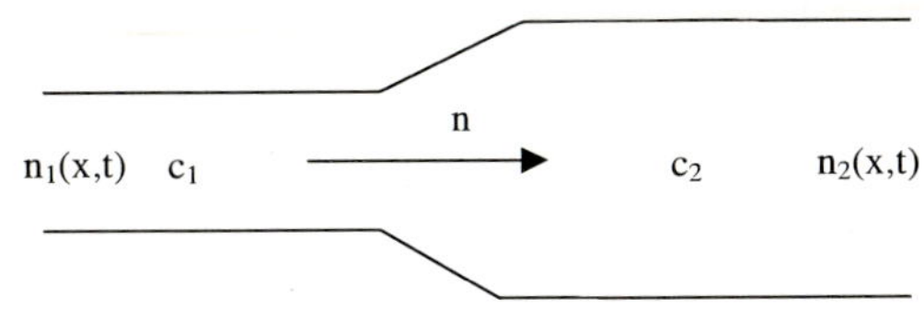

Fig. 9.

Physical system composed by two parts one on the left where the sound moves at a constant velocity c_1 and the other on the right where the sound moves at constant velocity c_2. In the physical system the velocity of the sound is not constant, but can be divided in two subsystems where the velocity is constant.

As we see in fig. 9 the sound wave enters in one tube with velocity c_1 and goes out with velocity c_2. The propagation of the wave of sound pressure $n(x,t)$ in all the tube or system is represented by the differential model

$$\left(\frac{\partial^2}{\partial x^2}-\frac{1}{c(x)^2}\frac{\partial^2}{\partial t^2}\right)n(x,t)=P\,n(x,t)=S \tag{18}$$

where S is the sound source. Because the velocity c(x) is function of the position x in the tube, the solution of the differential equation is difficult.. Because when the velocity is constant we can obtain the solution of the differential equation, we divide the system **in two subsystems** with constant velocity with

$$n_1(x,t)=\mu_1\,n(x,t),\quad n_2(x,t)=\mu_2\,n(x,t).$$

For the two variables we have the equations

$$\begin{aligned}\left(\frac{\partial^2}{\partial x^2}-\frac{1}{c_1^2}\frac{\partial^2}{\partial t^2}\right)\mu_1 n(x,t)&=\mathsf{P}_1\mu_1 n(x,t)\\&=\mathsf{P}_1 n_1(x,t)=\mu_1 S+\mathsf{S}_1^{ex}=\mathsf{S}_1+\mathsf{S}_1^{ex},\end{aligned} \tag{19}$$

$$\begin{aligned}\left(\frac{\partial^2}{\partial x^2}-\frac{1}{c_2^2}\frac{\partial^2}{\partial t^2}\right)\mu_2 n(x,t)&=\mathsf{P}_2\mu_2 n(x,t)\\&=\mathsf{P}_2 n_2(x,t)=\mu_2 S+\mathsf{S}_2^{ex}=\mathsf{S}_2+\mathsf{S}_2^{ex}.\end{aligned} \tag{20}$$

Because

$$\mu_1+\mu_2=1,\quad \text{we have}\quad \mu_2=1-\mu_1$$

and because

$$\mathsf{S}_1^{\mathrm{ex}}+\mathsf{S}_2^{\mathrm{ex}}=0,\ \mathsf{S}_2^{\mathrm{ex}}=-\mathsf{S}_1^{\mathrm{ex}}$$

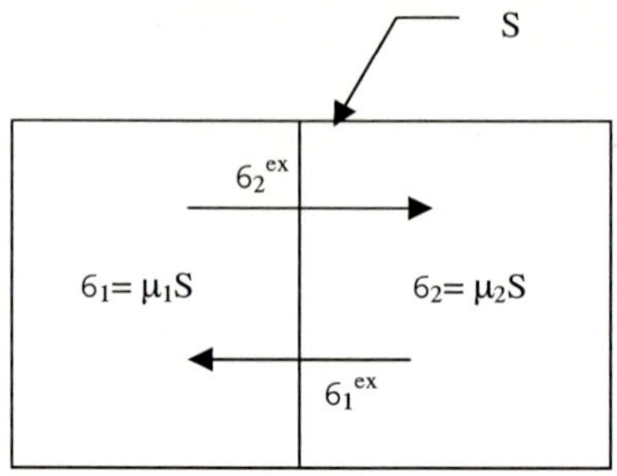

Fig. 10. internal and external sources to the subsystems

and

$$\mathsf{P}_1\mu_1\,\mathrm{n(x,t)} + \mathsf{P}_2(1-\mu_1)\,\mathrm{n(x,t)} = \mathrm{P\,n(x,t)}.$$

We have

$$\left(\frac{\partial^2}{\partial x^2}-\frac{1}{c_1^2}\frac{\partial^2}{\partial t^2}\right)\mu_1 n(x)n(x,t)+\left(\frac{\partial^2}{\partial x^2}-\frac{1}{c_2^2}\frac{\partial^2}{\partial t^2}\right)(1-\mu_1(x))\,n(x,t)=\left(\frac{\partial^2}{\partial x^2}-\frac{1}{c(x)^2}\frac{\partial^2}{\partial t^2}\right)n(x,t).$$

That can be writen in this way

$$\left(\frac{\partial^2\mu_1(x)n(x,t)}{\partial x^2}-\frac{\mu_1(x)}{c_1^2}\frac{\partial^2 n(x,t)}{\partial t^2}\right)+\left(\frac{\partial^2(n(x,t)-\mu_1(x)n(x,t))}{\partial x^2}-\frac{1-\mu_1(x)}{c_2^2}\frac{\partial^2 n(x,t)}{\partial t^2}\right)$$
$$=\left(\frac{\partial^2 n(x,t)}{\partial x^2}-\frac{1}{c(x)^2}\frac{\partial^2 n(x,t)}{\partial t^2}\right).$$

So we have

$$\left(-\frac{\mu_1(x)}{c_1^2}\frac{\partial^2 n(x,t)}{\partial t^2}\right)+\left(-\frac{1-\mu_1(x)}{c_2^2}\frac{\partial^2 n(x,t)}{\partial t^2}\right)=\left(-\frac{1}{c(x)^2}\frac{\partial^2 n(x,t)}{\partial t^2}\right)$$

and

$$\frac{\mu_1(x)}{c_1^2}+\frac{1-\mu_1(x)}{c_2^2}=\frac{1}{c(x)^2}.$$

In conclusion we have

$$\mu_1(x)=\frac{c_1^2}{c(x)^2}\left(\frac{c_2^2-c(x)^2}{c_2^2-c_1^2}\right). \tag{21}$$

When we take the solution in the vacuum as

$$n(x,t)=u(x)\exp(-i\omega t)\quad\text{where}\quad \omega=2\pi f$$

and

f *is the frequency of the wave.*

Because $\mathsf{S}_2^{\mathrm{ex}}=-\mathsf{S}_1^{\mathrm{ex}}$ we have

$$\left[\left(\frac{d^2}{dx^2}+k_1^2\right)\mu_1(x)-\mu_1(x)\left(\frac{d^2}{dx^2}+k^2\right)\right][\mathsf{G}(c_1)*(\mathsf{S}_1+\mathsf{S}_1^{ex})+\mathsf{G}(c_2)*(\mathsf{S}_2+\mathsf{S}_1^{ex})]=\mathsf{S}_1^{ex}$$

for

$$\mathrm{F(x)} = \frac{ik_1}{2} \int e^{-ik_1|x-x'|} \left[\mu_1(x')\, S(x')dx' + \frac{ik_2}{2} \int e^{-ik_2|x-x'|} [\mu_2(x')\, S(x')dx'\right.$$

we have

$$\left[\frac{\mathrm{d}^2\mu_1(x)}{\mathrm{dx}^2} - \mu_1(x)\mathrm{F(x)}\right] - \mu_1(x)\left[\frac{ik_1}{2}\int e^{-ik_1|x-x'|}\mathsf{S}_1^{ex}dx' - \frac{ik^2}{2}\int e^{-ik_2|x-x'|}\mathsf{S}_1^{ex}dx'\right] = \mathsf{S}_1^{ex}$$

where the Green function $\mathsf{G}(c)$ is

$$\mathsf{G}(c) = \frac{1}{2ik}\exp(-ik|x-x') \quad \text{and} \quad \mathrm{k} = \frac{\omega}{\mathrm{c(x)}},\ \mathrm{k}_1 = \frac{\omega}{c_1},\ \mathrm{k}_2 = \frac{\omega}{c_2}.$$

The product "$*$" is the convolution product.

$$u(x) = \mathsf{G}(c) * S \int \mathsf{G}(x,x')\, S(x')dx'.$$

The convolution product is the superposition of the fields (Green function) generated from the elementary sources inside S.

4 System Partition

Let us consider the system $\sum$ with the equations

$$\left[\frac{d}{dt}\mathrm{A} - f\right]\mathbf{n} = S_i \quad \text{where} \quad f\,\mathbf{n} = f_i\mathbf{n} = f_1(n_1, n_2, \ldots, n_q) \quad \text{and} \quad \frac{d}{dt}\mathrm{A}\,\mathbf{n} = \frac{dn_i}{dt}.$$

For the previous consideration we write

$$\left[\frac{d}{dt}\mathrm{A} - f\right]\mathbf{n} = \mathbf{P}\,\mathbf{n}. \tag{22}$$

So we have

$$\left[\frac{d}{dt}\mathrm{A} - f\right]\mathbf{n} = \mathbf{P}\,\mathbf{n} = \sum_{\mathrm{k}} \mathbf{P}_{\mathbf{k}}\,\mathbf{n}_{\mathbf{k}} = \sum_k \left[\frac{d}{dt}\mathrm{A} - f_k\right]\mu_{\mathrm{k}}\,\mathbf{n} = S. \tag{23}$$

For the previous equations and for the relation

$$\sum_k \mu_{\mathrm{k}}\mathbf{n} = \mathbf{n} \quad \text{or} \quad \sum_k \mu_{\mathrm{k}} = 1$$

we have

$$f\,\mathbf{n} = f\sum_k \mathbf{n}_{\mathbf{k}} = \sum_k f_k\mathbf{n}_{\mathbf{k}} \quad \text{where} \quad \mathbf{n}_{\mathbf{k}} = \mu_k\mathbf{n}.$$

Where we choose the functions f_k so that that we can solve the equations (24)

$$\mathbf{P_k}\,\mathbf{n_k} = \left[\frac{d}{dt}\mathrm{A} - f_k\right]\mu_k\mathbf{n} = \mathsf{S}_k + \mathsf{S}_k^{ex} \quad \text{so we have} \quad \mu_k\mathbf{n} = \mathbf{P_k}^{-1}(\mathsf{S}_k + \mathsf{S}_k^{ex}) \quad (24)$$

for

$$\sum_k f_k\mu_k\mathbf{n} = \sum_k \mu_k\mathbf{g}_k\mathbf{n} = f\,n.$$

We can obtain the transformation μ_k. Example :

$$\mu\mathrm{g}_1(\mathrm{n}) = \mathrm{f}_1(\mu\mathrm{n}), (1-\mu)\mathrm{g}_2(\mathrm{n}) = \mathrm{f}_2((1-\mu)\mathrm{n}) \quad \text{and} \quad \mu\mathrm{g}_1(\mathrm{n}) + (1-\mu)\mathrm{g}_2(\mathrm{n}) = \mathrm{f}((\mathrm{n})$$

so

$$\mu = \frac{f(n) + \mathrm{g}_2(n)}{\mathrm{g}_1(n) - \mathrm{g}_2(n)}.$$

In figure (11) we show the system partition

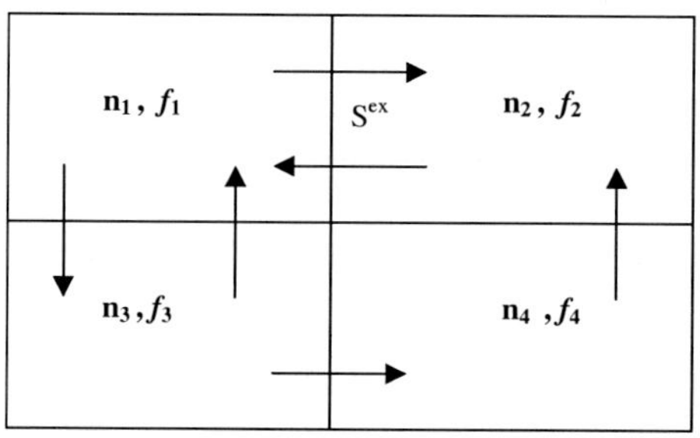

Fig. 11. system and subsystem functions and variables

For any part of the system we write the equation

$$\mathbf{P_k}\mathbf{n_k} = \mu_k S + \mathsf{S}_k^{ex} \quad (25)$$

Because we can solve for any subsystem the differential equation (25) we have

$$\mathbf{n_k} = \mathbf{P_k}^{-1}(\mu_k S + \mathsf{S}_k^{ex})$$

and we can create the feedback process (14) by which we can compute the sources $\mathrm{S}_{\mathrm{k}}^{\mathrm{ex}}$ and we can obtain the solution of the equation (22).+.

When the subsystem is isolated and $\mathrm{S}_{\mathrm{k}}^{\mathrm{ex}} = 0$, we have that

$$\mathbf{G_k} = \mathbf{P_k}\mathbf{n_k} - \mu_k S = 0$$

where G_k is the global variable for the k-th subset. In conclusion $\mathrm{S}_{\mathrm{k}}^{\mathrm{ex}}$ measures the change of the global variables for the subsystems.

References

1. Resconi G. M.Jessel [1986], A General System Logical Theory, International Journal of General Systems 12: 159–182, (1986)
2. P.Auger and G.Resconi [1990],Hilbert Space and Dynamical Hierarchical Systems, Intern. J. General Systems, vol. 16 pp.235–252
3. Maurice Jessel [1956], Une Methode pour etudier certains problemes d'interaction, Le journal de Physique et le radium tome 17, december 1956 page 1022
4. Resconi, G. C.Rattray, G.Hill [1999], The language of general systems logical theory (GSLT),Int.J.General Systems, vol.28 (4–5), pp. 383–416, 1999
5. Penna M.P.,Eliano Pessa,G.Resconi, [1996] General System Logical Theory and its role in cognitive psychology,Third European Congress on Systems Science Rome 1–4 October 1996
6. Tzvetkova G.V. G.Resconi,A [1996] Network recursive structure of robot dynamics based on GSLT, European Congress on Systems Science Rome 1–4 October 1996
7. Resconi G.,G.V.Tzvetkova, [1991] Simulation of Dynamic Behaviour of robot manipulators by General System Logical Theory,14-th International Symposium "Manufacturing and Robots" 25-27 June Lugano Switzeland,1991 pp.103–106
8. Resconi Germano, Gillian Hill, [1996] The Language of General Systems Logical Theory: A Categorical View, European Congress on Systems Science Rome 1–4 October 1996
9. C.Rattray,G.Resconi,G.Hill, GSLT and software Development Process, Eleventh International Conference on Mathematical and Computer Modelling and Scientific Computing, March 31-April 3,1997,Georgetown University, Washington D.C.
10. Mignani R. E.Pessa,G.Resconi,[1993] Commutative diagrams and tensor calculus in Riemann spaces, Il Nuovo Cimento 108B(12) December 1993
11. Petrov A.A. G.Resconi,R.Faglia and P.L.Magnani,A [1994] General System Logical Theory and its applications to task description for intelligent robot. In Proceeding of the sixth International Conference on Artificial Intelligence and Information Control System of Robots, Smolenize Castle,Slovakia,September 1994
12. Minati G. G.Resconi, [1996] Detecting Meaning, European Congress on Systems Science Rome 1–4 October 1996
13. G.A.Kazakov and G.Resconi,Influenced Markovian Checking Processes By General System Logical Theory,International Journal General System Vol.22 pp.277-296 (1994)
14. Saunders Mac Lane, Categories for Working Mathematician, Springer-Verlag New York Heidelberg Berlin 1971
15. A.Wayne Wymore,Model-Based Systems Engineering, CRC Press, 1993
16. Resconi G.A.W. Waymore, Tricotyledon Theory of System and General System Logical Theory, Eurocast'97
17. Sidney Yakowitz, Ferenc Szidarovszky, An Introduction to Numerical Computation, Macmillan Publishing Company, New York 1989

3 INTELLIGENT ROBOTS

Multiagent Approach to Intelligent Control of Robot

Witold Jacak and Karin Pröll

Dept. of Software Engineering,
Polytechnic University of Upper Austria at Hagenberg,
Hauptstrasse 99, A 4232 Hagenberg, Austria,
witold.jacak@fhs-hagenberg.ac.at, karin.proell@fhs-hagenberg.ac.at

Abstract. Within the last years the paradigm of intelligent software agents became an emerging topic in computer science research and development. Intelligent software agents are computational systems that populate complex dynamic environment, act and react there in an autonomous way in order realize a specific task based on a set of goals given to them. In multi agent systems (= MAS) intelligent software agents are grouped together and represent a network of problem solvers designed to achieve a common goal, which would be too large for a single centralized software agent. The member agents of such a multi agent system have a sufficient degree of decision-making autonomy and interact with other agents by explicit communication. Therefore the main fields of interest in multi agent based system design cover coordination, communication and negotiation based on uncertain data and knowledge of each member agent of the multi agent system. This article proposes s system theoretical approach towards the design and synthesis of a multi agent system for controlling a robotic agent.

1 Introduction

The practical applicability of multi agents systems covers various fields in computer science - like intelligent information systems, distributed manufacturing control systems or control systems for multirobot applications. This article presents the application of an multi agent based control system to the distributed control and internal communication in an autonomous robot enabling intelligent and flexible behavior of such a robot in partially known environment. We introduce the following controller components, which can be treated as independent control agents:
- *hardware agent* (robotic agent, or world object to be controlled), - *task understanding and interpreting agent*, - *action planning agent*,- *action execution agent*,- *action safety protection agent*, - *environment observation agent*, - *communication agent*, and - *negotiation agent*. The structure of such a control system is shown in Fig. 1.

The section 2 gives an overview of the formalism used for modelling the agents, the other sections describe some of the involved agents (action execution

F. Pichler, R. Moreno-Díaz, and P. Kopacek (Eds.): EUROCAST'99, LNCS 1798, pp. 185–200, 2000.

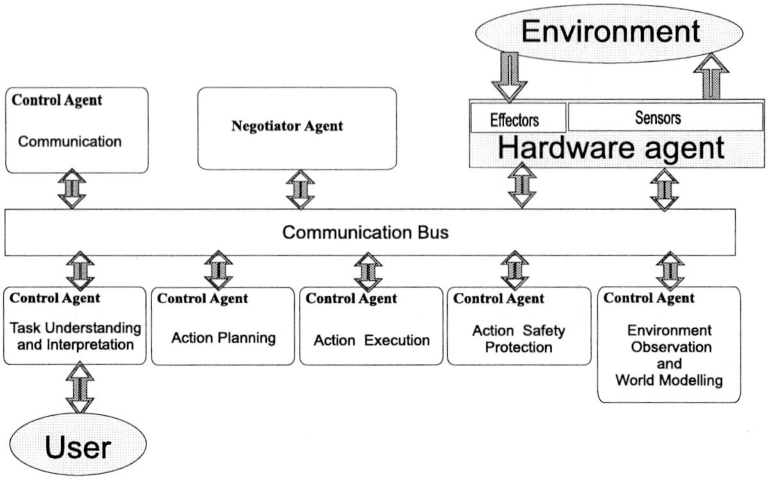

Fig. 1. Structure of multiagent based control system

agent, safety protection agent,communication agent and negotiation agent) of the multi agent system. The final section gives concluding remarks.

2 Basic Notions

2.1 Preliminaries to Multi Agent System

A multiagent system is the group of cooperating agents acting in common environment. Cooperation between agents have a different focus. The main task is to solve complex problems by using cooperation and communication mechanism with other agents or external resources [5,6]. Cooperation is used when a problem exceeds the capabilities of one individual agent or when other agents are available which can provide a proper solution or whose knowledge can be used by other agents. For the communication among agents communications channels have to be established and norms have to be defined in order to ensure that agents can exchange information.

The behavior of the overall multiagent system can be observed as the set of actions performed by each individual agent. Each agent follows its own specific goal. Between the goals of agents in a multiagent system can occur different dependencies. The set of all goals can change in the time. A goal set is quasi ordered by a priority relation, which determines the importance hierarchy of goals. The subset of goals which have the highest priority is called the primary goal set of the system.

2.2 Communication between Agents

Two agents communicate with each other by exchanging messages. They must have some common meeting point through which the message exchange is handled. The delivery of the messages is managed through a special communication agent (CA) of multiagent system [1,2,3].

The messaging mechanism uses protocol templates, messages and conversations .

Protocol templates. In order to simplify the mechanism of exchanging messages between agents individual messages are grouped into a set of related messages. These sets of messages are used to create *protocol template* definitions. Once a protocol template has been created it can be used to define an agent's interface. Interface definition is accomplished by the means of ports.

The protocol template contains a set of one or more parameters. For example the following protocol template parameters can be used:
conversation signature: Denotes the communicative act and the principal meaning of the protocol template.
sender: Denotes the identity of the sender, i.e. the identifier of the agent and its port (see Section 2.3) called *send_point*.
receiver: Denotes the identity of the intended recipient and contains the identifier of the receipting agent and its port (see Section 2.3) called $receive\ _point$.
Note that the recipient may be a single agent or a tuple of agent. This corresponds to the action of multicasting.
message: Denotes the message itself; equivalently denotes the object of the action.
priority: Indicates the level of significance of a specific message.

Based on the previous definitions a message is a data object that incorporates a specific protocol template.

$$Message = (Protocol\_template, data)$$

The exchange of messages is called a *conversation.* A conversation is an ongoing sequence of communicative acts exchanged between two (or more) agents relating to some ongoing topic of discourse. It takes place over a period of time that may be arbitrarily long but is bound to certain termination criteria.

2.3 Agent Architecture

An agent architecture is determined by agent's structure and behavior. The structure of an agent is defined as a tuple of four elements:

$$Structure = (C, I, Contracts, KB) \tag{1}$$

where C is the set of agent *components*, which their own input and output ports $C = \{Com_k | k \in K_{agent}\}$ (K_{agent} is the number of agent's components), I

(the interface) is the set of agent input and output relay ports, $Contracts$ is the set of connections between agent and other agents in multi-agent system and the KB (*knowledge base*) is the set of models and methods needed to decision making.
There are two components common to all agent architectures *a negotiation responsible component* and *a knowledge acquisition and management component.* The other components are agent's specific components, which are needed to realize the agent's goals.

Agent Knowledge base. The knowledge base of an agent builds formal representation of agent's specified data, which are needed for decision making. The decision making is performed on the basis of return values of the methods or functions coupled with knowledge base and called by entry or exit functions of the state machine of different agent's components. The general structure of knowledge base can be defined as

$$KB = (Formal\ Models, Data\ Objects, Methods) \tag{2}$$

where $Formal\ Models$ are the formal representations of stored knowledge, $Data\ Objects$ are the currently stored data, and $Methods$ is the set of procedures and functions, which are needed to data processing, reasoning and decision making The typical knowledge base contains four different models and their formal representations: a*world* model, a *social* model, a *mental* model and a *plant* model.

The *world model* represents knowledge about surrounding environments of an agent. The *social model* represents knowledge about other agent acting in the system. The types of agent, their typical tasks, and their purposes should be known and also communication norms needed for exchanging messages between the agents. The *mental model* represents the knowledge about risk by decisions making and their consequences. The *plant model* contains the knowledge about construction, properties and structure of an agent.

To processing the data of these basic models, functions and methods are grouped in a knowledge base methods set. The methods can be selected according to the following groups: processing and updating methods, reasoning methods, intention recognition methods, beliefs establishing methods, evaluation methods, learning methods, prediction and simulation methods.

Agent's Component. The component of an agent is agent's unit which is enable to perform specified actions based on messages from other components of agent and knowledge base of the agent. Each component has the following structure:

$$Com_k = (Ports_k, State\ Machine_k) \tag{3}$$

Component's port is a declaration (a reference) of the set of incoming and outgoing messages defined by their protocol templates. The ports can divided into two categories, peer ports and relay ports. Peer ports are used to connect the internal components of the agent and are not visible from the outside of the agent.

They are the part of the internal decomposition of the agent. Relay ports are the end ports of the agent and represent the interface of the agent for connecting the other agent [4].

Component state machine. The behavior of the component will be specified by an extended time-state machine.

$$State\ Machine = (M, S, A, \tau, t, g, e, x) \tag{4}$$

where:
• $M = M^{received} \cup M^{sent}$ is the set of incoming messages, and the outgoing messages.
• S is the set of elementary states of the component. A component is referred to as having a state when the effect of an input depends on the history of previous inputs. The state can be passive or active. The active state has finite advance time, defined by function τ.
• $\tau : S_k \to \Re$ is the time advance function of the state. Time advance of the passive state is equal to ∞ [8].
• A is the action set. The action represents the possible activities of the component, such as sending the message, calling the specified method or function from the knowledge basis methods.
• The guard function $g : E \times S \times Ports \to \{T, F\}$ must evaluate to true or false. This function defines an evaluation that must be performed when a message is received, to decide whether a state transition will be taken. The set E defines the event set of the component such as incoming events, internal events, and outgoing events. The internal event denotes the end of activity of the state s. For internal events the value of the guard function is always true.
• The trigger function $t : E \times S \times \{T, F\} \to S$ determines the state transition function. From the definition of the guard function it is easy to observe that for internal event the trigger function causes always the state transition.
• The entry function $e : S \to A$ decides on the entry action of the component. An entry action is performed when a state is entered by way of any transition. The action set contains different component actions enable sending the messages, calling the methods for knowledge base etc. For example the action set can be defined as $A = \{$ *Sent_ Message (·), Run_ Method (·) Change_Parameter (·)of_Method(·),...*$\}$
The entry function can be implemented as the decision making function, in form *if condition then action* where condition can based on the return values of previously called method.
• The exit function $x : S \to A$ generates the exit action. The exit action is taken when a state is vacated by way of any transition. The exit function generate such as the entry function the output event and can be implemented as the decision making function.

Agent interface and contracts. The agent interface is equal to so called *Relay_Interface* which is a class of input/output port references representing the ports that appear on the outside of the agent Each of the port reference is defined as a composite of two elements:

$$port = (port\_name, protocol\ template) \tag{5}$$

The contracts within the structure part of an agent class are partitioned into contracts involving bindings between agents and contracts involving connections which can be used by all agents in the system.
Binding contracts are defined by the set of binding bilateral contracts or multilateral contracts. Each bilateral contract must be created between two relay ports on the respective references, and each multilateral contract must be created between many relay ports on the respective references. The send point and receive points determinate the sender and receiver parameters of protocol templates.

Agent behavior. We refer to the internal operation of an agent over time as its behavior. The behavior of an agent will be specified by its state transition engine ($Behavior = State\ Engine$), which is the complex state machine over so called composite states. A state of an agent represents a period of time during which an agent is exhibiting a particular king of behavior. The states are complex states aggregated from groups of the component's states. For an agent's component activity we can ordered the composite state that represents this activity. A state engine has similar form as the component state machine, but uses only composite states and incoming/outgoing messages.

$$State\ Engine = (M, CS, Trans, Resp) \tag{6}$$

where
• M is the subset of the union of all messages sets of components, i.e. $M \subset \cup\{M_k | k \in K_{agent}\}$
• The set CS is the set of composite states of agent. The composite state is a complex state aggregated from group of the component's states. Let $St = \cup\{S_k | k \in K_{agent}\}$ be the union of the all states of all components. The aggregation relation $\alpha \subset St \times CS$ composes the agent state cs from agent's component states. The components states should be grouped into a composite state in the such a way that a composite state contains a lower level state machine of the component called subengine. ore exactly the composite state should be build from states of the one component, it means that $\alpha^{-1}(cs) \subset S_k$.
A transition represents an allowed path from one composite state to another. $Trans : M \times CS \rightarrow CS$. The transition can occurred only on the such pair (m, cs) where if the composite state cs is created from states of k-th component then the message m has to belong to a messages set of this component, i.e. $Trans(m, cs) = cs_{new} \wedge \alpha^{-1}(cs) \subset S_k \Rightarrow m \in M_k$.
• The response relation $Resp \subset CS \times M$ generates the possible output messages connected with the composite state. The message m can be response of the composite state cs only if there exists the state s of the adequate k-th component which causes with help of its entry or exit functions the action connecting with sent the output massage m.

3 Execution Agent

The first goal of the Execution Agent is to calculate the next path segment of robot motion depending on the type of input messages it receives. It selects the proper interpolation mode depending on the input message represents time trajectory in joint space, geometrical path of the effector or final pose. The functions for interpolation calculation can be retrieved form the knowledge base.
The second goal is to detect conflicts by monitoring the movement sequences. This has to be done by creating movement vectors which are used to monitor if the manipulator approaches the precalculated final point of the movement segment in a proper way. If it does not a conflict has occurred and the negotiation agent which is described in section 5 has to be contacted for conflict resolution.

Similarly to all agents the execution agent composes of general and specific components. The general components are the knowledge management and acquisition component and the negotiation responsible component. The execution agent has two specific components,which are called
○ *movement's path interpolation component* and
○ *the conflict detection component.*

3.1 Movement Path Interpolation Component

The task of the movement path interpolation component is calculating the next movement segment of the robot's manipulator depending on the kind of movement information coming from other agents. It also has to report if the destination point has reached, the motion of the robot was stopped or the robot is idle. The component performs the following functions to realize its task:
- Decision making about path interpolation mode
- Advance path planning
- Perform interrupts and track preference
- Tracing accomplishment

Decision making about path interpolation mode: Depending if the input message indicates **Path,Track** or **Destination** the message content includes a path specification in Cartesian space, a trajectory specification in joint space or a final pose specification via several points. In order to achieve the next movement sequence to the specified values a decision making of the proper interpolation mode has to take place. The methodology for computing the movement sequence according to the selected interpolation mode is represented in the self architecture model of the knowledge base of the action execution agent [11,13].

Advance path planning: If the input message indicates **Detour/Abandon** (sent by the collision avoidance component of the safety action agent which is described in section 4.3) the message contents includes the changes of robot configuration in order to circumvent an obstacle in the best way or by leaving the preplanned path to avoid collision with the obstacle. In the last case the advance path planning has to provide robot configurations which enable the manipulator to finally meet the goal point of the current movement segment. This can either be obtained by a manipulator movement directly to the end point of the movement segment or by motion to a point lying on the preplanned path from where it can continue tracking its original path.

Perform interrupts and track planning: If of any reason the current motion has to be stopped an interrupt is generated and a message indicating **Motion stopped** and the current robot configuration is sent to the other agents.

Tracing accomplishment: If the manipulator reaches the final point of the path segment the message **Destination reached** together with the current configuration data is sent.

3.2 Conflict Detection Component

The task of the conflict detection component is to monitor if the manipulator approaches its final point in the right way. If the distance to the final point increases during the observation periods or the movement keeps fluttering a conflict situation has occurred. The conflict detection component provides functions to monitor the manipulator's movements and to recognize if conflict situations have arisen. Fig. 2 presents a conflict situation. After the detection of such a conflict the negotiator agent which is described in Section 5 is contacted to provide a solution to the conflict In order to perform its task the conflict detection

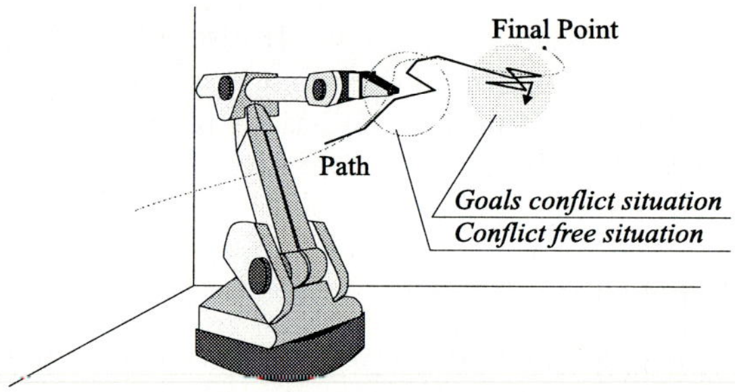

Fig. 2. Conflict detection based on movement vectors field

component uses three functions for conflict recognition.
- Movement vectors field creation and monitoring
- Opposite movement detection
- Goal attainability testing

Movement vectors field creation and monitoring: In order to observe the motion of manipulator, movement vector fields connecting each position of the manipulator on the current path are calculated within an observation period.

They are used for analyses of the actual motion of the manipulator by the following functions:

Opposite movement detection: By using heuristic computation the vector fields are used to detect opposite movement of the manipulator. They are several criteria which indicate an opposite movement, like very large angles between the movement vectors, vectors sums that increase too slowly, etc. They are all applied in order to test the current situation for conflict occurrence.

Goal attainability testing: The other possibility of a conflict occurs if the manipulator keeps moving away from is target during an observation period. This can be done by observing the distance between the endpoint of the movement vectors to the points of the preplanned path. If it keeps increasing the goal does not seem attainable.

In both cases the negotiation agent has to be contacted in order to provide a conflict resolution.

3.3 Knowledge Base and Knowledge Management and Acquisition Component of Execution Agent

The knowledge base of the Action Execution Agent represents three different models:
- Self Architecture Model: Here the robot kinematic model with its appropriate functions and methods is stored. A detailed description of this part of the knowledge base can be found in section 4.1. Besides the kinematic model a set of functions and methods is provided in order to calculate the interpolation mode depending on the input information.
-Social Model: The social model of the action execution agent knows about communication links to other agents. It is responsible to provide connections to other agents, f.e. a permanent connection to the action safety agent has to be provided, because the safety action keeps sending information about robot configuration changes which must be processed immediately by the action execution agent. A non permanent connection to the negotiation agent seems to be sufficient, it only has to be established in case of a conflict detection. After conflict resolution the connection can be disabled.
-Mental Model: The mental model represents a risk model which is applied in order to supersede path proposals of the negotiation agent. It enables the negotiation responsible component of the action execution agent to make decisions about tracing accuracy, adjustment of preplanned path and fine tuning of motion after the negotiation agent has provided a conflict resolution.

4 Safety Agent

The safety protection agent has to provide a collision free movement of the robot manipulator. Similarly to each agent, this agent has two fixed components:

∘ *negotiation responsible component* and ∘ *knowledge management and acquisition component* and two specific components:
∘ *danger recognition component*
∘ *collision avoidance component.*

4.1 Knowledge Base and Knowledge Management and Acquisition Component of Safety Agent

The world model. The world model of the knowledge base of the action safety protection agent is shown in Fig. 3. The knowledge represented here is the geometrical model of the robot environment [7,9]. The formal model used, describes the service space of the robot manipulator as triangle approximation. The points in the triangle net are coupled in triangle walls. The walls represent the data objects of the world model in knowledge base. This model is modified based on sensors data coming from external data bus and relay port of knowledge management and acquisition component.

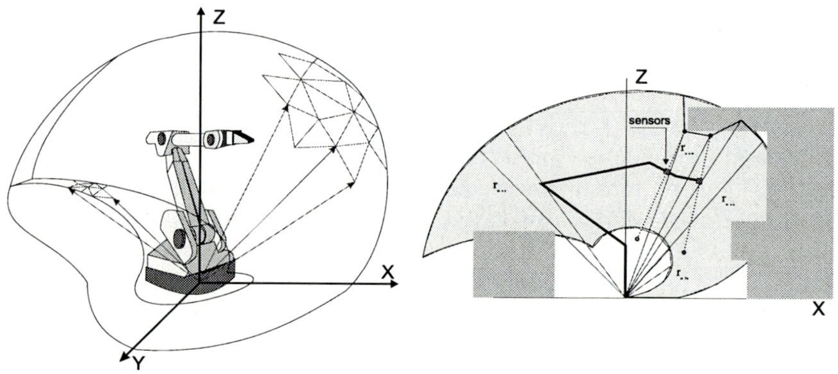

Fig. 3. World model and updating procedure of action safety protection agent

The self architecture model. The model contains the knowledge about construction, properties and structure of a hardware agent. The previously mentioned action safety protection agent needs the knowledge of the kinematical properties of the robotic agent in order to decide about collision avoiding mode and avoiding path. Therefore the forward and the inverse kinematic models of the robot should be known. These models can be created in different ways using different formal tools as e.g. symbolic computed models, numerical models or a neural network based model [7,10,15].

The planning of the new configuration is based on computation of robot kinematics and is computationally expensive. Therefore it is attractive to apply a neural network model of robot kinematics, stored in knowledge base of agent, which automatically generates safe configuration of the robot. This model uses a multilayer feedforward neural network with hidden units having sinusoidal activation functions [14,9,15].

The plant model of the robot kinematics is presented in Fig. 4.

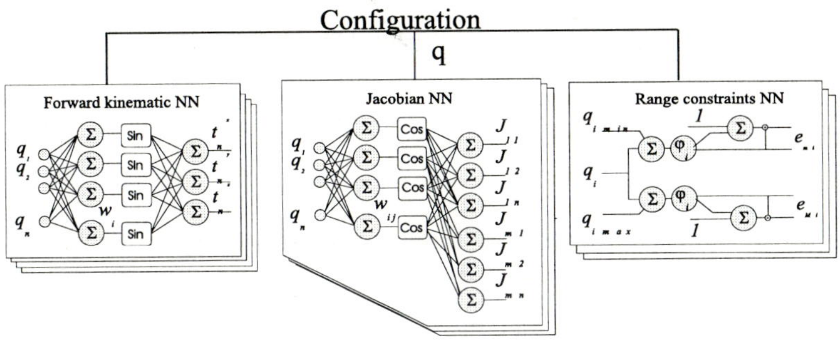

Fig. 4. Neural network modal of the robot kinematics

4.2 Danger Recognition Component

For preventive collision avoidance we propose the installation of ultrasonic sensors and a sensitive skin on each manipulator link and then use a neural network to estimate the proximity of the objects with the link of question. The resulting distances are compared with the local model of the robot environment to recognize the new obstacle and with the radius of the security zone ρ_ϵ. The security zone and sensors observations are shown in Fig. 5.

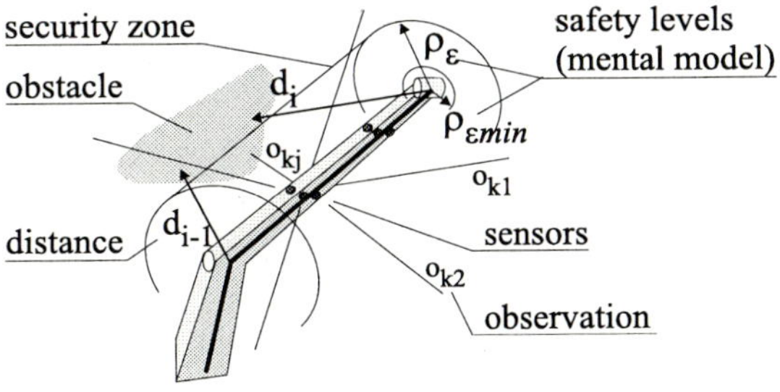

Fig. 5. Security zone of safety protection agent

In the case when some objects disturbing the trajectory execution are reported via system sensors.

When a manipulator is equipped with many sensors and these sensors are mounted on different manipulator links the problem arises how to fuse sensors readings to obtain useful information. The term "useful information" means, for example: what is the distance from the nearest obstacle to the manipulator body; what is "the safety" direction of manipulator movements; what is the type of visible obstacle (dynamic or static)?
The results of fusion action are vectors $d = (d_i | i = 1, .., n)$, represented the minimal distances to objects in robot environment for each joint of manipulator .

4.3 Collision Avoidance Component

The task of the collision avoidance component is a planning of the safe configuration of the robot's manipulator based on the information coming from the danger recognition component and from the action execution agent. To realize this task the component perform two general actions:
- decides about the obstacle avoiding mode, and
- calculates the changes of robot configuration robot, which avoid the obstacle in the possible best way.

Decision making about the avoiding mode. The outputs message **Danger** of the danger recognition component involves the vector.

$$\mathbf{d} = (d_i | i = 1, .., n)$$

Moreover the danger recognition component informs about the type of dangerous. In a case of previously known static obstacle the message **danger decreasing** is send, and in a case of new static or dynamic obstacle (which was not known in the world model stored in the agent's knowledge base) the message **danger decreasing** goes out.

The fuzzy rule based system decides now if the preplanned path should be abandon or if the manipulator should tray to circumvent an obstacle tracing the assumed path. In the first case, it is happen that the obstacle cannot be avoided if the end effector is to pursue its preplanned path. The manipulator must now be moved away to give the way to the obstacle.

In the second cases the robot must not abandon the preplanned trajectory and find the next collision free configuration. The second obstacle avoidance supposes that this position should be as near as possible to the preplanned path.

To calculate the safe configuration the planner uses information from the danger recognition component and combines it in the inverse kinematics computation, performed by *inverse calculation method* from agent's knowledge base (see section 4.1).

The obstacle avoidance can be achieved by introducing the additional errors for each joint, i.e. the errors between virtual points p_i and the joints positions $t_i(q)$. The virtual points are placed on the opposite side of the joint with respect to the obstacle (see Figure 6).

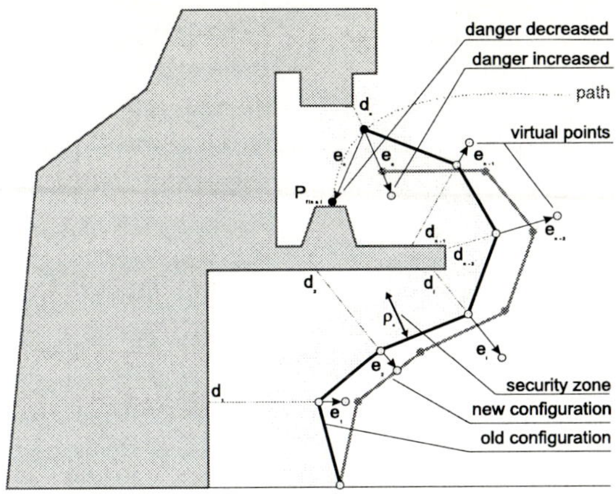

Fig. 6. The virtual point and security zone

Calculation of the changes of robot configuration robot. This action call the *inverse calculation method* from the knowledge base of the agent [7,12].

As a result of the method, the solution of the inverse kinematics with obstacle avoidance is obtained. Because computations performed in the loop are initialized with the desired positions of the manipulator's joints (virtual points), the real manipulator's configuration reflects the motion with tries to avoid the obstacles and perform the detour action. The *Detour* message denotes the state when the manipulator is trying to circumvent an obstacle and *Detour/Abandon* message denotes the abandon of the assigned motion path. *Deadlock* message is sent when the environment and the manipulator seem to be static.

The state machine of the collision avoidance component is presented in Fig. 7.

4.4 State Engine of Safety Protection Agent

The components states are grouped into a composite state in the such a way that a composite state contains a lower level state machine of the component called subengine. The states of the one component can be grouped in more as only one composite state. The state engine of the action safety protection agent is shown in Fig. 8.

5 Negotiation Agent

The situation forces negotiation only when there are conflicting goals of two agents.

Situations of conflict causes at first the theoretically test of achievability of the system primary goal of the multiagent system. The theoretically prove of

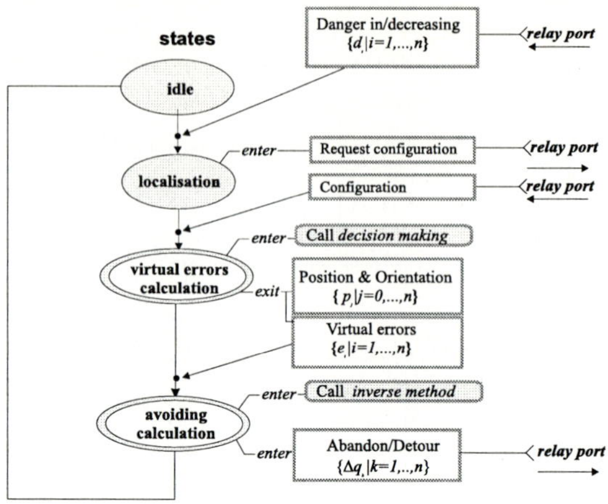

Fig. 7. State machine of collision avoidance component of safety protection agent

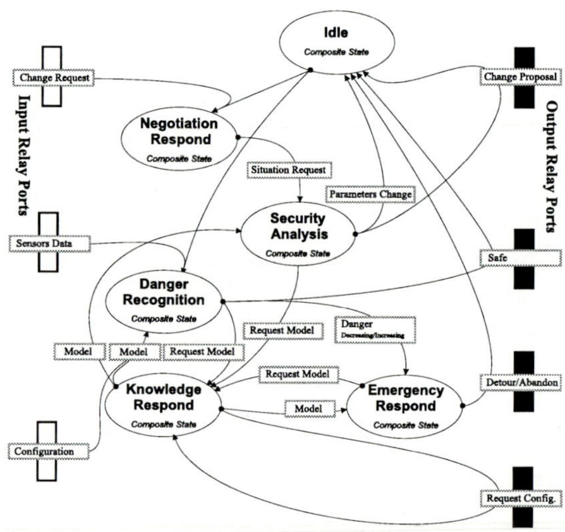

Fig. 8. State engine of action safety protection agent

the primary goal achievement of the system performs the negotiation agent, based on its own knowledge base and different models obtained from both safety protection and action execution agents.

More exactly, the negotiation agent creates current geometrical model of robot environment together with current position of the robot and simulates the robot motion from current position to the goal position. If this motion track is collision free, then the achievement of the primary goal is theoretically possible.

When a conflict situation appears but the achievement of goal of overall system is theoretically possible, then as the first strategy the non-symmetric compromise strategy will be used.

Non-symmetric compromise strategy: The negotiation agents selects the action execution agent as a predestinated winner. The action safety protection agent should decrease its safety level ρ_ϵ so long as it is acceptable by its mental model. Based on simulation's solution, the security zone of safety protection agent can be respectively changed. The parameter describing the range of the security zone will be decreasing. This change proposal is send with message *Change Request* to safety protection agent. When the risk estimation enables the modification of the security zone, the change is accepted. If no, the own proposal of zone changes is prepared by safety protection agent, and is send to negotiation agent with message *Change Proposal.*

In this strategy only one agent, namely safety protection agent is forced to make changes in quality of its individual goal achievement.

In case when this strategy of negotiation leads to the negative result, the symmetric compromise strategy is preferred.

Symmetric compromise strategy: The negotiation means a compromise for both parties and causes a degradation of their results. Using this strategy, the negotiation agent tries to decrease the safety level of the safety protection agent and to change the movement path realized by the execution agent, simultaneously.

Based on data involved in *Change Proposal* message, the local path of the motion should be changed, so that the modified security zone is obstacles free.

If this strategy leads to the negative result too, then the global conflict is recognized and conflict pragmatical resolve strategy o can be used.

Pragmatic resolve strategy: When the negotiation agent is absolutely sure that the achievement of the primary goal is possible, then the pragmatic resolve strategy leads to ignore of the goal and actions of one of the agents. In this case the negotiation agent decides to ignore the action safety protection agent, and takes over the responsibility from the safety protection agent.

When conflict appears such that the goals of the two agent are in direct conflict, which has the effect that only one can achieve its goal and theoretical test show that the achievement of the primary goal is impossible then only the symmetric compromise strategy is preferred. In this case the movement path tray to be replanned and the security level tray to be decreased.

6 Remarks

The article describes the formal models of a multi agent based control system for autonomous robots, which can be applied for the synthesis of a complex control system of a hardware agents in order to provide intelligent and flexible behaviour in partially known environment. This formalism is demonstrated by the synthesis of intelligent controller of robot manipulator

References

1. FIPA - Foundation for Intelligent Physical Agents,Agent Management,FIPA97 Specification Part 1, Version 2.0, Geneva, Switzerland, October 1998.
2. FIPA - Foundation for Intelligent Physical Agents,Agent Communication Language, FIPA97 Specification Part 2, Version 2.0, Geneva, Switzerland, October 1998.
3. Cohen, P. and Levesque, H., Communicative Actions for Artificial Agents, Proceedings of the First International Conference on Multiagent Systems, San Francisco, Cambridge, AAAI Press, pages 65-72, (1995).
4. Jacak W., Pröll K., Multiagent Approach to intelligent control of robot, Robotics and Applications RA99, IASTED International Conference, Santa Barbara, USA, 1999
5. Brenner W., Zarnekow R., Wittig H., Intelligent Software Agents, Springer-Verlag, 1998
6. Haddadi A., Communication and Cooperation in Agent Systems, Springer Verlag, 1995
7. Jacak W., Intelligent Robotic Systems, Kluwer Academic/Plenum Publishers, 1999
8. Zeigler B. P., Multifaceted modeling and discrete event simulation, Academic Press, 1984
9. Lee S., Bekey G. A., Application of neural networks to robotics: Advances in Control and Dynamics Systems, Academic Press, 1991
10. Sciavicco L., Siciliano B., A solution algorithm to the inverse kinematic problem for redundant manipulators,IEEE Trans. RA, RA-4, 1988
11. Brady M.(ed.) Robot Motion: Planning and Control, MIT Press, 1986
12. Jacak W. Discrete Kinematic Modelling Techniques in Cartesian Space for Robotic System, in: Advances in Control and Dynamics Systems, (ed.) C.T. Leondes, Academic Press,1991
13. Latombe J.C., Robot motion planning, Kluwer Academic Pub., Boston, Dordrecht, 1991
14. Lee S., Bekey G.A, Applications of neural network to robotics, in Control and Dynamic System: Advances in Robotic Systems, (ed.) C.T. Leondes, Academic Press Inc., San Diego, New York,1991,
15. Park J., Lee S., Neural computation for collision-free path planning, IEEE Inter. Joint Conf. on Neural Network, Vol. 2, 1990, pp. 229-232

Design of Competence Promoting Multi-Agent-Systems to Support the User in Fault Diagnosis of CNC-Machine Tools*

Regine Gernert[1] and Peter John[2]

[1] Technical University of Berlin, Institute for Human Factors, Steinplatz 1, D-10623 Berlin, Germany
Regine.Gernert@ipk.fhg.de

[2] Technical University of Berlin, Institute for Machine Tools and Factory Management, Pascalstr. 8-9, D-10587 Berlin, Germany
Peter.John@ipk.fhg.de

Abstract. The problem, which was addressed in this project, was the creeping loss of competence of the user when using conventional knowledge based systems. Therefore two main points were the aims of the project. At first it was concentrated on using the Multi-Agent architecture. The second aim was the development of design guidelines for competence promoting decision support systems. The promotion of competence was realised by offering different opinions about a problem to force decisions of the user. Two agents with different views on the diagnosis problem were implemented. Diagnosis strategies and data were analysed with regard to the different views and the integration into the system. At last methods to measure the effect on the competence of the user were developed and tested.

1 Introduction

Manufacturing systems in production technology become more and more complicated [3]. The use of computer based decision support systems is, therefore, necessary for fault diagnosis [4]. The earlier replacement of the human expert by knowledge based systems proved to be advantageous as well as disadvantageous from the human point of view. For this reason knowledge based systems were applied for supporting the user now. This kind of usage implies however the risk of creeping loss of his competence.

The promotion of user competence was, consequently, the aim of our research project. Design guidelines for developing competence promoting decision support systems were deduced. This goal influenced the system architecture as well. The archi-

* Sponsored by the German Research Foundation (DFG) Ti 188/6-1

F. Pichler, R. Moreno-Díaz, and P. Kopacek (Eds.): EUROCAST'99, LNCS 1798, pp. 201–208, 2000.

tecture of knowledge based systems did not meet well with this intention. So it was researched in the field of Multi-Agent-Systems (MAS) as basic architecture [2]. As the architecture is used for designing manufacturing systems in the maintenance application area, the project was a contribution to the research in holonic manufacturing systems.

It was a project of the Institutes for Human Factors and the Institute for Machine Tools and Factory Management of the Technical University of Berlin. It was sponsored by the German Research Foundation. The project was launched in co-operation with the Institute for production systems and design technology of the Fraunhofer community. Especially the machine equipment of it's production technology centre was used for the experiments. Also the experience of the maintenance staff was suited well for knowledge acquisition for the system.

2 Competence Promoting Multi-Agent-System

Point of emphasis was the usage of agent based knowledge. A multi-perspective view enabled the user to form his own opinion on the base of the various proposals given by the individual agents. Thus the user was deducing the resulting decision. Further point of emphasis was the distributed acquisition of knowledge according to the concept of Multi-Agents. To avoid the equalization of the knowledge it was acquired by one knowledge acquisition team per agent. This agent based acquisition prevented from standardizing and equalizing knowledge on an undesirable level as for example on the level of the lowest common denominator.

2.1 System Design

Essential part of the project was the implementation of a COMpetence Promoting Multi-Agent-System for Support (COMPASS). Application area of COMPASS was the fault diagnosis of CNC-machine tools.

The implementation of the multi-agent system took the following points into consideration:
- realization of competence promotion by offering different opinions about a problem to force decisions of the user;
- two agents of different views on the diagnosis problem;
- analyzing diagnosis strategies and data with regard to the different views and the integration into the system;
- self-developed methods to measure the effect on the user's competence.

The approach was based on the result in human factors research, that the multiple representation of knowledge was one of the reasons for higher performance of persons. So the working hypothesis was, that the multiple representation of knowledge within a system in accordance with the acquisition and use of the knowledge in different perspectives was leading to more performance and, as the cause of performance, to more competence. Offering different opinions and different solutions for a problem the system enabled the user to take the last decision.

For the first step implementation concentrated on two agents. The application area was fault diagnosis. Different perspectives on the diagnosis problem were modelled. Modelling included further different methods for fault diagnosis. So the system integrated different strategies of diagnosis to select by the system or the user.

From the human factors research point of view the evaluation of the hypothesis with tests was necessary. For measuring competence methods were developed or existing methods were used, for example existing questionnaires.

2.2 Competence Promotion

Main benefits of the integration of Multi-Agent Systems in diagnosis systems were modulisation of machine specific knowledge and the possibility of knowledge representation in different views. The agent related information allowed simple extension and modification of existing systems. The knowledge representation in different views showed the machine operator different solutions and left the final decision to him. In contrast to expert systems, which suggested only one solution, the machine operator had to decide himself. No loss of competence was occurred

The professional competence was improved by the preparation of expert knowledge from different perspectives. The technical competence was promoted by improving the use of technical equipment and enhancing the necessary knowledge of facts and procedures. The method based competence was supported by providing various methods for problem solving and strategic diagnosis. Adapting working methods to new tasks was trained.

2.3 System Architecture

The Multi-Agent System consisted of deliberative, adaptive, loose coupled agents. It was implemented with the integrated tool suite AgentBuilder [1]. The co-operation of the agents were controlled through a co-operation model. Each agent had it's own knowledge base, inference process and user interface. The system was equipped with a graphical user interface, multimedia elements and a CNC connection module (Fig. 1). The CNC connection module supported requests from the agents to the machine tool as well as the transfer of asynchronous messages from the machine tool. Access was supported by a multimedia graphical user interface with menu-supported error-selection. This interface offered acoustic and visual support for the elimination of errors as well as the representation of data transmitted between agent and inference process. The Multi-Agent System for support on diagnosis activities to CNC machine tools was not a black box with respect to the inference mechanism. The machine operator was not only informed about the communication between the agents but also about each step of the inference process. This was realized by display information on windows or store it in log files.

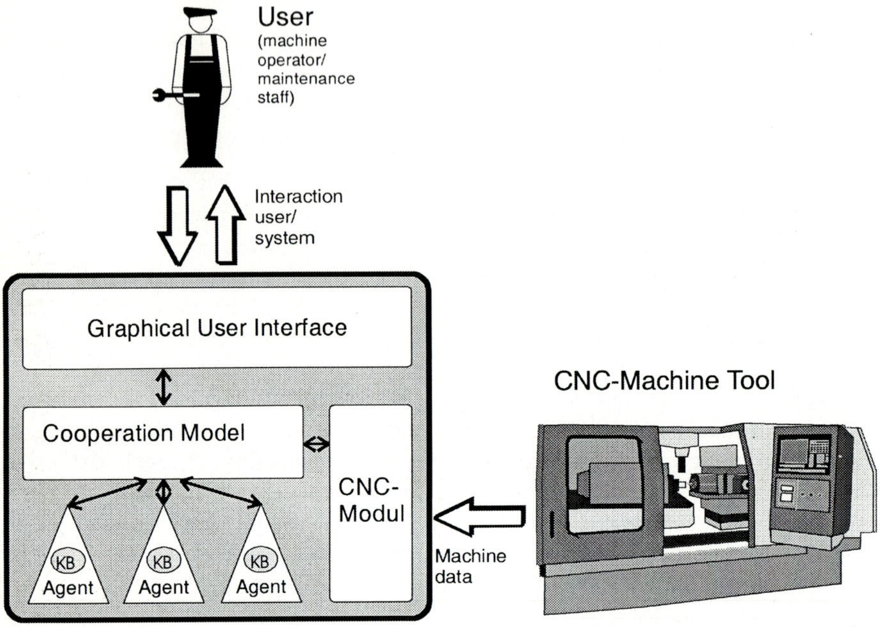

Fig. 1. Architecture of the Multi-Agent System

2.4 Diagnosis

Error messages of the CNC machine tool were received by the CNC module of the MAS and analysed by the agents. In addition the agents had the possibility of checking the state of the machine and requesting machine data. Each agent contacted the other agent, if its knowledge did not permit a further fault location. The language of the agents is based on the *Knowledge Query and Manipulation Language* (KQML) and uses the *Knowledge Interchange Format* (KIF).

For competence promotion different opinions about the problem were offered. Agents represented these opinions. The fields of competence and the diagnosis data

Table 1. Operations of horizontal and vertical agent

Horizontal agent	Vertical agent
In accordance with functional flow	in accordance with component structure
Within electrical, hydraulic or informational dependencies	within mechanical dependencies
In physically independent neighborhood	in physically local neighborhood

and strategies were divided into two agent views. The horizontal agent was responsible for fault location in horizontal direction. That means, for example, the agent operated in accordance with the functional flow (Table 1). In contrast to the horizontal agent the vertical agent worked in accordance with component structure.

The system supports the user in his diagnosis strategies by providing adequate information material. On the other hand, the system guides the user step by step in accordance with implemented diagnosis strategies. So the system executed error routines or delivered results of error routines of the CNC-machine. Information was provided in multimedia technology and included knowledge about the machine itself.

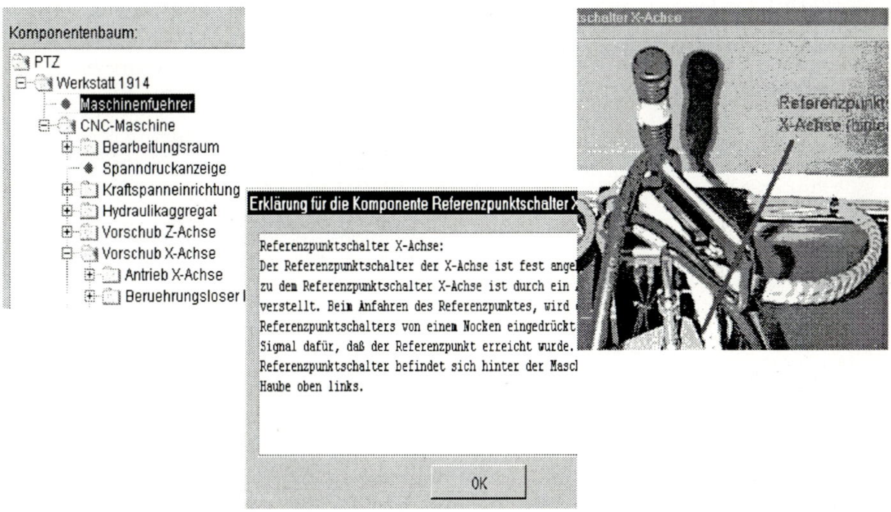

Fig. 2. Information material

The support of diagnosis strategies further integrated the consideration of diagnosis specific facts and methods like frequency of faults and deficits, historical information, dependent probabilities, sense perceptions, and signal flows.

General search strategies were also necessary for example the minimal effort approach, splitting, exclusion and uncertainty reduction. Special procedures in fault location were the reconstruction of errors and pattern matching of fault symptoms.

The system supported for example the user in his diagnosis strategies by providing adequate information material like component structure tree, explanations of components or pictures of parts of the machine (Fig. 2).

The horizontal agents offered for example the diagnosis strategy for following signal flows (Fig. 3, left hand side). It allowed the user to check symptoms of faults and to decide in each case, what fault is further analyzed. The diagnosis ended successful with finding the cause of the fault. At each step the system supported the user by displaying measures for check and correction on demand (Fig. 3, right hand side). One of the diagnosis strategies of the vertical agent was fault frequency. The fault frequency was attached to the component.

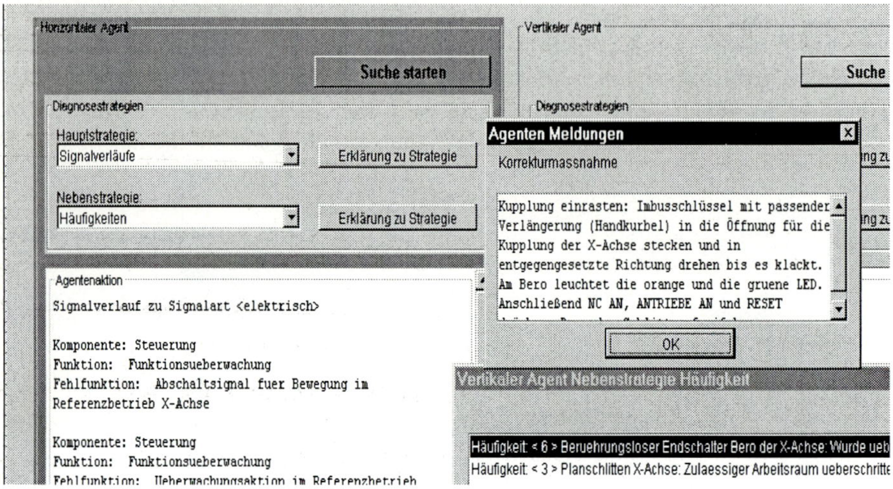

Fig. 3. Signal Flow

2.5 Competence Measurement

Development and test of methods for measuring competence promotion was of further interest. The professional and technical competence was measured by knowledge tests. So the increase of competence was proved by executing tests before and after the use of the multi-agent system. The method based competence was measured by carrying out simulation experiments with test persons. The test persons were 14 students with experiences in CNC-machine tools.

During the simulation experiment behaviors of the persons were observed and estimated based on the measured indicators. The experiment was verified by comparing the results of the group which used the system to a group which did not use it. The test ended with filling up a questionnaire concerning self assessment of the test persons.

The experiment started with the knowledge test. Facts about the machine and procedural knowledge about strategies of diagnosis were asked for. The number of successfully answered questions were increased after the intervention (Fig. 4). In that way the competence of the persons were also increased. If the intervention of the system or other reasons, for example the learning effect within the test situation, were the cause for the development of knowledge, couldn't be proved in this way. Therefore further studies are necessary.

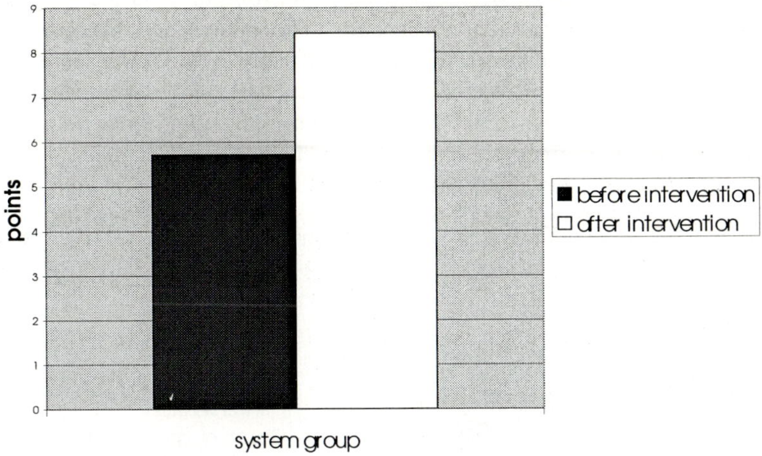

Fig. 4. Knowledge Test

During the simulation experiment the persons had be solved two tasks in fault location. The hypothesis was supposed, that with increasing competence the performance is also increased. Measured indicators of the performance were for example time and success for solving tasks. In Figure 5 the results are presented. The time was increased extremely. So the results were a good confirmation of the hypothesis. But the number of solved tasks were reduced. Before the intervention 10 of 14 tasks were solved. After the intervention only 7 persons finished the experiment successfully.

It was assumed, that the tiredness of the persons at the end of the test was a possible reason for this result. Nevertheless, all indicators of the performance together confirmed the theoretical assumptions.

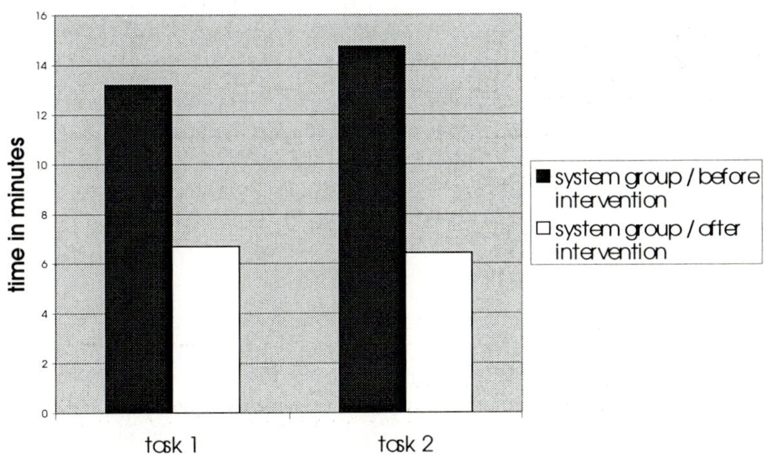

Fig. 5. Time for solving tasks

The measurement of competence was completed with filling up a questionnaire. The answers of the questionnaire were based on rating. The value 1 demonstrated a bad, 10 a good result. The questionnaire consisted of different scales. The scale ad hoc competence and emotional load were compared for example (Fig. 6).The hypothesis was assumed that with increasing ad hoc competence the emotional load is decreased. The evaluation confirmed this hypothesis. The contrast of the values were not very large, but the tend was essential.

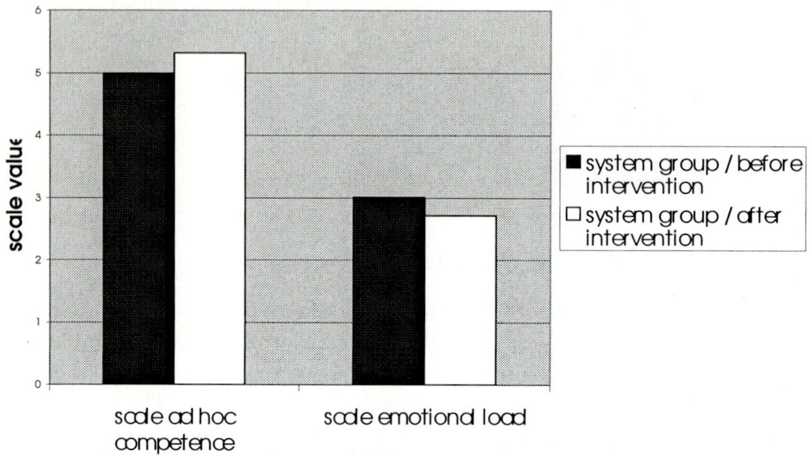

Fig. 6. Questionnaire about competence

2.4 Conclusions

The system is planned to be applied in industry. Manufacturer of equipment are interested in completing new machines with diagnosis systems on base of our technology. These are potential partners for co-operation. Supplier of machine tools are interested too. This is an traditional application area for diagnosis systems. For car industry the point of emphasis is the enlargement of competence with respect to social and organizational aspects. Intention of the management is the support of total productive maintenance.

References

1. http://www.agentbuilder.com
2. Bigus, J.; Bigus, J. (1998): Constructing intelligent agents with Java. Wiley, New York
3. Spur, G. (1996): Die Genauigkeit von Maschinen. Carl Hanser, München, Wien
4. Timpe, K.-P.; Rothe, H.-J. (1997): Wissenspsychologische Beiträge zur rechnerunterstützten Störungsdiagnose. In: Zeitschrift für wirtschaftlichen Fabrikbetrieb. München, 92(1997)5, S. 243-245

System Integration Techniques in Robotics

Libor Přeučil and Vladimír Mařík

The Gerstner Laboratory for Intelligent Decision-Making and Control,
Department of Cybernetics,
Faculty of Electrical Engineering,
Czech Technical University in Prague, Technická 2
CZ-166 27 Prague 6, Czech Republic
{preucil,marik}@labe.felk.cvut.cz

Abstract. Intelligent robots are systems which process on-line a vast volume of information consisting of both the data and knowledge. The data gathered by sensors are processed making use of knowledge incorporated in the software modules. To explore both the data and knowledge requires to integrate both of them on appropriate levels. As different the data usually represent the same part of the world, it is obvious to speak about data fusion than integration. On the other hand, the software modules are usually highly specialized and complementary to each other. Solving the problems of relevant data fusion and systematic software integration seems to be a crutial aspect of the robot design tasks. The paper describes core ideas and principles used for software and system integration in the GLbot (the **G**erstner **L**aboratory **Robot**) experimental platform, were the multi-agent approach has proved to be the very efficient one.

1 Introduction

Intelligent robots are usually understood as complex systems accomplishing tasks of mental and physical nature. As the former requires intelligent knowledge processing, the latter belongs to physical interacting with the environment. To achieve the required abilities a robot should be able:

- to percept and recognize the environment,
- to create and continuously modify the internal representation (model) of the environment,
- to make decisions about its own activity which guarantees robot's survival and leads to achieving of the given goals,
- to influence the environment: to manipulate objects and to accomplish movements,
- to communicate with a human operator.

The previous description formulates three main functional subsystems of each robot quite explicitly. The *perception subsystem* is responsible for preprocessing of sensor data in order to recognize basic elements of the environment and for integrating the partial descriptions into a (more or less complete and consistent) model of the environment (so called a world model WM).

F. Pichler, R. Moreno-Díaz, and P. Kopacek (Eds.): EUROCAST'99, LNCS 1798, pp. 209–223, 2000.

The *decision-making subsystem* is responsible for formulating global goals and for the global planning of activities to reach them. This subsystem mainly explores the world model created by the sensor subsystem.

The *action subsystem* performs a detailed planning of actions, generates instructions for the motion system and directly controls the motion. Again, it explores the world model and exceptionally some sensor data directly (e.g. tactile information can be used directly for assembling - without any complex processing in the decision making subsystem - to improve the finest motions and to make them very accurate).

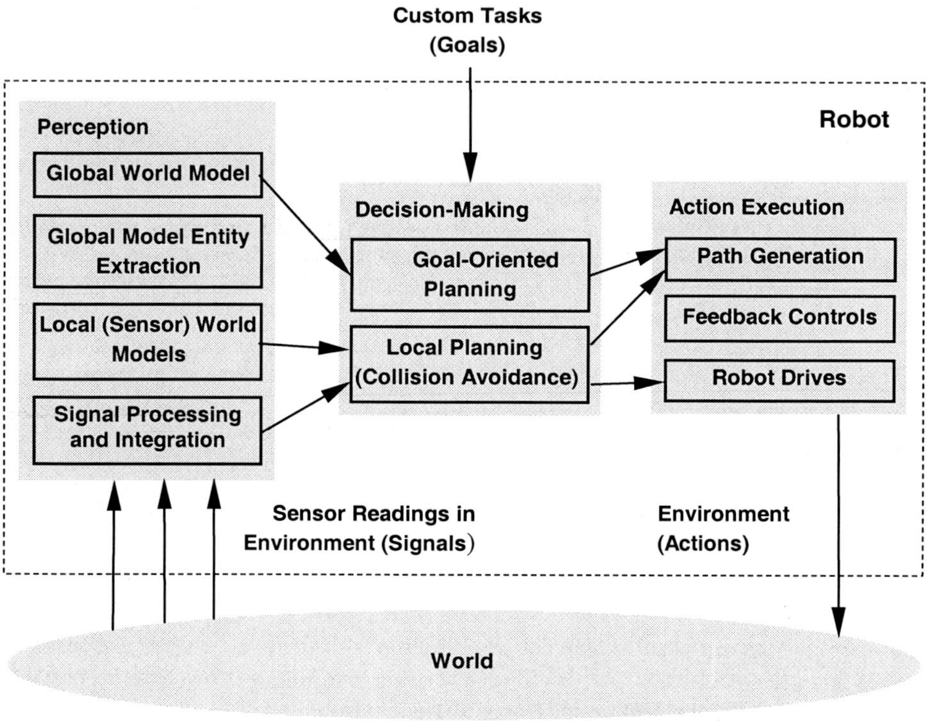

Fig. 1. Basic functional structure of an intelligent robot. Top-down position of each component stands for the level of abstraction.

As the robots have been developed to work in complex environments and to behave in a flexible and intelligent way, various kinds of information/control feedback loops of different nature can be found within the robots' architecture. This feedback explores different levels of data abstraction of the same data acquired by the same sensors. In the feedback loop on the lowest level, i.e. in the loop for solving emergency situations (e.g. collision avoidance), the direct sensor data are usually explored. On the top level, in the decision-making feedback loop, the global world model created through a multi-stage process of data processing/fusion is applied.

Each level of the feedback control requires a certain level of sensor data integration/fusion: The higher the feedback level, the more steps of data fusion are needed as a pre-requisite.

On the other hand: The more complex the task is to be solved, the more complex software accomplishing various feedback solutions should be implemented. Even the most complicated software can be split into a number of nearly independent software modules. These modules can be developed, implemented and maintained in a much simpler way than a monolithic solution. Only one problem remains: how to efficiently integrate these modules.

2 Intelligent Mobile Robots

To demonstrate the nature of the above mentioned techniques used in robotics, let's focus to a specific class of intelligent robots which are *self-guided autonomous vehicles* (SGV). Respecting the application field, the common task for this sort of robots is to conduct movements through 2D environments. The robot's typical goal is either to *reach a destination* or to *cover a region*. All such activities are closely related to major functions of the intelligent vehicle:

In the first place there is the *navigation task* [6] which provides the vehicle's position and orientation description at any world location. The self-navigation incorporates data acquisition by various sensor systems and their interpretation with respect to the level of internal knowledge contained within an *internal world model* (WM). This results in determination of the vehicle's position and orientation in the 2D world and in hypotheses of the world's properties expressed by the means of obstacle occurrence [17] and [20].

Knowing the actual position of the vehicle at certain moment and having the previously mentioned knowledge about the environment, *the activity planning* and *plan execution* can be resolved [1] and [17].

Generally, both the processes - the world data acquisition and planning - are carried out simultaneously and they influence each other. In the case that the vehicle's environment is not completely known in advance (before the execution of the planning task) the plan can only be estimated. Afterwards, during the plan execution and using interpreted measurements of the sensor system, the original plan estimate can be (and typically has to be) modified.

2.1 Main Functional Parts

From the basic concept introduced above one of the possible architectures of a SGV can come out comprising: Sensor data preprocessing and fusion, world model, fusion of different models, path planner and plan interpreter.

The *sensor processing and fusion* part fulfills two basic tasks. It is responsible for low level preprocessing operations of the raw sensor data such as noise reduction, dropout detection, etc. The methods used for information fusion range from simple statistical approaches to knowledge-based decision making techniques. The choice of a suitable approach depends in the first place on the abstraction level of information and on prior knowledge of the sensor response. This function results in general into two key functions of the robot's internal activity:

- A workspace updating mechanism which is responsible for overtaking long-term care for the global WM.
- On-line modification of the goal-driven behavior resulting into techniques of sensor-based control for collision avoidance.

Specific combinations of *sensing strategies* (global path planning strategies based on world modeling, local planning methods, border tracking strategies etc.) can be used for the mentioned tasks what leads to structuring of the system as shown in the Fig. 2.

The *world model* as a whole consists of two data-storing parts, the already mentioned *local world map(s)* and the *global world map* and one data-handling system. As the local maps contain temporary data about the vehicle's neighborhood, they are typically attached to a particular sensor. This concept enables optimization of the data (it is hard to speak about knowledge here) carriers with respect to the nature of the sensor in question. For example the GLbot system relies on the occupancy grid approach here as a 2D probability (occupancy) functions enable efficient storage and easy fusion of range measurements.

On the other hand, the global map contains long-term data covering the whole workspace. There is a plenty of concepts how to store overall knowledge. As the most efficient solutions to this task are strongly application dependent a useful categorization can be done with respect to the level of abstraction of the map content. From this point of view, an optimum in performance of such maps can be found for the data/knowledge representations which are sufficiently compressed (in the data quantity sense) but still do not require an obstacle recognition stage. To do the latter in unknown and unconstrained environments still belongs to hard-core problems. Sufficient level of data volume compression is mostly achieved via extraction of features giving a description of the workspace structure and shape.

The *scene feature finder* serves the local map data understanding by means of search for basic entities which set up the scene description. This function controls updating of both the global and local maps with found scene features and provides adaptability to changes in the environment.

The most common types of topological features are various approximations of obstacle boundaries, certain points-of-interest or specific regions. The GLbot system applies computationally cheap but fully functional solution using line-segment approximation of obstacle boundaries [10].

The *path planner* and the *plan interpreter* subsystems overtake responsibility for robot activity planning and plan execution. Although both these parts do not employ much of the data fusion methods we should mention the *collision avoidance* system at this point. It's linkage with the data fusion can be seen in usage of pre-processed (already partially integrated) data for a low-level feed-back for a sensor-based motion control.

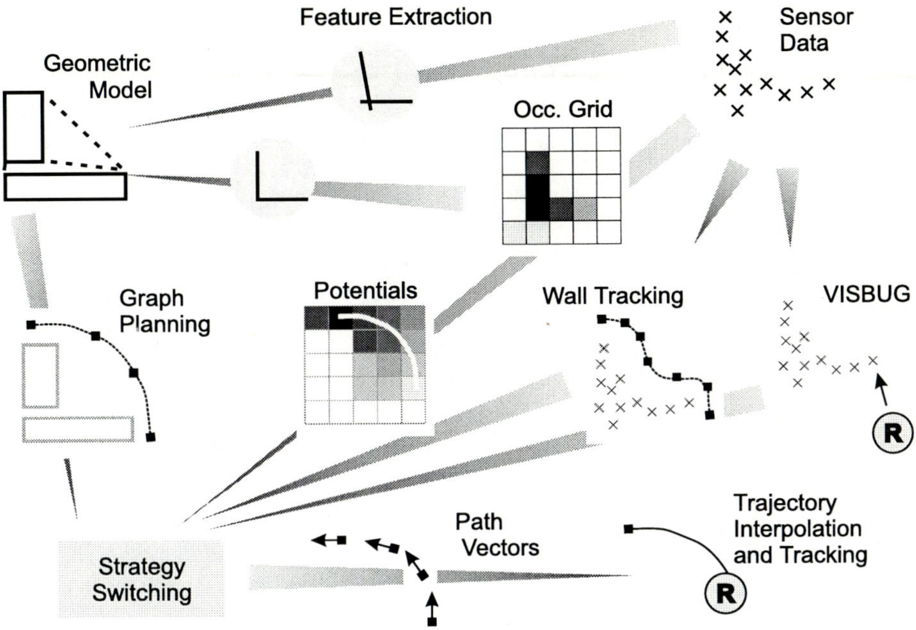

Fig. 2.. Example of a typical data abstraction levels and the data flow diagram within a self-guided autonomous vehicle

2.2 World Data Acquisition and the Sensor Subsystem

The most crucial processing phase in the proposed robot architecture (Fig. 1) is the world data acquisition process. With respect to performance of the used hardware, all the data on the environment can be gathered within certain limits on their quantity and/or rate. As it provides the only links from the real world to the vehicle, it is evident that the data acquisition can heavily influence the final performance of the whole system.

Respecting physical principles of the used sensors, more or less uncertainty in the measurements can be expected.

These two facts invoke the main tasks of the sensor subsystem - *data volume reduction* and *reliability improvement* - by preprocessing and fusion [2]. The incoming information, provided by the sensors, is in a pure signal form and can be either directly fused, or passed to further processing. This processing stage assigns a semantic context to raw signal data and converts it into higher level representations. Useful categorization of possible qualitative levels for data representation is given in the following figure.

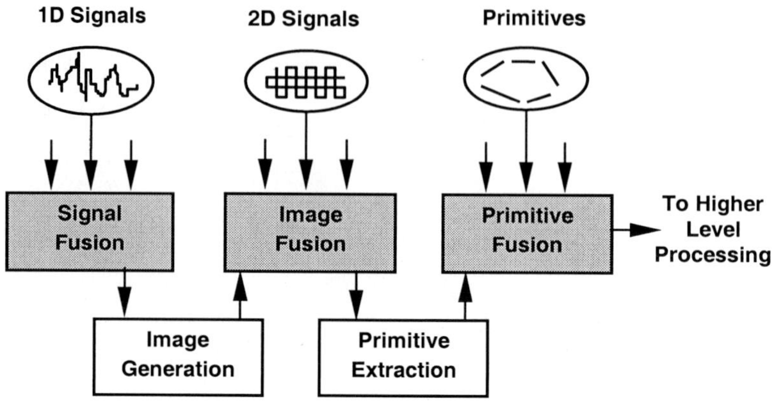

Fig. 3. Multi-level fusion concept. Each level integrates information of the same nature.

The *signal level* is a low level one as it usually represents single- or multidimensional signals. A suitable fusion method is signal estimation (filtration) which generally reduces the dispersion of the measured values. Fusion on this level can be used in real-time applications and can be considered as just an additional step in the global processing of signals.

On the *pixel level* the 2D data (images) are created and therefore it is used to fuse this sort of information. The fusion can be carried out by image estimation or pixel attribute combination [9], [10]. An example of this can be a task of fusing edge pixels (pixels attributed with high gradient value), this could help to detect syntactic features (e.g. line segments) for WM. The fusion process increases performance in solving image processing tasks [4].

The *syntactic feature (primitive) level* already represents the medium level and is based on features which can be extracted from signals and images [5]. An illustrating example: a fusion of all straight-line segments with similar direction at one world location (feature attribute). Suitable fusion methods incorporate the methods of geometrical and temporal correspondence and the feature attribute combination. As result, the fusion reduces processing time and increases feature measurement accuracy.

To be more specific: the GLbot system applies at this point two methods for feature extraction [9]. The goal is to interpret the data obtained from sonar measurements of the robot's environment as geometric map primitives. The extracted primitives can be used to update the robot position and the global map. This is the process how the robot can explore unknown environment. The first algorithm gathers sonar measurements into sensor-based map and using computer-vision techniques and morphological operations extracts geometric description of detected obstacles. The other method can be applied when the border tracking algorithm is activated. This method uses the raw sonar data as the input because obstacle border is well detected during the border tracking process.

2.3 Sensor Sources for Multiple Feedback Loops

Various input sensors in a multisensor mobile robot are used for different purposes, together setting up a perception subsystem. The perception provides three kinds of outputs.

One function is to support the collision avoidance subsystem by the information on region occupancy.

Another function provides the necessary data for updating of local maps which contain knowledge on the temporary neighborhood of the vehicle.

As shown in the Fig. 3, some of the sensors are used to match sensor data with the current content of the WM and afterwards to update the content of the WM to reflect the matching results. The other ones serve for landmark recognition which serves for robot position determination. The last kind of data exploration feeds the obstacle detection part which is responsible for the avoidance of collisions.

The degree of integration and fusion of sensor data required for each of these tasks can substantially differ.

The simplest approach to fusion in position location is trajectory integration. This calculates the vehicle's position from accumulated rotation and shift motions as determined by internal sensors such as an odometer and/or a gyro. Because of the nature of the always present inaccuracies of any sensors, the total error of position estimation accumulates as the vehicle moves. In order to reduce this cumulative error most of mobile robots periodically determine the location of some external landmark. Comparison of the position provided by the landmark with the one determined by internal sensors therefore keeps the location error within acceptable bounds.

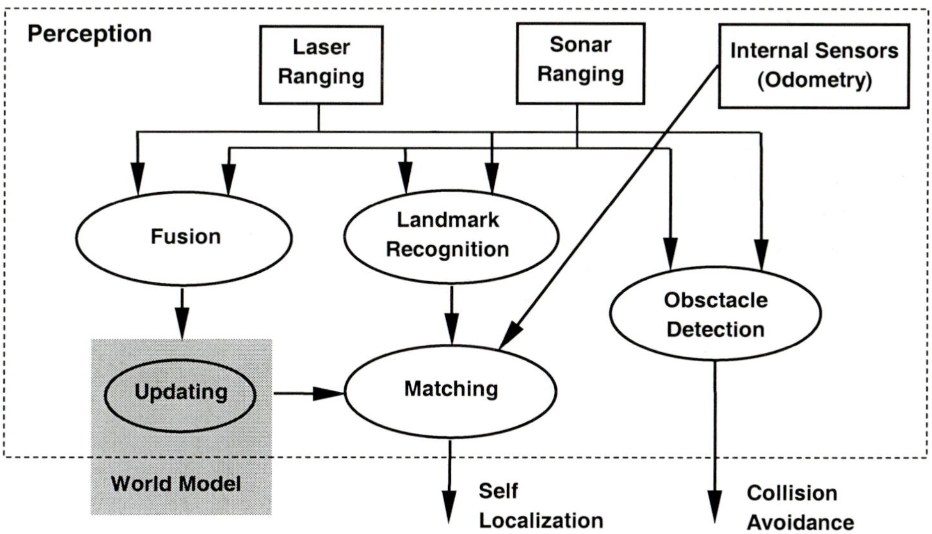

Fig. 4. Simplified perception function used by the GLbot system for navigation.

The WM data matching and updating task is another form of a fusion process which integrates particular sensor data with the world model. This requires that the sensor information and any associated measure of its uncertainty corresponds to the representation used in the world model. Only in that case the integration can take place. Information from different sensors of similar nature (e.g. laser range finder, stereo vision) can be fused before reaching the matching task in order to reduce the communication flow or complexity of the matching process.

Knowledge Handling. Nearly any entity (feature) segmented from the sensor data can be matched with the world model content. This can be done by matching of a feature derived from the sensor data to primitives obtained from the world map. The degree of similarity of the previous states if the hypothesis on the world state (primitives derived from sensor data) was a mismatch and/or uncertainty infiltrated from the sensor itself. Frequent successful matches vote for a world model updating.

The world model-to-sensor data matching and the model updating tasks require that the sensor information is self-evaluated by some measure of uncertainty. Only in that case some integration algorithms can be designed.

One possible integration approach leads towards building knowledge about space occupancies and certainty measures in the terms of their probabilities. Another area for this class of methods includes usage of a voice input by a robot. This is used e.g. for refinement of the already discovered knowledge or it supports conversions between symbolic and geometric world models (model fusion) through assistance.

3 Design Strategies

3.1 Sensing Strategies

The sensing strategies and namely the data fusion methods under consideration determine the applicability of the sketched concept of the robot. However, for each specific goal of the navigation task, a series of particular decisions concerning:

- particular integration of functions of the sensing subsystem,
- data representation levels to be explored for each functionality,
- data integration strategies (when and where to match different internal WMs?)

should be done by the designer. The robot perception subsystem is designed to fulfill three basic functions, namely:

- to match sensor data with the current content of the global WM, through which the local maps can be up-dated,
- to enable the robot's position and heading determination: a landmarking mechanism is used for this purpose,
- collision avoidance in the case of dynamically changing environment.

These functions cannot be implemented by quite independent, totally separated subsystems: some degree of integration of knowledge/data in the sketched functionality schema is, of course, required.

3.2 Workspace Representation and Planning Strategies

The knowledge about the robot's workspace is carried by the world map module containing the *world model* (WM) and is tightly bound with the global path *planner*. The WM part is responsible for storage and management of a database that stores primitives representing obstacles. This is called a *global map* and includes procedures which allow the other processes to conduct necessary operations with the map content. The most common operations are e.g. *addition* and *removal* of obstacles and/or their components (segments of borders), providing a *list of visible obstacle boundaries* from a given standpoint, etc. As a compromise between the data compression and problems appearing by extraction of map primitives, all world obstacles can be approximated by line segments. This simplification makes the search process for primitives (or a fusion of the sensor data into primitives) much easier as the model (the line segment) is known.

This brings an estimated error in obstacle representation which is acceptable for sufficient accuracy of the navigation and planning processes. On the other hand, the main advantage of the chosen representation is seen in avoidance of a recognition process which would be needed by direct segmentation of the whole obstacles. Building of the world model step-by-step from border segments copes with the visibility feature well. This allows updating of the world model from similar sensor aspects in smaller portions (and does not require discovery actions like „going around the whole obstacle"). The global map is stored as a simple visibility graph defined on vertexes of obstacles.

The planner is responsible for global path planning, e.g. it finds a free path from the starting point to the goal. The planning is done not in the configuration space but over the map visibility graph. Global plans consist of sequences of certain sub-goals. For suitable sub-goal points can serve corner points of obstacles after being slightly expanded to create a safety corridor along rigid borders. The plan can be generated by a standard A* search algorithm.

The both threads are integrated within one module because the path planning method strongly depends on the used map abstraction. Further execution of the obtained plan is usually done in cooperation with the *navigator* (local planner) and a *pilot* (execution driver) which form together the *collision avoidance* subsystem.

3.3 Grid Map and Local Path Planning

This process is usually divided into two tightly cooperating threads being responsible for sensor data integration and local map management.

The grid map receives measured data from the sonar system and the odometer (as position measurement) and integrates it into a *certainty grid* [10]. The second thread uses the grid map for local path planning making-use of the *potential field approach* [3]. The local (grid) planner receives a global path plan and serves for a first-instance method which attempts to refine the global plan. Reading the actual position of the robot the planner aims to find the safest path from the current position to the nearest sub-goal point on the global path. The algorithm also evaluates a measure of collision possibility as a sum of occupancy probabilities along the proposed local path. The possibility measure is passed to the decision-making process later. It chooses between the local planning (sensor-based guidance) methods of the grid-based and the border-tracking approaches.

3.4 Border Tracking

The border-tracking module is the second-instance method for refinement of global plans during the execution. This activity is invoked after the decision is done that for the local planning can not be applied the certainty grid planner. This situation happens when the robot path is „jammed" by an unexpected obstacle with no substitute path left and/or the sonar measurements are heavily ambiguous due to badly structured environment.

The central idea of this approach is to trace the border of the colliding obstacle unless any next local sub-goal is reached on the original path. The tracer uses a range scan provided by the range-finder (sonar or lidar). The first step is a search for the nearest obstacle followed by appropriate control action that brings the robot into parallel orientation with respect to the obstacle border (border approaching maneuver). For initial direction of tracking the closest one to the original path is chosen. The basic idea of the algorithm can be derived from the BUG and VISBUG algorithms [12]. The border following process can rely on measurements provided by the distance measurements from the robot's sides. Measurements in other directions serve for checking whether there is no smaller distance to any other obstacle than to the followed one.

4 System Integration in the GLbot

As we already mentioned, there are multiple path planning strategies used in the experimental GLbot [16]: the global path planning based on the global WM, the local path planning based on the grid map and the border tracking based on direct exploration of low-level data being supported by the landmark subsystem.

The data gathered from the (changing) environment and used for the local path planning should be - earlier or later - automatically integrated into the global WM. If this is not possible, the robot will be not able to learn in the conditions of learning environment and thus forced to handle every new change of the environment just only by the collision avoidance mechanism. The integration of the local data into the global WM has a crucial impact on integration of the navigation process as a whole.

This integration can be carried out - in principle - by applying one of the following two approaches. It is possible either to use range data and/or the grid map for extraction of primitives (edges and line segments) building the global WM or by direct matching single range scans - whenever is the collision avoidance system in active use - to the global WM.

As the former approach leads towards the application of image segmentation that has been found as highly problem-dependent and therefore not providing satisfying results here. The latter method avoids segmentation and uses the fact that an obstacle border becomes well determined during a border tracking process. The only task is to record the appropriate tracking curve and crack it into straight-line segments which can be immediately inserted into the world map. The collision-based updating process has been found as very robust.

Another activity of the discussed module is the support of the position and heading updating which aims to keep positioning errors within limited bounds. This has been based on matching single sonar range scans with the global map. The used position updating is a two-step algorithm: if the match is sufficiently good (the correlation coefficient exceeds a certain threshold), it implies examination whether it is not very distant (in the terms of rotation or translation). The latter condition decreases the danger of dropping the method into a local extreme caused e.g. by environment symmetries.

5 Functional Module Integration

The software equipment of an autonomous robot is typically organized as a set of independently developed modules [16]. E.g. the experimental mobile platform GLbot uses more than 20 simultaneous processes.

From the point of view of Artificial Intelligence the multi-agent approach can be used for software integration with advantage.

The multi-agent architecture of the set of modules enables to make structural changes (like adding a new module, changing the content of some of the modules) in a simple way.

Multi-agent systems (ideally) consist of a set of cooperating elements (=agents) which communicate among themselves in the "peer-to-peer" way using a specific ACL (Agent Communication Language).

Although there is no unified, widely accepted definition of an agent, it is possible to identify a minimal list of basic (nearly obligatory) features of the agents: autonomy, ability to cooperate, and ability to learn. In the case the software modules are equipped with these properties, they can be considered as agents creating a multi-agent community. Such a community of well internally organized and coordinated agents exhibits a strongly integrative behavior. In the other words: The software modules being extended into agents can be efficiently integrated using appropriate ACL. It is very often spoken about the so called *integration architecture* provided by the multi-agent approach. The ACL should meet the following requirements:

- the ACL must be general enough to enable communication among differently specialized agents,
- it must be open to enable enhancement of its expressing power,
- the ACL should make it possible to transfer different types and formats of data,
- the ACL should provide mechanisms to transfer data processing methods that will execute on various computing platforms hosting the agents.

All the current ACL specifications are based on the idea of declarative statements exchanged between the agents. The commonly discussed standard for this communication are based on the KQML philosophy [8] or FIPA standards. KQML is, in fact, an external language describing the expressions that compose the ACL (performatives). The internal language of describing the format of the expressions is the KIF (Knowledge Interchange Format). KIF is a prefix version of the first order predicate calculus with some features added for enhanced expressiveness. The entire inter-agent communication mechanism is based on a domain and task specific dictionary of terms that are well understood by the communicating agents.

The integration architectures usually consist of a small number of well-developed agents (=software modules) with a clearly defined highly specialized functionality. Each agent consists, as a rule, of a *functional body* (usually a stand-alone module with a well-defined functionality) and a *wrapper* which is responsible for agent's engagement in the agents' community (see Fig. 4).

The wrapper recognizes and translates relevant parts of the inter-agent communication into the instructions for the activity of the functional body and mediates the results of the body activity into the agents' community. Thus, the wrapper contains a representation of the agent's behavioral model. Plenty of such models have been developed recently, let's mention e.g. the twin-base model [7], or tri-base model [15].

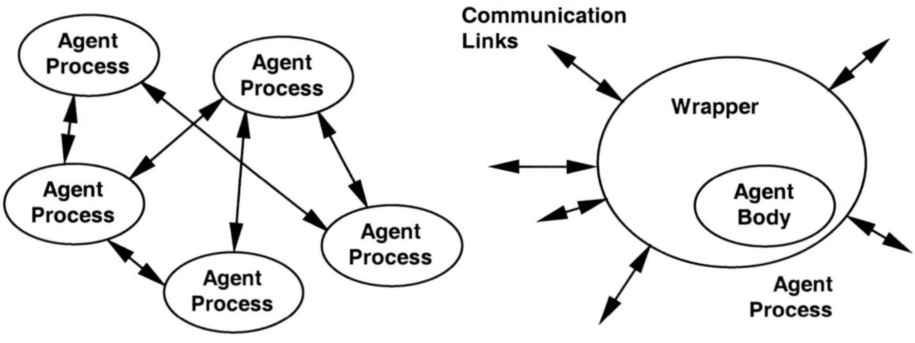

Fig. 5 Inter-agent communication (left) and an agent architecture consisting of the body and the wrapper (right).

The requirements on coordination and control in the integration architecture are typically very high, especially if it is aimed to reduce the volume of transferred data. Both the functionality and performance of a multi-agent community strongly depend on performance of each of the community members. The failure or loss of efficiency of each of the agents should be detected as soon as possible. It doesn't seem realistic that each agent should identify its own suspicious behavior. This can be more easily detected by an independent observer. That is why there has been introduced a new type of agents observing and processing just the information on behavior and mutual cooperation of a specific group of agents inside the community. As these "higher-level" agents contain and process mainly knowledge about behavior and activities of the other agents, they are called meta-agents or *super-agents* [19].

6 Conclusions

Many information feedback loops of diverse nature can be found within the particular robotics systems. These loops explore information based on sensor data gathered by sensors of different nature (the GLbot uses in the first place sonar, laser, odometer, TV camera, etc.).The raw data from specific sensors are usually pre-processed and then merged, fused or integrated with the data provided by the other sensors. If the data are gained from the sensors working on the same physical principles, it is possible to integrate them already on a lower level. The more diverse nature of sensor data, the higher level data fusion - as a part of the more global world model - is required.

The information feedback loops used e.g. in mobile robot platforms differ in the tasks they are targeted at. The schematic tree of typical data fusion approaches is sketched in the Fig. 3, but other choices are possible, too.

The experience of the Gerstner Lab team gathered in the development of the multi-agent system ProPlanT for production planning has been found useful in designing the multi-agent solution for the GLbot software integration [14], [15]. A simple *agentification process* converting separate modules into agents by adding wrappers has been applied.

In the first version, the behavioral models (located in the agents' wrappers) are surprisingly simple, but they can be significantly extended when much more sophisticated integration tasks would be solved. The efficiency and openness of the multi-agent solution was documented by different experiments when some functional modules (agents) were easily added or deleted. E.g., new agents as the agent for collision avoidance (IDA sensor-based planning) and a lidar-data based agent for WM updating have been added. On the other hand, the current GLbot setup enables easy suspension (or deletion) of old versions of agents without any danger for the GLbot global activities during these experiments. Of course, there are some constraints in the experiments, as not „to kill“ the „core“ agents which are crucial for the life of the GLbot (e.g. the pilot, the planner, or the local map updating agent).

Our experience shows that the standardization issues seem to be very important for software/algorithms reusability. When using a widely accepted communication standard (e.g. FIPA) as well as a good data organization standard (e.g. MIMOSA format), the exchange of software modules (agents) among different research teams and their simple integration into the global architectures will become quite natural.

The integration architecture of multi-agent systems has been proven as a suitable scheme for integrating of software modules of diverse nature. This architecture is open enough to enable a systematic growth of the robot software equipment and its maintenance.

Acknowledgement. This research has been carried out with the support of The Gerstner Laboratory for Intelligent Decision Making and Control (VS 96047) and the Decision Making and Control for Industrial Production (VZ MSM212300013) grants.

References

1. Allen, J., Hendler, J. and Tate A.(Eds.): Readings in Planning, Morgan Kaufmann Publishing Inc., Palo Alto (1990)
2. Abidi, M. A.: Data Fusion in Robotics and Machine Intelligence, Academic Press, New York (1992) 546 p.
3. Azarm, K. and Schmidt G.: Integrated Mobile Robot Motion Planning and Execution in Changing Indoor Environments, in: Proceedings of the IEEE International Conference Intelligent Robots and Systems, IROS'94, Germany, Vol.1 (1994) 298-305
4. Bray, Šonka, M., Hlaváč, V.: Image Processing and Machine Vision, Chapman-Hall Publ., 1993
5. Belknap, R., et.al.: The Information Fusion Problem and Rule-Based Hypotheses Applied to Aggregation of Image Events, in: Proceedings of the IEEE Conference on Computer Vision and Pattern Recognition, Miami Beach (1988) 227-234
6. Cesarone, J.: Navigation Problem in Manufacturing, Intelligent Design and Manufacturing (A. Kusiak, ed.), John Wiley & Sons, New York (1992) 469-490
7. Cao, W., Bian, Ch.-G. and Hartvigsen, G.: Achieving Efficient Cooperation in a Multi-Agent System: The Twin-Base Modeling, in: Cooperative Information Agents (Kandzia P., Klusch M. eds.), LNAI 1202, Springer-Verlag, Heidelberg (1997) 210-221
8. Finin, T, at al.: Specification of the KQML Agent Communication Language. Available from http://www.cs.umbc.edu/kqml/kqmlspec.ps (1995)
9. Kulich, M., Štěpán, P., Přeučil, L.: Feature Detection and Map Building Using Ranging Sensors. In: Intelligent Transportation Systems. Vol.1. Tokyo: The Institute of Electrical Engineers of Japan, Tokyo, Japan (1999) 201-206
10. Kulich, M., Štěpán, P., Přeučil, L.: Knowledge Acquisition for Mobile Robot Environment Mapping. In: Database and Expert Systems Applications. Vol.1. Springer-Verlag, Heidelberg:, LNCS No. 1677 (1999) 123-134
11. Lažanský, J.: Practical Applications of Planning Tasks. In: Advanced Topics in Artificial Intelligence, Lecture Notes in Artificial Intelligence, No.617, Springer-Verlag, Heidelberg (1992)
12. Lumensky, J. V. and Stepanov, A. A.: Path Planning Strategies for a Point Mobile Automaton Moving Admits Unknown Obstacles of Arbitrary Shape, Algorithmica (1987)
13. Matthies, L., et.al.: Integration of Sonar and Stereo Range Data Using a Grid-Based Representation, in: Proceeding of the IEEE Conference on Robotics and Automation (1988) 727-733
14. Mařík, V., Štěpánková, O. and Lažanský, J. : Role of Qualitative Reasoning in a Multi-Agent System, in: Computer Aided Systems Theory (Pichler F., Moreno-Diaz, R., Eds.), LNCS 1333, Springer-Verlag, Heidelberg (1997) 380-393
15. Mařík, V., Pěchouček, M., Lažanský, J., Roche, Ch.: PVS'98 agents: structures, models and production planning application. Int. Journal *Robotics and Autonomous Systems*, vol. 27, No. 1-2 (1999) 29-44

16.Štěpán, P., Král, L., Kulich, M., Přeučil, L.: Open Control Architecture for Mobile Robot. In: Proc. of 14th World Congress of IFAC. Exeter-England: Elsevier Science, Beijing, China (1999) 163-168

17.Suh, S. and Shin, K.: A Variational Dynamic Programming Approach to Robot-Path Planning with a Distance-Safety Criterion, IEEE Journal of Robotics and Automation, Vol.4, No.3 (1988) 334-349

18.Shapiro, S.C. (Ed.): Encyclopedia of AI, John Wiley & Sons Publ., New York (1990)

19.Zhong, N., Kakemoto, Y., and Ohsuga, S.: An Organized Society of Autonomous Discovery Agents, in: Cooperative Information Agents (Kandzia P., Klusch M. eds.), LNAI 1202, Springer-Verlag, Heidelberg (1997) 183-194

20.Zelinski, A., and Yuta S.: Reactive Planning for Mobile Robots Using Numeric Potential Fields, Intelligent Autonomous Systems IAS-3 (1993) 84-93

Multi-processor Design of Non-linear Robust Motion Control for Rigid Robots

Theodor Borangiu, Mitică Manu, and Virginia Ecaterina Oltean

"Politehnica" University of Bucharest, Faculty of Control and Computers
313, Spl. Independentei, sector 6, RO-77206
{borangiu, mitica}@icar.cimr.pub.ro, oltean@aii.pub.ro

Abstract. The research activities carried out in the framework of the proposed paper have been directed to the development of nonlinear control techniques, algorithms and architectures for the free motion of robots having rigid links, that provide the following global control performances: *global asymptotic stability*, *precision for time varying trajectory tracking*, *robustness* at external disturbances, nonlinear coupling between motion axes and uncertainties of the dynamic model of the manipulator with drives structure, for closed-loop robot systems with $n \geq 5$ degrees of freedom. Thus, there is proposed a solution for nonlinear motion control based on adaptive control techniques. The control method that is proposed provides high dynamic performances for the ensemble manipulator - drive systems - internal transducers - controller - user interface, that is: high displacement speed and tracking precision on desired time varying trajectories, rejection of disturbances of the type: variable masses carried by the arm, modeling errors, measuring noise. In the framework of this paper, the authors designed and simulated a library of procedures, algorithms and software control modules to be further implemented in a generic parallel multiprocessing architecture, as a robot controller. In this respect, the structural design was directed towards VME - based bus oriented multimaster structures.

1 Introduction

The "*motion controller*" unit for robot systems is composed from two parts:

- the *trajectory generator* (planner), the inputs of which are represented by the path specifications and constraints, and whose outputs are the time sequences of desired positions, velocities and accelerations of the n joints of the manipulator: $\boldsymbol{q}_{\mathrm{d}}, \dot{\boldsymbol{q}}_{\mathrm{d}}, \ddot{\boldsymbol{q}}_{\mathrm{d}}$;
- the *trajectory controller* (tracking module), the inputs of which are connected to the outputs of the trajectory generator, its outputs being the set of commands $\boldsymbol{u}$ for the n motors either as velocity signals if the actuators are considered as velocity-controlled generators (V-CG), or as torque signals if the actuators are considered as torque-controlled generators (T-CG).

Considering that, for free motions, there are no external forces acting on the end effector, and neglecting the static and dynamic friction forces in the joints, the generic

F. Pichler, R. Moreno-Díaz, and P. Kopacek (Eds.): EUROCAST'99, LNCS 1798, pp. 224–238, 2000.

dynamic nonlinear model of the system manipulator and drives is given by the equation:

$$\mathbf{B}(q)\ddot{q} + \mathbf{C}(q,\dot{q})\dot{q} + \mathbf{F}_v\dot{q} + \mathbf{g}(q) = \tau\,, \tag{1}$$

where $\mathbf{B}(\boldsymbol{q})$ is the inertia matrix of the system, $\mathbf{C}(\boldsymbol{q},\dot{\boldsymbol{q}})$ is the matrix that contains the Coriolis and centrifugal effects, $\mathbf{F}_v$ is the matrix of viscous forces in the joints, and $\mathbf{g}(\boldsymbol{q})$ represents the vector of gravitational forces in the system. For the plant model (manipulator and drives - MD), τ represents the vector of commands (active torques in the joints) that determine a certain evolution of the ($2n$ x 1)-dimensional state vector $[\boldsymbol{q}^{\mathrm{T}} \quad \dot{\boldsymbol{q}}^{\mathrm{T}}]^{\mathrm{T}}$, where $\boldsymbol{q},\dot{\boldsymbol{q}}$ are respectively the vectors of position and velocity in the robot's joints.

By extracting from $\mathbf{B}(q)$ the diagonal, time invariant coefficients representing the average joint inertia values - $\overline{\mathbf{B}}$, the controlled plant MD can be decomposed in two subsystems:

- one *linear* and *decoupled* system, having $\boldsymbol{u}$ (or τ) as input and the set of positions and velocities $\boldsymbol{q},\dot{\boldsymbol{q}}$ as output, as each component of τ influences through $\overline{\mathbf{B}}$ only the corresponding component of $\boldsymbol{q}_{\mathrm{m}}$ (the vector of angular positions of motors) and hence of $\boldsymbol{q}$;
- a *nonlinear* and *coupled* system having the set $\boldsymbol{q}_{\mathrm{m}},\dot{\boldsymbol{q}}_{\mathrm{m}},\ddot{\boldsymbol{q}}_{\mathrm{m}}$ as input and the signal $\boldsymbol{p}$ as output.

When direct drive systems are used, and/or when high velocities in the operational space are required, considering still the effects of the nonlinear coupling p as simple disturbances strongly reduce the dynamic performances of the MD system, especially the tracking error along the imposed trajectory.

Consequently, the motion control algorithms must rely on a certain degree of knowledge of the dynamic model of the manipulator: $\mathbf{B}(.)$, $\mathbf{C}(.,.)$, $\mathbf{F}_v$, $\mathbf{g}(.)$, or of its linear regressor model $\mathbf{Y}(.,.,.)$ that contains the set of pn parameters π that specify the mechanical structure, so that they can compensate directly the nonlinear coupling terms between the axes. This approach leads necessarily *to centralized multivariable nonlinear control structures.*

For this category of motion control techniques and algorithms, a real progress in worldwide research was possible in the last years, due to the availability of new data processing microstructures of high speed and storage capabilities, either at the level of specialized components (DSP, transputers), or at the level of new topologies of parallel, bus-oriented computing architectures.

In designing the nonlinear, modular, user configurable motion controller, the Lyapunov design method has been used in order to determine the stability propriety in equilibrium points, without solving the state equations. This method is based on associating an energy function to the autonomous nonlinear systems that express the manipulator's dynamics: the manipulator itself (its dynamic model) together with the actuating and transmission elements. There will be provided the condition that in each state of the global system, except for its equilibrium state, the energy decreases along the system's trajectories until the minimum is reached in the equilibrium state; thus, asymptotic stability (global or semi global) will be provided for the closed loop control system.

In what concerns the main objective of trajectory tracking, the adaptive control structure that has been developed provides an imposed path tracking accuracy (both in position and in velocity), even in the presence of uncertainty situations due to: modeling errors, numeric computation truncations, unknown transported masses, noise in joint position and velocity measurements.

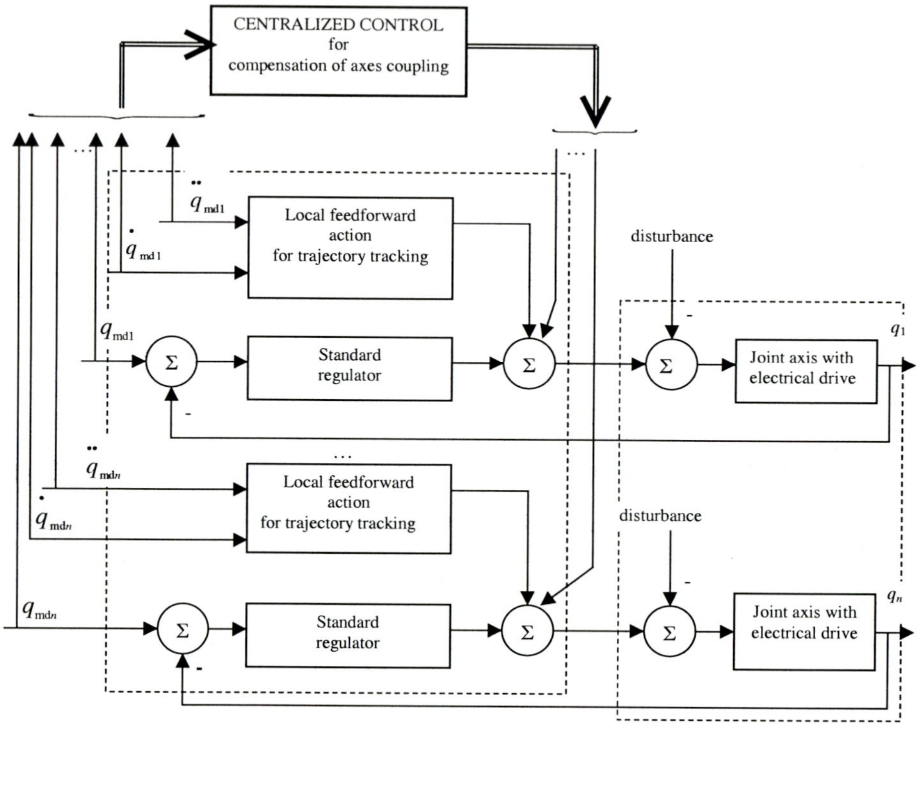

Fig. 1. Multi-microprocessor parallel pipe-lined architecture for advanced motion control

For the synthesis of the adaptive control, the authors designed a gradient – type algorithm for the updating of the estimates of model parameters, with additional facilities:

- avoiding the computation of the inverse of the estimated inertia matrix of the system (operation for which, in general, it is difficult to guarantee boundness, in the presence of model uncertainties);
- avoiding the dependence of the regressor associated to the parametric model on the current values of acceleration $\ddot{q}(t)$, in order to eliminate the difficulties of estimation, the time delays and the important errors that appear during the evaluation of this signal.

The structural synthesis of the adaptive controller uses an algorithm that is based on the passivity property of rigid manipulators, which simplifies the design procedure. With respect to this procedure, the control law contains two terms: the first one for compensating the nonlinear dynamics of the manipulator, and the second one for providing the linear PD action.

The utilization of this second term suggests a similitude of the adaptive controller with the PD one, thus increasing its robustness.

The proposed architecture, presented in the paper contains two functional areas, which are connected to the VME bus:

- a *centralized control zone*, built around a master one-board microcomputer unit equipped with mathematical coprocessor that cooperates with the vision processor in order to generate the desired trajectory data, to compensate the nonlinear couplings between the axes and to adaptively update the estimates of the manipulator's model parameters,
- a *decentralized control zone distributed among several intelligent slave processors*, that generate commands at the individual joint levels – the motion regulators, with local feedforward actions for accurate tracking of trajectory components for each joint. The two zones operate in parallel, in a pipelined configuration (fig.1).

2 Adaptive Control with Feedback Linearization

Adaptive motion control uses the linear parameterization property of manipulator dynamics models with respect to a suitable set of parameters included in the (p x 1) vector π:

- mass of links (1 per joint - link pair (JLP));
- components of the first moment of inertia (3 per JLP);
- components of the inertia tensor $\hat{\mathbf{I}}_i$ (6 per JLP);
- moments of inertia of rotor i (1 per JLP);
- viscous and static friction coefficients (2 per JLP).

The $p \leq 13n$ - dimensional parameter vector π is constant, as its components are invariant characteristics of the mechanical and actuating parts of the manipulator; $\boldsymbol{n}$ is the number of the degrees of freedom.

The parameterization property points out that, although the equations of motion of the JSDM are nonlinear, the parameters of interest such as link masses and moments of inertia, a.s.o., appear as coefficients of known function of the joint variables $\boldsymbol{q}$, $\dot{\boldsymbol{q}}$, $\ddot{q}$; by defining each coefficient as a separate parameter, a linear relationship results, so that the dynamic equations of the manipulator and drives can be written in the form:

$$\mathbf{B}(q)\ddot{q} + \mathbf{C}(q,\dot{q})\dot{q} + \mathbf{F}_v\dot{q} + \mathbf{g}(q) = \mathbf{u} = \mathbf{Y}(q,\dot{q},\ddot{q})\,\pi \tag{2}$$

$$\tau = u$$

in which the *regressor* **Y** is a (n x p) matrix of known functions, having superior triangular form:

$$\mathbf{Y} = \begin{bmatrix} \mathbf{y}_{11}^{\mathrm{T}} & \mathbf{y}_{12}^{\mathrm{T}} & \cdots & \mathbf{y}_{1n}^{\mathrm{T}} \\ \mathbf{0} & \mathbf{y}_{22}^{\mathrm{T}} & \cdots & \mathbf{y}_{2n}^{\mathrm{T}} \\ \vdots & & \ddots & \\ \mathbf{0} & & \cdots & \mathbf{y}_{nn}^{\mathrm{T}} \end{bmatrix} \tag{3}$$

where $\mathbf{y}_{ij}^{\mathrm{T}}$, $j \geq i$ (1 x 13) - dimensional block matrices and π is the p - dimensional parameter vector.

The choice of parameters in (2) is not unique; also, the dimension of the parameter space depends on the particular choice of parameters. In general, the degree of knowledge of the individual parameters π_i differs; the *estimate* of the exact, constant parameter vector π will be denoted by $\hat{\pi}$, and a measure of the knowledge about the system's parameters will be expressed by the *parameter of uncertainty* $\tilde{\pi}$:

$$\tilde{\pi} = \pi - \hat{\pi} \tag{4}$$

We assume that the estimates ($\hat{\mathbf{B}}, \hat{\mathbf{C}}, \hat{\mathbf{F}}_{\mathrm{v}}, \hat{\mathbf{g}}$) have the same functional form as (**B**, **C**, $\mathbf{F}_{\mathrm{v}}$, **g**) with estimated parameters, i.e.:

$$\hat{\mathbf{B}}(q)\ddot{q} + \hat{\mathbf{C}}(q,\dot{q})\dot{q} + \hat{\mathbf{F}}_{\mathrm{v}}\dot{q} + \hat{\mathbf{g}}(q) = \mathbf{Y}(q,\dot{q},\ddot{q})\hat{\pi} \tag{5}$$

with $\hat{\pi}$ the vector of estimated parameters. According to [1], the *inverse dynamics control law* for the system (1), (2) will be taken as :

$$\boldsymbol{u} = \hat{\mathbf{B}}(q)\left[\ddot{q}_{\mathrm{d}} + \mathbf{K}_{\mathrm{P}}(q_{\mathrm{d}} - q) + \mathbf{K}_{\mathrm{D}}(\dot{q}_{\mathrm{d}} - \dot{q})\right] + \hat{\mathbf{C}}(q,\dot{q})\dot{q} + \hat{\mathbf{F}}_{\mathrm{v}}\dot{q} + \hat{\mathbf{g}}(q) \tag{6}$$

Denoting by $\tilde{q} = q_{\mathrm{d}} - q$ the position tracking error, the ($2n$ x 1) vector $\boldsymbol{e}$:

$$\boldsymbol{e} = \begin{bmatrix} \tilde{q} \\ \dot{\tilde{q}} \end{bmatrix} \tag{7}$$

will be taken as state vector.

As compared with the robust control approach, which adds robustifying term $\boldsymbol{v}_{\mathrm{rob}} = \boldsymbol{w}$, to the control law (6), in order to compensate for the uncertainty ($\hat{\mathbf{B}}$, $\hat{\mathbf{C}}$, $\hat{\mathbf{F}}_{\mathrm{v}}$, $\hat{\mathbf{g}}$) in the dynamic model, the adaptive inverse dynamics method leaves the nominal control:

$$\boldsymbol{v}_{\mathrm{nom}} = \ddot{q}_{\mathrm{d}} + \mathbf{K}_{\mathrm{D}}\dot{\tilde{q}} + \mathbf{K}_{\mathrm{P}}\tilde{q} \tag{8}$$

input unchanged, but updates at a certain rate the estimates terms ($\hat{\mathbf{B}}, \hat{\mathbf{C}}, \hat{\mathbf{F}}_{\mathrm{v}}, \hat{\mathbf{g}}$). By substituting (6) in (3) and accounting for the model uncertainty:

$$\tilde{\mathbf{B}} = \mathbf{B} - \hat{\mathbf{B}},\ \tilde{\mathbf{C}} = \mathbf{C} - \hat{\mathbf{C}},\ \tilde{\mathbf{F}}_{\mathrm{v}} = \mathbf{F}_{\mathrm{v}} - \hat{\mathbf{F}}_{\mathrm{v}},\ \tilde{\mathbf{g}} = \mathbf{g} - \hat{\mathbf{g}} \tag{9}$$

it results:

$$\mathbf{B}(q)\ddot{q} + \mathbf{n}(q,\dot{q}) = \hat{\mathbf{B}}(q)\,[\ddot{q}_d + \mathbf{K}_D\dot{\tilde{q}} + \mathbf{K}_P\tilde{q}\,] + \hat{\mathbf{n}}(q,\dot{q}) \tag{10}$$

where for simplicity:

$$\hat{\mathbf{n}}(q,\dot{q}) = \mathbf{C}(q,\dot{q})\dot{q} + \mathbf{F}_v\dot{q} + \mathbf{g}(q) \tag{11}$$

Adding and subtracting the term $\hat{\mathbf{B}}(q)\ddot{q}$ in the left-hand side of (10), and in view of (9), it yields:

$$[\mathbf{B}(q) - \hat{\mathbf{B}}(q)]\ddot{q} + \tilde{\mathbf{n}}(q,\dot{q}) = \hat{\mathbf{B}}(q)\,[\ddot{\tilde{q}} + \mathbf{K}_D\dot{\tilde{q}} + \mathbf{K}_P\tilde{q}\,]$$

equivalent to:

$$\hat{\mathbf{B}}(q)\,[\ddot{\tilde{q}} + \mathbf{K}_D\dot{\tilde{q}} + \mathbf{K}_P\tilde{q}\,] = \tilde{\mathbf{B}}(q)\ddot{q} + \tilde{\mathbf{n}}(q,\dot{q}) \tag{12}$$

The right hand side of equation(12) is equal to $\mathbf{Y}(q,\dot{q},\ddot{q})\tilde{\pi}$ in view of (5), and assuming that $\hat{\mathbf{B}}$ is invertible, it gives:

$$\ddot{\tilde{q}} + \mathbf{K}_D\dot{\tilde{q}} + \mathbf{K}_P\tilde{q} = \hat{\mathbf{B}}^{-1}(q)\ \mathbf{Y}(q,\dot{q},\ddot{q})\tilde{\pi} \tag{13}$$

By denoting:

$$\hat{\mathbf{B}}^{-1}\mathbf{Y} = \Phi \tag{14}$$

it results:

$$\ddot{\tilde{q}} + \mathbf{K}_D\dot{\tilde{q}} + \mathbf{K}_P\tilde{q} = \Phi\,\tilde{\pi} \tag{15}$$

Next, the error dynamics system of equation (15) is re-arranged in state-space form:

$$\begin{aligned}\dot{\tilde{q}} &= \dot{\tilde{q}}\\ \ddot{\tilde{q}} &= -\mathbf{K}_D\dot{\tilde{q}} - \mathbf{K}_P\tilde{q} + \Phi\,\tilde{\pi}\end{aligned}$$

which, by adopting the same notations for the time-invariant matrices:

$$\mathbf{A} = \begin{bmatrix} \mathbf{0} & \mathbf{I} \\ -\mathbf{K}_P & -\mathbf{K}_D \end{bmatrix};\ \mathbf{B}_1 = \begin{bmatrix} \mathbf{0} \\ \mathbf{I} \end{bmatrix} \tag{16}$$

with the positive definite proportional and derivative gain (n x n) diagonal matrices $\mathbf{K}_P$, $\mathbf{K}_D$ included in **A**, becomes:

$$\dot{e} = \mathbf{A}\,e + \mathbf{B}_1\,\Phi\,\tilde{\pi}\,. \tag{17}$$

Our task is to derive an algorithm for updating the estimate $\hat{\pi}$ of the parameter vector π and, by choosing a positive definite quadratic form as Lyapunov function, candidate to prove that the overall system is globally asymptotically stable. To this purpose, a symmetric, positive definite matrix **P** is chosen and the Lyapunov equation

$$\mathbf{A}^{\mathrm{T}}\mathbf{Q}+\mathbf{Q}\,\mathbf{A}=-\mathbf{P} \tag{18}$$

is solved for **Q**. The unique solution is also symmetric, positive definite, which leads to the following choice for the Lyapunov function candidate:

$$\mathrm{V}(\boldsymbol{e},\tilde{\pi})=\boldsymbol{e}^{\mathrm{T}}\mathbf{Q}\,\boldsymbol{e}+\tilde{\pi}^{\mathrm{T}}\Gamma\,\tilde{\pi} \tag{19}$$

where Γ is a symmetric, positive definite (p x p) matrix. The time derivative of V along a trajectory of (8) is computed as:

$$\dot{\mathrm{V}}=\dot{\boldsymbol{e}}^{\mathrm{T}}\mathbf{Q}\,\boldsymbol{e}+\boldsymbol{e}^{\mathrm{T}}\mathbf{Q}\,\dot{\boldsymbol{e}}+2\tilde{\pi}^{\mathrm{T}}\Gamma\,\dot{\tilde{\pi}}$$

Substituting $\dot{\boldsymbol{e}}$ from (17) in the primary above form of $\dot{\mathrm{V}}$, and accounting for (18), it gives:

$$\dot{\mathrm{V}}=-\dot{\boldsymbol{e}}^{\mathrm{T}}\mathbf{P}\,\boldsymbol{e}+2\tilde{\pi}^{\mathrm{T}}[\Phi^{\mathrm{T}}\mathbf{B}_1^{\mathrm{T}}\mathbf{Q}\,\boldsymbol{e}+\Gamma\,\dot{\tilde{\pi}}\,] \tag{20}$$

In order to render $\dot{\mathrm{V}}$ negative definite, an obvious choice for the parameter update law is:

$$\dot{\tilde{\pi}}=-\,\Gamma^{1}\,\Phi^{\mathrm{T}}\mathbf{B}_1^{\mathrm{T}}\mathbf{Q}\,\boldsymbol{e}$$

or, in view of (4) and because π is a vector of time-invariant components:

$$\dot{\hat{\pi}}=\Gamma^{1}\,\Phi^{\mathrm{T}}\mathbf{B}_1^{\mathrm{T}}\mathbf{Q}\,\boldsymbol{e} \tag{21}$$

which makes the second term in the right-hand side of (20) zero, providing thus: $\dot{\mathrm{V}}(\boldsymbol{e})<0$, $\forall\boldsymbol{e}\neq\mathbf{0}$, $\dot{\mathrm{V}}(\mathbf{0})=0$, and $\dot{\mathrm{V}}$ is negative definite along all error system trajectories. Hence, the system is *stable* in the sense of Lyapunov, i.e., the origin $\boldsymbol{e}=\mathbf{0}$ is globally asymptotically stable for the nonlinear overall system whose error dynamics is expressed by equation (17).

Thus, $\boldsymbol{e}(t)$ and $\tilde{\pi}(t)$ are bounded. However, in order to prove rigorously that the tracking error does indeed converge to zero, several additional aspects must be considered:

1. It must be guaranteed that the function $\Phi=\hat{\mathbf{B}}^{-1}\mathbf{Y}$ remains bounded, which requires that the estimate $\hat{\mathbf{B}}(\cdot)$ never becomes singular, i.e. $\hat{\mathbf{B}}$ must be guaranteed to be invertible, possible by the use of projection in the parameter space. One way to ensure this is to impose to the estimate parameters $\hat{\pi}$ to be in compact, fixed region about the true parameters π. This implies that the border of the variance region for $\hat{\pi}$ must be *pre-defined*, which is equivalent to specify a priori the worst case estimates of the uncertainty $\tilde{\pi}$, $\sup\|\tilde{\pi}\|$. If the parameter estimate $\hat{\pi}$ should ever leave this fixed region, the estimation algorithm should be designed to automatically set $\hat{\pi}$ to the boundary of the region [2].
2. From (14) it results that Φ depends on the acceleration $\ddot{q}$ of the manipulator, and in order to execute the parameter updating according to the update law

(21), the acceleration needs to be measured, or at least evaluated indirectly by resorting to analog or digital filtering techniques applied to position (and velocity) measured signals $[\boldsymbol{q}^{\mathrm{T}} \quad \dot{\boldsymbol{q}}^{\mathrm{T}}]^{\mathrm{T}}$. Since $\boldsymbol{e}$ and $\tilde{\pi}$, and hence $\dot{\tilde{\boldsymbol{q}}}$, $\tilde{\boldsymbol{q}}$ and $\tilde{\pi}$ are bounded, the acceleration tracking error $\ddot{\tilde{\boldsymbol{q}}}$ and hence the acceleration $\ddot{\boldsymbol{q}}$ will be bounded by solving for $\ddot{\boldsymbol{q}}$ in equation (15). Thus, $\dot{\boldsymbol{e}}$ is bounded from (17) and from space-representation implying that $\boldsymbol{e}$ is uniformly continuous, and hence $\boldsymbol{e} \rightarrow \mathbf{0}$ as $t \rightarrow \infty$.

3. Suppose that related to the error dynamics equation (17), an output function $\boldsymbol{y} = \mathbf{C}\,\boldsymbol{e}$ is defined in such a way that the transfer function $\mathbf{C}[\mathrm{s}\mathbf{I}\text{-}\mathbf{A}]^{-1}\mathbf{B}_1$ is *strictly positive real* (SPR). Then, it can be shown, using the passivity theorem, that for a symmetric positive definite matrix $\mathbf{P}$, there exists a symmetric positive definite matrix $\mathbf{Q}$ satisfying the equations:

$$\mathbf{A}^{\mathrm{T}}\mathbf{Q} + \mathbf{Q}\,\mathbf{A} = -\mathbf{P}\,,\ \ \mathbf{Q}\,\mathbf{B}_1 = \mathbf{C}^{\mathrm{T}} \tag{22}$$

which bring the update parameter law (21) to the form:

$$\dot{\hat{\pi}} = \Gamma^{1}\,\Phi^{\mathrm{T}}\,\mathbf{C}\,\boldsymbol{e} \tag{23}$$

where $\Gamma = \Gamma^{\mathrm{T}}$ is positive definite.

4. The SPR property for the transfer function $\mathbf{C}[\mathrm{s}\mathbf{I}\text{-}\mathbf{A}]^{-1}\mathbf{B}_1$ also represents the first condition for the *convergence of parameter estimates* $\hat{\pi}$ to their true values π. Second, parameter convergence requires that the reference trajectory $\boldsymbol{q}_{\mathrm{d}}$ satisfies the condition of *persistency of excitation:*

$$\alpha\,\mathbf{I} \le \int_{t_0}^{t_0+T} \mathbf{Y}^{\mathrm{T}}(\boldsymbol{q}_{\mathrm{d}},\dot{\boldsymbol{q}}_{\mathrm{d}},\ddot{\boldsymbol{q}}_{\mathrm{d}})\mathbf{Y}(\boldsymbol{q}_{\mathrm{d}},\dot{\boldsymbol{q}}_{\mathrm{d}},\ddot{\boldsymbol{q}}_{\mathrm{d}})\mathrm{d}t \le \beta\,\mathbf{I} \tag{24}$$

for all t_0, where α, β and T are positive constants. Such a trajectory $\boldsymbol{q}_{\mathrm{d}}$ is considered to excite sufficiently the dynamic response of the system, so that the effects of the various parameters can be distinguished and can be used as a part of a "learning" algorithm to identify the system parameters, which can then be used in an inverse dynamics control law for tracking of any trajectory [9].

By exploiting the passivity of rigid dynamics, states that the mapping from the joint torque (input controls) $\tau = \boldsymbol{u}$ to the manipulator velocities $\dot{\boldsymbol{q}}$ is passive, i.e., there exists $\beta \ge 0$ such that:

$$\int_{0}^{T} \dot{\boldsymbol{q}}^{\mathrm{T}}(\xi)\tau(\xi)\mathrm{d}\xi \ \ge -\,\mathrm{H}(0) = -\beta \tag{25}$$

H(·) being the total energy of the manipulator at the current time, it is possible to derive more elegant and robust *passivity-based adaptive control algorithms* for manipulators. Such control algorithms, including the adaptive parameter update mechanism, are simple to design and implement as compared with the one above described.

Slotine and Li developed such adaptive control algorithms [7] based on a different approach as compared with that of inverse dynamics in the sense that, even with exact knowledge of the parameters, the control law does not linearize the equation of the motion of the robot.

This approach overcomes the main drawbacks of the adaptive inverse dynamics control previously described in that it does not require any measurement or estimation of the manipulator acceleration $\ddot{q}$, nor does it require inversion of the estimated inertia matrix $\hat{\mathbf{B}}$, thus making unnecessary the a priori specification of the variance domain for the parameter uncertainty $\tilde{\pi}$.

The passivity based control algorithm will be first presented in the case of perfect known parameters: $\hat{\pi} = \pi$.

Given the dynamic equation of the manipulator and drives system:

$$\mathbf{B}(\boldsymbol{q})\ddot{\boldsymbol{q}} + \mathbf{C}(\boldsymbol{q},\dot{\boldsymbol{q}})\dot{\boldsymbol{q}} + \mathbf{F}_v\dot{\boldsymbol{q}} + \mathbf{g}(\boldsymbol{q}) = \boldsymbol{u} \tag{26}$$

the choice for the inner loop control is now:

$$\boldsymbol{u} = \mathbf{B}(\boldsymbol{q})\dot{\boldsymbol{v}} + [\mathbf{C}(\boldsymbol{q},\dot{\boldsymbol{q}}) + \mathbf{F}_v]\boldsymbol{v} + \mathbf{g}(\boldsymbol{q}) + \mathbf{K}_D\boldsymbol{r} \tag{27}$$

where $\boldsymbol{v}$ and $\boldsymbol{r}$ are defined, in view of $\tilde{\boldsymbol{q}} = \boldsymbol{q}_d - \boldsymbol{q}$, as follows:

$$\boldsymbol{v} = \dot{\boldsymbol{q}}_d + \Lambda\tilde{\boldsymbol{q}} \tag{28}$$

$$\boldsymbol{r} = \boldsymbol{v} - \dot{\boldsymbol{q}} = \dot{\tilde{\boldsymbol{q}}} + \Lambda\tilde{\boldsymbol{q}} \tag{29}$$

with $\mathbf{K}_D$, Λ diagonal matrices of positive gains. The term $\mathbf{K}_d\,\boldsymbol{r}$ introduces a *linear PD action on the tracking error* $\tilde{\boldsymbol{q}}$; as for the input $\boldsymbol{v}$, this one weights the contribution that depends on velocity not only on the vase of the desired velocity $\dot{\boldsymbol{q}}_d(t)$, but also on the basis of the position tracking error $\tilde{\boldsymbol{q}}(t)$.

Substituting (27) into (26), and accounting for (28) and (29) it gives:

$$\mathbf{B}(\boldsymbol{q})(\dot{\boldsymbol{v}} - \ddot{\boldsymbol{q}}) + [\mathbf{C}(\boldsymbol{q},\dot{\boldsymbol{q}}) + \mathbf{F}_v](\boldsymbol{v} - \dot{\boldsymbol{q}}) + \mathbf{K}_D\boldsymbol{r} = \mathbf{0} \tag{30}$$

equivalent to:

$$\mathbf{B}(\boldsymbol{q})\dot{\boldsymbol{r}} + [\mathbf{C}(\boldsymbol{q},\dot{\boldsymbol{q}}) + \mathbf{F}_v]\boldsymbol{r} + \mathbf{K}_D\boldsymbol{r} = \mathbf{0} \tag{31}$$

where, from (29), $\dot{\boldsymbol{v}} - \ddot{\boldsymbol{q}} = \dot{\boldsymbol{r}}$. Taking as state-space vector for the system the error:

$$\boldsymbol{e} = \begin{bmatrix} \tilde{\boldsymbol{q}} \\ \dot{\tilde{\boldsymbol{q}}} + \Lambda\tilde{\boldsymbol{q}} \end{bmatrix} = \begin{bmatrix} \tilde{\boldsymbol{q}} \\ \boldsymbol{r} \end{bmatrix} \tag{32}$$

the choice of a Lyapunov candidate function that covers the entire system state $\boldsymbol{e}$, will be:

$$\mathrm{V}(\boldsymbol{e}) = \mathrm{V}(\tilde{\boldsymbol{q}},\boldsymbol{r}) = \frac{1}{2}\boldsymbol{r}^T\mathbf{B}(\boldsymbol{q})\boldsymbol{r} + \frac{1}{2}\tilde{\boldsymbol{q}}^T\mathbf{M}\,\tilde{\boldsymbol{q}} \tag{33}$$

This quadratic form (33) ensures $\mathrm{V}(\tilde{q}, r) > 0$, $\forall \tilde{q}, r \neq \mathbf{0}$ and $\mathrm{V}(\mathbf{0},\mathbf{0}) = 0$, that is V is positive definite for a positive definite (n x n) matrix $\mathbf{M}$. The time derivative $\dot{\mathrm{V}}$ of V along the trajectories of the system (31) is:

$$\dot{\mathrm{V}} = r^{\mathrm{T}} \mathbf{B}(q)\, \dot{r} + \frac{1}{2} r^{\mathrm{T}} \dot{\mathbf{B}}(q)\, r + \tilde{q}^{\mathrm{T}} \mathbf{M}\, \dot{\tilde{q}} \tag{34}$$

Substituting $\mathbf{B}(q)\, \dot{r}$ from equation (31) into equation (34) gives:

$$\dot{\mathrm{V}} = \frac{1}{2} \mathbf{r}^{\mathrm{T}} [\dot{\mathbf{B}}(q) - 2\mathbf{C}(q,\dot{q})]\, r - r^{\mathrm{T}} [\mathbf{F}_{\mathrm{v}} + \mathbf{K}_{\mathrm{D}}]\, r + \tilde{q}^{\mathrm{T}} \mathbf{M}\, \dot{\tilde{q}}$$

and by exploiting the skew-symmetry property of $\dot{\mathbf{B}} - 2\mathbf{C}$:

$$\dot{\mathrm{V}} = -r^{\mathrm{T}} \mathbf{F}_{\mathrm{v}}\, r - r^{\mathrm{T}} \mathbf{K}_{\mathrm{D}}\, r + \tilde{q}^{\mathrm{T}} \mathbf{M}\, \dot{\tilde{q}} \tag{35}$$

Using now the definition of r in equation (29), after some algebra , it results:

$$\dot{\mathrm{V}} = -r^{\mathrm{T}} \mathbf{F}_{\mathrm{v}}\, r - \dot{\tilde{q}}^{\mathrm{T}} \mathbf{K}_{\mathrm{D}}\, \dot{\tilde{q}} - \tilde{q}^{\mathrm{T}} \Lambda \mathbf{K}_{\mathrm{D}} \Lambda \tilde{q} + \tilde{q}^{\mathrm{T}} (\mathbf{M} - 2\Lambda \mathbf{K}_{\mathrm{D}})\, \dot{\tilde{q}} \tag{36}$$

In order to render $\dot{\mathrm{V}}$ negative definite, and in view of the negativeness of the first three terms in the right -hand side of equation (36), a convenient choice for $\mathbf{M}$ is:

$$\mathbf{M} = 2\Lambda \mathbf{K}_{\mathrm{D}} \tag{37}$$

for which $\dot{\mathrm{V}}$ has the final expression:

$$\dot{\mathrm{V}} = -r^{\mathrm{T}} \mathbf{F}_{\mathrm{v}}\, r - \dot{\tilde{q}}^{\mathrm{T}} \mathbf{K}_{\mathrm{D}}\, \dot{\tilde{q}} - \tilde{q}^{\mathrm{T}} \Lambda\, \mathbf{K}_{\mathrm{D}}\, \Lambda \tilde{q} \tag{38}$$

with $\dot{\mathrm{V}}(e) < 0$, $\forall e \neq \mathbf{0}$, $\dot{\mathrm{V}}(\mathbf{0}) = 0$. It follows that the origin $e = \mathbf{0}$ ($\tilde{q} = \mathbf{0}, r = \mathbf{0}$) of the state-space is globally asymptotically stable.

Moreover, from equations (33), (34) and (38), r is a bounded square integrable function, and from the definition (29) for r, it follows that e and $\dot{e}$ are bounded, and $e \to \mathbf{0}$ as $t \to \infty$. From the system dynamics (31), $\dot{r}$ is bounded - which implies that r and hence $\dot{\mathrm{V}}$ are uniformly continuous. Therefore, r converge to zero, which further implies that $\dot{e} \to \mathbf{0}$ as $t \to \infty$.

The previously obtained results will be used now to design a control law that is adaptive with respect to the vector of parameters π. It is considered that the estimate $\hat{\pi}$ of the parameters of the computational model ($\hat{\mathbf{B}}, \hat{\mathbf{C}}, \hat{\mathbf{F}}_{\mathrm{v}}, \hat{\mathbf{g}}$) of the robotic dynamics differ from the exact set π, i.e., there exists a nonnull *parameter uncertainty* $\tilde{\pi}$ reflected in the *model uncertainty* ($\tilde{\mathbf{B}}, \tilde{\mathbf{C}}, \tilde{\mathbf{F}}_{\mathrm{v}}, \tilde{\mathbf{g}}$) according to equations (9). Then, the control law is chosen as:

$$u = \hat{\mathbf{B}}(q)\dot{v} + [\hat{\mathbf{C}}(q,\dot{q}) + \hat{\mathbf{F}}_{\mathrm{v}}]v + \hat{\mathbf{g}}(q) + \mathbf{K}_{\mathrm{D}} r = \mathbf{Y}(q,\dot{q},\ddot{q},v,\dot{v})\hat{\pi} + \mathbf{K}_{\mathrm{D}} r \tag{39}$$

Because $\dot{\boldsymbol{v}} = \ddot{\boldsymbol{q}}_{\mathrm{d}} + \Lambda\dot{\tilde{\boldsymbol{q}}}$, the regressor **Y** does not depend on the manipulator acceleration $\ddot{\boldsymbol{q}}$, but only on $\dot{\boldsymbol{v}}$, that is on the acceleration of the reference trajectory $\ddot{\boldsymbol{q}}_{\mathrm{d}}$, and on the velocity tracking error $\dot{\tilde{\boldsymbol{q}}}$; this avoids problems related to acceleration measurement or indirect evaluation by velocity filtering.

Substituting the control input $\boldsymbol{u}$ of equation (37) into the system equation (26), gives by a similar procedure as above:

$$\mathbf{B}(\boldsymbol{q})\,(\dot{\boldsymbol{v}} - \ddot{\boldsymbol{q}}) + [\,\mathbf{C}(\boldsymbol{q},\dot{\boldsymbol{q}}) + \mathbf{F}_{\mathrm{v}}]\,(\boldsymbol{v} - \dot{\boldsymbol{q}}) + \mathbf{K}_{\mathrm{D}} =$$
$$\tilde{\mathbf{B}}(\boldsymbol{q})\,\dot{\boldsymbol{v}} + [\tilde{\mathbf{C}}(\boldsymbol{q},\dot{\boldsymbol{q}}) + \tilde{\mathbf{F}}_{\mathrm{v}}]\,\boldsymbol{v} + \tilde{\mathbf{g}}(\boldsymbol{q}) = \mathbf{Y}(\boldsymbol{q},\dot{\boldsymbol{q}},\boldsymbol{v},\dot{\boldsymbol{v}})\,\tilde{\boldsymbol{\pi}}$$

equivalent to:

$$\mathbf{B}(\boldsymbol{q})\,\dot{\boldsymbol{r}} + [\,\mathbf{C}(\boldsymbol{q},\dot{\boldsymbol{q}}) + \mathbf{F}_{\mathrm{v}}]\,\boldsymbol{r} + \mathbf{K}_{\mathrm{D}}\,\boldsymbol{r} = \mathbf{Y}(\boldsymbol{q},\dot{\boldsymbol{q}},\boldsymbol{v},\dot{\boldsymbol{v}})\,\tilde{\boldsymbol{\pi}} \tag{40}$$

The Lyapunov candidate function will be modified, as compared to the choice in equation (33) and in view of the result obtained in (37), to the form:

$$\mathrm{V}(\tilde{\boldsymbol{q}},\boldsymbol{r},\tilde{\boldsymbol{\pi}}) = \frac{1}{2}\,\boldsymbol{r}^{\mathrm{T}}\mathbf{B}(\boldsymbol{q})\,\boldsymbol{r} + \tilde{\boldsymbol{q}}^{\mathrm{T}}\,\Lambda\mathbf{K}_{\mathrm{D}}\,\tilde{\boldsymbol{q}} + \frac{1}{2}\,\tilde{\boldsymbol{\pi}}^{\mathrm{T}}\,\Gamma\,\tilde{\boldsymbol{\pi}} \tag{41}$$

with Γ a symmetric, positive definite matrix. For this new function, the time derivative of V along solution trajectories becomes:

$$\dot{\mathrm{V}} = \boldsymbol{r}^{\mathrm{T}}\mathbf{B}(\boldsymbol{q})\,\boldsymbol{r} + \frac{1}{2}\,\boldsymbol{r}^{\mathrm{T}}\,\dot{\mathbf{B}}\,(\boldsymbol{q})\,\boldsymbol{r} + 2\tilde{\boldsymbol{q}}^{\mathrm{T}}\,\Lambda\mathbf{K}_{\mathrm{D}}\,\dot{\tilde{\boldsymbol{q}}} + \tilde{\boldsymbol{\pi}}^{\mathrm{T}}\,\Gamma\,\dot{\tilde{\boldsymbol{\pi}}} \tag{42}$$

One substitutes $\mathbf{B}(\boldsymbol{q})\,\dot{\boldsymbol{r}}$ from equation (40) into equation (42), and exploiting the skew symmetry of ($\dot{\mathbf{B}}$ -2**C**), after some algebra, it results:

$$\dot{\mathrm{V}} = -\,\boldsymbol{r}^{\mathrm{T}}\mathbf{F}_{\mathrm{v}}\,\boldsymbol{r} - \dot{\tilde{\boldsymbol{q}}}^{\mathrm{T}}\,\mathbf{K}_{\mathrm{D}}\,\dot{\tilde{\boldsymbol{q}}} - \tilde{\boldsymbol{q}}^{\mathrm{T}}\,\Lambda\mathbf{K}_{\mathrm{D}}\,\Lambda\tilde{\boldsymbol{q}} + \tilde{\boldsymbol{\pi}}^{\mathrm{T}}\,[\Gamma\dot{\tilde{\boldsymbol{\pi}}} + \mathbf{Y}^{\mathrm{T}}(\boldsymbol{q},\dot{\boldsymbol{q}},\boldsymbol{v},\dot{\boldsymbol{v}})\,\boldsymbol{r}\,] \tag{43}$$

The adaptation law for the estimate $\hat{\boldsymbol{\pi}}$ of the parameter vector is chosen so that the last term in the right-hand side of equation (43) vanishes, i.e.:

$$\dot{\hat{\boldsymbol{\pi}}} = \Gamma^{1}\;\mathbf{Y}^{\mathrm{T}}(\boldsymbol{q},\dot{\boldsymbol{q}},\boldsymbol{v},\dot{\boldsymbol{v}})\,\boldsymbol{r} \tag{44}$$

in view of equation (4) and of π being constant. This choice for the adaptive parameter updating law leads to:

$$\dot{\mathrm{V}} = -\,\boldsymbol{r}^{\mathrm{T}}\mathbf{F}_{\mathrm{v}}\,\boldsymbol{r} - \dot{\tilde{\boldsymbol{q}}}^{\mathrm{T}}\,\mathbf{K}_{\mathrm{D}}\,\dot{\tilde{\boldsymbol{q}}} - \tilde{\boldsymbol{q}}^{\mathrm{T}}\,\Lambda\mathbf{K}_{\mathrm{D}}\,\Lambda\tilde{\boldsymbol{q}} \tag{45}$$

for which $\dot{\mathrm{V}} < 0$, $\forall\tilde{\boldsymbol{q}}, \boldsymbol{r} \neq \mathbf{0}$, and $\dot{\mathrm{V}} = 0$ for $\tilde{\boldsymbol{q}} = \boldsymbol{r} = \mathbf{0}$. The error trajectories globally and asymptotically converge to the origin $\boldsymbol{e} = \mathbf{0}$ ($\tilde{\boldsymbol{q}} = \mathbf{0}$, $\boldsymbol{r} = \mathbf{0}$) of the state-space $\boldsymbol{e} = [\tilde{\boldsymbol{q}}^{\mathrm{T}}\,\boldsymbol{r}^{\mathrm{T}}]^{\mathrm{T}}$ which implies convergence to zero of $\tilde{\boldsymbol{q}}$, $\dot{\tilde{\boldsymbol{q}}}$ and boundness of $\hat{\boldsymbol{\pi}}$.

Fig. 2 represents the block scheme of the adaptive feedback linearization control with PD action.

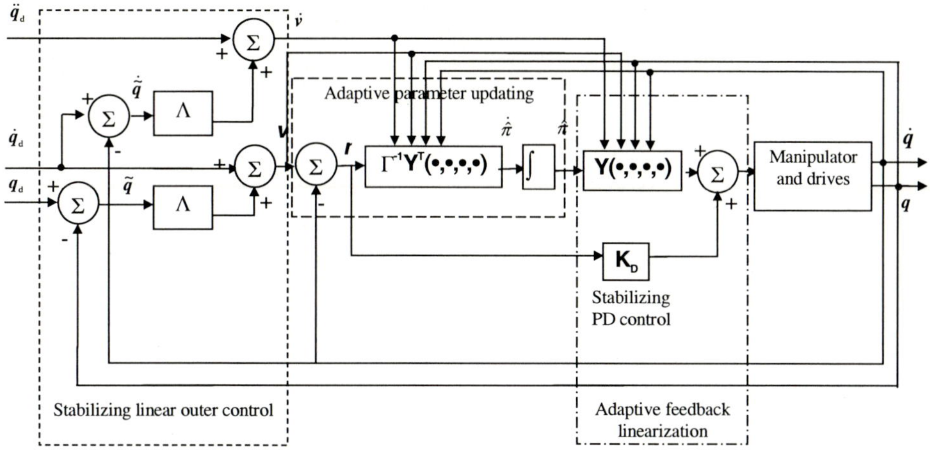

Fig. 2. Passivity-based adaptive control with PD action

It can be recognized that the adaptive control law that has been designed:

$$u = \mathbf{Y}(q,\dot{q},v,\dot{v})\,\hat{\pi} + \mathbf{K}_D(\dot{\tilde{q}} + \Lambda\tilde{q}) \tag{46}$$

highlights the following actions:

- an *adaptive feedback linearization* that uses the term $\mathbf{Y}\hat{\pi}$ for compensation nonlinear dynamic effects;
- a *stabilizing PD control action* exerted by the term $\mathbf{K}_d\, r$ on the trajectory tracking composite error: $\dot{\tilde{q}} + \Lambda\tilde{q}$.

The global adaptive feedback uses parameter estimates $\hat{\pi}$ that are *updated* by an *adaptive law* of *gradient type* (44) to their asymptotic values π.

A number of refinements to this basic approach have been proposed. Using, for example, the *reference trajectory* instead of the measured joint positions and velocities in both the control and parameter update laws:

$$u = \mathbf{Y}(q_d,\dot{q}_d,\ddot{q}_d)\ \hat{\pi} + \mathbf{K}_P\tilde{q} + \mathbf{K}_D\dot{\tilde{q}} \tag{47}$$

with:

$$\dot{\hat{\pi}} = \Gamma^{-1}\ \mathbf{Y}^T(q_d,\dot{q}_d,\ddot{q}_d)\ (\dot{\tilde{q}} + \Lambda\tilde{q}) \tag{48}$$

it is possible to obtain:

- exponential stability of the tracking error in the case of known parameters;
- asymptotic stability in the adaptive case of computed parameter estimates;
- convergence of the parameter estimation error to zero under persistence of excitation of the reference trajectory.

The passivity-based adaptive motion controller can be generally formalized with the following generic control law and adaptation law:

$$\boldsymbol{u} = \hat{\mathbf{B}}(\boldsymbol{p})\boldsymbol{v} + [\hat{\mathbf{C}}(\boldsymbol{p}, \boldsymbol{p}_1) + \mathbf{F}_{\mathrm{v}}]\boldsymbol{v} + \hat{\mathbf{g}}(\boldsymbol{p}) + \mathbf{K}_{\mathrm{p}}\varepsilon + \mathbf{K}_{\mathrm{D}}\nu + \mathbf{K}_{\mathrm{f}}\|\varepsilon\|^2\gamma \tag{49}$$

$$\dot{\hat{\pi}} = \Gamma^{-1}\,\mathbf{Y}^{\mathrm{T}}(\boldsymbol{q}_{\mathrm{d}}, \dot{\boldsymbol{q}}_{\mathrm{d}}, \ddot{\boldsymbol{q}}_{\mathrm{d}}) \tag{50}$$

where:

- $\boldsymbol{r} = \dot{\tilde{\boldsymbol{q}}} + \Lambda\tilde{\boldsymbol{q}}$ with $\tilde{\boldsymbol{q}} = \boldsymbol{q}_{\mathrm{d}} - \boldsymbol{q}$ and Λ diagonal matrix of positive gains, no-necessarily time-invariant;
- $\hat{\mathbf{B}}$ (.), $\hat{\mathbf{C}}$ (.,.), $\hat{\mathbf{F}}_{\mathrm{v}}$, $\hat{\mathbf{g}}$ (.): estimated counterparts of the components ($\mathbf{B}$, $\mathbf{C}$, $\mathbf{F}_{\mathrm{v}}$, $\mathbf{g}$) of the manipulator dynamic model;
- Γ is a (p x p) positive definite matrix of adaptation gains;
- $\mathbf{Y}$(.,.,.) is a (n x p) regressor matrix of known functions for the manipulator dynamic model;
- $\boldsymbol{p}$, $\boldsymbol{p}_1$, $\boldsymbol{v}$, $\boldsymbol{v}$, ε and ν are signals that depend on the particular passivity-based adaptive controller to be used.

It must be pointed out that the adaptation law (50) for parameter estimate updating is of gradient type, driven by trajectory tracking error.

As for the control law (49), two parts can be distinguished: the first one (containing the first three terms in the right hand side of (49)) is responsible for *compensating the non-linear dynamics* of the manipulator, while the second one (containing the last three terms in the right hand side of (49)) represents a *linear PD action.*

The fact that the parametric error $\tilde{\pi}$ depends both on the tracking error $\tilde{\boldsymbol{q}}$ and on the prediction error $\dot{\boldsymbol{q}}$ motivated Slotine to propose as alternative adaptation mechanisms the so-called *adaptation law of composite parameters* [8]. These laws extract information about the true parameter set from the prediction and tracking errors, and have the form:

$$\dot{\hat{\pi}} = \Gamma^{-1}(t)\,\{\mathbf{Y}^{\mathrm{T}}(\boldsymbol{q}, \dot{\boldsymbol{q}}, \boldsymbol{v}, \dot{\boldsymbol{v}})\,\boldsymbol{r} + \mathbf{W}^{\mathrm{T}}(\boldsymbol{q}, \dot{\boldsymbol{q}})\,\mathbf{R}\,\varepsilon_{\mathrm{f}}\} \tag{51}$$

where $\mathbf{R} = \mathbf{R}^{\mathrm{T}} > \mathbf{0}$ is a weighting (n x n) matrix. This weighting matrix will be used to evaluate at what extended influences the prediction error in the composite law (51) with respect of the tracking error.

In order to generate the matrix $\Gamma(t)$ of adaptation gains, various techniques can be used. The most general algorithm is the *cushioned-floor method* [8]:

$$\Gamma^{-1}(t) = \alpha(t)\,\{\Gamma^{-1}(t) - \Gamma^{-1}(t)\,\mathbf{K}_0\,\Gamma^{-1}(t)\} - \Gamma^{-1}(t)\,\mathbf{W}^{\mathrm{T}}(\boldsymbol{q}, \dot{\boldsymbol{q}})\,\mathbf{R}\,\mathbf{W}(\boldsymbol{q}, \dot{\boldsymbol{q}})\,\Gamma^{-1}(t) \tag{52}$$

where:

- $\Gamma^{-1}(0) = \Gamma^{-1^{\mathrm{T}}}(0) = \mathbf{0}$;

- $\mathbf{K}_0 = \mathbf{K}_0^{\mathrm{T}} > \mathbf{0}$
- $\alpha(t) = \alpha_0 \left(1 - \dfrac{\left\|\Gamma^{-1}(t)\right\|}{\mathrm{k}_0}\right)$, $\alpha_0 \geq 0$, $\mathrm{k}_0 > 0$ is a variable omission factor.

This method allows to impose an upper-margin to the adaptation law, of the form $\Gamma^{-1}(t) < \mathbf{K}_0^{-1}$.

3 Conclusions

The approach that has been considered in the research refers to the category of nonlinear, multivariable control, with original contributions in the following respect:

1. Development of a *robustifying algorithm* for the linear stabilizing control in robust control structures, by using Look-Up-Table correspondences, off-line computed, and dependent on the joint space configuration.
2. Development of an adaptive control algorithm based on the *passivity property* of rigid robot systems, having a *new adaptive law of gradient type with dependence on the parameters of the desired trajectory* $\boldsymbol{q}_d$, for updating the regressor model parameters π.
3. Development of *multiprocessor control architecture with acceleration of the feedforward centralized compensations* for the nonlinear coupling between the first three principal axes of the anthropomorphic and SCARA-type robots.

It can be observed that, unlike for inverse dynamics controllers based on feedback linearization, in passivity-based adaptive controllers the adaptation loop affects the compensator $\mathbf{Y}(.,.,.,.)$ of the nonlinear dynamic model as long as it does not interact with the feedforward loop.

Another useful feature of passivity-based adaptive controllers is their reduced computational effort, which makes them particularly attractive for implementation purpose.

The passivity of the adaptive controller can be interpreted like follows: due to the presence of the gradient-type adaptation law the closed-loop system can be decomposed in three interconnected blocks, as represented in the Fig. 2. One of them - the parameter updating adaptive law - is passive, as results from:

$$\begin{aligned} \left\langle -\mathbf{Y}(.)\tilde{\pi}, \boldsymbol{r} \right\rangle &\equiv -\int_0^T (\mathbf{Y}(.)\tilde{\pi}(\xi))^{\mathrm{T}} \boldsymbol{r}(\xi)\mathrm{d}\xi \\ = -\int_0^T \tilde{\pi}^{\mathrm{T}}(\xi)\Gamma \frac{\mathrm{d}}{\mathrm{d}t}\{\tilde{\pi}^{\mathrm{T}}(\xi)\}\mathrm{d}\xi &\geq -\frac{1}{2}\tilde{\pi}^{\mathrm{T}}(0)\Gamma\tilde{\pi}(0) = -\beta(0) \end{aligned} \tag{53}$$

in view of equations (50), (4), and $\pi =$ const.

As a concluding remark, the research and design activities that are presented in the framework of the paper aim to produce a complete documentation for the future implementation of a powerful, generic robot controller to be used as research platform for advanced robot control.

References

1. Craig, J. J., Hsu, P., S. S. Sastry: Adaptive Control of Mechanical Manipulator. Int. J. of Robotics Research, vol. 6, (1987) 16-28
2. Craig J. J.: Adaptive Control of Mechanical Manipulators. Addison Wessley, Reading, MA, (1988)
3. Kreuzer, E.J., Lugtenburg, J.-B., Meissner, H.-G.: Industrieroboter. Technik, Berechnung und Anwendungsorientierte Auslegung. Springer-Verlag, Berlin, (1994)
4. Liu, G., Goldenberg, A.A.: Robust Control of Robot Manipulators Based on Dynamics Decomposition. IEEE Trans. on Robotics and Automation, vol. 13, no. 5, oct. (1997) 783-789
5. Qu, Z.: Robust Control of Nonlinear Uncertain Systems Under Generalized Matching Conditions. Automatica, 29(4), (1993) 985-998
6. Sciavicco, L., Siciliano, B.: Modeling and Control of Robot Manipulators. Mc Graw-Hill, (1996)
7. Slotine, J. J. E. and W. Li: On the adaptive control of robot manipulators. Int. J. Robot Res., vol. 6, no. 3 (1987) 49-59
8. Slotine, J. J. E., W. Li: Composite Adaptive of Robot Manipulators. Automatica, Vol. 25, (1989) 509-519
9. Spong, M. W., M. Vidyasagar: Robot Dynamics and Control. John Wiley and Sons, N.Y., (1989)

Mobile Robot Path Planning Among Weighted Regions Using Quadtree Representations

Jozef Vörös

Slovak Technical University,
Faculty of Electrical Engineering and Information Technology
Department of Automation and Control
Ilkovicova 3, 812 19 Bratislava, Slovakia
voros@nov1.kar.elf.stuba.sk

Abstract. An approach to the mobile robot path planning among heterogeneous regions is presented. Proper weight factors are associated with the regions of different kind (rock, sand, grass) to characterize the amount of the difficulty required to travel through that particular region, as compared to traveling over a flat, smooth surface (corresponding to a minimum weight). The weighted regions are represented in the form of modified matrix quadtree. The technique of distance transform is extended to the weighted regions and applied to the robot path planning. Illustrative examples are included.

1 Introduction

Application areas of robots in flexible manufacturing and assembly systems but also in agriculture, forestry and other services demand use of intelligent robots. Planning a collision-free path is one of the fundamental requirements for a robot to execute its tasks [1]. The objective is to navigate the robot from its start position to a goal position in the presence of a given set of obstacles and not to collide with any of them. In the standard path planning approaches all the obstacles in workspaces are assumed to be impassable (solid).

Much work has been done on solving the path planning problem between distinct locations in an environment with impenetrable obstacles. However, not all applications can be treated in this way [2]. Such an example would be the path planning for a mobile robot traveling over various terrains. The terrains may include regions that are indeed inaccessible (such as rocks), but may also include regions that just slow the movement down (such as sand, grass). Such regions can be modeled by assigning a particular weight factor to each of them. The weight of a region represents the amount of the difficulty or loss required to travel through that particular region, as compared to traveling over a flat, smooth surface (corresponding to a minimum weight).

Using quadtree representations [3] in mobile robot path planning in many applications may simplify the algorithms, because large areas of obstacles or those of free space can be represented by small number of square blocks with different dimensions [4]. The matrix form of quadtree enables compact description of the scene

F. Pichler, R. Moreno-Díaz, and P. Kopacek (Eds.): EUROCAST'99, LNCS 1798, pp. 239–249, 2000.

with weighted regions. This is used in the presented approach to the path planning among weighted regions.

The proposed approach is based on the distance transform technique and uses an artificial path planning wave to efficiently search the quadtree in order to find the weighted shortest path. The wave is used to compute a weighted distance transform of the quadtree for a given goal position. The weighted length of a path through such a region is determined by the distance of the path (in the chosen metric) times the weight of the specific region. Then from any starting point in the environment, the shortest path to the goal is traced by following the path of steepest descent. The sequence of free region quadrants includes at least a collision free path of the mobile robot. The weighted distance transform of the quadtree can be effectively performed using the recently proposed repetitive neighbor finding strategy [5].

First, the so-called matrix quadtree is described and modified for the representation of mobile robot environment, which is heterogeneous. Then the distance transform map is extended for quadtrees with weighted regions and the algorithm for path planning will be described. Finally, illustrative examples show the path planning results.

2 Quadtree Representation

A quadtree representation of a binary image is a tree whose leaves represent square areas or quadrants of the image and are labeled with the color of the corresponding area, i.e., BLACK, WHITE, or GRAY (both black and white). The quadtree R corresponding to a $(2^N \times 2^N)$-dimensional binary image can be characterized as a column vector of all the GRAY node descriptions [6]

$$R = [P_1, P_2, P_3, \dots, P_L]^T, \tag{1}$$

where L is the number of GRAY nodes. A GRAY node at level k of quadtree representing a $(2^k \times 2^k)$-dimensional region can be described as an ordered quadruple

$$P_i = (Q_{i1}, Q_{i2}, Q_{i3}, Q_{i4}), \tag{2}$$

where Q_{im}, m = 1, ... , 4, denote its four children at level k-1.

If Q_{im} is a leaf or terminal node, its value can be assigned one of two distinct non-positive integers for the corresponding color; e.g., it is proper to use

$$Q_{im} = \begin{cases} -w & \text{for BLACK quadrat,} \\ 0 & \text{for WHITE quadrat.} \end{cases} \tag{3}$$

If Q_{im} is a nonterminal node, there must be a quadruple later in (1) describing the corresponding quadrant. Hence, it is appropriate to substitute

$$Q_{im} = r, \qquad \text{for GRAY quadrant,} \tag{4}$$

where r = 2,3,..., refers to the position (row) in vector (1) where the quadruple P_r corresponding to this GRAY quadrant is described. As the quadruples (2) consist of four integers, the quadtree description (1) is an (L x 4)-matrix, consequently the name "matrix quadtree". The matrix quadtrees are minimum and top-down descriptions of binary images.

3 Weighted Regions

A quadtree is a recursive decomposition of a 2D environment into uniformly sized regions. The nodes of a standard quadtree can be classified according to the above definition as either (i) free nodes, representing obstacle-free areas; (ii) obstacle nodes, representing regions full of obstacles, which cannot be traveled through; or (iii) mixed nodes, representing regions of both obstacle and free space.

An important feature of matrix quadtree description is that the substitutions for terminal nodes given by (3) can significantly extend the informational content of tree representations. Besides their main role, to give information on the shapes and locations of 2D regions or objects, they allow to store further relevant information on areas considered. Namely, different negative integers can be used to distinguish among more objects in a scene represented by one tree, to assign colors, labels for connected areas, to store distance data, or another information related to the corresponding nodes.

Hence the matrix quadtree description is proper also for effective representation of a scene with weighted regions. Different negative integers can be used to assign the weights of given regions. We can define:

$$Q_{im} = \begin{cases} -w & \text{for weighted regions,} \\ -1 & \text{for free regions,} \\ 0 & \text{for forbidden regions,} \end{cases} \tag{5}$$

where w > 1, for the matrix quadtree representation of weighted regions.
For example, the 2D scene in Fig. 1 is described by the matrix quadtree given in Table 1. In this matrix, appearance of a positive integer, say *r*, means that the subquadrant associated with the given column is nonterminal and is described in the r-th row of matrix. All negative and zero entries denote terminal nodes, i.e., 0's denote the forbidden regions (C) (impassable or solid obstacles), -1's are for free regions (weight 1), -2's and -3's are for the regions with the weights 2 (A) and 3 (B), respectively.

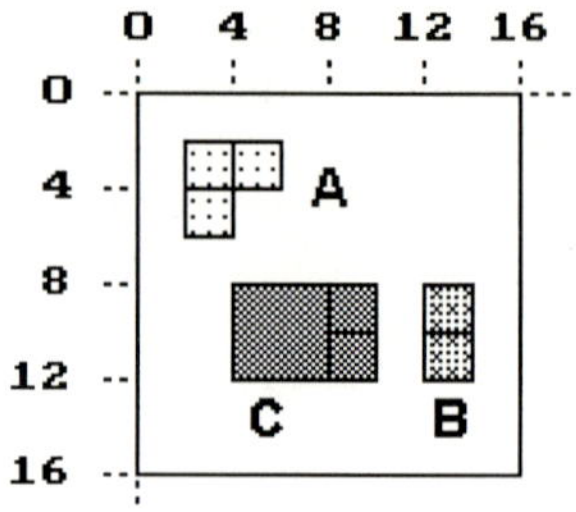

Fig. 1.

Table 1.

P_i / Q_{ij}	0	1	2	3
1	2	-1	3	4
2	5	6	7	-1
3	-1	0	-1	-1
4	8	9	-1	-1
5	-1	-1	-1	-2
6	-1	-1	-2	-1
7	-1	-2	-1	-1
8	0	-1	0	-1
9	-3	-1	-3	–1

4 Distance Transforms

Distance transforms (also known as distance maps) have been used in robotics for a variety of path planning and collision-avoidance applications [1]. A distance transform is the minimum cost to reach a particular location from a starting position.

Computational implementations of these distance maps invariably involve discretized of grid-based representation of the domain over which the distance map is defined. This distance map is represented in an array with each pixel containing an integer representing the distance between it and the given goal pixel. A main drawback of such discretized representations, however, is the large amount of memory required to store it. Larger spaces and/or finer resolution, of course, increase this usage exponentially.

A more memory-efficient method than a raw binary array is the quadtree data structure [3]. The distance transform for a quadtree can be defined as a function/map that yields for each free quadrant in the tree the distance (in the chosen metric) to the quadrant containing the given goal point G.

The quadtree distance transform uses an artificial wave starting in a given quadrant G to efficiently search the quadtree in order to find the shortest paths to other quadrants. Note that the wave is not an actual physical wave but an artificial computation technique. The wave propagates outward from G. The boundary of the propagating wave at any point in time is called the wavefront. It encloses the area that has been searched by the wave at a particular point in time. When the wave assigns a distance transform value to a particular quadrant, that quadrant is said to be covered by the wave. If all the free quadrants are covered, the transformed quadtree can be used to find the shortest path from any free quadrant to the quadrant G.

The essence of most methods for computing distance transforms is to approximate the Euclidean distances between pairs of pixels by propagating distance information locally and adding increments that represent distances between neighboring pixels [7], [8], [9]. In the case of quadtree representation the distance transform is computed rather at squares of different size than at pixels [10], [11], [12]. Therefore the definition of distance should be oriented to the nodes of quadtree and to the type of neighborhood chosen.

For the sake of simplicity, the neighbors in 4 main directions will be considered only, i.e. the pairs of edge neighbors in the horizontal and vertical directions. The distance between two neighboring nodes of quadtree can be defined in different ways. In this paper the distance corresponds to the size of neighboring node, hence the distance transform maps the sum of neighboring nodes' sizes from the reference/goal node (quadrant) to the respective one. Hence the global distances in the scene are approximated by propagating local distances, i.e., distances between neighboring quadrants.

To illustrate the process of quadtree distance transform, assume the reference (goal) node is the uppermost one on the left of the scene in Fig. 2. It is reasonable to associate/map its size (8) as the starting value of following transformation. Then for its two edge neighbors the values are computed as increments to the starting node value, i.e., 8 + 8, and so on. Naturally, the incrementation is performed only once for every node.

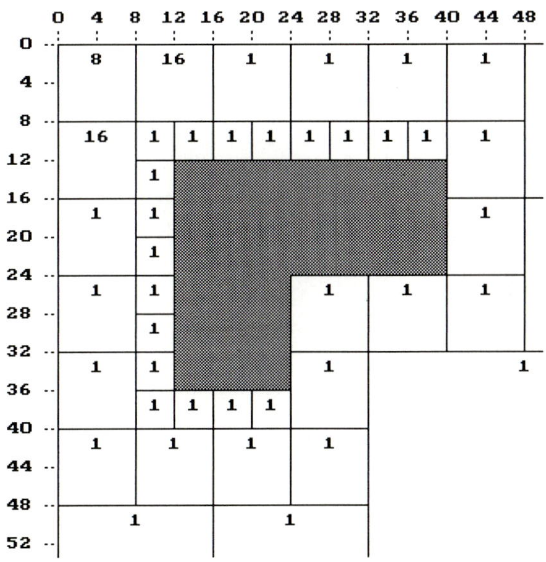

Fig. 2.

The distance transform computation algorithm requires to examine the four neighbors of each free node. This can be effectively performed using the recently proposed repetitive neighbor finding strategy [5]. It is worthy to note that the neighbor finding algorithm used in this case can cope also with the neighbors of nodes with smaller sizes.

5 Weighted Distance Transforms

Let a path P from S to G (the positions of the start and goal) consists of a sequence of line segments (p_sp_1, p_1p_2, ... , $p_{g-1}p_g$), where p_s and p_g are the positions of S and G, respectively. Let L(a,b) denote the unweighted path length between locations *a* and *b* (i.e., through a region with a weight 1) corresponding to the sum of the lengths (in some metric) of the line segments making up the path. Let w(R) be the weight of a particular region R, where $0 \leq w(R) < \infty$. The term weight and cost, when referring to a particular region, will be used interchangeably throughout this paper. The weighted length of a segment *cd* in R is defined as L(c,d)*w(R). Thus, the weighted length of a path P from G to S is simply the sum of the weighted lengths of the line segments making up P. Of all possible paths from S to G, the path of minimum total cost corresponds to the weighted shortest path between the two locations. There can be any number of weighted regions in the workspace, and each weight can be of any nonnegative value.

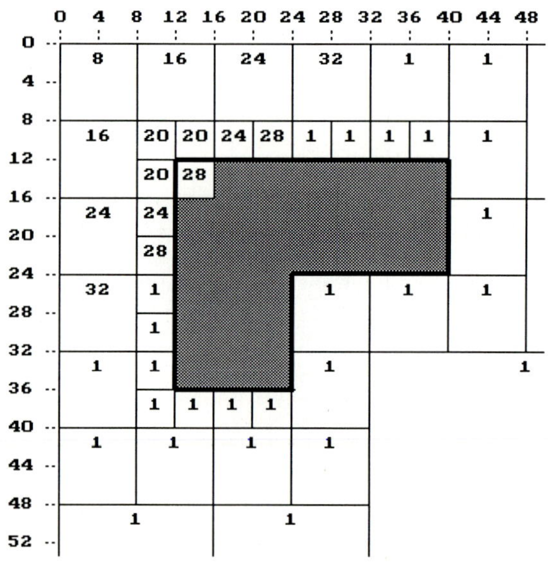

Fig. 3.

The weighted distance transform can be converted into the quadtree representations by copying the above ideas. If the distance wave reaches a quadrant with weight, then the increment added to the previous value is simply multiplied by this weight forming the weighted local distance. This can be seen in Fig. 3, where the first quadrant of a weighted region has been reached. If the weight factor of this region is 2, the corresponding distance increment has to be multiplied by 2, therefore this node gains the transform value 20+2*4=28.

In this way the weighted distance transform of the whole workspace can be performed. Naturally all the weighted regions are included into the distance transform except the forbidden ones. Note that in the case of weighted path planning the start and/or goal points can be positioned also in the weighted regions.

6 Path Planning

In the case of general weighted region, the robot environment is partitioned into a set of regions, each of which is associated with a given weight factor. A path through a weighted region assumes a cost that is determined by the defined distance of the path in that region and that region's weight factor.

We are interested in generating a shortest path between two distinct locations in an environment consisting of weighted regions. We base the optimality of the weighted path on the above mentioned definition of distance and the weighted distance transform of all the free and weighted regions.

Let the robot's 2D environment (workspace) be represented by a matrix quadtree using substitutions (5). It means that all the weighted regions are marked by the weight factor (with minus sign), free regions have the factor 1 (in the matrix -1) and the forbidden areas are assigned as 0.

Given the goal (G) point in the workspace, first we determine the corresponding quadtree leaf node G, representing the region of the space containing this point. For this goal node G the weighted distance transform of workspace is performed in the way described above. It consists of marking every free or weighted node with its distance or weighted distance from the goal node. The matrix quadtree enables direct storing of these extra data as they can be put into the corresponding matrix entries as integers with minus sign. Now, from any starting point in the environment, the shortest path to the goal can be traced by following the path of steepest descent.

Beginning with the node corresponding to the start point S, the transformed quadtree representation is used for finding an optimal node adjacent to the node being processed, i.e., the neighbor node with the least distance value is chosen as optimal. This distance optimal node is included into the path list. The process continues with searching an optimal neighbor for this node and is repeated up to reaching the neighbor node identical with the node G.

The result of applying the proposed algorithm to the transformed quadtree representation of workspace is the list of nodes from the quadtree that determines a sequence of adjacent free or weighted squares (ordinarily of varying sizes). This space includes at least one weighted shortest path between the start and goal nodes, passing through free and weighted regions. If desired, an optimal path through these squares can be computed.

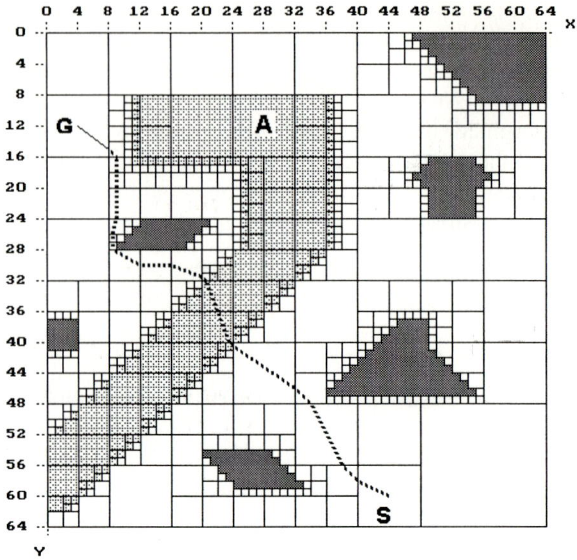
0 4 8 12 16 20 24 28 32 36 40 44 48 52 56 60 64
X
Y
A
G
S

Fig. 4.

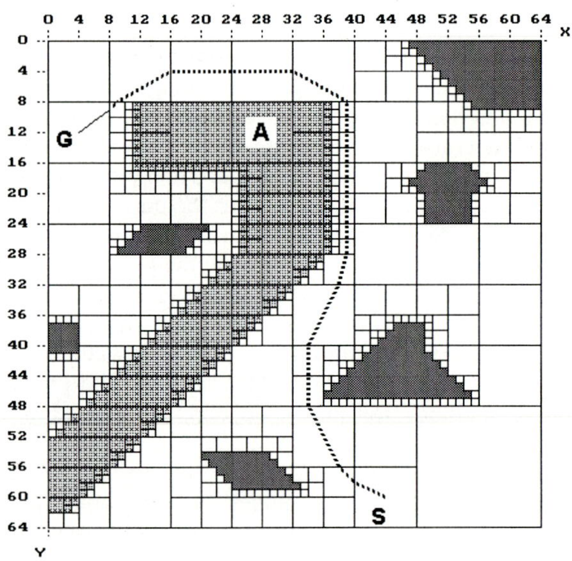
0 4 8 12 16 20 24 28 32 36 40 44 48 52 56 60 64
X
Y
A
G
S

Fig. 5.

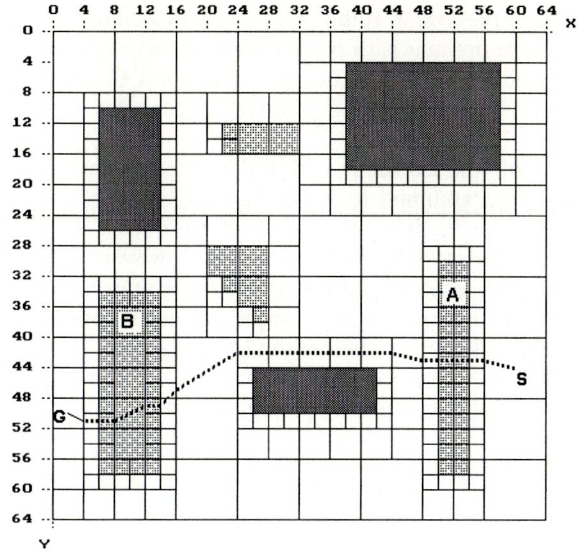

Fig. 6.

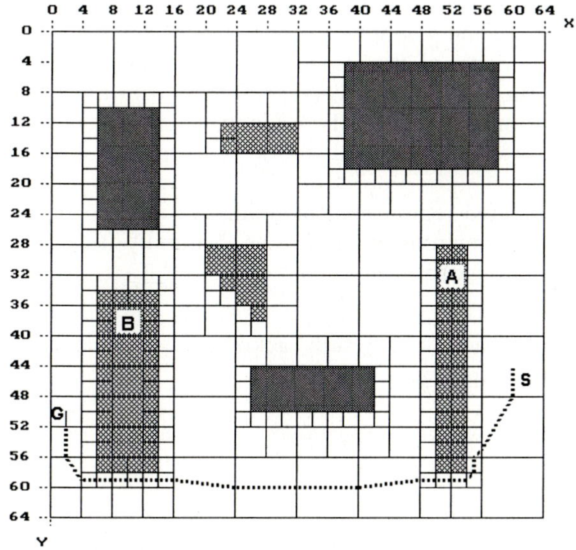

Fig. 7.

Now some computer simulation results of solving the path planning in different weighted robot workspaces are presented. In the first example the 2D scene has been considered according to the Fig. 4. If the region A is associated with the weight factor 2, the distance transform optimal path passes through this region as it is shown by the dotted line between the start point S and the goal point G. However, if the weight factor of the region A is 3, the resulting path will bypass this region as it is shown in Fig. 5.

In the second example the 2D scene with more objects is presented - Fig. 6, where the regions A and B are assumed to have the weight factor 2 while all the other regions are forbidden. The resulting path between the quadrants S and G is drawn by the dotted line. The path passes through A and B. However, if the weight factors of regions A and B are 3, the resulting path bypasses these regions as shown in Fig. 7.

Finally note that although the resulting paths are not so optimal as in the case of more elaborated methods [2], nevertheless the computational efforts by this approach are not so extensive and the results are satisfactory. And also, an extension to 8-neighbor case can improve the optimality of path planning.

7 Conclusion

The hierarchical nature of quadtree representation makes it popular in mobile robot path planning because it is very efficient for representing large areas of obstacles and free space. This has been extended to the representation of scenes with weighted regions. For such representations the distance transform has been introduced to form the basis of effective path planning in terrains with different types of obstacles.

The proposed path planning approach requires less computational effort, as always the largest parts (quadrants) of the free space or weighted regions are investigated. The distance transform of the robot workspace (based on the new neighbor finding strategy) can be performed very quickly. Computer simulation results of solving the path planning in different weighted robot workspaces illustrate the proposed approach.

The presented weighted region path planning for mobile robots can be used for different heterogeneous environmental terrains. Naturally it can be simply extended for the 3D cases, assuming octree representations.

Acknowledgment. The author gratefully acknowledges financial support from the Slovak Scientific Grant Agency.

References

1. Latombe, J.C.: Robot Motion Planning. Kluwer Academic Publishers, Boston (1991)
2. Szczerba, R.J. et al: Planning Shortest Paths among 2D and 3D Weighted Regions using Framed-subspaces. Int. J. Robotics Research 17 (1998) 531-546
3. Samet, H.: The Design and Analysis of Spatial Data Structures. Addison Wesley, Reading, MA (1990)
4. Vörös, J.: Simple Path-planning Algorithm for Mobile Robots using Quadtrees. In: Kopacek, P. (ed.): Preprints IFAC Workshop on Human-Oriented Design of Advanced Robotics Systems, Wien (1995) 197-202

5. Vörös, J.: A Strategy for Repetitive Neighbor Finding in Images Represented by Quadtrees. Pattern Recognition Letters 18 (1997) 955-962
6. Vörös, J.: Top-down Generation of Quadtree Representation using Color-Change-Code. Computers and Artificial Intelligence 13 (1994) 91-103
7. Borgefors, G.: Distance Transformations in Digital Images. Computer Vision, Graphics, and Image Processing 34 (1986) 344-371
8. Borgefors, G. et al: Parallel Distance Transforms on Pyramid Machines: Theory and Implementation. Signal Processing 21 (1990) 61-86
9. Piper, J., Granum, E.: Computing Distance Transformations in Convex and Non-convex Domains. Pattern Recognition 20 (1987) 599-615
10. Shaffer, C.A., Stout, Q.F.: Linear Time Distance Transforms for Quadtrees. CVGIP: Image Understanding 54 (1991) 215-223
11. Jung, D., Gupta, K.K.: Octree-based Hierarchical Distance Maps for Collision Detection. Journal of Robotic Systems 14 (1997) 789-806
12. Samet, H.: Distance Transform for Images Represented by Quadtrees, IEEE Trans. Pattern Analysis and Machine Intelligence 4 (1982) 298-303

Matrix Model of Robot in Matlab - Simulink

František Šolc

Department of Control, Measurement and Instrumentation
Brno University of Technology
Božetěchova 2, 612 66 Brno, Czech Republic
solc@dame.fee.vutbr.cz

Abstract. The article describes construction of a mathematical model of a robot The method is based on a matrix calculus. The advantage of the described method is in unified approach which allows systematic construction of mathematical model of mechanical part of a robot together with its electrical drives. Matrix engine of the MATLAB-SIMULINK then allows easy simulation of the complete robot. The article also discuss classical way of model construction and gives its comparison with the matrix approach.

1 Introduction

It is well known that the realistic mathematical model of an industrial robot is a complicated non-linear system. Until now, methods for modelling of robots and design of their control systems were rather complicated. The aim of this paper is to show that the above mentioned problem could be simplified with the use of MATLAB-SIMULINK[1] software. MATLAB which stands for matrix laboratory is a well known computing environment for high-performance numeric computation with matrices. SIMULINK is built on the MATLAB numeric computation system, offering to the analyst and/or designer immediate access to a comprehensive selection of mathematical and engineering techniques for further analysis of an investigated dynamic system. The advantages of the software are shown in this article on example of a robot modelling and simulation.

2 Classical Model

The mathematical model of the mechanical part of the robot is generally expressed in the form of a set of differential equations which are usually written in matrix form:

$$\mathbf{H}(q)\ddot{q} + \mathbf{h}(q, \dot{q}) = \mathbf{T} \tag{1}$$

[1] MATLAB and SIMULINK are registered trademarks of the MathWorks, Inc.

F. Pichler, R. Moreno-Díaz, and P. Kopacek (Eds.): EUROCAST'99, LNCS 1798, pp. 250–255, 2000.

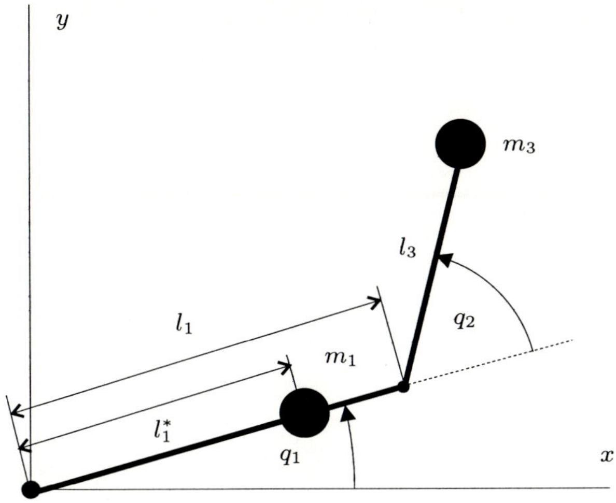

Fig. 1. Kinematic scheme of a manipulator

E.g. for very simple planar manipulator from Fig. 1 the equations are, [1]:

$$
\begin{aligned}
(m + 2\,m_3\,l_3\,l_1\cos q_2)\,\ddot{q}_1 + (m_3\,l_3^2 + m_3\,l_3\,l_1\cos q_2)\,\ddot{q}_2 - \\
- m_3\,l_3\,l_1\,\dot{q}_2\,(\dot{q}_2 + 2\,\dot{q}_1)\sin q_2 = T_1 \\
(m_3\,l_3^2 + m_3\,l_3\,l_1\cos q_2)\,\ddot{q}_1 + m_3\,l_3^2\,\ddot{q}_2 + m_3\,l_3\,l_1\,\dot{q}_1^2\sin q_2 = T_2 \\
m = m_1\,l_1^{*2} + m_3\,l_1^2 + m_3\,l_3^2
\end{aligned}
\tag{2}
$$

Knowledge of dynamic model of mechanical part only is not sufficient for control synthesis. One must know also model of actuators which develop driving torques and forces. General dynamic model of a simple DC drive is described in state variable form (3) and matrix form (4). Torques and forces from (1) and/or (2) can be considered as disturbing (reactive) forces acting on individual drives of course via gearing. That is why coordinates $\mathbf{q}$ of the manipulator, coordinates θ_m and ω_m of drives, torques T of the manipulator and torques M of the drive are related by gear ratio N, see eq. (5).

$$
\begin{aligned}
\dot{\theta}_{mk} &= \omega_{mk} \\
\dot{\omega}_{mk} &= \frac{1}{J_m}(C_m\,i_k - B_m\,\omega_{mk} - M_{mk}) \\
\dot{i}_k &= \frac{1}{L_m}(u_k - C_e\,\omega_{mk} - R\,i_k)
\end{aligned}
\tag{3}
$$

$$
\dot{\mathbf{x}}_{mk} = \mathbf{A}_{mk}\,\mathbf{x}_{mk} + \mathbf{b}_{mk}\,u_k + \mathbf{f}_{mk}M_{mk} \tag{4}
$$

where

$$
\mathbf{x}_{mk} = [\theta_{mk}\;\;\omega_{mk}\;\;i_k]^T, \quad \theta_{mk} = N q_k, \quad M_k = T_k/N_k, \quad k = 1,2 \tag{5}
$$

Denoting state vector of the complete system $\mathbf{x}_r = [q_1 \;\; \omega_1 \;\; i_1 \;\; q_2 \;\; \omega_2 \;\; i_2]^T$ where ω means velocity of respective joint co-ordinate we can substitute from equations (2) and (5) to equations (3) and write complete equations of the robot (manipulator with drives) in form

$$
\begin{aligned}
\dot{q}_1 &= \omega_1 \\
\dot{\omega}_1 &= \frac{1}{NJ_m}(C_m i_1 - NB_m\omega_1 - \frac{1}{N}[(m + 2\,m_3\,l_3\,l_1\cos q_2)\,\dot{\omega}_1 + \\
&\quad + (m_3\,l_3^2 + m_3\,l_3\,l_1\cos q_2)\dot{\omega}_2 - m_3\,l_3\,l_1\omega_2(\omega_2 + 2\omega_1)\sin q_2) \\
\dot{i}_1 &= \frac{1}{L_m}(u_1 - C_e N\omega_1 - R\,i_1) \\
\dot{q}_2 &= \omega_2 \\
\dot{\omega}_2 &= \frac{1}{NJ_m}(C_m i_2 - NB_m\omega_2 - \frac{1}{N}[(m_3\,l_3^2 + m_3\,l_3\,l_1\cos q_2)\,\dot{\omega}_1 + \\
&\quad + m_3\,l_3^2\,\dot{\omega}_2 + m_3\,l_3\,l_1\omega_1^2\sin q_2]) \\
\dot{i}_2 &= \frac{1}{L_m}(u_2 - C_e N\omega_2 - R\,i_2)
\end{aligned}
\tag{6}
$$

While equations for $\dot{q}, \dot{i}$ are ordinary state equations the other two equations for $\dot{\omega}$ are not true state equations and create "algebraic loop" which should be resolved to bring the equations to form

$$\dot{\omega} = f(\mathbf{x}_r) \tag{7}$$

Fortunately the algebraic loop is linear with respect to $\dot{\omega}$ and can be rewritten in form

$$
\begin{aligned}
\dot{\omega}_1 &= A_1\dot{\omega}_1 + B_1\dot{\omega}_2 + C_1 \\
\dot{\omega}_2 &= A_2\dot{\omega}_2 + B_2\dot{\omega}_1 + C_2
\end{aligned}
\tag{8}
$$

where

$$
\begin{aligned}
A_1 &= -\frac{1}{N^2 J_m}(m + 2\,m_3\,l_3\,l_1\cos q_2) \\
B_1 &= -\frac{1}{N^2 J_m}(m_3\,l_3^2 + m_3\,l_3\,l_1\cos q_2) \\
C_1 &= \frac{1}{NJ_m}[C_m i_1 - NB_m\omega_1 + \frac{m_3\,l_3\,l_1}{N}\,\omega_2(\omega_2 + 2\,\omega_1)\sin q_2] \\
A_2 &= -\frac{1}{N^2 J_m} m_3\,l_3 \\
B_2 &= -\frac{1}{N^2 J_m}(m_3\,l_3^2 + m_3\,l_3\,l_1\cos q_2) \\
C_2 &= \frac{1}{NJ_m}(C_m i_2 - NB_m\omega_2 - \frac{m_3\,l_3\,l_1\,\omega_1}{N}\sin q_2)
\end{aligned}
\tag{9}
$$

Solution of equations (8) gives us rather complicated state variable equations

$$\dot{\omega}_1 = \frac{-A_2C_1 + B_1C_2 + C_1}{A_1A_2 - A_1 - A_2 + 1 - B_1B_2}$$
$$\dot{\omega}_2 = \frac{-A_1C_2 + C_2 + C_1B_2}{A_1A_2 - A_1 - A_2 + 1 - B_1B_2} \qquad (10)$$

which can be used for classical modeling of the robot. Of course situation will be much more complicated for manipulator with greater number of degree of freedom.

3 Matrix Model

Matrix access to robot modeling is discussed in [2]. Considering equations (6) and relation (5) one can write model of a drive in form

$$\dot{\mathbf{x}}_k = \mathbf{A}_k\mathbf{x}_k + \mathbf{b}_k u_k + \mathbf{f}_k T_k \qquad (11)$$

where $\mathbf{x}_k = [q_k \;\; \omega_k \;\; i_k]^T$ and individual matrices are

$$\mathbf{A} = \begin{bmatrix} 0 & 1 & 0 \\ 0 & -\frac{B}{J_m} & \frac{C_m}{NJ_m} \\ 0 & -\frac{NC_e}{L_m} & -\frac{R}{L_m} \end{bmatrix}; \quad \mathbf{b}_k = \begin{bmatrix} 0 \\ 0 \\ \frac{1}{L_m} \end{bmatrix}; \quad \mathbf{f}_k = \begin{bmatrix} 0 \\ -\frac{1}{N^2J_m} \\ 0 \end{bmatrix} \qquad (12)$$

Model of both drives together can be written again in matrix form

$$\dot{\mathbf{x}}_r = \mathbf{A}_d\mathbf{x}_r + \mathbf{B}_d\mathbf{u} + \mathbf{F}_d\mathbf{T} \qquad (13)$$

where complete state vector of the robot $\mathbf{x}_r = [\mathbf{x}_1^T;\; \mathbf{x}_2^T]^T$ and matrices $\mathbf{A}_d, \mathbf{B}_d$ and $\mathbf{F}_d$ are diagonal block matrices consisting of the respective matrices of individual drives, thus

$$\mathbf{A}_d = diag(\mathbf{A}_1, \mathbf{A}_2), \quad \mathbf{B}_d = diag(\mathbf{b}_1, \mathbf{b}_2), \quad \mathbf{F}_d = diag(\mathbf{f}_1, \mathbf{f}_2) \qquad (14)$$

The vector of joint co-ordinates $\mathbf{q}$ may be obtained from the complete state vector with the help of multiplication by a separation block matrix $\mathbf{T}_q = diag(\mathbf{k}_q)$, where $\mathbf{k}_q = [1 \;\; 0 \;\; 0]$. Thus $\mathbf{q} = \mathbf{T}_q\mathbf{x}_r$. Similarly, the vector of velocities of joint co-ordinates may be obtained in the same manner, but now the separation matrix is $\mathbf{T}_\omega = diag(\mathbf{k}_\omega)$, where $\mathbf{k}_\omega = [0 \;\; 1 \;\; 0]$, thus:

$$\dot{\mathbf{q}} = \mathbf{T}_\omega\mathbf{x}_r, \quad \ddot{\mathbf{q}} = \mathbf{T}_\omega\dot{\mathbf{x}}_r \qquad (15)$$

Now we can write equation (1) in form

$$\mathbf{H}(\mathbf{x}_r)\mathbf{T}_\omega\dot{\mathbf{x}}_r + \mathbf{h}(\mathbf{x}_r) = \mathbf{T} \qquad (16)$$

substituting for $\mathbf{T}$ from (16) to (13) we come to matrix "algebraic loop" which can be solved more easily than in the classical case. We come to model

$$\dot{\mathbf{x}}_r = [\mathbf{I} - \mathbf{F}_d\mathbf{H}(\mathbf{x}_r)\mathbf{T}_\omega]^{-1}[\mathbf{A}_d\mathbf{x}_r + \mathbf{B}_d\mathbf{u} + \mathbf{F}_d\mathbf{h}(\mathbf{x}_r)] \tag{17}$$

or

$$\dot{\mathbf{x}}_r = \mathbf{A}_c(\mathbf{x}_r) + \mathbf{B}_c\mathbf{u} \tag{18}$$

where

$$\mathbf{A}_c = [\mathbf{I} - \mathbf{F}_d\mathbf{H}(\mathbf{x}_r)\mathbf{T}_\omega]^{-1}[\mathbf{A}_d\mathbf{x}_r + \mathbf{F}_d\mathbf{h}(\mathbf{x}_r)]$$
$$\mathbf{B}_c = [\mathbf{I} - \mathbf{F}_d\mathbf{H}(\mathbf{x}_r)\mathbf{T}_\omega]^{-1}\mathbf{B}_d$$

which can be easily implemented in MATLAB-SIMULINK thanks to its matrix calculation engine. Model in MATLAB-SIMULINK is shown in Fig. 2. MATLAB functions *Bc*, *suma* and *Ac* are listed in Appendix.

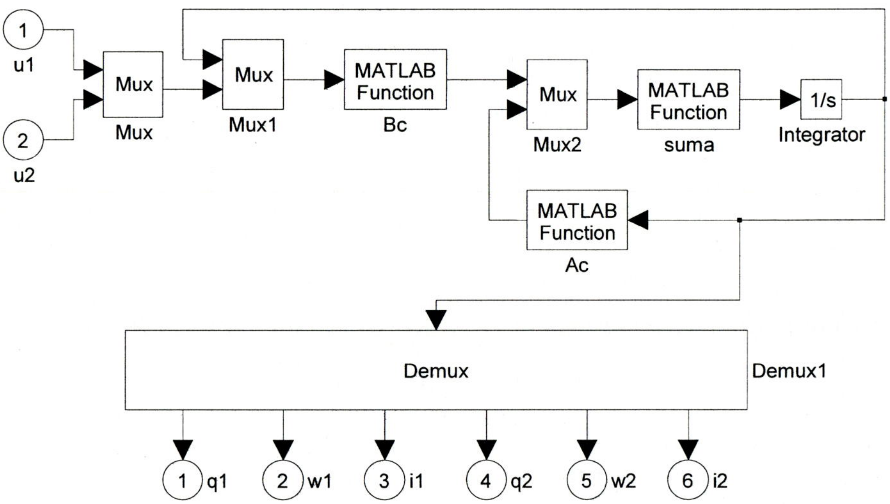

Fig. 2. Simulink scheme of the robot

The described principle can be used for systems with greater degree of freedom easily. Equations (17) and (18) stay true for such a case . User of the model simply fills in the matrices which prevents making errors significantly. A disadvantage of the model is lower speed of simulation run as inversion of matrices must be done in each integration step.

Acknowledgement

The work was supported by research projects J22/98:262200012
and J22/98:260000013 and by grant GACR 102/98/0552

References

1. Spong, M. W., Vidyasagar, M.: Robot Dynamics and Control. J. Wiley, N. Y. (1989)
2. Vukobratovic, M., Stokic, D.: Applied Control of Manipulation Robots. Springer-Verlag (1989)

Appendix

$\mathbf{A}_c$ - matrix of complete system model of 2 joint robot with drives:

```
function y=ac(x)
global A B T F m1 m3 l1 l1d l3
H=[m1*l1d^2+m3*l1^2+m3*l3^2+2*m3*l3*l1*cos(x(3))   m3*l3^2+m3*l3*l1*cos(x(3))   0
   m3*l3^2+m3*l3*l1*cos(x(3))                              m3*l3^2              0
           0                                                0                  m3];
h=[-m3*l1*l3*x(4)*(x(4)+2*x(2))*sin(x(3))
   m3*l3*l1*x(2)^2*sin(x(3))
       m3*9.81                          ];
INV=inv(eye(3)-F*H*T);
y=INV*(A*x+F*h)
```

$\mathbf{B}_c$ - matrix of complete system model of 2 joint robot with drives:

```
function y=bc(x)
global A B T F m1 m3 l1 l1d l3
H=[m1*l1d^2+m3*l1^2+m3*l3^2+2*m3*l3*l1*cos(x(3))   m3*l3^2+m3*l3*l1*cos(x(3))   0
   m3*l3^2+m3*l3*l1*cos(x(3))                              m3*l3^2              0
           0                                                0                  m3];
INV=inv(eye(3)-F*H*T);
u=[x(7)
   x(8)
   x(9)];
y=INV*B*u;
```

Function *suma*:

```
function y=suma(x)
for i=1:6
  y(i)=x(i)+x(i+6);
end
```

4 MODELING AND SIMULATION

Integrating Two Dynamic Models of Business-Logistics Plant

R. Sato

Inst. of Policy and Planning Sciences, University of Tsukuba,
Tennodai, Tsukuba, Ibaraki 305-8573, Japan
rsato@sk.tsukuba.ac.jp

Abstract. In order to engineer business processes a model is needed that allows us to analysis and design of the dynamic properties of the processes. This paper shows a unified framework for data flow diagrams (DFD, for short) and Petri nets in modeling business processes. Introducing the concept of business transaction Petri net, we provide a firm mathematical basis for a comparison of a such Petri net to a business process modeled by a DFD. It will be turned out that the behavior of a business transaction Petri net is simulated by the corresponding business transaction system.
The resulted framework makes it possible that the experience with DFD's will be applicable to Petri nets in modeling business processes, and vice versa.

1 Introduction

Control engineering has been developed to meet the demand for industrialization. Starting from PID control, it continues to grow producing the concepts of state space method and optimal control. The model employed is, mainly, a set of linear continuous differential equations. Differential equations are often used to approximate the original models or to implement control algorithms into, firstly operational amplifiers, and then a computer. Typical examples include a chemical plant for refinement of crude oil, many kinds of electric circuits, and inverted pendulum. Now a computer can be seen as an interface device between mechanical, electrical, and/or human actuator and information. Such a computer can even be implemented in a VLSI chip and be embedded in miscellaneous industrial products.

If we look out at organizational process such as design and development, production, procurement and sales and distribution rather than products themselves, then computers turn out to be also important components of such business processes. That is, most of businesses use computers and network in their business operation. Also used are ERP (enterprise resource planning) packages, e.g., SAP R/3, Oracle applications, Baan, and so on. In this way, our economies continue to invent and develop business processes that consist of business tasks to convert raw materials into finished products at customers through production. This flow of materials is called logistics, and a database-based information system serves as a control device for a company-wide virtual plant formed by its logistics.

In order to control an inverted pendulum, first the physical law governing the pendulum and related things are described by differential equation. Then, using the

F. Pichler, R. Moreno-Díaz, and P. Kopacek (Eds.): EUROCAST'99, LNCS 1798, pp. 259–273, 2000.

concept of state, stability and state observer, a feedback control is devised with operational amplifiers. This is a typical methodology to control physical objects.

The same approach is applicable to attain suitable dynamic properties in business processes. For example, using data flow diagrams (DFDs) or event-driven process chain (Scheer, 1994; Keller, et.al., 1998), the entity types like customer order or production order and the activities like accept or issue orders are extracted from business process. Entities correspond to business documents. Dynamic aspects of a business process such as lead-time and through put did not been paid much attention, because there seems to be no good way to calculate them accurately. Dynamic models should have explicit relation to DFD and/or other descriptive modeling tools. The situation is explained as Fig. 1.

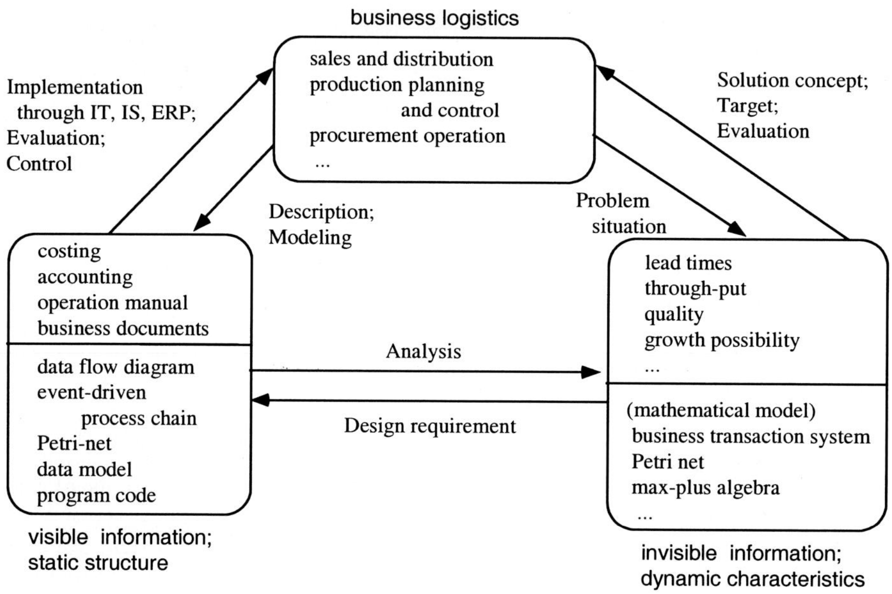

Fig. 1. Engineering framework for business process

The dynamic model of business processes is developed as a discrete-event system, called a business transaction system [6] (BTS, for short). It describes that the activities in an organizational logistics process are executed in parallel, and that the integrity and synchronization of the process as a whole is attained through the business documents that connect activities. Those documents can be either paper- or database package-based. B.P. Zeigler (1976, 1984) has originated the DEVS formalism mainly for modeling discrete-event phenomena, and the formulation of business transaction systems is in a multicomponent DEVS. That is, each of the activities in a business transaction system is a DEVS and the whole is also a DEVS.

A data flow diagram (DFD) is a simple graphical network of data flows, business processes, data stores and external organizations. The data flow is a labeled arrow, the process a round rectangle, the data store an open rectangle, and the sink or source a rectangle. Since the DFD is convenient for modeling business systems, most information systems methodologies have adopted it. Sato and Praehofer (1997) has shown that a business system modeled by DFD's is a business transaction system. If a DFD has the following two characters, then is called a BTS. First, there is a data store between two processes. Second, each source is connected to a process via data store.

Establishing the relation between a BTS to a timed Petri net, we can calculate an agility index of a business process. The index is known as the cycle time of a timed Petri net (Baccelli et.al., 1992). Certain BTS shows periodic behavior in the sense that the same state of a BTS is observed periodically. A BTS has a description of state transition function, and its state consists of the inventories of business documents and the list of respective times to finish operations of activities in the BTS. The timed Petri net corresponding to a business transaction Petri net (BT-PN, for short).

According to Sato (1999a), this paper will provide a formal relation between a business transaction system and a corresponding timed Petri net called a business transaction Petri net.

2 Business Transaction Petri Net

The components of a business transaction are depicted by a data flow diagram. In this section we define a business transaction Petri net (a BT-PN, for short) that has an isomorphic dynamics to a business transaction system.

A Petri net is defined as follows (Murata, 1992).

Definition 1. Petri net
A Petri net is a quintuple (P, T, F, W, μ_0), where

- $P = \{p_1, p_2, ..., p_x\}$: a finite set of places;
- $T = \{t_1, t_2, ..., t_y\}$: a finite set of transitions;
- F: a set of arcs between transitions and places;
- $W: F \to \{1, 2, ...\}$: the assigned weight with an arc;
- $\mu_0: P \to \{0, 1, 2, ...\}$: initial marking of the places.
- $P \cap T = \phi$, where ϕ is the empty set.

In the above definition, x and y represent the number of places and transitions, respectively. For a marking μ_0, $\mu_0(p_i)$ represents the number of tokens in the place, p_i. Therefore, μ_0 is a function from P to the set of non-negative integers. If a Petri net is given without any marking then it is called a Petri net structure, or a Petri net graph. For a Petri net graph J and a marking μ_0, we have a Petri net (J, μ_0).

A Petri net has a diagrammatic representation. A circle represents a place. A bar represents a transition. Directed arcs (arrows) connect the places and the transitions. An arc directed from a place pi to a transition tk defines the place to be an input to the transition. An output place is defined by an arc from a transition to the place. Multiple inputs and outputs are possible with multiple arcs. An arc has its weight shown by the function W. When the weight of an arc is 1, then it is not shown in the diagram of a Petri net, and vice versa.

A marking shows a deployment (an allocation) of tokens in the places of a Petri net. That is, a marking m is a function $\mu:P \rightarrow \{0, 1, 2, ...\}$. When a transition fires, tokens are transferred from the input places to the output places according to their weights. The fact that a place p is an input of a transition t is denoted by $p \in I(t)$, and p is called an input place of t. Similarly, $p \in O(t)$ means that the place p is an output place of the transition t. The firing rule that shows when and how transitions fire and corresponding tokens are transferred. Let a transition t be arbitrary. Let w(p, t) be the weight of the arc from a place *p* to *t*, and w(t, p') the weight of the arc from t to a place p'.

General firing condition: If for any input place p of t the number of tokens in p is greater than or equal to w(p, t) then the transition t is said to be enabled (by the input tokens to fire).

Transferring tokens: If an enabled t fires, then w(p, t) tokens in an input place p are removed and w(t, p') tokens are produced in an output place p'. If t has multiple input places and multiple output places, respective numbers of tokens in the input places are removed and produced in the output places depending on their weights.

If there are not enough tokens in an input place of a transition then the transition is not enabled and cannot fire. Firing a transition changes the marking of a Petri net from, say, μ to μ'.

In order to explicitly deal with state transition along time, we introduce the concept of a timed Petri net according to Baccelli et al (1992). Any place of a timed Petri net has its own holding time. For a token in a place, if the elapsed time from its entry is equal to exceeds the holding time, then it can be used to fire any transition that has the place as an input place. Such tokens are called matured. A transition of a timed Petri net will fire if it is enabled by matured tokens.

We assume that the holding time of a place does not vary among tokens. That is, any token in a place is available to be used to fire the output transition if it stayed longer than or equal to the holding time of the place. (Different places usually have different holding times.)

A business transaction Petri net is a subclass of the class of timed Petri nets, which satisfies additional conditions defined below.

Definition 2. Business Transaction Petri Nets
A timed Petri net is called a business transaction Petri net (BT-PN, for short) if satisfies the following five conditions:

(1) A part of BT-PN is consists of two transitions and a place between them, |→o→|, is called an activity unit of the BT-PN and the place a busy place. The input places of an activity unit are those of the input transition of the busy place. The output places of an activity unit are defined similarly. The weight of the arcs coming into and going out of the busy place is one.
(2) Any place that is not a busy place has one input arc and one output arc, and is called a connecting place. The weights of input arcs and output arcs are positive integers. For an activity unit there is a place of which input transition is the output transition of the busy place and of which output transition is the input

transition of the busy place. This place for an activity unit is called an idle place (of the unit).

(3) Any arc that goes into an activity unit goes into the input transition of the busy place.

(4) On any path in a BT-PN, activity units and connecting place appears alternatively. In this sense, it is bipartite.

(5) Holding times of a BT-PN are positive integers, while those of connecting places zero.

For any place p of a BT-PN, the weight of the arc coming into p is written by w_{pi}. Similarly, the weight of the arc going out of p is denoted as w_{po}. For example, if p is a busy place then $w_{pi} = w_{po} = 1$.

3 State Transition of Business Transaction Petri Net

We describe the state transition mechanism of a business transaction Petri net. That is, the precise description of state transition of a BT-PN will be defined in this section.

Let $J = <P, T, F, W>$ be an arbitrary Petri net structure. All of the transitions as well as the places of J will be numbered. Let n_A is the number of activity units of J that are numbered as k, $1 \leq k \leq n_A$. For a busy place k, the input transition is represented by t^i_k, and the output transition t^o_k. That is , $T = \{t^i_k, t^o_k \mid 1 \leq k \leq n_A\}$. Let n be the number of the places (i.e., the cardinality of P is n). Let P_A be the set of the busy places and P_C the set of connecting places. Then $P = P_A \cup P_C$ and $P_A \cap P_C = \phi$ hold and the cardinality of P_A is n_A.

Let N be the set of natural numbers, and N^n be the set of vectors with n integer components. An element μ of N^n can be thought of a marking for J. In this case, μ (k) represents the number of tokens in the place numbered by k. Let R be the set of reals, and $R^\infty = R \cup \{\infty\}$. R^{nA} denotes the set of vectors with n_A real components.

The function lag: $N^n \times P_A \rightarrow R^\infty$ gives the holding time for the busy places. That is,

$$\text{lag}(\mu, q_k) = \begin{cases} \text{lag}_k, & \text{if } \mu(q_k) = 1 \\ \infty, & \text{otherwise.} \end{cases}$$

The value $\mu(q_k)$ denotes the holding time of the busy place numbered by k. Any initial marking is assumed to satisfy the following condition:

(Initial condition)
For any busy place $q_k \in P_A$ and arbitrary marking $\mu \in N^n$, $lag(\mu, q_k) - e_k > 0$ holds, where e_k is the elapsed time from the entry of the last token currently held in the busy place q_k.

If no token is in q_k, then the initial condition is satisfied, since $lag(\mu, q_k) = \infty$.

The state transition function $\delta: N^n \times R^{nA} \to N^n \times R^{nA}$ in the following part of this section. First we define the function dueP: $N^n \times R^{nA} \to \Pi(T)$, where $\Pi(T)$ is the power set of T. For arbitrary $\mu \in N^n$ and $e \in R^{nA}$, $dueP(\mu, e) = \{t^o{}_{i1}, t^o{}_{i2}, ..., t^o{}_{im}\} \Leftrightarrow$ for each k, $1 \le k \le m$, the following (1) and (2) hold:

(1) $I(t^o{}_{ik}) = \{q_{ik}\}$. That is, $t^o{}_{ik}$ is an output transition of a busy place q_{ik}.

(2) $lag(\mu, q_{ik}) - e_{ik} = \min_{[q_a \in P_A]}\{lag(\mu, q_a) - e_a \neq \infty\}$.

The function $dueP(\mu, e)$ gives the transitions to fire at the nearest future. The needed time for the transitions $dueP(\mu, e)$ is given by the function $taP(\mu, e)$. That is,

$$taP(\mu, e) = \min_{[q_a \in P_A]}\{lag(\mu, q_a) - e_a \mid 0 < lag(\mu, q_a) - e_a \neq \infty\}.$$

We define the first half of the whole state transition function δ. The function $\delta_1: N^n \times R^{nA} \to N^n \times R^{nA}$ is defined as follows. Let $\mu_0 = \mu$, and for each j, $1 \le j \le m$,

$\mu_j(p) =$

$\mu_{j-1}(p)$ - w_{po}, if $I(t^o{}_{ij}) = \{q_{ij}\}$;

$\mu_{j-1}(p) + w_{pi}$, if $p \in O(t^o{}_{ij})$;

$\mu_{j-1}(p)$, otherwise.

Let $\mu^1 = \mu_m$, and define e^1 such that $e^1(q_k) = e_k + taP(\mu, e)$ holds for each $q_k \in P_A$. Define δ_1 such that $\delta_1(\mu, e) = (\mu^1, e^1)$. This definition for δ_1 represents that matured tokens in enabled busy places are removed and tokens are produced in the idle places that correspond to the busy places, and that the time $taP(\mu, e)$ till the firing of the output transitions of the busy places with matured tokens are added to the elapsed times of the tokens in the other busy places.

We now define δ_2 for the second half of δ. The function δ_2 will represent that firings of all possible enabled transitions that are inputs of some busy places. Define a function $st(\mu)$ to show the enabled input transitions of busy places. For arbitrary $\mu \in N^n$, $st(\mu) = \{t^i{}_{k1}, t^i{}_{k2}, ..., t^i{}_{kr}\} \Leftrightarrow$ For each h, $1 \le h \le r$, the following (1) and (2) hold:

(1) $O(t^i{}_{kh}) = \{q_{kh}\}$. That is, $t^i{}_{kh}$ is an input transition of a busy place q_{kh}.

(2) For arbitrary p, if $p \in I(t^i{}_{kh})$ then $\mu(p) \ge w_{pi}$ holds.

The above condition (2) assures that the transitions of idle places with tokens can only be in st(μ). Let $\mu'_0 = \mu$, and for each j, $1 \le j \le r$,

$\mu'_j(p) =$

$\mu'_{j-1}(p) - w_{po}$, if $p \in I(t^i_{k_j})$;

$\mu'_{j-1}(p) + w_{pi}$, if $p \in O(t^i_{k_j})$;

$\mu'_{j-1}(p)$, otherwise.

A function g is used to represent the resultant μ'_r gained from μ. That is, $g(\mu) = \mu'_r$. The final marking is represented by the function h that is defined as follows:

$h(\mu) =$

μ, if $st(\mu) = \phi$;

$h(g(\mu))$, if $st(\mu) \neq \phi$.

It can be shown that the definition of h is well-defined. Define $e^2 \in R^{nA}$ based on μ and $e \in R^{nA}$ as follows: for each $q_k \in P_A$,

$e^2(q_a) =$

0, if $O(t^i_a) = \{q_a\}$ and $t^i_a \in \bigcup_{[p = 0, \infty]}\{st(g^p(\mu))\}$ hold;

$e(q_k)$, otherwise.

Define δ_2 such that $\delta_2(\mu, e) = (h(\mu), e^2)$.

Now we define the state transition function $\delta: N^n \times R^{nA} \rightarrow N^n \times R^{nA}$ of a BT-PN. For arbitrary μ and e, define $\delta(\mu, e) = \delta_2(\delta_1(\mu, e))$. The function δ has the following interpretation for a PT-PN: Given a marking $\mu \in N^n$ and a distribution e $\in R^{nA}$ of elapsed times of tokens in busy places with respect to μ, fire the matured activity units and then start activity units as much as possible, updating the elapsed times of tokens in busy places. Then the resultant marking and updated list of the elapsed times are represented by $\delta(\mu, e)$.

4 Business Transaction System to Simulate Business Transaction Petri Net

A business transaction system that simulate a given BT-PN is defined in this section. This section provides a formal definition, while the next section describes an example of this definition.

The concept of business transaction systems has been introduced in Sato and Praehofer (1997) and Sato (1997) as a general model of routine business processes. Its characteristics are

(1) A data flow diagram (DFD, for short) is a static model of a business process. Data flows are described as a file system that is modeled by a kind of the Entity-Relationship model called TH model.
(2) The whole dynamic structure is a discrete-event system that is formulated as a multicomponent DEVS.

DEVS is a systems theoretic formalism for discrete-event systems, which is originated by Zeigler (1974). Each activity in a DFD is model as a DEVS, and then they are connected by the file system to constitute a whole DEVS.

We construct a BTS that simulates a given BT-PN in the sense that Fig. 2 commutes.

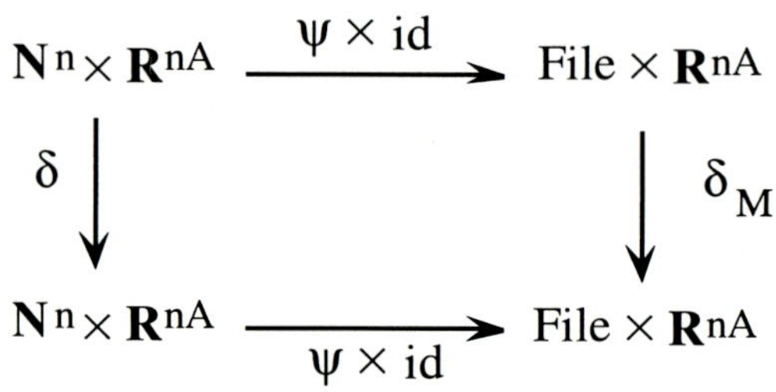

Fig. 2. Commutative diagram

Given a BT-PN, the function $\psi: \mathbf{N}^n \rightarrow \text{File}$ in Fig. 2 natural correspondence from a marking to a file structure. The function δ_M is the state transition function of the corresponding BTS.

To define a BTS, we need to define its components. They are activities, respective conditions to start activities, and the information sharing mechanism. Formally, they are described by the following sets and functions.

(1) The set of activities:
$\{G_a = \langle S_a \times R^{\infty}, \delta_a, ta_a \rangle \mid 1 \leq a \leq n_A\}$
(1) The function to describe staring condition for activities decided by the file system:
$f_{FA}: \text{File} \rightarrow \Pi(\{1, 2, \ldots, n_A\})$
(2) The function to describe updating the file system triggered by activities:
$f_{FileValue}: \text{File} \times \Pi(\{1, 2, \ldots, n_A\}) \rightarrow \text{File}$

This section provides the definition of these set and functions for a BT-PN.

As indicated above, an activity unit of a BT-PN corresponds to an activity of a BTS. Ga is a DEVS. In the following, a busy place q_a corresponds to an activity *a* of a BTS.

We define the file system of a BTS. The set KS of controlled entity types virtually represents the set of table names in a relational database. Let $KS = P - P_A = P_C$. That is, all of the places but busy places correspond to tables names in the file system. In the definition of a BTS an element p of KS is virtually a table, so it has its defining attributes. Define the attributes of p as {#FT(p), number_of_tokens}, where #FT(p) is the primary key of the table for p. Since we think of non-colored tokens, this kind of table has only one line to show that a place p has some tokens. This completes the definition of the file system. We employ f_μ to show the file contents function that corresponds to a marking μ. Then, $f_\mu(p)$ is the number of tokens in a place p. That is, $f_\mu(p) = \mu(p)$ holds for each p. This correspondence between makings and file contents are represented by the function ψ:$N^n \rightarrow$ File, $\psi(\mu) = f_\mu$.

Both of the functions f_{AK} and f_{KA} specify a subset of KS for each activity as follows. Take Fig. 3 as an example.

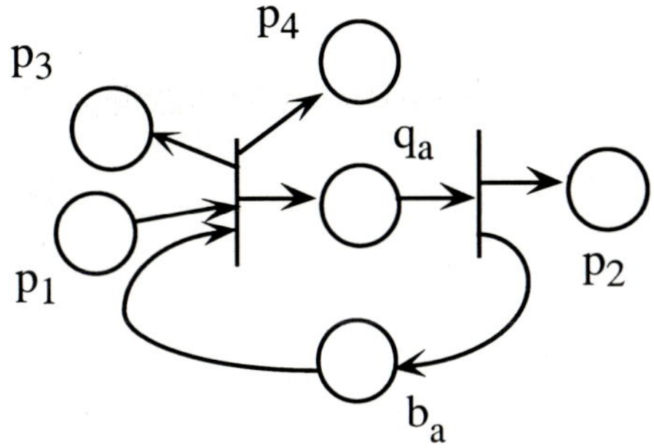

Fig. 3. Activity unit and connected places

We define b_a, $p_1 \in f_{KA}(a)$, b_a, $p_2 \in f^F{}_{AK}(a)$, and p_3, $p_4 \in f^S{}_{AK}(a)$. That is, the fact that p_3 and p_4 are the output places of the input transition of the activity unit is shown by $f^S{}_{AK}(a) = \{p_3, p_4\}$ while the fact that b_a and p_2 are the output places of the output transition of the activity unit is shown by $f^F{}_{AK}(a) = \{b_a, p_2\}$. We define the function f_{AK} such that $f_{AK}(a) = f^S{}_{AK}(a) \cup f^F{}_{AK}(a)$. The input places to the input transition of the activity unit is shown by $f_{KA}(a)$. For a general activity unit those three functions are defined in the same manner. Therefore, the places connected to an activity unit is represented by $f_{AK}(a) \cup f_{KA}(a) \cup \{q_a\}$.

Let $S = \{(f_\mu, e_1, e_2, ..., e_{n_A}) \mid f_\mu \in \text{File and } (e_1, e_2, ..., e_{n_A}) \in R^{n_A}\}$. The set S is the state space of a BTS (Sato and Praehofer, 1997).

Let $S_a = \text{File}|_{f_{AK}(a) \cup f_{KA}(a)}$; that is, the restriction of File to $f_{AK}(a) \cup f_{KA}(a)$. Therefore the restriction $f^a{}_\mu$ of f_μ to $f_{AK}(a) \cup f_{KA}(a)$ consists of the set S_a; That is, $S_a = \{f^a{}_\mu \mid f_\mu \in \text{File}\}$.

Define the maturity (time-advance) function ta_a: $S_a \rightarrow R^\infty$ for G_a by $ta_a(f^a{}_\mu) = lag(f'_\mu, q_a)$, where f'_μ is any file contents function except that $f'^a{}_\mu = f^a{}_\mu$ holds. Notice that the time to maturity of token is infinity for an activity place without any tokens. The definition of the function ta_a is well-defined (Sato and Praehofer, 1997).

We define the function f_{FA} to show the enabled transitions:
$a \in f_{FA}(f_\mu) \Leftrightarrow$ For each p, if $p \in f_{KA}(a)$, then $f_\mu(p) \bullet w_{pi}$ holds.

This definition says that $a \in f_{FA}(f_\mu)$ holds iff all of the input tables of the activity a have enough matured tokens to commence the activity. Notice that $a \in f_{FA}(f_\mu)$ assures that the activity is not in busy state. We say that an activity of a BTS is said to be "ready to start at $f^a{}_\mu$," if and only if there is a distribution f'_μ of tokens such that $f^a{}_\mu = f'_\mu|_{f_{KA}(a)}$ and $a \in f_{FA}(f'_\mu)$ holds.

For an activity a, in order to define the state transition function of, a we need two functions δ_{aO} and δ_{aI} that are defined below. The function δ_{aO} shows the updating of the file system of the BTS to represent transference of tokens in a BT-PN with respect to finishing activities. When we apply this function to $f^a{}_\mu$, it is needed to hold the fact that $f^a{}_\mu(b_a) = 0$.

$\delta aO{:}File|_{f_{AK}(a) \cup f_{KA}(a)} \rightarrow File|_{f_{AK}(a) \cup f_{KA}(a)}$,

$\delta aO(f^a\mu)(p) =$

$f^a\mu(p) + wpo$, if $p \in f^F AK(a)$ and $f^a\mu(ba) = 0$;

$f^a\mu(p)$, otherwise.

The function δ_{aI} shows the updating of the file system of the BTS to represent transference of tokens in a BT-PN with respect to starting activities.

$\delta_{aI}{:}File|_{f_{AK}(a) \cup f_{KA}(a)} \rightarrow File|_{f_{AK}(a) \cup f_{KA}(a)}$,

$\delta_{aI}(f^a{}_\mu)(p) =$

$f^a{}_\mu(p) - w_{pi}$, if $p \in f_{KA}(a)$ and a is ready to start at $f^a{}_\mu$;

$f^a{}_\mu(p) + w_{pi}$, if $p \in f^S{}_{AK}(a)$ and a is ready to start at $f^a{}_\mu$;

$f^a{}_\mu(p)$, otherwise.

Define $\delta_a = \delta_{aI} \cdot \delta_{aO}$. That is, first apply δ_{aO} and then apply δ_{aI}.

We next define the file updating function $f_{FileValue}{:}File \times \Pi(\{1, 2, \ldots, n_A\}) \rightarrow File$.

For any a, define as follows:

$f_{FileValue}(f_\mu, \{a\})(p) =$

$\delta_{aI}(f^a{}_\mu)(p)$, if $p \in f_{AK}(a) \cup f_{KA}(a)$;

$f_\mu(p)$, otherwise.

For arbitrary $\{a_{i_1}, a_{i_2}, \ldots, a_{i_r}\}$, define as follows:

$f_{FileValue}(f_\mu, \{a_{i_1}, a_{i_2}, \ldots, a_{i_r}\}) =$

$f_{FileValue}(\cdots f_{FileValue}(f_{FileValue}(f_\mu, \{a_{i_1}\}), \{a_{i_2}\}), \ldots, \{a_{i_r}\})$.

Based on the functions defined above, the state transition function δ_M of the BTS is constructed in a normal procedure provided in Sato and Praehofer (1997). The time advance function $ta{:}File \times R^{nA} \to R^{\infty}$ is defined by

$$ta(f_\mu, e_1, e_2, ..., e_{nA}) = \min_{[a \in \{1, 2, ..., n_A]}\{ta_a(f_\mu) - e_a\}.$$

Define the function $due{:}File \times R^{nA} \to \Pi(\{1, 2, ..., n_A\})$ as follows:

$$due(f_\mu, e_1, e_2, ..., e_{nA}) = \{a_{i_1}, a_{i_2}, ..., a_{i_m}\} \text{ iff}$$

for each k, $1 \bullet k \bullet m$, $ta(f_\mu, e_1, e_2, ..., e_{nA}) = ta_{a_{i_k}}(f^{a_{i_k}}{}_\mu) - e_{i_k}$.

We next define the function f_F that represents the mechanism to start all of the possible activities start in turn from a file value:

$f_F(f_\mu) =$

f_μ, if $f_{FA}(f_\mu) = \phi$;

$f_F(f_{FileValue}(f_\mu, f_{FA}(f_\mu))$, otherwise.

The dispatching function $f_{Dispatch}$ represents the resultant file system when all of the possible activities start in turn for the file value: $f_{Dispatch}(f_\mu) = f_{FileValue}(f_\mu, f_{FA}(f_\mu))$.

The elapsed times $e_1, e_2, ..., e_{nA}$ for each activity is defined as follows. For each *i*,

$e'_i =$

0, if $i \in due(s) \cup_{[k=0, \infty]} f_{FA}[f_{Dispatch}{}^k (f_{FileValue}(f_\mu, due(s)))]$;

$e_i - ta(f_\mu, e_1, e_2, ..., e_{nA})$, otherwise,

where $f_{Dispatch}{}^k(f_\mu)$ represents *k* times application of $f_{Dispatch}$ to f_μ, and $s = (f_\mu, e_1, e_2, ..., e_{nA})$.

Now the state transition function δ_M is defined:

$\delta_M(f_\mu, e_1, e_2, ..., e_{nA}) = (f_F[f_{FileValue}(f_\mu, due(f_\mu, e_1, e_2, ..., e_{nA}))], e'_1, e'_2, ..., e'_{nA})$.

This completes the definition of the BTS that corresponds to a given BT-PN. The commutativity in Fig. 2 is shown in the next section.

5 Commutativity between Business Transaction Petri Net and Business Transaction System

A diagrammatic explanation for the commutativity between a business transaction Petri net and business transaction system is provided in this section. A simple Kanban-system is an example. Since a business transaction system is a general model of business processes, the commutativity we will show is applicable to business processes that are described by data flow diagrams.

A Kanban systems is a just-in-time control system. Kanbans are cards and used between two sequential processes. In the production process controlled by kanbans, one kanban is attached to the container for small quantity of specific part or products. When such part in a container is used up by a processing or assembly, then the attached kanban is put off into its holding place. When prescribed number of kanbans are accumulated in the holding place for the kanban then an production-order is issued to the preceding production/assembly process. In this way, production orders for various part propagate upstream to suppliers, while materials and part flow downstream to customers. The market demand for product becomes the order to the final process and is the only input to the whole system. The kanban system is also called as Toyota system (Monden 1983).

In Fig. 4 a supplier is modeled as an activity that accepts production order for part and delivers the corresponding amount of part to the manufacturer in some time. Tokens in Fig. 4 represent inventories of materials, parts, or products.

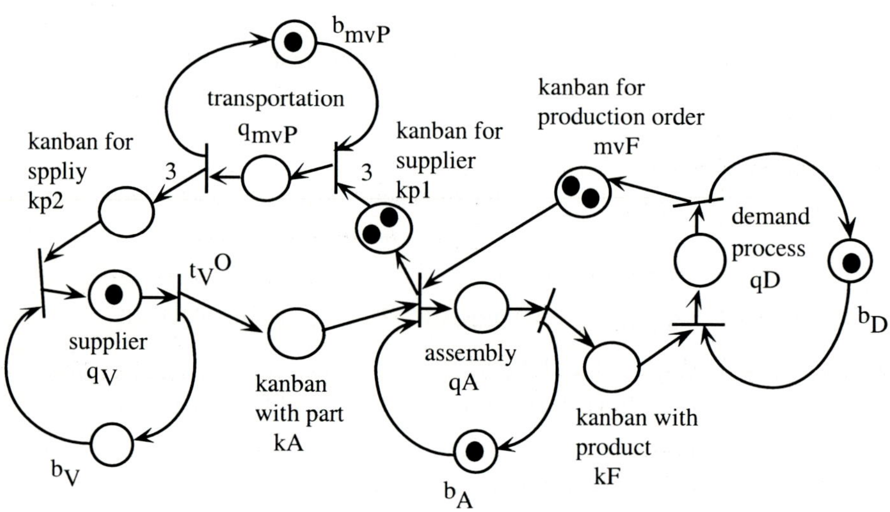

Fig.4. A BT-PN on Kanban system and initial marking

From the state depicted in Fig. 4 we show how the state transition by d will bring new state in turn. The process of the calculation consists of five steps: from (PN1) till (PN5).

(PN1) Because $t_V{}^o$ is the only enabled transition, we have $deuP(\mu, e) = \{t_V{}^o\}$.

(PN2) Let $(\mu^1, e^1) = \delta_I(\mu, e)$. Since (μ^1, e^1) is the state just after transition $t_V{}^o$ was fired, μ^1 is the marking just after the supplier has unloaded.

(PN3) Now, $st(\mu^1) = \{t_A{}^i\}$ holds. That is, $t_A{}^i$ is enabled.

(PN4) Starting from the firing of $t_A{}^i$, δ_2 gives the final result of successive firings of possible transitions. Let $(\mu^2, e^2) = \delta_1(\mu^1, e^1)$. Then, we have $st(g(\mu^1)) = \{t_{mvP}{}^i\}$. Since $st(g(g(\mu^1))) = \phi$, we have $\mu^2 = g(g(\mu^1))$.

(PN5) Here we have that $\bigcup_{[p=0, \infty]}\{st(g^p(\mu))\} = st(\mu^1) \cup st(g(\mu^1)) = \{t_A{}^i, t_{mvP}{}^i\}$ hold, it holds that $e^2(q_A) = e^2(q_{mvP}) = 0$ in addition to the fact that $e^1(q_V) = 0$.

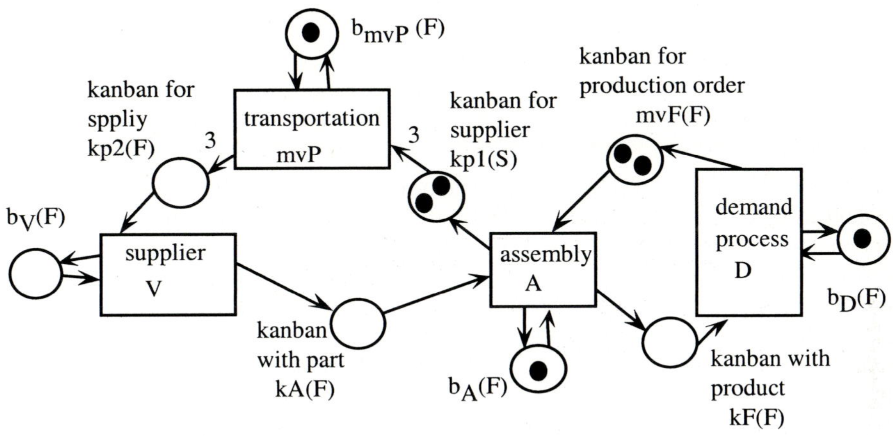

Fig. 5. The BTS correcponding to Fig. 4

We will construct a BTS corresponding to the BT-PN of Fig. 4, and show how the BTS makes its state transition. They are depicted in the steps from (BTS1) till (BTS5).

(PN1) The BTS corresponding to the BT-PN in Fig. 4 is Fig. 5 with the marking, f_μ. In Fig. 5, k_A(F) means that k_A is an element of $f^F{}_{AK}(V)$, and $k_{p_1}(S)$ means that k_{p_1} is an element of $f^S{}_{AK}(A)$.

(PN2) The initial condition on Fig. 4 implies that $due(f_\mu, e) = \{V\}$ and $f_{FA}(f_\mu) = \phi$ hold.

(PN3) We have μ^1. For the μ^1, the file content function satisfies $f_{\mu 1}(k_A) = 1$ and $f_{\mu 1}(b_V) = 1$. And for other place p, it holds that $f_{\mu 1}(p) = f_\mu(p)$.

(PN4) We calculate the value f_F $[f_{FileValue}(f_\mu, due(f_\mu, e))] = f_F(f_{\mu 1})$. Since $f_{FA}(f_{\mu 1}) = \{A\}$, the definition of f_F says that $f_F(f_{\mu 1}) = f_F$ $[f_{FileValue}(f_{\mu 1}, \{A\})]$. That is, $f_F(f_{\mu 1}) = f_{g(\mu 1)}$ holds. the value $f_{g(\mu 1)}$ represents the situation just after the commencement of the activity, A. At this stage we have $f_{FA}(f_{g(\mu 1)}) = \{mvP\}$. Thus, the activity mvP is started. The value, $f_F(f_{g(\mu 1)}) = f_F(f_{FileValue}(f_{g(\mu 1)}, \{mvP\})$ is decided. Comparing this to the BT-PN, we have $f_F(f_{g(\mu 1)}) = f_F(f_{g(g(\mu 1))}) = f_F(f_{g^2(\mu 1)})$. There exists no activity that can be commenced in $f_{g^2(\mu 1)}$. That is, $f_{g^2(\mu 1)} = \phi$ holds. Therefore, we have that $f_F[f_{FileValue}(f_\mu, due(f_\mu, e))] = f_{g^2(\mu 1)}$.

(PN5) Let s be (f_μ, e). Since $\{A, mvP\} = due(s) \cup_{[k=0, \infty]} f_{FA}[f_{Dispatch}^k(f_{FileValue}(f_\mu, due(s)))]$ holds, the elapsed time for both A and mvP becomes 0.

The above steps provide the commutativity of Fig. 2. For a BT-PN there exists a BTS that simulates in the sense of Fig.2.

6 Conclusion

Based on the model of business transaction system, the dynamics of business process as a logistics plant can be analyzed. The theory of max-plus algebra[1, 9] for business transaction can be introduced for that purpose, introducing the corresponding business transaction Petri net. Using a business transaction system and the business transaction Petri net, we can investigate how feedback control schema produces periodic behavior and how bottleneck activity effects. Since this paper has provided a firm theoretical basis for two models to such business process engineering, a thorough development of a kind of business process algebra and a methodology to design business process equipped with the algebra is a topic of future research.

Peterson (1981) pointed out that modeling through Petri net had not been developed yet. The equivalence between DFD and Petri nets shown in this paper can be thought of a step toward that direction in modeling business processes.

Acknowledgment. The author is grateful to the Grant-in-Aid for Scientific Research (no. 11630119) of the Japan Society for Promotion of Science for the partial support for the research.

References

[1] Baccelli, F.L., Cohen, G., Olsder, G.J., and Quadrat, J.P (1992) Synchronization and Linearity -- An algebra for discrete event systems. John Wiley.

[2] Keller, G and Teufel (1998) T.:R/3 Process Oriented Implementation: Iterative Process Prototyping, Addison -Wesley.

[3] Monden, Y. (1983)*Toyota Production System - Practical approach to production management.* Industrial Engineering and Management Press, Institute of Industrial Engineers.

[4] Murata, T. (1989), "Petri Nets: Properties, Analysis and Applications," Proceedings of the IEEE, 77-4, pp.541-580.

[5] Peterson, J. (1981), Petri net theory and the modeling of systems, Prentice-Hall.

[6] Sato, R. and Praehofer, H.(1997): A discrete event model of business system - A Systems Theoretic Foundation for Information Systems Analysis: Part 1, IEEE Transactions on Systems, Man, and Cybernetics , 27-1, pp.1-10.

[7] Sato, R. (1997): Meaning of Dataflow Diagram and Entity Life History - A Systems Theoretic Foundation for Information Systems Analysis: Part 2, IEEE Transactions on Systems, Man, and Cybernetics, 27-1, pp.11-22.

[8] Sato, R. (1999a): Modeling business processes through DFDs and Petri nets, Journal of Japan Society for Management Information, 8 - 1, pp.1 - 15. (in Japanese)

[9] Sato, R.(1999b): On control mechanism for business processes, Journal of Japan Society for Management Information, 8 - 1, pp.17 - 28. (in Japanese)

[10] Sato, R, Praehofer, H, and Pichler, F.(1995): Realization theory of general discrete event systems and the uniqueness problem of DEVS formalism, ISEP Discussion Paper Series No.625, University of Tsukuba.

[11] Sato, R. (1998).: Realization theory of discrete-event systems and its application to the uniqueness property of DEVS formalism, (submitted).

[12] Scheer, A.W.(1994): Business Process Engineering (2nd edtion), Springer.

[13] Zeigler, B. P.(1976),Theory of modeling and simulation. John Wiley.

[14] Zeigler, B. P.(1984) Multifaceted modeling and discrete event simulation. Academic.

Assembly Reengineering Model

Dragica Noe and Peter Peternel

Laboratory for Handling, Assembly and Pneumatics,
Faculty of Mechanical Engineering, University of Ljubljana
Aškerčeva 6, 1000 Ljubljana, Slovenia
dragica.noe@fs.uni-lj.si

Abstract. Market demands and internal needs forced companies to permanent adaptations. It exists a need for a systematic approach to the perpetual business process improvement and company growth. Reengineering of assembly is one of the possibilities, and involves activities on assembly technology, logistic, assembly system development and product redesign. For that reason, the adequate tools for modelling of the assembly process and system have been developed. Activities needed for reengineering process are part of such model. The reengineering process is finished and started again by evaluation phase of reengineering.

1 Introduction

An enterprise is a dynamic structure, it has to adapt itself permanently to the laws and demands of the market and adjust its technological and sociological abilities. Competition narrows sales possibilities and enforces reduction of production costs. Customers demand new products in many varieties and short delivery times, short development times, high and constant quality and low prices. New materials and technologies are emerging, and society demands environment friendly products. Internally enterprises encounter rigid organisation, high labour costs, inability to manage production costs, material flow and quality, rejects, limited investment founds, and incompetent production management. Numerous new philosophies and organisation models should provide improvements, but practical results of these models are not encouraging.

A few years ago production rationalisation was a proper approach to improve enterprises marketing ability, nowadays this is reengineering. Production rationalisation was aiming on technology improvement, rising productivity, and cost reduction on the operational level. Reengineering means complete renewal of the enterprise, on technological and organisational level, management of logistics, information flow, and employment as well as development of new products and redesign of actual products and programs.

Reengineering of assembly means: »Fundamental reconsideration and radical renewal in assembly organisation and technology, design and redesign of products, logistics, education of workers for dramatical improvement of all success parameters like cost, selling amount and market share, adaptation to market and environmental

F. Pichler, R. Moreno-Díaz, and P. Kopacek (Eds.): EUROCAST'99, LNCS 1798, pp. 274–281, 2000.

demands, customer satisfaction, aiming on general profit improvement for all employees.«

Until now only a general model of reengineering for business – production process and information systems has been known. Reengineering of assembly as a part of business-production process hasn`t been established yet. Reasons for that are: differences in enterprise organisation, human resources and their knowledge, level of success and others.

In an assembly process as an integral and crucial part of production process the problems on realisation level are mostly connected with organisational decisions, logistics, technology, products, and all kind of resources. In comparison with optimisation and process ratioantionalisation the assembly reengineering process is a simultaneously changing and evaluating of the assembly system, material and information flow as well as the improving of human recourses skills connected with the whole production process.

2 Assembly Proces Modelling Tools and Methodes

When crossing the route from existing assembly system to a new one in a reengineering process, the following objectives and constraints have to be considered: Modelling formalisms, Evaluation criteria, Reference model (Fig 1).

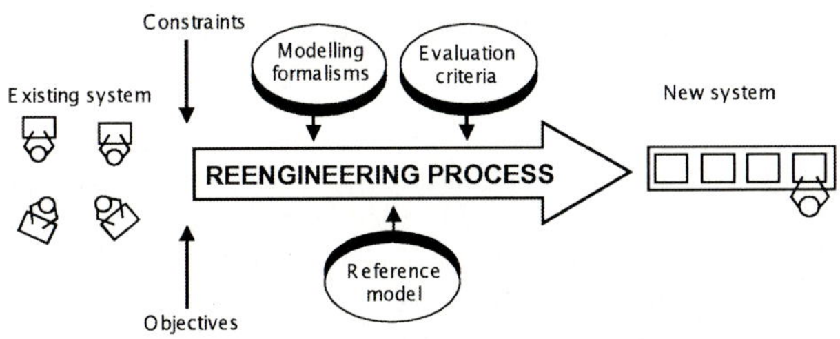

Fig. 1. Reengineering process form existing assembly system to new one

A number of modelling tools already exist, like CIM-OSA (Kosanke 91), IDF 0, SADT and GRAI method (Rolstadas 95), MO^{2}GO (Mertins 94) and are being used in the reengineering of business process management systems. Some of them can also be used in assembly reengineering process. Especially some ideas of GRAI and presentation of activities with IDF 0 will also be used in project presented in this paper. In planning process relation models can also be used for presenting the assembly operation information.

Concerning the reengineering process for crossing from existent assembly system toward a new one, we can notice similarity, three kinds of decisions have to be observed: realizational, structural and conceptual. And on conceptual level the question "what?'" has to be answered without concerning the organisational or technical frames of assembly process. Questions "who?, when? where?" are concerning organisational point of view and are mostly structural nature of decisions and "how?" lead us to realisation phase of reengineering process and is concerning only technical constrains of the studied case and the choice of equipment (Fig. 2).

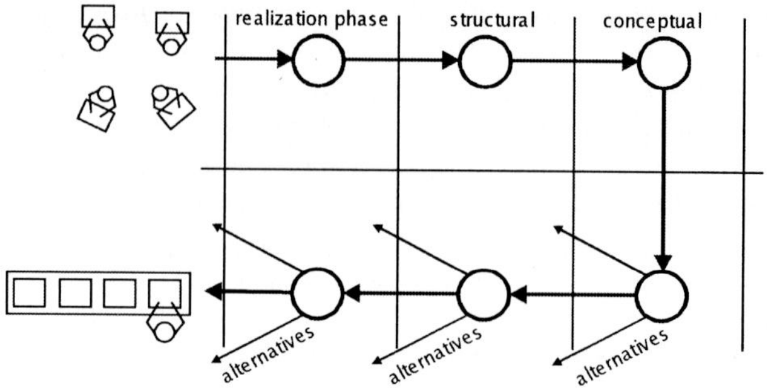

Fig. 2. Decisions circle when redesigning an existing assembly system

In a new assembly system planning process on all, conceptual, structural and realizational levels, more alternatives or choices have to be estimated with performance indicators. There are some modelling formalisms used on conceptual and structural levels for describing and analysing the function of existing assembly system, material flow, information flow and decisions making process in reengineering and planning process as entity models, relations models, activity models.

3 Company Reference Model and Assembly Reengineering Process

The production system can be divided on three level concerning decisions making and reengineering process implementation. On the *strategic level* the company efficiency is evaluated and the long- and short-term plans are discussed and accepted. *Tactical level* is level of performing the strategic decisions, analysing the existing assembly systems, planning a new one, and preparing demands and orders in form of drawings, reports for *operational level*, where the assembly process is executed (Fig. 3)

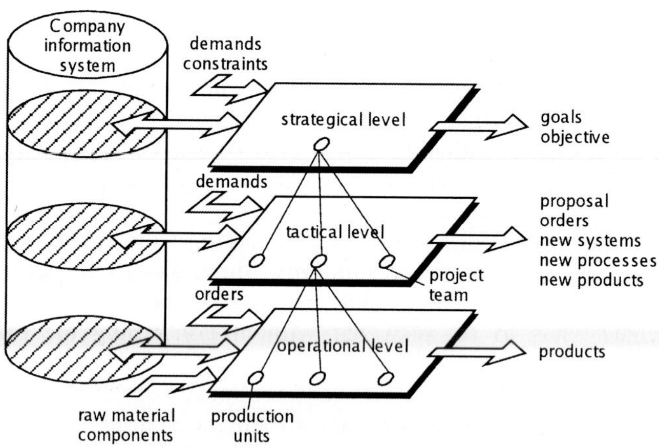

Fig. 3. Production system reference model

All three levels are supported by a suitable information system, which enables access to all relevant data, for each decision and process level. The structure of production system shows us that we have on strategic level a management team which forms product oriented teams on tactical level. Such teams are responsible for individual reengineering projects and are responsible for one or more assembly units on operational level.

The top management group sets up goals and demands for reengineering process based on market demands and conditions as well as conditions in production process. The acceptance of decisions are supported by evaluation criteria or variables for market, environment and production system, evaluation as follows:

- profit (costs, product market price)
- market demands on product (volume, new product, new material, product variants)
- profitability or investment pay back (time and intensity)
- response and company flexibility (velocity, intensity, stability)
- environment (pollution, energy, resources)
- market share.

The decisions of management group are in form of demands like: to develop a new product, to improve the productivity, to reduce all kinds of costs, to eliminate the pollution etc.

The decisions done by team responsible for reengineering process on tactical level are based on following influenced parameters:

- product development (time, variety, group)
- planning and development of assembly system (development time, optimal solutions, support of planning tools)
- assembly system controlling (respond, overview)
- material flow controlling (overview, effectiveness)
- total quality assurance

and

- capacity of assembly system
- assembly system flexibility
- quality, disturbances, rejects
- assembly process costs (energy, space, investment, labour costs)
- assembly system reliability and availability
- material flow and stocks
- human resources.

In the process of assembly reengineering most of work is done on tactical level. On this level proposals are prepared, systems are developed and realisation of the reengineering project is managed. Installation and implementation of the new process are realised on the operational level.

4 Activity Model of Assembly Process Reengineering

The activity model of assembly reengineering process for the transfer from actual assembly process or assembly system to a new one was developed. Activity model allows us to follow individual working phases and shows also on which production level they are performed (Fig.4).

Assembly reengineering process consists of following activities:

- *Analysis* of the existing production process and by its results *determination* of demands for renewal of production process on structural level.
- Evaluation and analysis of existing assembly process respectively assembly system and comparison with demands, designing a proposal of a new assembly process or system is a result of the first activity on tactical level.
- The proposal has to be *verified* on strategic level.
- Planning and designing of the new assembly system are the next activities.
- Installation activities involve the activities on operational level.
- The new or renewed assembly process or assembly system has to be evaluated and the process of reengineering can process again in term of constant improvement of assembly system.
-

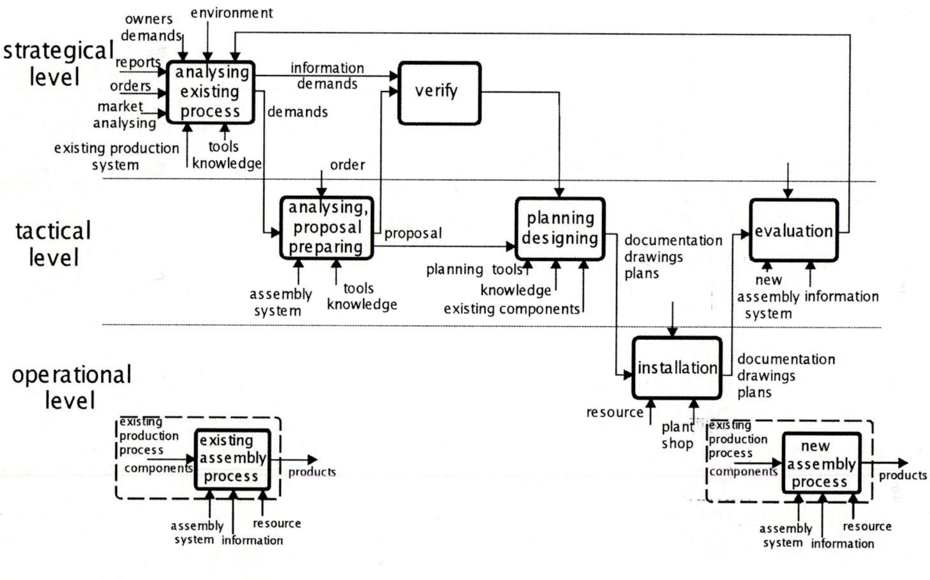

Fig. 4. Assembly reengineering process model

5 Temporary results

The assembly reengineering process has recently been implemented in two small companies.

Example 1 (Fig 5):

Company producing metal parts for windows and the window ruler assembly was reengineered.

Goal was to increase market share. This could be accomplished by assembly costs reduction and by increasing the production volume. For that reason the new automated flexible assembly system was developed.

Example 2 (Fig. 6):

Company producing components for domestic appliances. The pressure switch assembly was reengineered.

Goals were: labour cost reduction, intermediate stocks reduction, through put time reduction.

This was accomplished by development of a new assembly system.

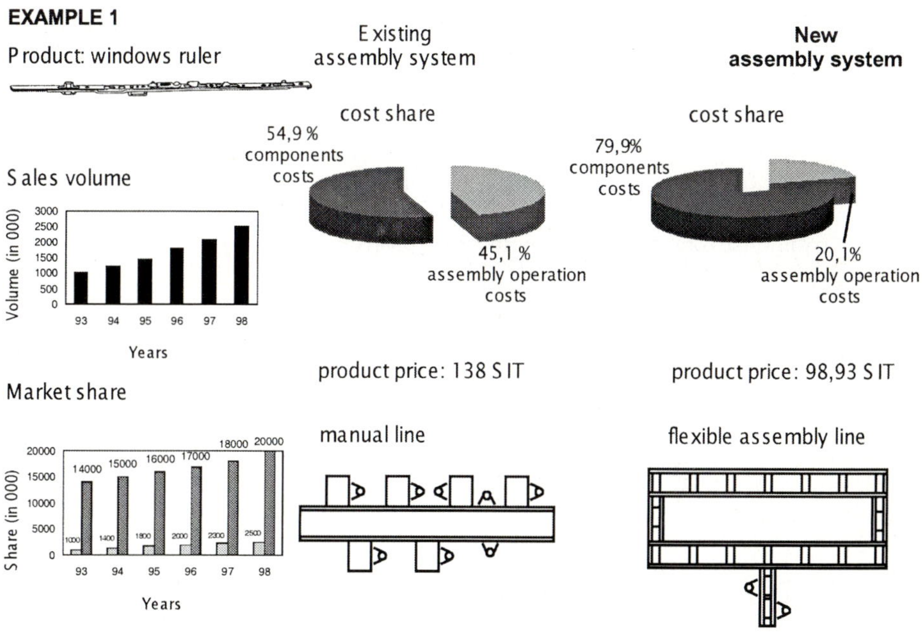

Fig. 5. Example 1

EXAMPLE 2

Product: pressure switch

COSTS SHARE	OLD	NEW
COMPONENTS COSTS	61 SIT	62 SIT
ASSEMBLY COST	64 SIT	32 SIT
PRODUCT COST	125 SIT	94 SIT

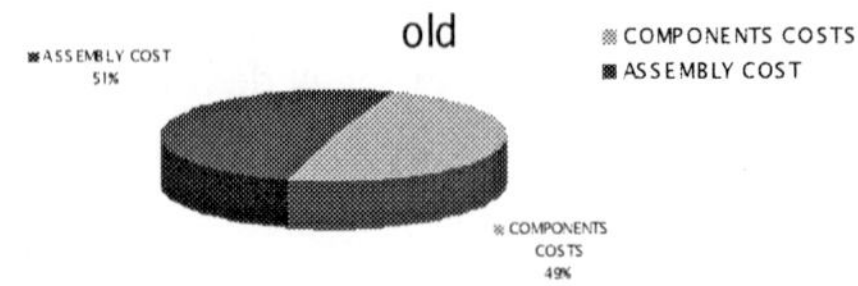

Demands:
- Labour cost reduction by 10 %
- Capacity flexibility integration
- Stocks reduction 50 %
- Through put time reduction by 20 %

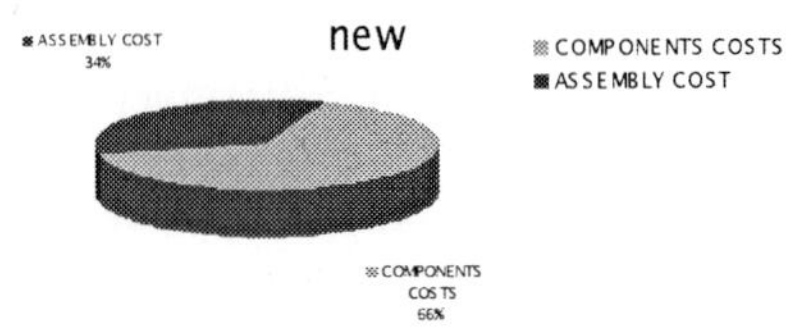

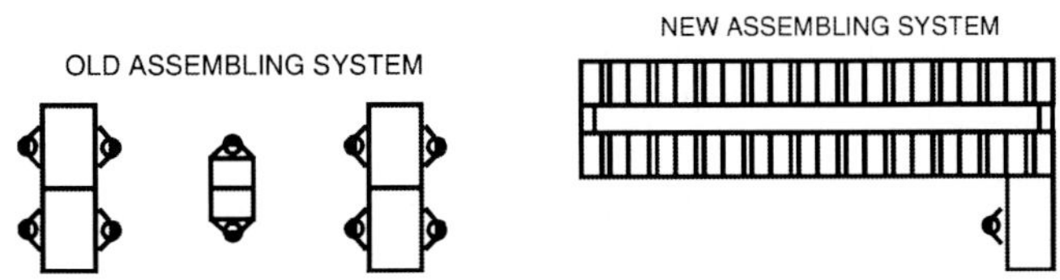

Fig.6. Example 2

Conclusion of several realised reengineering projects is (Fig. 7):

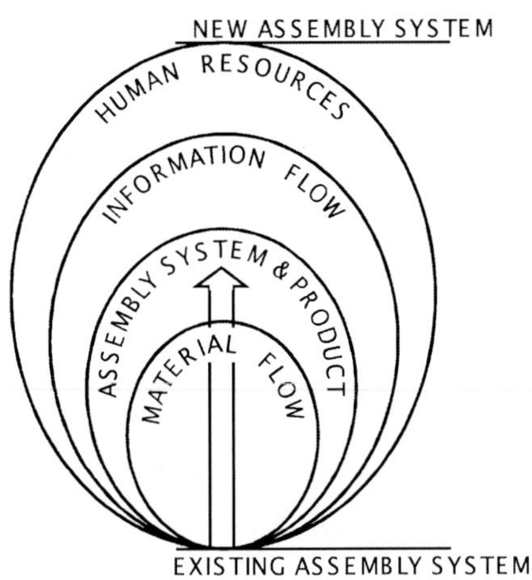

Fig. 7. Activity circles of improvement

Assembly reengineering starts with activities on all four activity circles: material flow, assembly system, information flow and human resources. Optimal solutions are first achieved in material flow circle, than in assembly system, followed by information circle and finally in human resources.

6 Future Work

The research work on improving the assembly process started two years ago, supported by the Ministry of Technology and Science and with co-operation of different small and medium sized Slovenian companies.

- The whole activity model has to be verified in real environment.
- Some assembly reengineering oriented tools for reliable verifying of the effectiveness of assembly process have to be developed.
- Newly developed tools are implemented for assembly systems performances determination and analysing.
- More emphasis has to be given to the development of user friendly assembly process and system planning process.
- Different tools have to be used for material flaw planning and analysing.
- All decision making places have to be connected with effective information system and give the management on strategical and also on tactical level more adequate information in the real time.
- The permanent educational program has to be implemented for all participants in the production process.

References

1. Kosanke, K.: The European Approach for an open system architecture for CIM Computing and Control Journal, (1991)
2. Marcotte, F.: Organizational/decisional view. FOF/GRAI/200, (1990)
3. Mertins, K., Jochem, R., Jakel, F.-W.: Reengineering und Optimierung von Geschaftsprozessen. ZwF 89 (1994) 10. Carl Hanser Verlag, München, 479-481
4. Rolstadås, A. Enterprise modelling for Competitive Manufacturing. Control Eng. Practice, Vol. 3, No. 1, (1995), 43-50
5. Vallespir, B. The GRAI Integrated Method: A technical – economical methodology to design manufacturing systems. Internal report. GRAI – Laboratory of Automatics and Productics (LAP), University BORDEAUX I, FRANCE

Design for Disassembly and Recycling for Small and Medium Sized Companies for the Next Generation

Harald Zebedin

Institute for Handling Devices and Robotics,
Technical University of Vienna,
Favoritenstrasse 9-11,
A-1040 Vienna, Austria
zebedin@ihrt1.ihrt.tuwien.ac.at

Abstract. As we left the 20th century the industry has paid particular attention to the ecological impacts of new facilities, energy use, number of used materials and several other very important environmental issues. Now many impacts to design and recycle facilities as „ life cycle " (planning, design, development, supply, manufacture, use, recycling) have not been systematically and comprehensively studied and analysed. Engineers, operators, professionals and other people who works in this area together maybe in different firms should understand the environmental and interactions of management, team behavior and methods because their everyday decisions carry substantial social implications.

1. Introduction

A rapidly developing market and an increasing availability of new productions technologies speed up the development of products and their corresponding means. To respond competitive pressure, companies invested heavily in production automation and computer integrated manufacturing (CIM). Therefore the global marketplace is changing rapidly, companies have to answer to the requirements of customers.

This paper focuses on the need for creation of a closed loop system integrating research in the area of design for disassembly and end of life cycle equipment disposition. The manufacturer must respond quickly to rapidly changing markets driven by customer based valuing of products and services. Low cost intelligent manufacturing systems are necessary to realise this in an economic way. Hence the producers have to rethink their whole organisation. Therefore it is obligatory to reorganise the product development.

The paper explores closed loop (including reuse) and open loop recycling issues and presents some quantitative data on the impact of design for disassembly and recycling on the recycling process.

F. Pichler, R. Moreno-Díaz, and P. Kopacek (Eds.): EUROCAST'99, LNCS 1798, pp. 282–288, 2000.

2. Design for Disassembly and Recycling

The main goal of Design for Disassembly (DFD) is to impact the economics of electronics recycling in a number of important ways. Techniques for Design of disassembly are driven by equipment disposition, recycling/reuse technology. Practical design for disassembly can only be formulated within the work of standard equipment, demanufacturing techniques and the economic realities of these techniques. In end of life management of considering product disposal, the designer becomes a more and more important role to consider final disposition processes at end of life, thus "Design for Recycling" becomes part of the over all design plan. The incorporation of ***Design for Recycling*** into the design process is an critical step towards the creation of more recyclable electronic products. The greatest impact on product design has both the costs of the recycling process and the logistics. The design can dictate both the methodology used to recycle the product and the profit generated from the recycling process.

2.1 Design for Disassembly

In the field of automation and robotics a very high standard have been reached in the last decade of this century. But the focus was only on "**assembly problems"**.

When we have a look at the characteristics of a product, it can be described in three main phases of living: production, utilisation and disposal. The closed product life cycle (chain) is one of the two main streams called "traditional" product life chain (such as production, distribution and use) called **"Eco-design**". The other one, re-using and recycling products, components and materials (such as take back, re-use and recycling) is called "**End-of-life Management**". Therefore in the future disassembling will be the most important process.

The number of products to be disassembled will increase dramatically, so all companies should be aware of this problem. This is mainly caused by the environmental legislation of many countries, which leads towards a life cycle strategy that involves different concepts:

- product recycling
- rising waste disposal costs
- demand of the consumer and the corporate image
- public perception

Therefore these topics becoming more and more a competitive factor for companies, especially for small and medium sized, all over the world.

The different products to be recycled and also disassembled, such as computers, printers, telephones and other electronic devices, it is necessary to automate this aim. High flexibility and "Low - Cost" of disassembly processes will be necessary.

The automation potential will be one of the most important productivity factors for this new production process.

The automation of disassembling is becoming a new challenge for engineering where the main goals are:

- reduce the costs of disassembling
- extend the performance of the disassembly process
- optimise the different recycling processes
- extend the life cycle of a product
- create a human working environment in disassembly factories.

2.2 A New Kind of Organisation

In the current market not the rise of quantity, but those of variations is significant. Nowadays the life of the products is often shorter than the time to develop, to design and to produce it. These are the reasons why traditional sequential product development will not be able to fulfil these requirements. This procedure includes a lot of iterations loops between the single development steps and has periods of dead time. But the market wants small quantities of high quality and highly customised products at low unit costs with high responsiveness. The producer must respond quickly to rapidly changing markets driven by customer based valuing of products and services. Based on all these demands and also with the new industry of disassembling with all its influences the working conditions of the employees can suffer dramatically. That is also because the human is not in the centre anymore and only money is taken into account. To deal with all these problems the producers have to rethink their whole organisation, not only for the people also for the Management, the Team Behavior, the Method and the Environment.

Significant in this area is the flow of information. Therefore it is essential that the information between the components (CAD/CAM/CAP) will be controlled by one system. Only when all components work together, when there is a continuos flow of information and when there is a simultaneous execution the planning process will be efficient. This means that beginning with the first concept for the construction of the product you shall also develop a concept for the production and disassembling of the product, the layout of the production and the disassembly cell, to determine the necessary working and auxiliary material, the recycling processes and so on and estimate all possible occurring impacts on mankind. An integrated CIM-System is obligatory for every enterprise. But not every small and medium sized company had enough money to install such a CIM concept. So we at the Institute for Handling Devices and Robotics (IHRT) try to find solution to maintain small and medium sized companies.

Based on the flow of information between the single components, which start to work together in a more specific phase of the product development, it is also obligatory to organise the co-operation between the different departments. Hence it is necessary to form a project team (Fig. 1) which operates together from the beginning in a methodical way.

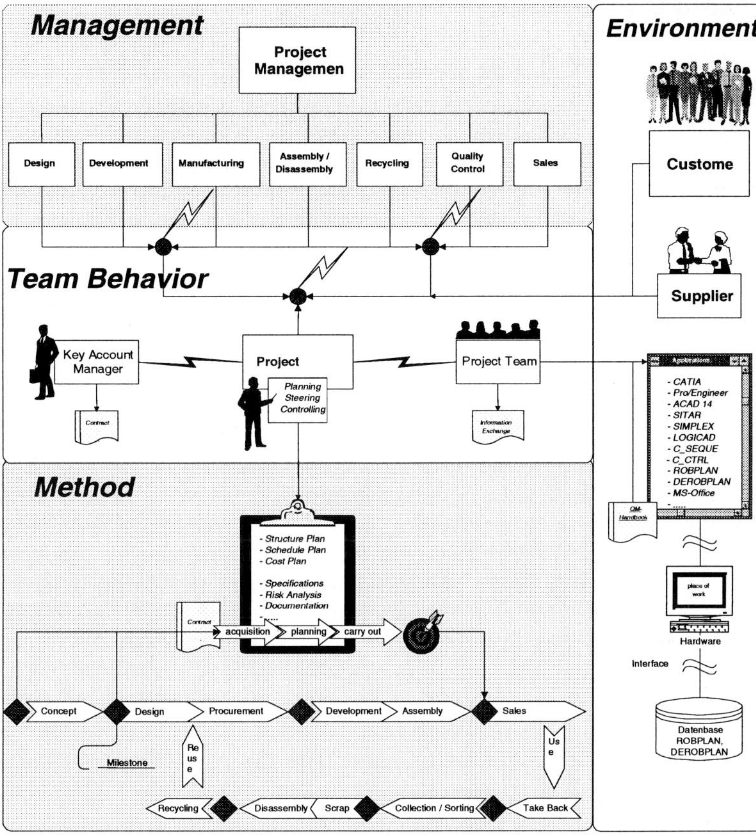

Fig. 1. 4 – Key Factors of a Project Team

Here it is important that the members of the team extend their knowledge also in the other fields. During these phases it is necessary that the project team operates strongly together. Design, manufacturing, disassembling and recycling planning must take place parallel. The next phase of the development is not initiated until all parties have reached agreement and have identified a global and economically satisfactory solution.

Starting from the next phase it is possible, and for future development essential, to divide the product into several independent modules. These modules can be created according to their function, material, manufacturing and/or recycling process or as recommend here to their assembly and disassembly structure. For the employees, the enterprise, the environment and also logistically this will be the best solution, also because these groups can easily be organised for manufacturing and recycling processes. Hence, resulting from the large impact of the design especially the design for disassembly on the working environment, the next chapter should give a view of different software solution especially for small and medium sized.

3. Applications for Design of Disassembly

Robotized disassembly automation has been growing up dramatically in the last years. Especially for small and medium sized companies, there is a great demand for flexible modular disassembly cells usually equipped with components like robots, grippers and tools, storage devices and part feeders, and other auxiliary devices. To guarantee a high flexibility and efficiency, an appropriate distribution and a context sensitive supply of information between all components must be ensured. A number of modular software system developed at the Institute for Handling Devices and Robotics (IHRT) of the Vienna University of Technology will be described (DEROBPLAN, ACAD Simulation packages).

3.1 Derobplan (Planning of Robotized Disassembly Cells)

There are many similarities to the planning of assembly systems, the application area of ROBPLAN will be extended now to planning of robotized disassembly cells – DEROBPLAN (Fig.2).

Product Database: In addition to the database fields designed for planning system ROBPLAN, the product database has to include all necessary information to build disassembly families. For the complete description of the product, the database has to contain also a detailed description of the part connections (how to solve them, possible tools for solving the connections). One key aspect for the disassembly process is to know the „constitution" (age, damage, etc.) of the product and expected complications caused by this conditions.

Symbol Database: The pre-defined assembly icons from the Symbol Database have to be extended with representative disassembly steps.

Component Database: The description of available cell components has to be expanded with tools and systems necessary for disassembling. Especially, a detailed description of available sensor systems and their characteristic data has to be included to the Component Database.

Planning Database: Equivalent to the planning process of an assembly cell, the results of the design phase are written into the Planning Database. As an extension for the disassembling process, the proceeds of the disassembling can be calculated (and compared to the cost of the cell) and included to this database in order to proof the efficiency of the planned cell. As an additional result, the planning database also includes the calculated disassembly depth.

Parts Database: Besides the four 'original' databases there planning system DEROBPLAN includes a fifth one, the 'Parts Database'. This database defines the current industrie's need of reusable and/or remanufacturable parts. Via connection to the world-wide-web (virtual market place) this information is permanently updated and represents the actual market demand. This information (together with information available from the 'Product Database') allows calculation of the required disassembly depth.

Modeling the Emergence of Social Entities

Gerhard Hanappi

Institute of Economics, University of Technology of Vienna
Argentinierstraße 8/175
A-1040 Vienna, Austria
Hanappi@pop.tuwien.ac.at

Abstract. This paper proposes a new approach to a rarely investigated topic: How can the emergence of new agents in social life be modeled. It proceeds in two steps. First a set of general simulation features needed to embed the notion of emergence are developed. Then it is shown how these challenges could be met by the use of a certain simulation environment called HE (Human Evolution). It turns out that not only specification and information structures used in HE are radically different from other contemporary approaches, but that the nature of results too leads researchers in a new - in my opinion highly prosperous - direction.

1 Introduction

Simulation usually is seen as a tool that enables researchers to observe the dynamics of highly interactive, complex systems that could not be studied in direct experiments. If the essential variables and the essential relations between them are modeled adequately, then the essence of the trajectories of the real processes should be close to the trajectories produced by the simulation - neglecting for a moment the difficulties with chaotic systems. The toolbox of simulation techniques thus clearly is one of the most important scientific methods for the social sciences - there is no controlled experiment in this area and the force of abstraction of the thought experiments of singular scientists is less powerful, at least with respect to capacity and rigidity, than computer simulation.

These facts acknowledged, models of social processes (indeed already models of biological systems) entail a new challenge that cannot be found in physical systems: The entities in these systems are not simply given, they emerge and vanish again, and the relations between them are not fixed and waiting to be discovered, they rather are temporary valid constellations based mainly on intentional behavior of the modeled entities. In short, social system evolve. Despite its evidence, most contemporary social modeling circumvents this challenge by focussing only on time spans where the setting is stable. This paper collects and evaluates some ideas that try to grasp the emergence of social entities. In principle, three perspectives have to be interwoven: some historical instances of the emergence of social entities (1) have to be confronted with the existing simulation attempts (2) to find out which modeling problems emerge - so that finally something for the development of a simulation language for social modeling can be learned (3). As with any modeling exercise, the importance of a

F. Pichler, R. Moreno-Díaz, and P. Kopacek (Eds.): EUROCAST'99, LNCS 1798, pp. 289–294, 2000.

simulation tool derives from its capacity to go beyond a thorough meditation on already observed behavior, to produce something new and perhaps surprising. Note that modeling the emergence of entities involves itself innovation of modeling techniques.

2 On Emergence

The basic element of a model of any society is a **social entity**. History shows that at any moment in time a set of social entities for a given geographical area can be identified - but this set changes over time. A descriptive model thus can only start at a specified point in time, making an assumption on the prevailing, essential social entities. In other words, the basic assumption of methodological individualism has to be rejected. These social entities create, modify and exchange models of the society. They do so by the use of a ***language***, which they share. The development of this language is a necessary ingredient of human societies, since it enables memory, consciousness and the experience of time. Again the language used at a certain point of time in history, of course stripped to its bare essentials, has to be assumed if one models a human society. A major difficulty that arises in models of language is the ability of language elements not only to refer to non-language items, but also to other language elements: they are sometimes self-referential. In fact any model of language is just an example of a self-referential language. In the sphere of language, where language elements only refer to other language elements, rules for the use of these elements will have to be agreed upon. These rules are the **syntax of this language**, they are a convention made by social entities. But using the language, using models, often implies the influence of models on the choice of action outside the sphere of language. Models thus might refer to non-language elements. This relation is called the **first order semantics of a language**. Extending the argument, language elements refering to a language element with first order semantics are said to have second order semantics, and so on. So while the communication of social entities is somehow bound to their non-language environment via first order semantics, compare [1], the growth of social entities relative to their environment implies a growing dominance of higher order semantics. Sharing higher order semantics thus enables social entities to locate certain informational areas which they commonly 'inhabit'.

But social entities not only communicate within themselves and with each other, they also are exposed to a historically determined **metabolism** - an exchange with their non-living environment that enables their physiological survival. Perhaps the most significant property of almost every strand of economic theory - though it is rarely made explicit - is the assumption that social entities use language and model-building mainly as a supporting tool for this metabolism.

In a sense, the most extreme position was that of Friedrich Hayek, who held that the prices in a fully developed market system are all that is needed to guarantee the optimal support of the metabolism of a social entity: The price structure incorporates everything that is needed for optimal allocation of ressources, see [2], it replaces all other language - in everyday language 'money speaks'. Prices in these theories are anonymously determined outcomes of auctions with given participants, the informational link between the social entity and its environment is as small and efficient as possible. The role of prices and their physical carriers, money signs, thus

becomes a kind of 'natural' system, quite independent and not contributing to the development of social entities whose language, now cut off from their metabolism, for them still is their characteristic property. In short, Hayek-type approaches cannot contribute to the current proposal, since they loose sight of the endogenous dynamics of social entities. More recent economic theory for the greater part still sticks to the assumption that the core problem of the metabolism can be cast as 'allocation of scarce ressources', and that the price signals for allocation auctions can be directly derived from predetermined preference orders of individuals. From this perspective any evolution of social entities is clearly out of scope. Nevertheless in the last two decades, model-building of social entities was taken more seriously in economic theory. It is treated under the label of 'expectation formation'. From simple adaptive updating rules to sophisticated solution of differential equation systems many variants of language use have been postulated and tested in their interaction. The simplest case, and in a sense the extreme counterposition to Hayek, is the one assumed by the school of rational expectations: The true model of the social metabolism is the market model, every individual knows it and solves it to express its preference order with price signals. This indeed loads the burden of total information on each single entity instead of freeing it from expectation formation as was the case in Hayek's view. On the other hand the equivalence of the actual core problem of the social metabolism and the models individuals build, together with the assumed capacity to derive from these models exactly the same 'correct' price signals as in Hayek's world, makes the rational expectations approach as pointless with respect to the emergence of social entities as Hayek's.

The most fruitful theoretical development for the purpose at hand seems to be the theory of strategic games. From the perspective of **cooperative game theory**, the set of social entities at a point in time can be interpreted as a set of coalitions[3]. But contrary to the usual assumptions in this body of theory, expected payoffs are computed by the bounded rationality models of the social entities. Furthermore payoffs are also subject to the working of the different metabolisms of the social entities. For any given coalition structure the choices of actions do change the achieved rewards. This is so because, contrary to mainstream economic theory, it has to be insisted that interaction takes place long before any auction process leads to equilibrium levels[1]. Allocation of scarce ressources therefore is just one aspect in a broader process of dynamic adaption, learning, communication, manipulation (see [5]) and intended optimization. Clearly any such system will run through equilibrium positions - either in the sense of supply equals demand or in the sense that expected values equal actually reached values - only sporadeously, induced by dynamic adjustments, general equilibrium being an extremely unlikely state. It is a model of **general disequilibrium**[2]. While the disequilibrium property generates continuous oscillations, it does not imply continuous emergence of new social entities. As already mentioned above a new coalition structure of lower level entities, i.e. emergence and vanishing of higher level entities, will only occur if a critical mass of old coalitions due to a continuous change in expected payoffs breaks up. In that way continuous change provokes sudden discrete events. More precisely, to model these ideas two

[1] This point is developed more carefully in [4].

[2] General disequilibrium models are not the current fashion in economics, though they gain ground, e.g. [6].

game theoretic tools are combined: On the one hand the cooperative, coalition game given in ***characteristic function form*** is used, with the value of each coalition transformed into an ***expected*** value, which in turn is computed by an explicitly given (bounded rationality) model. An important aspect of this model is the use of time preferences, discounting expected future utilities. Contrary to the assumptions in standard economic theory, time preference of a social entity certainly changes over time. On the other hand instead of the use of an equilibrium concept like the core, elements of Stephen ***Brams 'Theory of Moves'*** (see [7]) are used to analyze a bimatrix game where each coalition has the choice either to continue or to break up. The second player in each of these games is always the complementary set of existing coalitions. To include the empirically most important attitudes towards risk, the utilities in the bimatrix games are based on mean-variance considerations. It fits well into the spirit of this approach that Brams' theory insists on the importance of inherited settings and on the force of traditional behavior.

From this perspective the hierarchy of social entities will come into turmoil only if the traditional structure is judged to be inferior by at least one coalition. The breakup of coalitions then sets free smaller social entities open for ***new combinations***[3]. Clearly there is much room for cumulative effects and surprising new social entities.

3 Towards a Prototype for Social Simulation

Stating the challenges that a new approach is thought to master and to design a simulation prototype that meets these challenges are two different things. With the simulation tool HE, for a more thourough description compare [9], in principle ***all aspirations can be adequately incorporated*** - though at the price of rather ***demanding specification inputs*** from the side of the model-builder. In this prototype of a simulation environmet metabolisms of social entities are represented explicitely and are connected to each other as well as to an explicitely modeled environment. Metabolisms themselves might change over time due to endogenous forces, endogenous technical progress being just a specific example. Moreover communication between social entities is included in the simulation too. Learning thus does not only occur as some kind of Bayesian learning from the success of the interaction with the environment. There is also the possibility that social entities transfer models to others - be it to teach them something correct about their environment, or be it to maniplulate them to act favourably for the sender. Social entities thus 'speak' by the use of models. If one considers the importance of learning from other social entities relative to the learning processes that derive from direct interaction with the physical environment, it immediately becomes clear how important that feature of the simulation prototype is.

Moreover the necessary computations to recognize possible break-ups of coalition structures - see the previous section - could not be incorporated within some reasonable constraints on information processing capacities if there was not the possibilty of communicating models. Indeed in many cases the communicated models can substitute the correct and empirically tested models completely! Social entities

[3] This is the term Schumpeter used in [8] to characterize innovations.

can base their actions on quite strange, more or less commonly believed models that escape any empirical validation - and they will continue to use them as long as long as a significant majority of metabolisms produces satisfactory results. This might sound strange to economists but surely is trivial for ethnologists.

To enable the implementation of such a simulation environment it is necessary, or at least wise, to restrict the possible information processing abilities of modeled social entities right from the start. Indeed with communication at its disposal a social entity will store parts of its knowledge outside itself to avoid overflows of its memory. The design of the prototype thus evidently leads to the implementation of external 'model stores', that in HE are called 'media'. With a common language the same medium can be used by different entities and different media can be compared by the same entity. In that way a new type of learning from external sources emerges.

The great advantage of a simulation run of this type, once it is set up, is that a wealth of interesting results on a historically and empirically valid level can easily be generated. It differs from the currently flourishing 'Artificial Life' approaches[4] in that it does not aim at spartanic behavioral traits of ever larger numbers of entities. While there is a necessity to keep things simple, it would mean to overstate this technical need if one substitutes the use of models of modeled social entities by simple stimulus-response functions. If reactions are not dealt with in a somewhat more sophisticated way, it is the implementation of learning that suffers. More detailed information on the algorithmic side of the prototype can be found in a compagnon paper to this contribution [9].

4 Conclusion

It should be evident by now, that HE can provide results that are substantially different from those of other types of formal analysis. Though they are closer to empirical observations, they usually will be rather sensitive to changes in the rich set of specifications. In a sense they are more like metaphors for possible interpretations of the past or forecasts of future developments. Their very nature is to guide exploration rather than to present themselves as a unique, discovered law governing reality.

With respect to the central question of this contribution, the emergence of social entities, this means that HE can help to ***discover prospective future social entities***. Or, it can help to single out those social entities that are prone to be eliminated. This highly sensitive tool thus clearly transforms the world-view of the social scientists in a fragile, quantitative result that will induce the latter to regard his or her own role in the modeling process as active part: The observer is part of the model. In that way normative and objective findings by the use of HE can be digested as being permanently in a state of flux. With every simulation run and the implied newly emerging entities, the model-builder is confronted with the question if these expected entities are judged as good (and should be furthered) or as bad (and should be suppressed). The judgement itself, of course, should be derived explicitely from current preference orders of model-builders, a further hint at self-reference that not

[4] For an overview see [10], for an interesting recent example see [11].

only comes back to the discussion of language, but also shows that self-reference tends to accumulate beyond every scope.

References

1. Hanappi H., Hanappi-Egger E.: Norbert Wiener's Cybernetic Critique of the Information Society -An Update, International Conference Cybernetics 99, Gran Canaria, Spain, February 8-12, Proceedings forthcoming with Springer, Heidelberg (1999)
2. Hayek F.: The Use of Knowledge in Society, American Economic Review (1945)
3. Owen G.: Game Theory, Academic Press, New York (1982) 143-192
4. Takayama A.: Mathematical Economics, Cambridge University Press, Cambridge (1985)
5. Hanappi H.: Evolutionary Economics, Avebury Press, Aldershot (1994) 21-41
6. Duménil G., Lévy D.: The Economics of the Profit Rate, Edward Elgar, London (1993) 111-196
7. Brams S.: Theory of Moves, Cambridge University Press, Cambridge (1994)
8. Schumpeter J.: Business Cycles, McGraw-Hill, New York (1939)
9. Hanappi H.: HE - A Simulation Environment for Socio-Economic Dynamics, International SASE Conference, Vienna, July, forthcoming with Gordon & Breach (1998)
10. Langton C. (ed.): Artificial Life, MIT Press, Massachussetts (1997)
11. Caldas J., Coelho H.: The Origin of Institutions: socio-economic processes, choice, norms and conventions, Journal of Artificial Societies and Social Simulation (JASSS), vol.2, no.2 (1999)

Simulating Social Grouping: An Interactive Team-Building Tool (ITBT)

Edeltraud Hanappi-Egger

Vienna University of Technology
Institute of Technology Design and Assessment
Argentinierstr.8/187
A-1040 Vienna
Edeltraud.Hanappi-Egger@tuwien.ac.at

Abstract. The paper presents the concept of an interactive team-building tool for simulating social grouping. It shows how decisions whom to assign which task can be supported from a game-theoretic perspective. Thus the group member are modeled as social entities following their preferences and being able to resist against work distributions they do not want. The features of ITBT are deduced from the decision situation of a team leader being in charge for the work distribution and are described in more detail in the paper.

1 Introduction

The theme of the presented paper is how to support team leaders in their decision making processes on team constellation. As known from several organization theory contributions *grouping* is an essential part in organizations. In particular if the organization can be characterized by the necessity of building temporary project groups (see [1]), whereby *groups* refer to those parts in an organizational structure that support collaboration in terms of reaching and fulfilling corporate objectives (see [2]). Organization theory has treated the coordination problem in several different ways: While institutional economics and transaction cost analysis [3, 4] focus on the role of conflicting sub-goals in strict hierarchies, the clan paradigm [5] stresses the role of sharing values as way to obtain organizational cohesion. Nevertheless they do have in common that the assignment of people to tasks seem to be a coordination problem, even though it is accepted that there are several soft factors influencing the according decisions.

Contrary to these approaches the presented interactive team-building tool starts from the assumption that it is necessary to take group dynamics into account and to allow for the modeling of subjectivity, a term explained in more detail in the following paragraphs.

There is no doubt that the constellation of teams plays a crucial role in the economic field: the division of labor requires the distribution of several qualifications, therefore a group has to consist of people with according abilities in order to fulfil the group's task. Evidently experience, routines and knowledge of the single group members are parts of one's qualification profile. Normally the labor

F. Pichler, R. Moreno-Díaz, and P. Kopacek (Eds.): EUROCAST'99, LNCS 1798, pp. 295–301, 2000.

process can be divided into single steps, each of them requiring certain qualifications. If the people have these abilities with no overlapping, it is rather clear, who has to do what. There is no space for discussion. Egger [6] argues that such settings can hardly be called 'cooperative'.

As soon as there are redundant abilities in a group, the question emerge to whom assign certain tasks, since more people can do the same job. Assumed that people do have ways to resist against work distributions they don't want, this decision becomes crucial. In other words the assignment cannot be treated as a scheduling job, but is highly determined by group dynamics: phenomena such as power games, individual utility functions, conflicting interests, different preferences and the like do influence the performance of working teams. If people are dissatisfied with the jobs they have to do, the performance will decline.

Following this basic idea computer support for social grouping has to take into account all the mentioned aspects as well as the problem of subjectivity. This means that a person in charge for the work distribution has to make decisions in a highly uncertain decision situation: Many of the mentioned aspects can just be anticipated very vaguely. In order to offer a decision making support, ITBT simulates the subjective view of him/her on the decision situation. By doing so a feedback loop is implemented: The results of the simulation serve as evaluation for the view or they can be used to compare it with reality initiating learning processes.

2 The Decision Situation

From a team leader's point of view the distribution of tasks belongs to his/her job and has to be done. Furthermore it is a necessity in terms of division of labor and has to follow certain company' objectives. This means that the decision situation on one side is determined by the company's objectives, on the other side it is influenced by the group members and according dynamics.

If the members of the organization are treated as entities with no possibility to react on the proposed assignments the decision situation can be modeled as *optimization problem* resulting in a work schedule following the company's objectives (see e.g. [7]). As soon as it is assumed that the (human) group members do have their own view and ways to deal with proposed work distributions in a destructive way (in other words as soon as it is assumed that they do have certain power), their behavioral alternatives have to be considered in the decision situation (for the discussion on the role of conflicts see also [8]). This leads to the approach that there is an interaction between the person in charge for the work distribution and the group members as well as among the group members. Evidently the latter ideas lead to a game-theoretic modeling view: The decision (of a team-leader) whom to assign which tasks depends also on the expected actions of the group members (see [9, 10]). Clearly these expected actions of the team members depends on their utility functions – that are usually at least partially hidden from the team leader.

The concept of the utility function in the presented case refers to a preference order which each team member has with respect to certain tasks. This preference order reflects many different influences such as expected salary, reputation, career opportunity, but also status and power. Egger and Hanappi [11] distinguish between *operational* and *inter-relation* level of the decision making: while the former means the "objective matter-of-fact" issue – namely in terms of qualification and company's objectives – the latter refers to the "backstage", the group dynamic space (for a more detailed discussion see also [12]).

The following figure presents the decision situation from a team-leader's point of view.

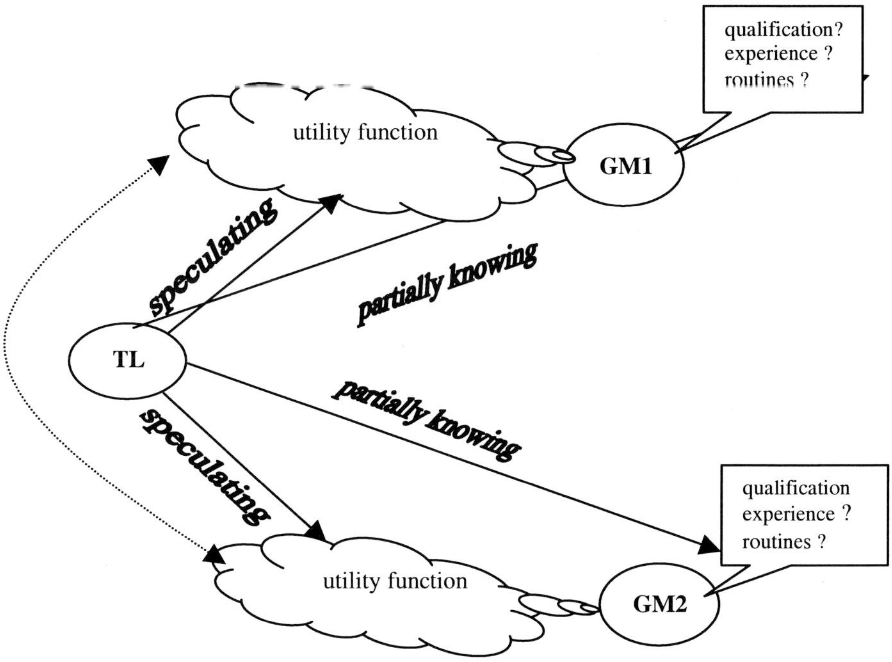

Fig. 1. Decision situation of a team leader representing visibility of qualification and utility functions of the group members

Group member 1 as well as group member 2 (GM1, GM2) has a qualification profile which is partially known by the team-leader (TL), the person in charge of the team-constellation, even if the assessment of experience and routine might be uncertain. The 'visible' part of qualification consists of educational degrees, certificates of training courses and the like. Those documents (together with a CV) offer an impression of the consumed education. In connection with information on former working places and duration an imagination of the abilities of a group member can be built, even though the real know-how cannot be assessed by these data. The aspects concerning the "hidden" space of a group member can be defined as utility function (including expected utility from certain activities, preferences (concerning tasks and/or roles), interests (such as income, status, power, labor intensity)). All these parameters of the utility function stay hidden. Thus the team leader can only speculate. Even though working experience with people will expose some aspects of one" utility function (you probably can assess if one is interested in making career or in social appreciation and the like), but it is difficult to catch more of the inter-related facets.

Note, that the relation between the group members might be of informal type (expressible in form of a status-relation) or formal (expressible in form of a power-relation externally given). *Status* refers to the group value a member is ascribed by others. As a consequence a group member interested in getting higher status has to act strategically: E.g. One could offer to undertake jobs not wanted by others, even though it ranges at the end of the own preference list.

Power refers to formal ways to restrict others' action space. Informal power represents ways of coping with unwanted situations, e.g. spending more time than necessary for the completion of a task.

It is assumed that there is a correlation between status and power: If the status of a certain team member is high enough it usually leads to a change in the power structure: Informally this group member will gain power, formally it will probably be useful in the eyes of a manager to adapt the power structure to the observed group situation.

The team-leader has the problem to choose group members according to requirements of the job. If there exist redundant job qualifications group dynamics in terms of co-operation between the team members can be considered, too. Therefore the team leader has to make the choice on the basis of his/her imagination (i.e. model) of the group members.

In order to improve the ability of a team leader to assess the constellation of group members, a simulation tool shall help to evaluate the assignment decisions made.

3 The Features of the Interactive Team-Building Tool

In order to support the team leader in this decision situation an interactive tool for team-building is designed which will simulate the decision situations as presented in figure 1. The main idea of ITBT is to allow for simulating effects of certain team constellations by linking the operational level with the inter-relational level. In other words the group situation in terms of company's circumstances as well as group specific internal dynamics is implemented. Assignments a team leader made are evaluated on the basis of the utility function of the group members. Therefore the evaluation of the decisions has to refer to the specific group specification. Note that this group specification represents the view of the team leader. Thus it is highly subjective and might not necessarily be a 'good' representation of the group. 'Good' means an accurate model of reality. Since the team leader is always uncertain concerning the utility functions of the group members, he/she shall learn to reflect on the own view. This is exactly the purpose of ITBT.

In brief the ITBT consists of two main features, the group specification and the group member module which are presented in the following. Those features are linked to each other by the decision module.

Group Specification: The group specification asks the team leader to model his/her view of the general situation of the group: The group task is represented as a Petri

Net showing the logic of the sub-tasks as well as the qualification required. The objective of the company has also to be expressed in terms of e.g. shortest completion time.

The number of the group members as well as their position within the group's hierarchy is made explicit.

Group Member: A group member is simulated by a program: His/her profile can be determined randomly or specifically consisting of qualification, preference list of tasks, experience-index, social type (interested in power), and "resistance"-index expressing the power potential (i.e. how strong is the group member in acting destructively).

There might exist several group members who are related to each other by a net of power respectively status. Power refers to the potential to force somebody else to follow a rule, while status represents the value of the person in the eye of the others. It is assumed that permanently growing status will lead to changes in the power structure (formally or informally).

Moreover the interaction of utilities of group members can be specified: the utility of group member 1 might be perceived by member 2 and might enter member 2's utility function (positively or negatively), and vice versa.

The group member specification allows for describing the different involved group members,- as many people are participating in the labor process as many programs are generated.

Decision-Module: Group specification and group members serve as framework for the Decision-Module. The team leader is asked to make assignments and the simulation is started. The results are interpreted as criterion for the quality of the decisions. I.e. how 'good' is the view of the team leader, which group members did he/she describe accurately, whose utility function did he/she catch rightly and so forth.

The interplay of the above mentioned features can be presented as follows:

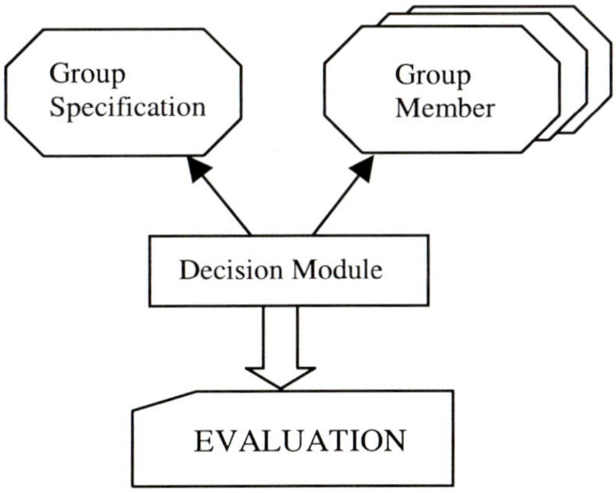

Fig. 2. The interplay of the components of ITBT: group specification, group member and decision module resulting in an evaluation of the team leader's view

The interactive team-building tool serves as a learning device for people dealing with the described kinds of decisions. There is a particularly interesting issue of modeling social processes: As figure 1 shows the team leader is in a decision situation with a high degree of uncertainty concerning his/her model of team members. As a consequence he/she can only speculate on the utility function of the people. In order to assess his/her ability to catch reality, a tool is needed serving as mirror and allowing for reflection on the own view. ITBT acts as such a mirror in several ways:

- Firstly, the user is forced to make his/her (very subjective) view of the decision situation explicit.

- Secondly, the user is asked to express his/her model of the group members involved into the decision situation.

- Thirdly, the tool allows for investigating the view by simulating certain decisions and assessing the results.

- Fourthly, the tool highlights the influence of certain team constellations on the work schedule.

ITBT shall improve decision making processes concerning work distribution by considering the operational as well as the inter-relational level in working teams. The applied learning method could be described in [13] terminology as "inner-feedback" that is reflective feedback to the user about his/her mental model. It implies judging the assumptions and increasing the understanding of the problem being studied.

4 Conclusion

Work processes change, at least in several organizational settings. Teamwork, flexibility, temporal project groups and the like are some notions describing the new ways of working together. These developments lead to critical decision situations for people being in charge for work distribution. It is well-known that certain team constellation may lead to unsatisfying results, while others might function very well. In order to provide team leaders with a learning tool, ITBT simulates the interdependence of group members based on the concept of utility functions. It shows which aspects play crucial roles in cooperative settings and it serves as mirror for the team leader.

References

1. Bedeian A.G. and Zammuto R.F.: Organizations: Theory and Design, The Dryden Press, Orlando (1991).
2. Migliarese P., Paolucci E.: Cooperation Support Through the Use of Group Decision Systems, Journal of Organizational Computing, 3(2) (1993) 215-243.
3. Williamson O.: Markets and Hierarchies: Analysis and Antitrust Implications, Free Press, New York (1975).
4. Williamson O.: The Governance of Contractual Relations, Journal of Law and Economics, 22, (1974) 233-261.
5. Ouchi W.: Markets, Bureaucracy, and Clans, Administrative Sciences Quartely, 25(1), (1980) 129-141.
6. Egger E.: Computer Supported Cooperative Work: The Bargaining Aspect, Peter Lang Verlag, Wien, Bern, New York (1996).
7. Hanappi-Egger E.: Modeling Group Decision-Making: Some Important Aspects for System Design, in: European Journal of Work and Organizational Psychology, Special Issue on 'The Introduction of IT in Organizations' (1996).
8. Hay D.A. and Morris D.J.: Industrial Economics and Organizations, (1991) Oxford.
9. Edwards P.K. and Scullion H.: The Social Organization of Industrial Conflict, Blackwell, (1982) Oxford.
10. Hanappi-Egger E., Hanappi H.: Simulating Working Groups' Bargaining on Task-Distribution, in: Research on Cases and Theories, Special Issue on Complex Problem Solving: Methodological Support for Societal Policy Making, Vol. 2 (1998).
11. Egger E. and Hanappi H.: Multi-criteria Decision Making in Groups: A Game-theoretic Model, in: Proc. of the International Symposium on Human Interaction with Complex Systems, Greensboro (1994).
12. Weick K.: Sources of Order in Underorganized Systems. Themes in Recent Organizational Theory, in: Lincoln Y.S. (Ed.), Organizational Theory and Inquiry, Beverly Hills (1985).
13. Engerström Y: Learning, Working and Imagining. Twelve Studies in Activity Theory, Orientakonsultit, Helsinki (1990).

Sociological Aspects of Data Acquisition and Processing

I must Create a System, or be enslav'd by another Man's;
I will not Reason and Compare;
my business is to Create.

W. Blake, Jerusalem

In memory of

Professor Jerzy Jaron

Ryszard Klempous, Barbara Lysakowska, and Jan Nikodem

The Wrocław University of Technology,
Institute of Engineering Cybernetics,
27 Wybrzeże Wyspiańskiego Street, 50-370 Wrocław, Poland

Abstract. In this article, we want to demonstrate that data acquisition and processing are not only technical and information problem, but that they have to fulfil some sociological aspects, too.
The data acquisition and processing is the part of a cybernetic system. However, nowadays it is too often identified only with an information processing system itself. Such a contraction is not harmful. However, by perpetuating it we lose a fundamental sense of the activity – it becomes incomprehensible and 'art for art's sake'.

1 Introduction

One of the main tasks exacted by a man on computer systems is data acquisition and processing. It is the result of a natural characteristic of human beings: the curiosity of the environmental world. Nowadays, we observe such a rapid increase of data acquisition and processing amount that it is often called an information explosion.

Taking a part in emerging challenge, we work on new, faster computers and algorithms. Thus we collect more data and process it faster. In that untamed race, we have to decompose complex problems into simpler tasks in order to solve them (*Descartes, Discours de la methode 1637*). Focusing on their solution, we can clearly see only fragmentary goals, forgetting and losing the main purpose of our activity.

2 Teleologics of Cybernetic Systems

Cybernetics is the science concerning goal-oriented process solutions, fulfilling their demanded progress [11]. The goal-oriented sense of purpose is a paradigm of cybernetics [6], pointing the primordiality of contemplation over an activity.

F. Pichler, R. Moreno-Díaz, and P. Kopacek (Eds.): EUROCAST'99, LNCS 1798, pp. 302–306, 2000.

Cybernetics is an interdisciplinary science that can combine rational and empirical methodologies. The second methodology uses to a considerable extent the notion of conceptualism. Therefore it does not negate the rationalism.

Looking back in the history of cybernetics development, which is a rather new scientific discipline, indicates that its roots are in ancient times.

Descartes had been searching for similarities in the machines, organisms, or communities. Looking back further – the Boole's algebra or the 'Yi-king' binary logic – are the foundations for the modern data processing system. The ancient abacus was undoubtedly the early stage of computer [10], whereas the numerical Babylonian knowledge gives the beginning for computer science rather than for mathematics [8].

Referring to the names of researchers from the beginning of this discipline; [11] talks about "control and communication" (with the defined purpose but not only for the action necessity). Ashby [1], while building a Dispensive and Multistable System, accurately specifies that multistability is "the ability to reach balance by the whole system through the change of a goal". Finally the 'Shanon mouse', which learns by a trial and error method, realizes the fixed goal.

The researches, changing in kaleidoscopic-like fashion, are mathematicians, philosophers or physiologists, mainly humanists having vast technical knowledge. These people in particular, bringing their impression on cybernetics, finally gave to cybernetics its teleological character. Cybernetics without this character is a science without the fundamental principles.

3 Creative Data Acquisition and Processing

Data acquisition and processing are the processes that have been in the centre of importance for ages. The history of our civilisation's progress shows us how differently they can be realized. For example the Egyptians took note of events occurring in the sky and in nature. Thereafter, they drew conclusions in accordance with the principle 'similar after similar'. The collections of such notes from the periods long enough had made possible to predict many phenomena [9]. Today, for example, we similarly forecast weather. The empirical methodology (we talk about it here) is not without merit. Induction techniques, experimenting or concluding on the basis of data collection, clearly contradicts this.

On the other hand, rationalism, especially in the course of building algorithms of data processing problems solution, is that part of cybernetics that has been inherited from mathematics. Looking back at history of science development related to cybernetics suggests one more characteristic feature of that process: co–existence and common diffusion of the two trends i.e., the practicians–inventors and theoreticians–prophets. We can mention here many surnames, for example Pascal calculators – to Leibnitz's theories, the counting machine of Schickard – to Keppler's Rudolphian Tables, analytic machine of Babbage – to Boolean's algebra, computers of Konrad Zuse – to the machine model of Turing, computers of Atanasoff – to Wiener's memorandum.

The data acquisition and processing are contemporary typical tasks for computer systems.

Computer science, having its roots in mathematics, is based on the algorithms development that can solve numerical and symbolic problems. It technically realizes the developed algorithms. The widespread use of a computer improves the possibility of the algorithms' technical realization. Simultaneously, computer science leaves the structures and schema of proving that stem from mathematics. "The most crucial feature of the applied algorithms is a fact, that most of them haven't got any proof of correctness and nobody cares to fulfil it" [9].

The commonness of a computer in data acquisition and processing is not detrimental. However one must not give in to the pressure of an activity (vide motto). We have to remember the purpose it serves and to look for motivation of an activity, to justify and to compare the results obtained with our expectations.

So, how to understand the 'creative data acquisition and processing' is the matter of our consideration.

Pascal constructing his 'computing machine' had shown how to connect empiricism with abstraction. Leibnitz in 1671 had directly written "it's not proper for eminent people to waste their time on slavish work, on calculations, which could be done by anybody with the use of machines".

It is impossible to determine how long Kepler would have worked in Linz on Rudolfian's Tables showing the planets' movement, if he had used Nepper's 'sticks' and not Schickard's 'calculus machine'. No doubt, however, that the machine had sped up only calculations and the astronomer's intellect had been the 'creative factor'.

The similar suggestions occur when we analyze the works on the analytical machine of Babbage or the logic machine of Turing. Also in Wiener's works we cannot find anything presuming that the main component is a machine. Creating general theories of communication and data processing, Wiener punctuated moral and social conditions what were written in his work [12]. "Using of human beings only like sources of energy seems like their degradation and dooming to galleys, but forcing people in the factories to the repeatable jobs with the use of only one-millionth part of their intellectual potential, is equal to dooming them to the same degradation. Unfortunately, it seems simpler to build factories, which demand a very small amount of human abilities, than to build the world, where people could entirely develop".

4 Multistability and Feedback Theory

Leibnitz, while giving the lecture at the Paris Academy of Science in 1702 on the binary arithmetics, was clearly pointing at the tract "Yi–king" (3000 BC) as his source of inspiration. That esoteric binary language relies on two fundamental rules: "Yang" – primary and "Yin" – complementary, which stimulate each other's progress. The accurate conditions of progress need the co-existence and balance of these two elements.

We also know the European roots of feedback described by Bertalanffy [3]. Nikolaus von Kues (XV–th century) introduced the concept of coincidentia oppositorum, "antagonisms, battles between unique parts of the whole", which still create entity of higher category.

Creative data acquisition and processing need that gentle balance between a computer and a man operating it. This is the balance between speed (and also the primitive nature) of acting and intellect with association variety.

A computer is a tool for a man as fundamental as a microscope is for a biologist. It is the tool that enlarges man's capabilities [7]. But, it is a mad dream to expect that the microscope substitute human eyes; it only intensifies them. To get such an intensification, it is necessary to provide both 'sage's eyes and a glass'. Only working together can they give impressive results. As a matter of fact, that is Aristotle's "the whole is more than a sum of its parts".

The data acquisition and processing is similar – it does not exist only for itself but first of all for the interaction between the two activities. Since data acquisition develops our knowledge about what the surroundings are, data processing uses that knowledge to create what the activity should be (feedback). Only the balanced co-existence of these two elements tells us what things are and how they work.

When building systems of data acquisition and processing, we should use both analytic and synthetic methods. This means that with specifications how the system works, we should give the specifications of goals it realizes and describe the system's purpose.

There are many situations when goaloriented methods allow for the aggregation of information. Our task is the realization of goal–oriented feedback in a data acquisition and processing system. Thus we must exhibit active rather than passive behavior as elements of the system. Ashby [7] points that the information needs rapidly increase if we liberate interactions between the system's elements. He also proposes that "after we have won a battle in common interaction allowance, we should learn the moderation in its usage". One who realizes the feedback in a system of data acquisition and processing, should control this level of mentioned above with moderation, ensuring the information system's stability.

5 Conclusions

The computer has been introduced to a civilisation with a strict aim, perhaps not always clearly articulated by its creators. That aim is the intellectual progress of a man and our civilisation. Ashby states directly [2] that "a computer is the amplifier of a man's intellect". One should not forget it, especially when there are proposed new applications of those tools. It was never the intention of Wiener to build "new factories, new galleys", where the tool is a computer forcing people to do the repeatable jobs with the use of only one-millionth part of their intellectual potential.

The existence of feedback is a necessary condition for the system's stability [2]. In data acquisition and processing, the role of feedback is played by us. Our activities cause data to be separated (having no designation) from information after the interpretation transformation. Wiener, while discussing information says: "the process of getting and using information is the process of our adaptation to various circumstances of surroundings and active life in it".

When we destroy the feedback connection, we induce instability. Thus, when we eliminate a man from data processing and acquisition process, we irrevocably establish instability of information systems.

With the use of computers, we can collect and process more and more data, we gather them, enlarge our archives and computer memories, but we cannot know if anybody can analyse such vast material. Maybe just like the XIXth century was called the century of steam, our XXth century will be called the century of huge amount of information that nobody needs.

Synthetic vs. analytic, and a man vs. machine dichotomies have similar meanings. The first characterizes the whole while the second singularity, local events. Dichotomy provides equifinality rather than stability of a system [4], which proves homeostatic conditions of it (Ashby).

The reason for a violent expansion of communication processes is the disturbance of feedback between the man and the machine. It yields to the instability of the contemporary information system, which can be seen nowadays as information explosion. We are the subjects of data acquisition and processing, since the information is for us. Thus, we must take an active part in that process.

The information system has the homeostatic property. Taking into account quantum limitation of the information processing speed [5], we can be sure that a machine will never be more intelligent than a man who has built it. There is, however, a pessimistic version: the man after he builds the machine, will bring himself to the degradation of his intellect lower than the machine's level.

We have no choice; 'be optimistic' is our imperative.

References

1. Ashby W.R.; Design for a Brain, Chapman and Hall 1952, London
2. Ashby W.R.; Introduction to Cybernetics, John Willey & Sons Inc., New York 1956
3. von Bertalanffy L.; Nikolaus von Kues, Munchen, Muller 1928
4. von Bertalanffy L.; General Systems Theory; in Main Currents in Modern Thought, vol 71, 1955
5. Bremermann H.J.; Quantal Noise and Information; in 5-th Berkeley Symp. On Mathematical Statistics and Probability, vol.4, pp.15-20, 1967
6. Jaron J.; The Goals Space Of the Cybernetical System; 3-th Intern. Congress of Cybernetics and Systems, Bucharest 1976.
7. Klir G.J. (editor); Trends in General Systems Theory, John Willey & Sons Inc., New York 1972
8. Knuth Donald E.; Ancient Babylonian Algorithms, Communication Of the ACM 1972
9. Kordos Marek; Lectures on Mathematics History (in Polish); Editions WSiP, Warsaw 1994
10. Ligonniere Robert; Prehistoire et histoire des ordinateurs; Editions Robert Laffont, S.A., Paris 1987
11. Norbert Wiener; Cybernetics or Control and Communication in the Animal and the Machine, Hermann & Cie, Editeurs, Paris 1948
12. Wiener N.; The Human Use of Human Beings; Cybernetics and Society, Eyre and Spottiswoode, London 1954 (Revised Edition).

Efficient Concurrent Simulation of DEVS Systems Based on Concurrent Inference*

M. Cabarcos, R.P. Otero, and S.G. Pose

Dept. of Computer Science. University of Corunna.
15071 Campus Elviña s/n, A Corunna, Spain
Tel: +34-981-167000 ext 1276 Fax: +34-981-167160
{cabarcos,otero,silvia}@dc.fi.udc.es
http://www.dc.fi.udc.es/ai/

Abstract. This paper studies the concurrent simulation of DEVS systems by using their encoding into a formalism for dynamic systems called Generalized Magnitudes (GM). When represented with GMs, the internal parallelism of a DEVS model is foregrounded and the concurrent inference techniques developed for the GMs formalism can be applied to speed up simulation. This approach is used for both atomic and coupled DEVS models, and does not require the modeler's intervention. The internal structure of a typical DEVS control system is identified, so that the scheduling algorithms can be adapted for further efficiency improvement.

1 Introduction

Discrete Events Simulation Systems (DEVS) is a formalism intended for discrete event systems modeling [8,9]. It is based on Systems Theory, and provides a methodology for modular and hierarchical modeling. Previous studies have established the relationship between DEVS and *Generalized Magnitudes* (GMs), a formalism for reasoning in dynamic domains. A first translation from DEVS models to GMs was presented in [5] and was extended allowing simulation of DEVS models in [4]. Besides, a scheme for making concurrent the execution of GM's inference method was introduced in [1]. In this paper we relate these works and study the applicability of the concurrent inference scheme to the concurrent simulation of DEVS systems.

The GMs formalism can be used both for the encoding of the internal behavior of each atomic DEVS model and the representation of the interactions among components of a coupled DEVS model. The resulting representation in GMs has a modular shape which shows the internal parallelism of the system, and allows for fine-grain parallelization. We will recall the interesting properties of the GMs formalism for DEVS modeling, including homogeneous representation

* This work was supported in part by the Government of Galicia (Spain) under grant ref. XUGA10501B98, and in part by the Government of Spain, grant ref. PB97-0228.

F. Pichler, R. Moreno-Díaz, and P. Kopacek (Eds.): EUROCAST'99, LNCS 1798, pp. 307–318, 2000.

of transitions (internal and external), the partition of the transition function— which is introduced in each variable definition— and the integration of the time advance function into the expression of the transition function.

In [1], a scheme for aplying concurrency to the GMs inference process was presented. It was developed for speeding up the inference in a multiprocessor environment with shared memory, without requiring user intervention. Several scheduling algorithms were studied, all of them based on simple heuristics derived from structural properties of the expressions involved in the inference.

These concurrency algorithms for GMs systems can, therefore, directly be applied to atomic or coupled DEVS systems encoded in GMs. Since the GMs formalism is used for representing the whole DEVS model, the scheme exploits not only the parallelism among coupled components, but also the internal parallelism of each component.

This paper is organized as follows. In the next section we present a survey of the DEVS and the GMs formalisms, and how DEVS models can be translated into GMs. The scheduling techniques developed for concurrent evaluation of systems modeled in the GMs formalism are explained in section 3. Section 4 will show how these techniques can be applied to the particular features of DEVS systems in order to speed up the simulation. In section 5 we focus on the concurrent simulation of coupled DEVS systems. Finally, we discuss some related work and present our conclusions.

2 Relating the Two Formalisms

A DEVS system is defined by seven parameters $(X, Y, S, \delta_{ext}, \delta_{int}, \lambda, ta)$, where:

- X is the set of external events.
- Y is the set of outputs of the system.
- S is the set of sequential states through which the system can pass. In DEVS systems there are usually two special state variables called `phase` and σ representing the global state of the system, and its corresponding time, respectively. The σ variable allows defining the time advance function `ta`.
- δ_{ext} is the external transition function, that defines the next state of the system given the current state and an external event.
- δ_{int} is the internal transition function, that yields the next state of the system given the current state and without need of any external event.
- λ is the output function, that provides the outputs the system yields before an internal transition occurs.
- ta is the time advance function, that assigns to each state the amount of time that must elapse to reach the next internal transition.

Therefore, a DEVS system evolves along time due to either an internal or an external transition. External transitions are due to external events, and internal transitions happen when a certain amount of time (which depends on the current state) has elapsed since the last transition of the system. Just before an internal transition happens, the system yields the corresponding outputs. A collision occurs when an internal and an external transition happen simultaneously; in this case, a `select` function chooses between them.

```
------- initial conditions -------

initial phase: WAIT
initial sigma: tmin(P1)
initial checkstate: P1

------- external transition -------

when receive value on sensor-port
        case of: phase
        WAIT    :       hold-in EARLY 0
        WINDOW  :       if value = expected (Pi)
                    then hold-in SEND-COMMAND 0
                    else hold-in ERROR 0

------- internal transition -------

case of: phase
        WAIT    :       hold-in WINDOW window (Pi)
        WINDOW  :       hold-in LATE 0
        SEND-COMMAND: set checkstate = next (checkstate)
                hold-in WAIT tmin (next(Pi))
        ERROR   :       passivate

------- output function -------

case of: phase
        EARLY   :       send "(Pi) input arrived too early" to error-port
        SEND-COMMAND:   send control command(Pi) to command port
        ERROR   :       send "(Pi) error in sensor value" to error-port
        LATE    :       send "(Pi) input arrived too late" to error-port
        else    :       send null message
```

Fig. 1. DEVS specifications of a typical control system

Figure 1 shows an example of a generic control DEVS system taken from [8]. The system processes the inputs received from a sensor, and takes one of several checkstates depending on it. The inputs `X` must arrive inside a time window (`tmin`) in order to be considered valid, and there is an `expected` function that decides if the input is expected in the current checkstate. If an input arrives out of the time window, the system generates the corresponding error message. The internal and external transitions obtain the next state, that is captured by the `phase` state variable, and set the lapse of time `ta` (time advance function) for the next internal transition. The `ta` value plus the time of the last transition `tlast` provides the time value for the next internal transition `tnext`. When the elapsed time `e` equals the `ta` value, an internal transition occurs. Finally, an output `Y` is produced just before an internal transition is done.

As for the *Generalized Magnitudes* formalism, it is a formalism for dynamic systems with a modular representation of the system. The basic unit of knowledge representation is the *generalized magnitude* (GM), defined by a name, the set of possible values it can take, and a *Knowledge Expression* (KE) that determines what (single) value is assigned to the GM in each evaluation of knowledge.

The evaluation of the system's knowledge takes place at discrete time instants, although located in a continuous time basis. Each evaluation starts when a GM becomes *pertinent*, that is, when it receives a new value. A GM can receive an new value due to external events, due to references to the special temporal identifier `now`—that represents the current time— or due to the evaluation of its KE. Together with pertinence, the system applies inertia for the non-pertinent GMs, so that they preserve their values from the previous evaluation.

Once introduced the basic concepts of the GMs formalism, we present a summary of how a DEVS system can be modeled under this formalism (see [5, 4] for a more detailed study):

- The external events will be implemented as GMs without knowledge expression.
- The outputs of the system will be implemented as final GMs, i.e., GMs which are not referred in the KE of any GM.
- The state variables will be represented by GMs. We will distinguish the GM `phase` for representing the state of the system.
- A transition, either external or internal, corresponds with an evaluation of the knowledge in the system, i.e., with the evaluation of the KE of all the pertinent GMs at a given moment. External transitions are directly translated as consequences of the input events. In order to implement the internal transitions, it is necessary to define a GM `tnext` relating the next time point where an internal transition will take place.
- The output function is implemented in the knowledge expressions of the GMs that represent the outputs of the system.
- The Time Advance Function (`ta`) is another GM which depends on the global state of the system, and determines the time for the next internal transition (GM `tnext`). For the sake of simplicity, we will omit the σ state variable on which the `ta` function depends, and we will use instead directly a GM called `ta`.

The GMs formalism forces an homogeneous representation of DEVS models. The input, output and state variables are represented as GMs, and both the internal and external transition functions are represented as knowledge expressions.

The translation of the previous DEVS example into the GMs formalism, using the equivalences presented before, is shown in Figure 2[1]. The `phase` GM obtains the state of the system for a given previous state, transition, and input, as well as the `output` GM represents the output of the system, and the GMs `e`, `tlast` and `tnext` contain the elapsed time, the time of the last transition and the time for the next transition, respectively.

3 Concurrent Evaluation in the GMs Formalism

As explained before, [1] has studied the concurrent execution of systems represented in GMs in a shared memory multiprocessor. This study was based on the modular structure of the knowledge representation scheme provided by the GM formalism, and it allowed to make the parallelization process transparent for both designers and users. Several scheduling policies to achieve an efficient parallel execution were shown for a broad spectrum of expert systems.

The knowledge execution is done as an ordered evaluation of the KE of the system. A knowledge expression consists of references to other GMs and constant values, combined with boolean, mathematical and conditional operators, and the `previous` operator for accessing the previous value of a GM. Each reference of a GM in the KE of another one establishes a *dependence* between them. Obviously, the order of evaluation is restricted in the sense that an expression cannot be evaluated until all the values for its dependences are available. The global ordering can be represented by a *dependences net*, which is defined as a directed graph where each node represents a GM, and each arc $< A, B >$ represents a dependence, i.e., a reference to GM A in the knowledge expression of GM B. The result is a graph that captures the ordering constraints imposed by the references among GMs in their knowledge expressions.

The dependences net can be arranged in *layers*, defined as a set of GMs that do not have direct or indirect dependences among them. For our purposes, we define for each GM a layer number as the length of the longest path from that GM to any final GM. The layer *width*, i.e., the number of GMs in a layer, is a measure of the quantity of work that can be done concurrently.

The concurrent evaluation scheme for GMs tries to use all the available processors. Each processor takes a GM from a shared *ready queue* and evaluates it. Next, it marks all the GM's outgoing arcs in the dependences net as solved. If a solved arc makes ready a GM, this GM is introduced in its turn in the ready queue. With this simple scheme all the processors share their work, and synchronization is mainly needed for accessing the central ready queue.

Together with this scheme, we apply the concept of *pertinence* previously defined. The non-pertinent GMs will preserve their values from the previous evaluation. Therefore only those GMs that depend directly or indirectly on the inputs must be processed. The scheme drives, in this way, the processors to compute exclusively the GMs whose values may change. This yields a significant

[1] See [4] for a more detailed explanation.

```
GM phase (Initial Value: WAIT)
Knowledge Expression:
WINDOW      if previous(phase) = WAIT and internal;
LATE        if previous(phase) = WINDOW and internal;
WAIT        if previous(phase) = WINDOW and external and
                                 expected(value);
PASSIVE     if previous(phase) = WINDOW and external and
                                 unexpected(value);
EARLY       if previous(phase) = WAIT and external;

GM chstate (Initial Value: P1)
Knowledge Expression:
next(previous(chstate)) if phase = WAIT;

GM Output
Knowledge Expression:
"send command"   if phase = WAIT;
"error message"  if phase = PASSIVE;
"late message"   if phase = LATE;
"early message"  if phase = EARLY;

GM e (Initial Value: 0)
// time elapsed since last transition
Knowledge Expression:
timeof(now)-previous(tlast);

GM tlast (Initial Value: initial-absolute-time)
// time of the previous phase change
Knowledge Expression:
timeof(previous(phase));

GM tnext (Initial Value: initial-absolute-time)
// time of the next transition
Knowledge Expression:
tlast+ta;

GM external (Initial Value: false)
// indicates if there is input
Knowledge Expression:
true        if timeof(now)=timeof(value);
false;

GM internal (Initial Value: false)
// indicates if there is an internal transition
Knowledge Expression:
true        if e=prev(ta);
false;
```

Fig. 2. Translation into GMs formalism

speedup of the evaluation, since it is known that the subnetwork actually affected by change is usually a small part of the whole network.

The dependences net is also fundamental for the task scheduling during the inference process. The aim is to finish the inference as soon as possible, making use of all the available processors. To this end, it is necessary to select in each moment the most appropriate GM for execution, from the ready ones. This selection can be done following different criteria, as for example the layer number, the number of dependences, and the estimated evaluation time. These criteria, as well as their combinations, and their performance results with different number of processors were studied in [1].

The layer number criterion selects the GM with highest layer number. The aim is to process GMs belonging to the longest paths first, because they are supposed to delay the processing of the whole net. On the other hand, the number of dependences criterion selects firstly the GM with more outgoing arcs, because it will likely provide more work to the processors. Another criterion combines the two preceding ones, and selects the GMs with highest layer number, and among these, the one with more outgoing arcs. Finally, several criteria include the selection by estimated evaluation time, which is based on the idea that a GM with high evaluation time yields a higher delay for all the GMs depending on it.

The better simulation results were obtained by the scheduling algorithms that included the criteria of estimated evaluation time or the layer number. For instance, the HLFET [1] scheduling algorithm combined these two criteria, by selecting the GM with highest evaluation time for any path to any final GM. This criterion reached the highest efficiency: up to 80% for generated KBs, and up to 90% for a reduced set of real KBs. The second algorithm regarding its performance, was the one including the layer number as first criterion, and the evaluation time as the second.

4 Concurrent Simulation of DEVS Systems Represented in GMs

This section is devoted to analyse the structure of the dependences net for a typical DEVS control system represented in GMs. The selection of the scheduling policy and its improvements will be done based on this analysis.

Figure 3 represents the dependences net resulting from the translation of a generic DEVS control system into the GM formalism. It is easy to identify all the parts of the system: inputs (X_i), outputs (Y_i), state variables (S_i, for instance, `chstate` in Figure 2), state (`phase`), the select function (`select`), and several GMs intended for time management (`now` for clock, `e` for elapsed time, `ta` for time advance, `tlast` for time of last transition, and `tnext` for time of next transition). The select function depends on two subfunctions: one that detects when an internal transition occurs (`internal`), and another for detecting external transitions (`external`). The boxes stand for sets of GMs whose number and structure may vary from one DEVS model to another.

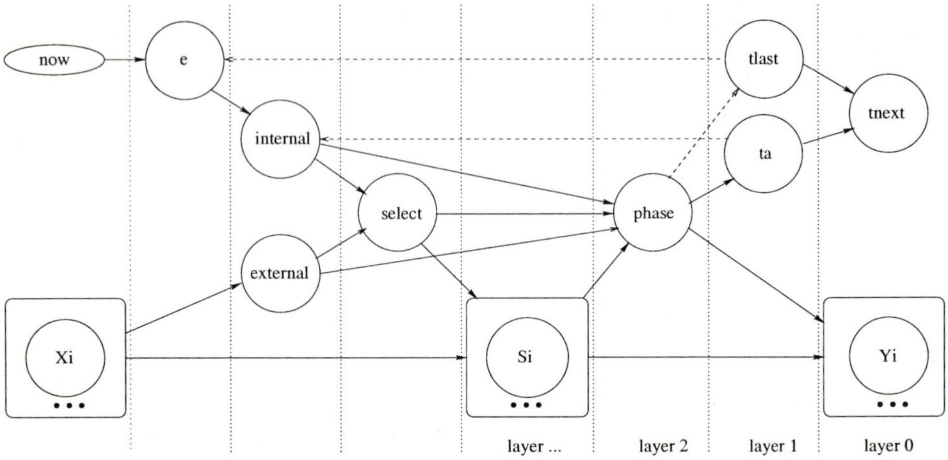

Fig. 3. A generic DEVS system modeled using the GM formalism

The general structure of these systems can also be identified. Three parts may vary for each particular DEVS model depending on the desired behavior: X_i, S_i and Y_i. But the rest of GMs corresponding to time, state and transition management will always be present.

The dependences among GMs are represented by two different kinds of arrows: regular arrows for usual dependences, and dashed arrows for those dependences modified by the `previous` operator. In the picture the different layers can be identified, each one containing a set of independent GMs.

The structure of this dependences net has the shape of a big X, where the `phase` GM is located in the cross-point. This means that the amount of work for processors will decrease from the starting points to the middle point, and just after it, it will grow to the final points.

Note that since the dashed arrows represent references to the `previous` operator, which obtains the value of a GM in a previous evaluation, they do not impose restrictions for evaluating GMs, because all the previous values are available when a new execution starts. Therefore, these arrows will not be considered as dependences for the concurrent evaluation. This is also the reason why they do not constitute cyclic references in the structure (for example `internal-phase-ta`).

In Figure 4, a concurrent execution of the previous example using three processors is presented. We use the layer number scheduling algorithm, which selects from the ready GMs those with highest layer number. Initially, the ready queue contains `e`, `external` and S_i. Applying the algorithm, `e` and `external` are selected by processors 1 and 2 due to their layer numbers. Meanwhile, the third processor starts the evaluation of elements from the S_i subnet which do not depend on `select`. The execution of `e` makes ready `internal`, which is selected next. After this, the three processors focus on the S_i subnet. The execution continues as represented in the diagram.

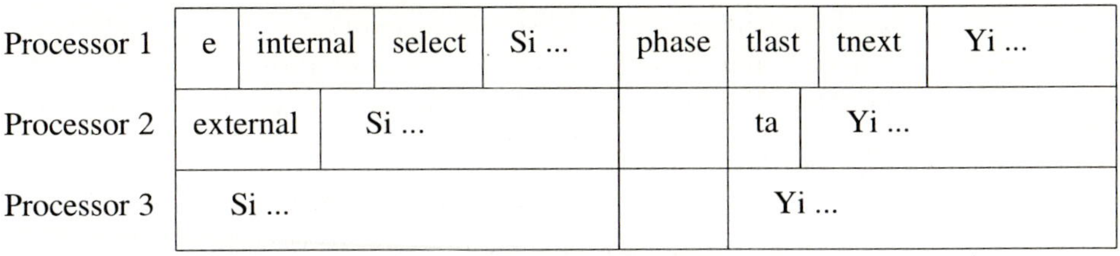

Fig. 4. A possible schedule for the simulation of a DEVS model using two processors

We improve the concurrency by evaluating at the same time the common and the specific parts of the system. The initial layers of the structure have few GMs, so little work could be done concurrently. But a detailed study of the structure shows that the initial layers can be done concurrently with part of the S_i subnet, particularly with those GMs that do not depend on the `select` GM. The situation is analogous in the final part of the net, where the `ta`, `tlast` and `tnext` GMs can be evaluated concurrently with the outputs Y_i.

The dependences net of the example presents several difficulties for its concurrent simulation. The most obvious one is that it has a layer with only one GM: the GM named `phase`. Therefore, only one processor will be busy when this layer is in progress. In order to reduce the impact of this forced break in the processor usage, we try to occupy all the processors just until the evaluation of `phase` begins. This is done by delaying the processing of GMs with lowest layer number as much as possible. The layer number criterion is adequate for this problem, because it drives the execution of the net by evaluating the set of GMs layer by layer.

5 Concurrent Simulation of Coupled DEVS Models Represented with GMs

The coupled DEVS model is represented in the GMs formalism by a set of GMs for each atomic DEVS model plus additional GMs that represent the interface among them [5]. The connections are made by modifying the KE of the input GMs of the DEVS models that receive outputs from another model. The KE contains now the name of the output GM of the model from which it receives the value. This representation introduces a restriction for the modeler, because he(she) should not reference GMs of different subsystems, apart from those representing connections between subsystems.

The concurrent simulation of the coupled DEVS model does not involve special difficulties when represented in the GMs formalism. The fine-grain concurrent evaluation scheme developed for the GMs formalism can be applied directly. Several atomic DEVS models will be evaluated concurrently, following the scheme explained in the previous section. In Figure 5 a coupled model with five atomic models A-E is represented. The A and C models can be executed sharing all the processors, as there are no connections between them. And when one model finishes its evaluation, the following one will start its evaluation. In the example, if model A requires more time for its evaluation than C, then the execution of model D will start even though model A is still being evaluated.

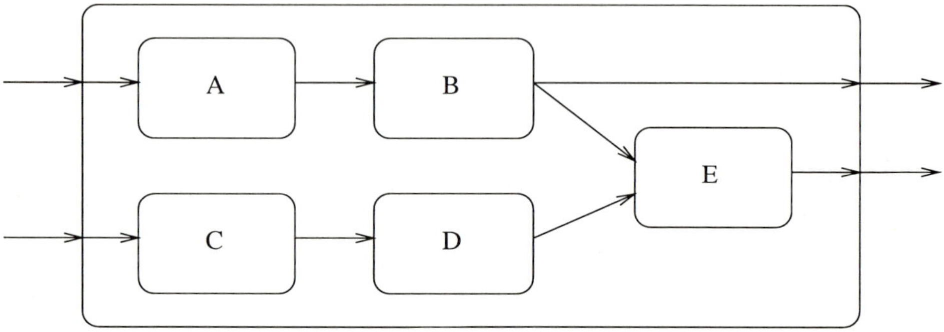

Fig. 5. A coupled DEVS system

The mechanism for concurrent evaluation of GMs does not actually distinguish different atomic DEVS models. It processes each component's GMs as if they belonged to one single DEVS system. A large dependences net is built, and the concurrent execution uses available processors for progressing through this net. The ready queue will contain pieces of work (GMs) belonging to different atomic DEVS models, and the processors will execute it using the scheduling algorithms presented in previous sections.

The simultaneous occurrence of multiple external and internal transitions is managed taking into account the influences among components. Multiple inputs can be introduced giving values to several X_i, making them available for the external transition. However, if multiple internal transitions are scheduled, an implicit order is established. Firstly, the components whose inputs are not affected by the outputs of other scheduled components are processed. Then, although in the same evaluation, the rest of components are processed since they have all their inputs available. For instance, assume that in Figure 5 components A and B have simultaneous internal transitions. In that case, component A starts its evaluation first, and component B is evaluated afterwards. Note that component A has an internal transition, while component B has an internal transition together with an external one (due to the output of A).

If there are cyclic dependences among components, the current translation into the GMs formalism will assign the value `unknown` to all the GMs, because the system does not have enough information to solve the cycle. Of course, this problem can always be overcome by introducing a general `select` function for the coupled DEVS model. Each component's select function would use the value of this function in order to determine what component has been selected for executing its internal transition.

6 Discussion and Related Work

The Revised DEVS formalism presented in [2] increases the parallelism of the formalism allowing multiple input events and simultaneous internal transitions. The `select` function is replaced by a `confluent` transition function that leaves up to the modeler the description of the system behavior when a collision between the internal and external transitions occurs. Among the possible instances of this function, it is proposed that the external transition is done just after the internal one.

The GMs formalism also allows exploiting this type of parallelism. For example, it is possible to give values to several input GMs at the same time for representing simultaneous input events. Components with simultaneous internal transitions can be concurrently executed if they have no dependences among them. Otherwise, the dependences net will guide the concurrent evaluation.

The distributed simulation of DEVS models has been studied in [6]. In this work by Praehofer et al., an algorithm which uses conservative and optimistic strategies is presented. It computes input time estimates for each atomic component in order to know when event processing is safe. Our work differs from theirs in several aspects. For instance, we simulate only conservatively, our parallelization deals with finer grains, and we consider the possibility of parallelizing the execution of one atomic DEVS model. Besides, instead of requiring the designer to decompose the whole model into clusters to run on different processors, our approach does all the work in a transparent way, with no designer or user intervention.

The identification of the structure for DEVS models represented in GMs yields interesting benefits. One of them is that the scheduling algorithms can be modified for improving performance when a system of this structure is being executed. A similar approach is presented in [3,7], where the specific properties of each task are exploited in the verification and validation process of a knowledge based system. This proposal allows for a more detailed verification that includes task specific properties. Furthermore, it allows the verification of the knowledge-based system against a specification of the intended behavior and, consequently, repairing actions can be suggested.

7 Conclusions

We have shown how the choice of the GMs formalism for representing DEVS systems allows for their efficient and concurrent simulation. On the one hand, the modular representation of knowledge allows an efficient simulation thanks to the identification of the part of the model affected by change. On the other hand, it exposes the internal parallelism of the DEVS model, that can be exploited by the concurrent execution techniques developed for the GMs formalism.

The concurrent execution scheme presented in this work is powerful enough to make an efficient parallelization of both atomic and coupled DEVS models. It exploits the internal parallelism of the model, and therefore, it does not require any intervention of the designer or the user. Finally, the identification of the structure of a generic control DEVS model allows for efficient scheduling algorithms.

Several lines remain open. The concurrency techniques could be improved if several evaluations were allowed to occur at the same time, but a detailed estimation of the time for the next transition must be done in order to avoid wrong computations. Besides, the identification of structural patterns in the dependences nets for specific systems could provide new efficiency improvements by means of adaptations of the scheduling algorithms.

References

1. M. Cabarcos, M. Otero-Díaz, P. Cabalar, and R. P. Otero. Efficient concurrent execution of medtool expert systems. In *Conference on Artificial Intelligence Applications (EXPERSYS'96)*, Paris, October 1996.
2. A. C. Chow and B. P. Zeigler. Parallel DEVS: A parallel, hierarchical, modular modeling formalism. In *Winter Simulation Conference Proceedings*, Orlando, Florida, 1994.
3. F. Cornelissen, C.M. Jonker, and J. Treur. Compositional verification of knowledge-based systems: a case study in diagnostic reasoning. *Lecture Notes in AI*, 1319:65–80, 1997.
4. R. P. Otero, A. Barreiro, P. Cabalar, and D. Lorenzo. Discrete event simulation in an environment for temporal expert systems. *Lecture Notes in Computer Science*, 1030:271–282, 1996.
5. R.P. Otero, A. Barreiro, H. Praehofer, F. Pichler, and J. Mira. STIMS-MEDTOOL: Integration of expert systems with systems modelling and simulation. *Lecture Notes in Computer Science*, 763:347–356, 1994.
6. H. Praehofer and G. Reisinger. Distributed simulations of DEVS-based multiformalism models. In *AIS'94*, Gainesville, FL, December 1994.
7. F. van Harmelen and A. ten Teije. Validation and verification of conceptual models of diagnosis. *Proceedings of the Fourth European Symposium on the Validation and Verification of Knowledge Based Systems (EUROVAV'97)*, pages 117–128, June 1997.
8. B. P. Zeigler. DEVS representation of dynamical systems: Event-based intelligent control. *Proceedings of the IEEE*, 77(1):72–80, 1989.
9. B. P. Zeigler, T.G. Kim, and H. Praehofer. *Theory of Modelling and Simulation.* Academic Press, New York, 2 edition, 1999.

Simulation of Gaussian Processes and First Passage Time Densities Evaluation*

E. Di Nardo[1], A.G. Nobile[2], E. Pirozzi[3], L.M. Ricciardi[4], and S. Rinaldi[4]

[1] Dipartimento di Matematica, Via N.Sauro 85, Potenza, Italia
dinardo@unibas.it,
[2] Dipartimento di Matematica e Informatica, Via S. Allende, Salerno, Italia
nobile@unisa.it,
[3] Dipartimento di Informatica, Matematica, Elettronica e Trasporti, Via Graziella, Reggio Calabria, Italia
pirozzi@ns.ing.unirc.it,
[4] Dipartimento di Matematica e Applicazioni, Via Cintia, Napoli, Italia
{ricciard,rinaldi}@matna2.dma.unina.it

Abstract. Motivated by a typical and well-known problem of neurobiological modeling, a parallel algorithm devised to simulate sample paths of stationary normal processes with rational spectral densities is implemented to evaluate first passage time probability densities for time-varying boundaries. After a self-contained outline of the original problem and of the involved computational framework, the results of numerous simulations are discussed and conclusions are drawn on the effect of a periodic boundary and a Butterworth-type covariance on determining quantitative and qualitative features of first passage time probability densities.

1 Introduction

The purpose of the present paper is to focus on the determination of the probability density function (pdf) of the random variable (rv) representing the instant when for the first time a dynamic system enters a preassigned critical region of the state space. Our attention will be devoted to the instance in which such a region collapses into one single point, while the system is described mathematically by a one-dimensional stochastic process. This is a typical first passage time (FPT) problem, of the kind that has been massively approached by us in the past to model single neuron's firing problems. To pinpoint the relationship existing between an FPT problem and the mentioned neurobiological example, a few considerations are in order.

As is well known, neurons are subject to ceaseless stimulations due to the bombardment of inhibitory and excitatory signals that in many cases impinge

* This work has been performed within a joint cooperation agreement between Japan Science and Technology Corporation (JST) and Università di Napoli "Federico II", under partial support by National Research Council (CNR) and by Ministry of University and of Scientific and Technological Research (MURST).

F. Pichler, R. Moreno-Díaz, and P. Kopacek (Eds.): EUROCAST'99, LNCS 1798, pp. 319–333, 2000.

on them according to random time sequences. If the sum of the net excitation is such that a change in the membrane potential higher than a certain threshold voltage occurs, then an explosive process starts, that ends in an electrical event called a "spike". The resting potential of the neuronal membrane is then restored within a few milliseconds. This situation corresponds to an FPT problem for the associated stochastic process $X(t)$ representing the membrane potential between two consecutive neuronal spikes. The initial voltage, namely the reset value following a spike, is often assumed to be equal to the resting potential $X(t_0) = x_0$. The threshold potential is denoted by the deterministic function of time $S(t)$, with $S(t_0) > x_0$. Then the theoretical counterpart of the interspike interval is the FPT rv defined as

$$T = \inf_{t \geqq t_0} \{t : X(t) > S(t)\}, \qquad X(t_0) = x_0 < S(t_0) \ . \tag{1}$$

Therefore, the reciprocal relationship between the firing frequency and the interspike interval naturally leads to the problem of characterizing the pdf of T, namely the function

$$g[S(t), t|x_0, t_0] = \frac{\partial}{\partial t} P(T \leqq t) \ . \tag{2}$$

The birth of this research area, on quantitative grounds, stems out of a classical paper [7] in which the authors invoked a diffusion process as responsible for the fluctuations of the membrane potential under the assumption of numerous simultaneously and independently acting input processes. They were able to show that, by suitably choosing the parameters of the model, the experimentally recorded interspike interval histograms of numerous units could be fitted to an excellent degree of approximation by means of the FPT pdf of a Wiener process. However, despite the excellent fitting of some data, this neuronal model has been the object of various criticisms [13], particularly because of the absence of the well-known spontaneous exponential decay of the neuron's membrane potential occurring between successive spikes. Ever since, alternative stochastic diffusion models have been proposed in the literature, aiming at refinements and embodiments of other neurophysiological features (see [9], [10] and references therein). Among these, the most famous is undoubtedly the Ornstein-Uhlenbeck (OU) model (cf., for instance, [9]).

Differently from the mentioned Wiener model, FPT problems for the OU process are in general very complicated so that cumbersome computations have traditionally been performed in the literature to come to some reliable evaluation of the statistics of the firing time. A breakthrough was provided in [2] where an efficient procedure was devised to solve a nonsingular Volterra integral equation in the unknown FPT pdf, thus obtain accurate numerical evaluations. Such a procedure holds for the general case of time-varying boundaries, for arbitrary diffusion process and, as recently shown, also for Gauss-Markov (GM) processes (cf. [5]).

Diffusion models and, more generally, GM processes, rest on the strong Markov assumption, which implies the possibility of making use of various existing analytic methods for the FPT pdf evaluation. However, it is conceivable that, particularly if a dynamical system is subject to strongly correlated stimulations, the Markov assumption is inappropriate, so that models based on non-Markov stochastic processes ought to be considered. Similarly to what has been done for GM and OU models, one can thus challenge an approach to correlated Gaussian processes.

The difficult here is due to the lack of effective analytical methods for obtaining manageable closed-form expressions for the FPT pdf. Henceforth we shall set $t_0 = 0$ and denote by $\{X(t), t \geqq 0\}$ a one-dimensional non-singular stationary Gaussian process with mean $E[X(t)] = 0$ and covariance $E[X(t)X(\tau)] = \gamma(t-\tau) = \gamma(\tau - t)$ such that $\gamma(0) = 1, \dot{\gamma}(0) = 0$ and $\ddot{\gamma}(0) < 0$. Furthermore, let us denote by $S(t) \in C^1[0,\infty)$ the boundary (or threshold) with $S(0) > x_0$. The FPT pdf of $X(t)$ through $S(t)$ is given by (cf. [11])

$$g[S(t), t|x_0] = W_1(t|x_0) + \sum_{i=1}^{\infty}(-1)^i \int_0^t dt_1 \int_{t_1}^t dt_2 \cdots \int_{t_{i-1}}^t dt_i W_{i+1}(t_1, \ldots, t_i, t|x_0) \ , \tag{3}$$

with

$$W_n(t_1, \ldots, t_n|x_0) = \int_{\dot{S}(t_1)}^{\infty} dz_1 \cdots \int_{\dot{S}(t_n)}^{\infty} dz_n \times \prod_{i=1}^{n} [z_i - \dot{S}(t_i)]\, p_{2n}[S(t_1), \ldots, S(t_n); z_1, \ldots, z_n|x_0] \ , \tag{4}$$

where $p_{2n}(x_1, \ldots, x_n; z_1, \ldots, z_n|x_0)$ is the joint pdf of $X(t_1), \ldots, X(t_n), Z(t_1) = \dot{X}(t_1), \ldots, Z(t_n) = \dot{X}(t_n)$ conditional upon $X(0) = x_0$. Unfortunately, the functions (4) appearing in (3) cannot be simplified because the involved multiple integrals cannot be calculated. Since (3) is a Leibnitz series for each fixed $t > 0$, estimates of the FPT pdf can in principle be obtained; indeed, its partial sum of order n provides a lower or an upper bound to g depending on whether n is even or odd [12]. However, also the evaluation of such partial sums is extremely cumbersome because of the complexity of the involved terms, except for the first-order term which has the following closed form expression (cf. [11]):

$$\begin{aligned} W_1(t|x_0) &= \frac{|\Lambda_3(t)|^{1/2}}{2\pi[1-\gamma^2(t)]} \exp\left\{-\frac{S^2(t)}{2[1-\gamma^2(t)]}\right\} \\ &= \left\{\exp\left(-\frac{\sigma^2(t|x_0)}{2}\right) - \sigma(t|x_0)\int_{\sigma(t|x_0)}^{\infty} \exp\left(-\frac{y^2}{2}\right)\right\} \ , \end{aligned} \tag{5}$$

where $\Lambda_3(t)$ is the covariance matrix of $X(0), X(t), \dot{X}(t)$,

$$\Lambda_3(t) = \begin{pmatrix} 1 & \gamma(t) & \dot{\gamma}(t) \\ \gamma(t) & 1 & 0 \\ \dot{\gamma}(t) & 0 & -\ddot{\gamma}(0) \end{pmatrix} \tag{6}$$

and

$$\sigma(t|x_0) := \left(\frac{1-\gamma^2(t)}{|\Lambda_3(t)|}\right)^{1/2} \left\{\dot{S}(t) + \frac{\dot{\gamma}(t)[\gamma(t)S(t) - x_0]}{1-\gamma^2(t)}\right\} . \tag{7}$$

Note that for all $t > 0$ the first-order approximation $W_1(t|x_0)$ provides an upper bound to the FPT pdf in (3), but is a good approximation of g only for small values of t. In the sequel we shall assume that the covariance $\gamma(t)$ is of the Butterworth type (B-2), i.e.:

$$\gamma(t) := E[X(t+\tau)X(\tau)] \equiv \sqrt{2}\, e^{-\alpha t} \sin\left(\alpha t + \frac{\pi}{4}\right) \quad t \geqq 0, \quad \alpha \in \mathbb{R}^+ . \tag{8}$$

Being the simplest type of an oscillatory covariance that carries a concrete engineering significance, (8) has often been invoked to approximate other covariance functions of interest for applications (cf. [14]).

Figures 1–3, obtained via (5), refer to a stationary Gaussian process originating at $x_0 = 0$ with zero mean and covariance (8). Figure 1 shows plots of $W_1(t|x_0)$ with $\alpha = 1$ in (8) for various constant boundaries. We note that as the boundary increases $W_1(t|x_0)$ decreases for all $t > 0$ to approach a constant value as t diverges. Figures 2 and 3 show plots of $W_1(t|x_0)$ as functions of t with $\alpha = 2\pi$ for the periodic boundaries $S(t) = 1 + \sin(2\pi t/Q)$ and $S(t) = 1 + \cos(2\pi t/Q)$, respectively, and for different values of Q. In all cases as t increases the function $W_1(t|x_0)$ becomes periodic with the same period of the boundary.

The computational complexity of equations (3) and (4) apparently urge that alternative analytic procedures be used in order to obtain information on the FPT pdf. A complementary, effective approach is instead the object of the present paper. Here, we implement a simulation procedure in order to disclose the essential features of the FPT pdf for a class of Gaussian processes and specified boundaries. Our approach relies on a simulation procedure [1] by which sample paths of the stochastic process are constructed and their first crossing instants through the assigned boundary are recorded. The underlying idea, originally proposed by Franklin [6], can be applied to any Gaussian process having spectral densities of a rational type. Since the sample paths of the simulated process are generated independently of one another, the simulation procedure is particularly suited for implementation on supercomputers, so it has been resumed recently. Such a procedure has been extensively implemented by us in order to explore the possible different shapes of the FPT pdf as induced by the oscillatory behaviors of covariances and boundaries.

Our goal has been is characterize the FPT pdf in terms of parameters of the covariance in the single periodic boundary problem. The motivation has been partially suggested by some results obtained in [8], where it is shown there that for a certain class of Markov processes, the FPT pdf through periodic boundaries exhibits damped oscillations reflecting the periodicity of the boundary. We have thus been naturally led to investigate whether such a behavior also occurs for Gaussian processes. Section 2 contains a sketch of the theoretical model while Section 3 indicates how the simulation procedure has been vectorized. Section 4 reports the results of the analysis of the simulations, as well as conclusions on the effect of the periodic components of the covariance and of the boundary in determining the qualitative features of the corresponding FPT pdf.

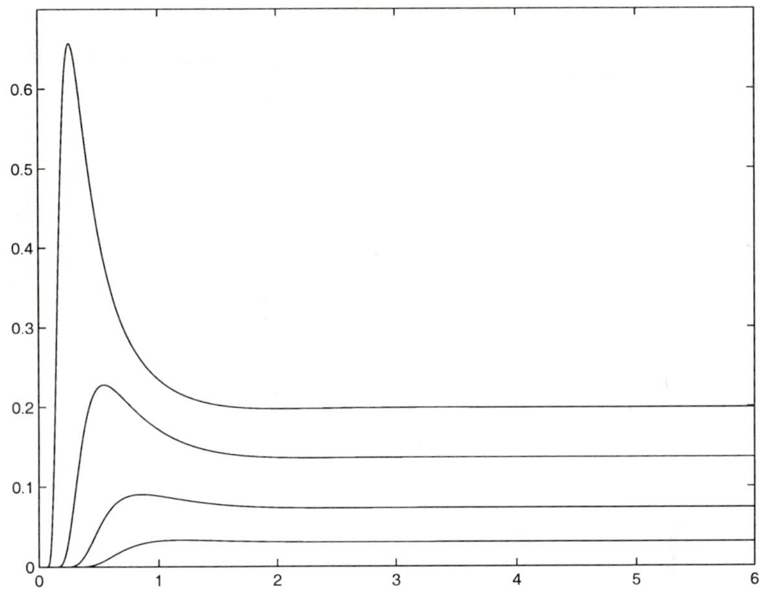

Fig. 1. Plot of $W_1(t|x_0)$ as function of t for a stationary Gaussian process with zero mean and covariance function (8) with $\alpha = 1, x_0 = 0$ and $S(t) = 0.5, 1, 1.5, 2$ (top to bottom).

2 The Stochastic Model

Let us consider the linear filter

$$X(t) = \int_0^\infty h(s)\, W(t-s)\, ds \tag{9}$$

where $h(t)$ is the impulse response function and $W(t)$ is the input signal. By Fourier transforming, (9) yields

$$\Gamma_X(\omega) = |H(\omega)|^2 \Gamma_W(\omega) \tag{10}$$

where $\Gamma_W(\omega)$ and $\Gamma_X(\omega)$ are the spectral densities of input $W(t)$ and output $X(t)$, respectively, and where $H(\omega)$ denotes the Fourier transform of $h(t)$. Equation (10) is suggestive of a method to construct a Gaussian process $X(t)$ having a preassigned spectral density $\Gamma_X(\omega) \equiv \Gamma(\omega)$. It is indeed sufficient to consider a white noise $\Lambda(t)$, having spectral density $\Gamma_W(\omega) \equiv 1$, as the input signal and

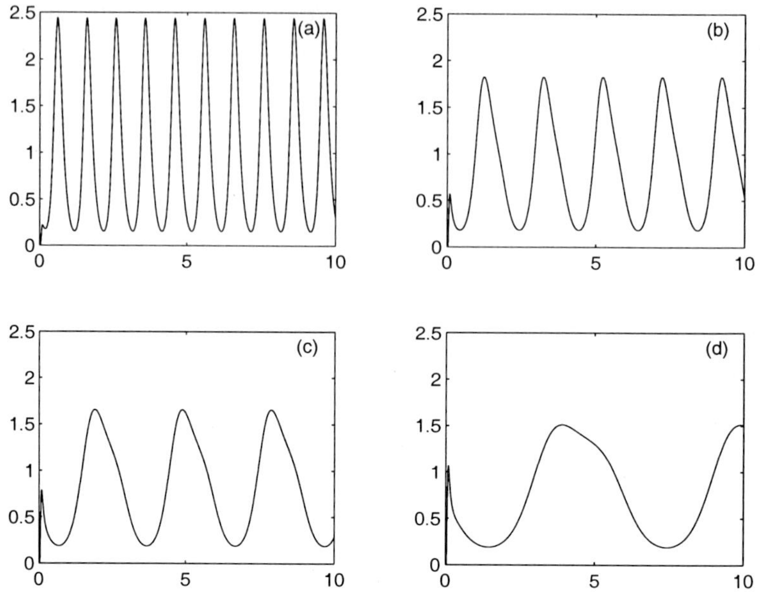

Fig. 2. Plot of $W_1(t|x_0)$ as function of t for a stationary Gaussian process with zero mean and covariance function (8) with $\alpha = 2\pi$, $x_0 = 0$, $S(t) = 1 + \sin(2\pi t/Q)$ for $Q = 1$ in (a), $Q = 2$ in (b), $Q = 3$ in (c) and $Q = 6$ in (d).

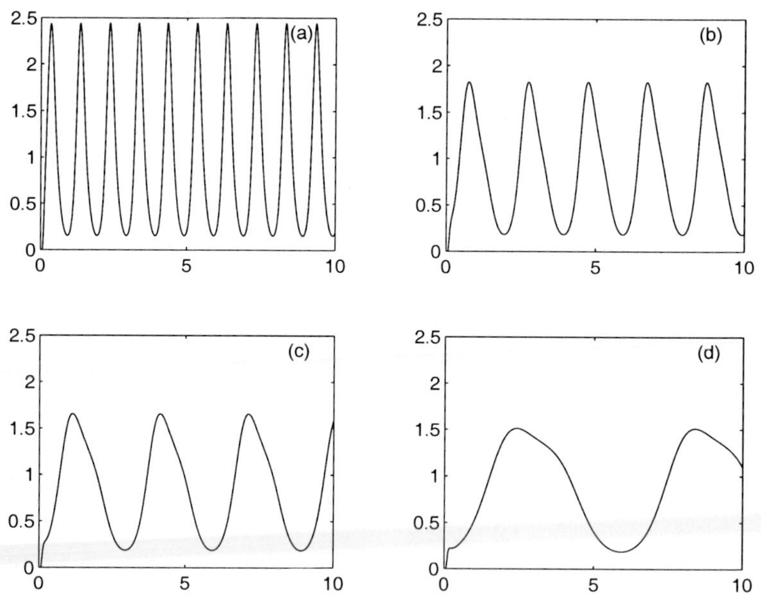

Fig. 3. Plot of $W_1(t|x_0)$ as function of t for a stationary Gaussian process with zero mean and covariance function (8) with $\alpha = 2\pi$, $x_0 = 0$, $S(t) = 1 + \cos(2\pi t/Q)$ for $Q = 1$ in (a), $Q = 2$ in (b), $Q = 3$ in (c) and $Q = 6$ in (d).

then select $h(t)$ in such a way that $|H(\omega)|^2 = \Gamma(\omega)$. Referring to the covariance function (8), the spectral density of $X(t)$ is given by

$$\Gamma(\omega) := \int_{-\infty}^{\infty} \gamma(t)\, e^{-i\,\omega\, t}\, dt = \frac{8\alpha^3}{\omega^4 + 4\alpha^4} \ . \tag{11}$$

Choosing the following real-coefficients polynomials

$$P(z) = 2\sqrt{2\alpha^3} \qquad Q(z) = z^2 + 2\alpha z + 2\alpha^2 \ , \tag{12}$$

from (11) it results $\Gamma(\omega) = |P(i\omega)/Q(i\omega)|^2$. Setting $H(\omega) = P(i\omega)/Q(i\omega)$, from (10) it follows (cf. [3])

$$X(t) = \frac{P(D)}{Q(D)} \Lambda(t) \tag{13}$$

where $D = d/dt$. To calculate the output signal $X(t)$ it is thus necessary to solve first the differential equation $Q(D)\phi(t) = \Lambda(t)$ to obtain the stationary solution $\phi(t)$, and then evaluate $X(t) = P(D)\phi(t)$. Equation (13) is equivalent to

$$\frac{d\mathbf{v}(t)}{dt} = C\,\mathbf{v}(t) + \mathbf{y}(t) \ , \tag{14}$$

where

$$\mathbf{v}(t) = \begin{pmatrix} \phi(t) \\ \phi'(t) \end{pmatrix}, \quad C = \begin{pmatrix} 0 & 1 \\ -2\alpha^2 & -2\alpha \end{pmatrix}, \quad \mathbf{y}(t) = \begin{pmatrix} 0 \\ \Lambda(t) \end{pmatrix} \ . \tag{15}$$

The simulation procedure aims to construct sample paths of the process $X(t)$ at the instants $t = 0, \Delta t, 2\Delta t, \dots$ where Δt is a positive constant time increment. Solving (14) at these instants, one obtains

$$\mathbf{v}(0) = B\,\tilde{\mathbf{\Lambda}}^{(0)}, \quad \mathbf{v}[(k+1)\Delta t] = e^{C\Delta t}\mathbf{v}(k\Delta t) + B\tilde{\mathbf{\Lambda}}^{(k)} \ , \tag{16}$$

where $\tilde{\mathbf{\Lambda}}^{(k)}$ for $k = 0, 1, \dots$ is a sequence of i.i.d. standard normal rv and

$$e^{C\Delta t} = (d_{ij}) = \frac{e^{-\alpha\Delta t}}{\alpha} \begin{pmatrix} \alpha[\cos(\alpha\Delta t) + \sin(\alpha\Delta t)] & \sin(\alpha\Delta t) \\ -2\alpha^2 \sin(\alpha\Delta t) & \alpha[\cos(\alpha\Delta t) - \sin(\alpha\Delta t)] \end{pmatrix} \ , \tag{17}$$

$$B = \begin{pmatrix} \sqrt{m_{11}} & 0 \\ \dfrac{m_{12}}{\sqrt{m_{11}}} & \sqrt{m_{22} - \dfrac{m_{12}^2}{m_{11}}} \end{pmatrix} \ , \tag{18}$$

$$m_{12} = \frac{d_{12}^2}{2}, \quad m_{22} = \frac{1 - d_{22}^2}{4\alpha} - \alpha\, m_{12}, \quad m_{11} = \frac{m_{22} - \alpha d_{12}^2 - d_{11}d_{22}}{2\alpha^2} \ . \tag{19}$$

In this way, the equation $X(t) = P(D)\phi(t)$ becomes

$$X(k\Delta t) = 2\sqrt{2\alpha^3} v_1(k\Delta t), \quad k = 1, 2, \dots \tag{20}$$

by which the sample paths of the process are constructed independently of one another.

3 The Simulation Procedure

According to (1), we denote by T the rv representing the FPT through the boundary $S(t)$ with the condition that $S(t)$ has never been previously crossed. Without loss of generality, in the sequel we shall assume $X(0) = 0$ at the initial time $t_0 = 0$. Our aim is to estimate the FPT density (2) for a stationary Gaussian process with zero mean and covariance (8). To this purpose, the stopping criterion of a simulated path is the occurring of the first crossing of the boundary. The instant when such a crossing takes place is recorded, the sample path is then discarded and the simulation is carried afresh from the initial state $X(0) = 0$ and the initial time $t_0 = 0$.

Let us denote by $NPMAX$ the number of paths to be simulated and by $NCOUNT$ the counter indicating the number of paths reset to the initial state. The length of vector loops is represented by the integer $NPARA$, which depends on the employed supercomputer: by means of (20) at each step the states of the $NPARA$ simultaneously simulated sample paths are determined. The procedure for simulating the sample paths of $X(t)$ and for recording the boundary crossing instants can be summarized as follows:

STEP 0. Input $\alpha, NPMAX, NPARA, \Delta t$.
STEP 1. Parameters computation by sequential instructions:
- $P(z)$ and $Q(z)$ coefficients in (12);
- elements of matrix $e^{C\Delta t}$ in (17);
- elements of matrix B in (18);

STEP 2. $NCOUNT := 0;\ T := 0;$
STEP 3. generate $NPARA$ standard Gaussian numbers;
STEP 4. for $i := 1$ to $NPARA$ do
- compute the initial state vector $\mathbf{v}(0)$ in (16);

STEP 5. if $(NCOUNT > NPMAX)$ stop;
STEP 6. generate $2 * NPARA$ standard Gaussian numbers;
STEP 7. for $i := 1$ to $NPARA$ do
- $T(i) := T(i) + \Delta t;$
- compute the vectors $\mathbf{v}(i)$ in (16);
- compute $X(i)$ in (20);
- calculate the values of the boundary $S(i)$;
- if $X(i) \geqq S(i)$ then
  $FLAG(i) := 1,$
  resetting the vector $\mathbf{v}(0)$
- if $X(i) < S(i)$ then
  $FLAG(i) := 0;$
  $NCOUNT := NCOUNT + FLAG(i);$
  $T(i) := [1 - FLAG(i)] * T(i);$

STEP 8. go to Step 5.

By using the recorded instants in the array $T(\cdot)$, the unknown FPT pdf $g[S(t), t] \equiv g[S(t), t|0, 0]$ can be estimated by constructing histograms with a suitable choice of the number of classes. The error due to the introduced discretization time is not larger then Δt. The error induced by the Monte Carlo method is of the order of $1/\sqrt{NPMAX}$, so that $NPMAX$ must be large (for our simulations $NPMAX = 10^7$.)

4 Computational Results

Along the lines indicated in the previous section, we have implemented a parallel FORTRAN90 program on a 128-processor Cray T3E/1200 supercomputer, based on MPI language for parallel processing. The simulation procedure is structured in such a way that after establishing the form of the covariance and of the boundary, the choice of the involved parameters can be made. Input variables are the total number of sample paths to be simulated, the number of those that must be simultaneously simulated and the time discretization step. From the recorded array $T(\cdot)$ of crossing instants, histograms are then constructed to estimate the FPT pdf $\tilde{g}(t)$.

By making use of this simulation procedure, extensive computations have been performed by us to gain some insight on the behavior of FPT pdf through varying boundaries of the form

$$S(t) = S_0 + A \sin\left(\frac{2\pi t}{Q}\right) , \tag{21}$$

$$S(t) = S_0 + A \cos\left(\frac{2\pi t}{Q}\right) , \tag{22}$$

where S_0, A and Q are positive constants. The results of the simulations indicate that in the case of the constant boundary obtained by setting $A = 0$ in (21) and (22), the FPT pdf $\tilde{g}(t)$ is rigorously unimodal (see Fig.4), in apparent contrast with the oscillatory behavior of the covariance function (8). In general, due to the relation $P = 2\pi/\alpha$ holding between the period P of the oscillatory part of the covariance and the decay constant α, for large values of α one obtains highly damped oscillations of the covariance so that such oscillations do not affect the shape of FPT pdf. On the other hand, as α decreases the damping effect reduces but the period P of the oscillation increases; the effect of such oscillations is therefore invisible in the shape of $\tilde{g}(t)$ because practically most of simulated sample paths have crossed the boundary before the first oscillation has been completed. Hence, as α increases the mode decreases while the height of the peak increases. This means that as α increases the whole probability mass approaches the $y-$axes, so that the corresponding FPT pdf goes to zero faster and faster. We note that as α increases, the covariance function (8) approaches that of an OU-type diffusion process. Hence, its sample paths are characterized by very rapid fluctuations, so that boundaries close to the equilibrium point are almost instantaneously crossed.

It should be stressed that a multimodal behavior of the FPT pdf may be expected if the assumption of constant boundary is removed. Indeed, the performed simulations support the conjecture that for Gaussian processes, the behavior of FPT pdf shows some similarities to that discovered for diffusion processes [8]. In particular one finds that, after an initial phase on which we shall soon say something more, for the case of periodic boundaries (21) and (22), the FPT pdf $\tilde{g}(t)$ becomes multimodal: its successive relative maxima occur after time intervals equal to the period Q of the boundary and the amplitudes of such maxima decrease in time.

The simulation results have showed that the parameter

$$\lambda := \frac{Q}{P} = \frac{\alpha Q}{2\pi}, \tag{23}$$

i.e. the ratio between the period Q of the boundary (21) or (22) and the period P of the oscillating component of the covariance function (8), plays a fundamental role in determining the behavior of FPT pdf.

Let's begin to analyze the behavior of estimated FPT pdf in the presence of boundary (21). Figure 5 refers to the case $\lambda \geqq 1$. We note that initially a peak shows up whose amplitude increases with λ. (In Fig. 5(a), where $\lambda = 1$, such a peak is too small to be noticed.) In Fig. 5(b), where $\lambda = 2$, the presence of the initial peak can be observed; such a peak becomes progressively more pronounced as λ increases (see Fig. 5(c) and 5(d)). It should be noted that as the initial peak increases the amplitudes of the successive peaks progressively decrease because of the conservation of the total probability mass. Furthermore, comparing Fig. 5 with the same cases shown in Fig. 2, we note that the function $W_1(t|0)$ in the neighborhood of the origin gives a good approximation of the FPT pdf of $\tilde{g}$, and hence also of g. As λ decreases, the initial peak of $\tilde{g}$ closed to the origin tends to decrease, to disappear for $\lambda < 1$ (see Fig. 6(a)).

Qualitatively analogue results are obtained for periodic boundaries of type (22) that are initially decreasing. For $\lambda < 1$, as in the presence of boundary (21) the estimated FPT pdf $\tilde{g}(t)$ exhibits a multimodal behavior with maxima occurring after time intervals equal to the period of boundary (compare Fig.6(a) with Fig.6(c)). The relative maxima of $\tilde{g}$ in the case of boundary (22) are greater than those of $\tilde{g}$ in the case of boundary (21). For $\lambda > 1$, differently from the previous case of boundary (21), the presence of the initial peak is less evident because of the initial decrease of boundary (22) (compare Fig.5 with Fig.7 and Fig.6(b) with Fig.6(d)). In particular, Fig. 7 shows that as λ increases the mode increases while the absolute maxima of $\tilde{g}$ decreases. As for the previous boundary, by comparing the curves of Fig. 7 with the corresponding ones in Fig. 3, we note that the function $W_1(t|0)$ in the neighborhood of the origin again gives a good approximation of the FPT pdf of $\tilde{g}$, and hence also of g.

For the boundaries (21) and (22), Fig. 8 shows estimated FPT pdf with $\lambda = 10$ and $\lambda = 30$. As is to be expected, whenever λ becomes large — i.e. when the boundary's period is much larger then that of the periodic term of the covariance — such pdf becomes unimodal. Indeed, one expects to observe a behavior similar to that characterizing constant boundaries because for very large values of λ the boundary remains practically constant during the time interval in which the covariance is appreciably non vanishing.

It should be pointed out that parameter λ plays also a different role. Indeed, FPT pdf corresponding to the same value of the ratio $\lambda = Q/P$ can be mapped into one another by a suitable transformation. To see it, let us consider the stationary normal process $X(t)$ with zero–mean and covariance $\gamma(t) := E[X(\tau)X(\tau+t)]$; furthermore let us assume that $P[X(0)=0]=1$. Let us now consider the transformed process $Y(t^*) = X(t^*/c)$. Process $Y(t^*)$ is itself normal, stationary, with zero–mean and initial state $Y(0)=0$. The covariance function of $Y(t^*)$ is given by

$$\widehat{\gamma}(t^*) = E[Y(\tau+t^*)Y(t^*)] = E\Big[X\Big(\frac{\tau}{c}+\frac{t^*}{c}\Big)X\Big(\frac{t^*}{c}\Big)\Big] = \gamma\Big(\frac{t^*}{c}\Big)\ , \tag{24}$$

with $\widehat{\gamma}(0)=1$. Hence, if $X(t)$ has covariance (8), then $Y(t^*)$ has a B2-covariance with parameter $\widehat{\alpha} = \alpha/c$ and period $\widehat{P} = 2\,\pi/\widehat{\alpha}$ of the oscillating component. The FPT problem for $X(t)$ through $S(t)$ is thus changed into the FPT problem for $Y(t^*)$ through the boundary $\widehat{S}(t^*) = S(t^*/c)$. Hence, the FPT pdf $g_Y[\widehat{S}(t^*), t^*|0]$ of the process $Y(t^*)$ is related to FPT pdf (2) as follows:

$$g_Y[\widehat{S}(t^*), t^*|0] = \frac{1}{c}\, g\left[S\left(\frac{t^*}{c}\right), \frac{t^*}{c}\,\middle|\, 0\right]\ . \tag{25}$$

We remark that if the boundary $S(t)$ is periodic with period Q, also the boundary $\widehat{S}(t^*)$ is periodic with period $\widehat{Q} = c\,Q$, so that

$$c = \frac{\widehat{Q}}{Q} = \frac{\alpha}{\widehat{\alpha}} = \frac{\widehat{P}}{P} \tag{26}$$

and

$$g_Y[\widehat{S}(t^*), t^*|0] = \frac{Q}{\widehat{Q}}\, g\left[S\left(\frac{Q}{\widehat{Q}}t^*\right), \frac{Q}{\widehat{Q}}t^*\,\middle|\, 0\right]\ . \tag{27}$$

In conclusion, the FPT pdf for a preassigned pair $\widehat{P}, \widehat{Q}$ can be determined via (27) from the FPT pdf corresponding to the pair P, Q provided that $\lambda = Q/P = \widehat{Q}/\widehat{P}$. As an example, let us consider the FTP problem with $\alpha = 2\pi$ and $S(t) = 1 + \sin(2\pi t)$ and let g denote the corresponding pdf. Due to (27), since $Q/P = \widehat{Q}/\widehat{P}$, the FPT pdf g_Y for $S(t) = 1 + \sin(2\pi t/3)$ and $\alpha = 2\pi/3$ is obtained from g as follows:

$$g_Y\left[1+\sin\left(2\pi\frac{t}{3}\right), t|0\right] = \frac{1}{3}\, g\left[1+\sin\left(2\pi\frac{t}{3}\right), \frac{t}{3}\,\middle|\, 0\right]\ . \tag{28}$$

Graphs (a) and (b) of Fig. 9, obtained via our simulation procedure, "experimentally" indicate the validity of such relation. Fig. 9(c), obtained by simulation as well, refer to the FTP problem with $\alpha = 2\pi/6$ and $S(t) = 1 + \sin(2\pi t/6)$. Relation (27), again expected to hold, turns out to be experimentally confirmed, as one can see by comparing Figs. 9(a) and 9(c). We incidentally note that all this further stresses the reliability of the adopted simulation procedure.

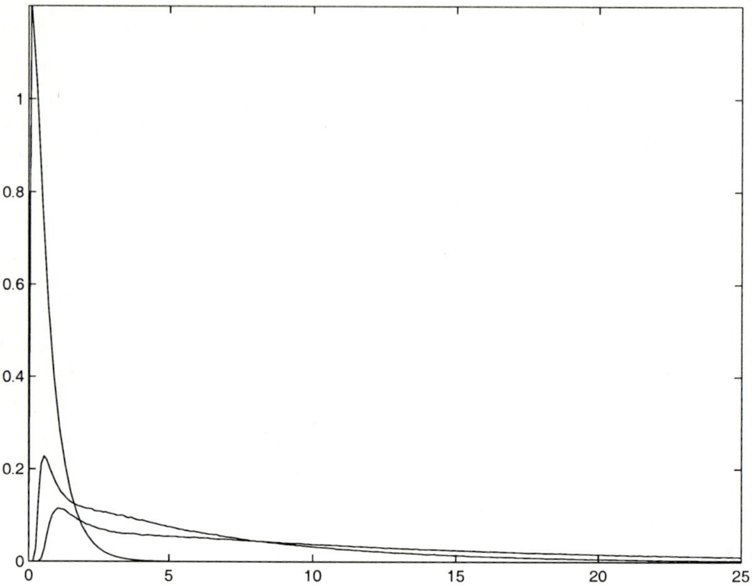

Fig. 4. Plot of $\tilde{g}(t)$ in the presence of boundary (21) and (22) with $A = 0, S_0 = 1$ for $\alpha = 0.5, 1, 5$ (bottom to top)

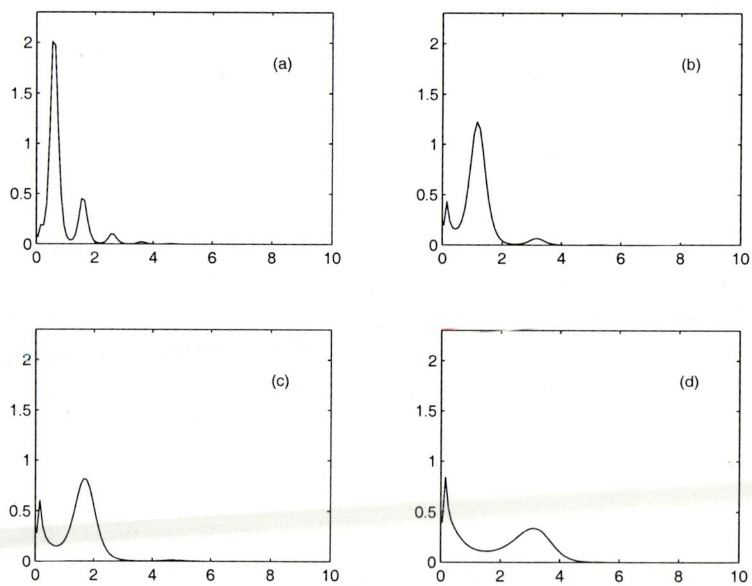

Fig. 5. Plot of $\tilde{g}(t)$ in the presence of boundary (21) with $A = 1, S_0 = 1$ and $\alpha = 2\pi$ for $Q = 1$ in (a), for $Q = 2$ in (b),for $Q = 3$ in (c) and for $Q = 6$ in (d).

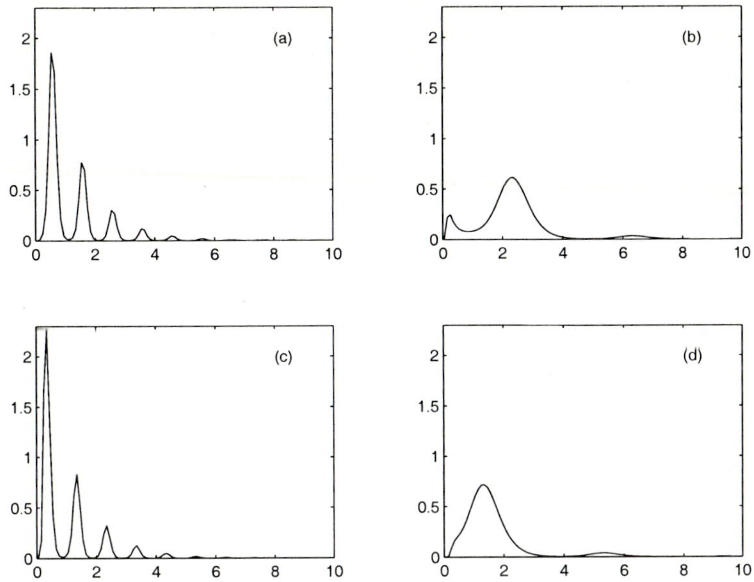

Fig. 6. Plot of $\tilde{g}(t)$ in the presence of boundary (21) with $A = 1, S_0 = 1, \alpha = \pi$ for $Q = 1$ in (a) and for $Q = 4$ in (b). Plot of $\tilde{g}(t)$ in the presence of boundary (22) with $A = 1, S_0 = 1, \alpha = \pi$ for $Q = 1$ in (c) and for $Q = 4$ in (d).

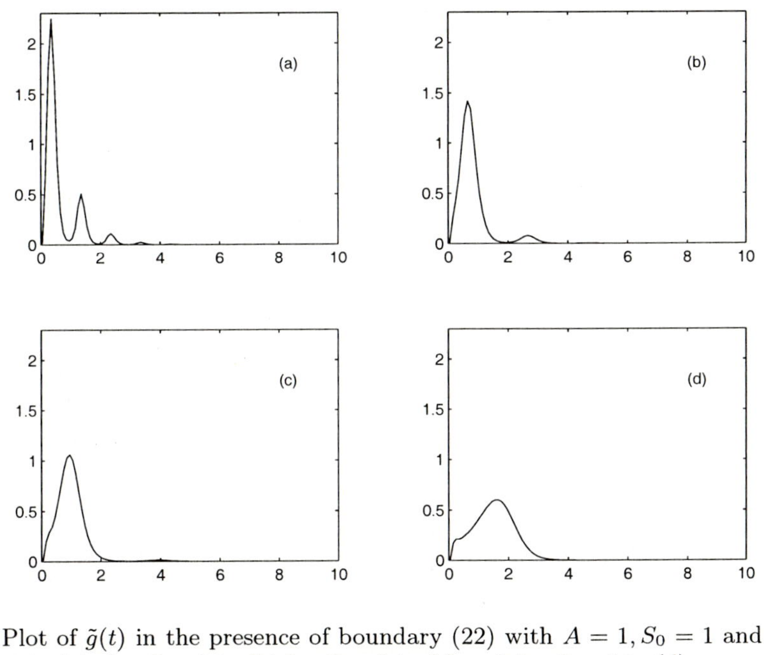

Fig. 7. Plot of $\tilde{g}(t)$ in the presence of boundary (22) with $A = 1, S_0 = 1$ and $\alpha = 2\pi$ for $Q = 1$ in (a), for $Q = 2$ in (b),for $Q = 3$ in (c) and for $Q = 6$ in (d).

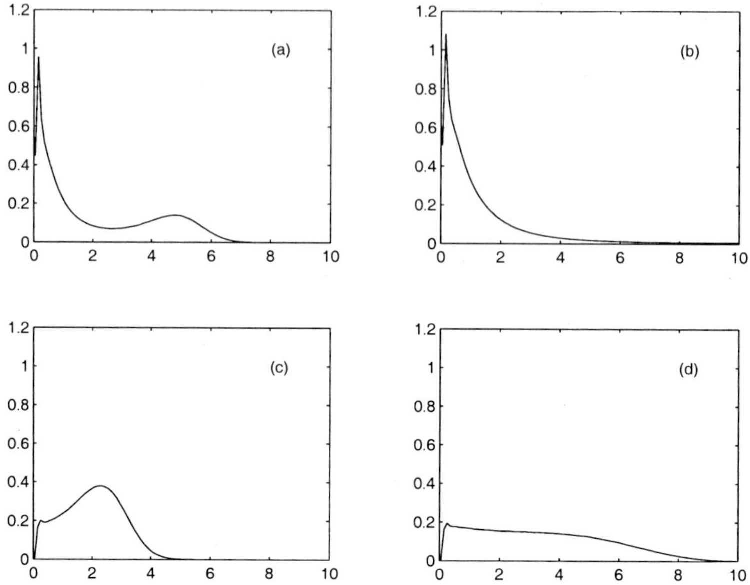

Fig. 8. Plot of $\tilde{g}(t)$ in the presence of boundary (21) with $A = 1, S_0 = 1, \alpha = 2\pi$ for $Q = 10$ in (a) and for $Q = 30$ in (b). Plot of $\tilde{g}(t)$ in the presence of boundary (22) with $A = 1, S_0 = 1, \alpha = 2\pi$ for $Q = 10$ in (c) and for $Q = 30$ in (d).

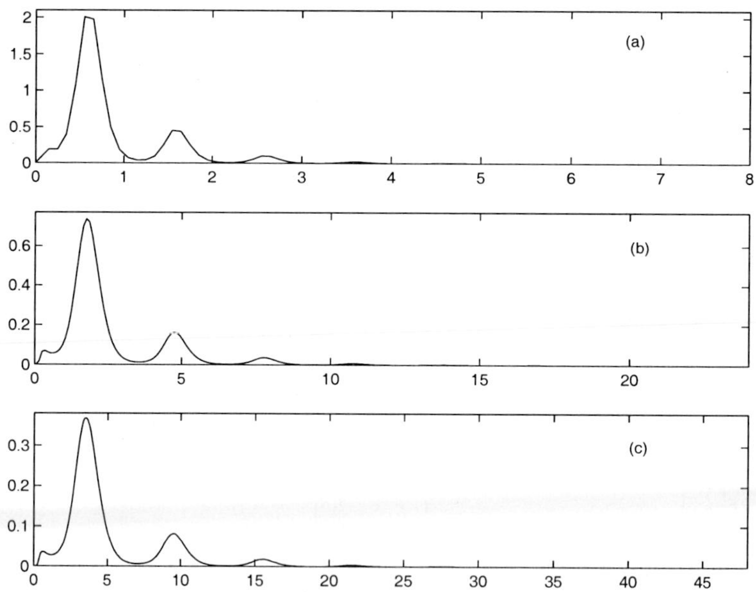

Fig. 9. Plot of $\tilde{g}(t)$ in the presence of boundary (21) with $A = 1, S_0 = 1$ for $P = Q = 1$ in (a), for $P = Q = 3$ in (b),for $P = Q = 6$ in (c).

5 Concluding Remarks

In the foregoing, a simulation procedure for evaluating FPT pdf's for correlated stationary Gaussian processes has been sketched. Some indications about its features in the case of both constant and oscillatory boundaries have been provided when the covariance is of the Butterworth type. Use of such a procedure is valuable for the investigation of the time evolution of a variety of dynamical systems subject to random perturbations. In particular, this is certainly the case of single neuron's activity behavior modeled by stochastic processes of a non-Markov Gaussian nature, for which the existing literature is scarce and fragmentary (cf. [4]). Although a specific reference has been made here to the neurobiological case, FPT problems for correlated Gaussian processes bears a great relevance in a variety of other fields, including population dynamics and risk-assessment, for which analytical methods are lacking or ineffective.

References

1. Buonocore A., Di Crescenzo A., Iardino F., Nakamura A., Ricciardi L.M., Rinaldi S.: A vectorized simulation procedure for computations of first crossing time densities of normal processes with oscillatory covariances, Rapporto Tecnico N. 1/93, "Progetto Finalizzato Sistemi Informatici e Calcolo Parallelo " del CNR (1992)
2. Buonocore A., Nobile A.G., Ricciardi L.M.: A new integral equation for the evaluation of first-passage-time probability densities. Adv. Appl. Prob. **19** (1987) 784–800
3. Davenport W.B., Root W.L.: An introduction to the theory of the random signals and noise. Mc Graw-Hill (1958)
4. Di Nardo E., Nobile A.G., Pirozzi, E., Ricciardi L.M.: On a non-Markov neuronal model and its approximations. BioSystems **48** (1998) 29–35
5. Di Nardo E., Nobile A.G., Pirozzi E., Ricciardi L.M.: A computational approach to first-passage-time problem for Gauss-Markov processes. Preprint **5**, Universitá degli Studi della Basilicata (1998)
6. Franklin J.N.: Numerical simulation of stationary and non stationary Gaussian random processes. SIAM Review **7** (1965) 68–80
7. Gerstein G.L.,Mandelbrot B.: Random walk models for the spike activity of a single neuron. Biophys. J. **4** (1964) 41–68
8. Giorno V., Nobile A.G., Ricciardi L.M.: On the asymptotic behavior of first-passage-time densities for one-dimensional diffusion processes and varying boundaries. Adv. Appl. Prob. **22** (1990) 883–914
9. Ricciardi L.M.: Diffusion models of single neurons activity. Neural Modeling and Neural Networks (F. Ventriglia, ed.) Pergamon Press, Oxford (1994) 129–162
10. Ricciardi L.M.: Diffusion models of neuron activity. The Handbook of Brain Theory and Neural Networks (M.A. Arbib, ed.) The MIT Press, Cambridge (1995) 299–304
11. Ricciardi L.M., Sato S.: A note on first passage time for Gaussian processes and varying boundaries. IEEE Trans. Inf. Theory **29** (1983) 454–457
12. Ricciardi L.M., Sato S.: On the evaluation of the first-passage-time densities for Gaussian processes. Signal Processing **11** (1986) 339–357
13. Tuckwell H.C.: Introduction to Theoretical Neurobiology, Vol. 2. Nonlinear and stochastic theories. Cambridge University Press (1988)
14. Yaglom A.M.: Correlation Theory of Stationary Related Random Fluctuations. Vol: I Basic Results. Springer (1987)

Distributed Simulation with Multimedia Interface

Pedro Corcuera, Mario Garcés, Eduardo Mora, and Marta Zorrilla

Department of Applied Mathematics and Computer Science,
University of Cantabria
Av. de los Castros s/n 39005 Santander. Spain
{pcorc, mgarc, emora, mezorr}@ macc.unican.es

Abstract. The training of the personnel plays a major role in the security and reliability of many industrial processes. In order to fulfil this objective it is necessary to develop applications that support distributed interactive simulation with an interface near the real world. This paper describes such type of application during the development of a training simulator prototype for a nuclear power plant. The simulator is based on a client/server architecture that allows the execution on different machines in a network and many users to participate in the same simulation. The interface was designed to support the interaction of the operators with the simulator through touch screens with high fidelity displays of the control room developed using the component technology. The main features of the simulator are the distributed execution of the models using inexpensive hardware and the flexibility of design and maintenance of the interface.

1 Introduction

An important factor in the security and reliability of industrial processes is the training of the operators that intervene in them. Such training should cover stages of normal operation and of emergency in order to prepare the personnel in the pursuit of the procedures of emergency and mitigate the consequences of an anomaly in the production process [1]. This last factor is particularly important in the nuclear sector, where it is recognised that human factors are key elements in order to assure a safe operation [2]. In order to fulfil this objective, the problem has traditionally been solved by means of the construction of computing full-scope simulators that reproduce physically the control room, supported by expensive and proprietary systems. The main advantage is that they reproduce exactly all the factors that intervene in the man-machine interface of the process.

The advances in the computing power of microprocessors, as well as the development of protocols to communicate over software components, has made it possible to apply Client/Server architecture to simulation. It enables the simulator user interface to run in one computer (client) and the simulator software in another (server). Moreover, the development of multimedia techniques and devices, allow the replica of real panels by soft panels composed of indicators and operational switches incorporated by means of computer graphics, touch screen interactivity and audio characteristics.

F. Pichler, R. Moreno-Díaz, and P. Kopacek (Eds.): EUROCAST'99, LNCS 1798, pp. 334–342, 2000.

This paper describes the design and development of a training simulator for a nuclear power plant with almost the same functions and scope of simulation as a full scope simulator. The interaction of the operators with the simulator is carried out through touch screens. The specifications of the simulator were high fidelity of the power generation system, flexibility, extendibility and recording of response data. The first specification fulfils the technical requirements of the simulator with sufficient accuracy. The second and third specifications mean that it must be easy to change freely the arrangements of instruments, switches, indicators, etc., and that it must be possible to add new systems or displays to the simulator easily. The last specification enables the simulator to record data concerning human behaviour in response to malfunctions.

The paper is organized as follows. In section 2 we describe the environment where the simulator will be executed: the RNI's OpenSim NT™ simulator executive, paying attention in those aspects more related to this project. In section 3 we introduce the object oriented component technology and analyze the soft panels development process. In section 4 we present the simulator's architecture. We describe how the soft panels are integrated onto the simulation environment, the modules that comprise the simulator and the chosen configuration for an effective user interaction. Finally, in section 5 we present a complete prototype that is currently available and expose the main conclusions derived from its development and the operators feedback.

2 Simulation Environment

The simulator is based on the real time simulator executive called OpenSim NT™ of RNI Technologies [3]. The PC/Windows NT technology has increased competition and decreased cost, allowing the use of standard, off the shelf hardware and software for functions that, in the past, were performed by proprietary, specialised hardware and software. It allows, in conjunction with the physical models of the reference plant, to verify the requirements included in the ANSI 3.5 for nuclear plant simulators. As an open environment based on software standards it simplifies the task of adding new modules to extend its capabilities. This feature has been widely applied to the development of this simulator.

The OpenSim NT™ software architecture consists of servers, a client library, clients, and an operating system interface library. Client applications normally access the various services provided by the servers contained in the client library, that is designed to allow the use of separate processes executed on different machines in a network. The operating system interface library provides a portable application interface to several process control services used in real-time applications which are not uniformly implemented in Unix and non-Unix operating systems. The use of threads allows the distribution of processes over multiple processors.

A single server, called the directory server, provides the location and Remote Procedure Call (RPC) numbers of the servers which comprise a project, along with the root point of the project's directory hierarchy. An OpenSim NT project is logically divided into the development system and the run-time system. The processors which make up the run time system can communicate with each other using the shared memory and semaphore services, and between themselves and the development system processors using the network services based on the standard ONC RPC/XDR to create the client/server environment. To aid in the maintenance of the simulator is provided a database manager that uses a library written to the ODBC standard and allows to use any commercial database with this driver.

The module named simulator control server, provides a control and monitoring interface to the simulator executive and I/O processing systems. This server executes in the run-time system because it requires both memory mapped access to the global partitions and use of the semaphore service to control the executive system. The services provided by the simulator control server are available to any client which includes the simulator control (this provides control functions including freeze, run, reset, snapshot, module in/out control, module query, I/O override manipulation, and I/O query) and the operator action monitor (this provides a way for clients to obtain control panel device manipulation data from the simulator executive system).

The components of the typical simulator configuration includes a simulator computer, a system server, the instruction station and the site network. The simulator computer has to be preferred a multiprocessor PC because it support the real time execution of the simulator models. The system server is a PC that contains the simulator database, initial condition files, simulation model source code and instructor data files. From the instructor station the simulation models running on the simulator computer may be monitored and manipulated and instructor inputs and training scenarios may be inserted. A fast ethernet network is used for the simulator system.

The models of the simulator should be developed and tested using a engineering workstation in the form of a static library. Normally the Fortran language is used for this purpose and the Matlab or ACSL environment to design the control systems.

OpenSim NT™ also includes a broad set of utilities designed to perform varied tasks like FORTRAN and C code precompiling, initial conditions generation, database labels searching, simulator executive linking or execution time control. The modules that constitute the simulator use one of this tools, a Dynamic Data Exchange (DDE) server, to obtain and/or modify simulator data in run-time. This utility is enclosed to allow OpenSim to be accessed by third party applications through Windows DDE protocol. The DDE conversation is established between the client module and an intermediate DDE server which communicates directly with OpenSim using RPC. This data exchange method simplifies the development process of our modules avoiding RPC programming.

3 Soft Panels Development

The simulator being developed is a multimedia application that thoroughly implements that concept. As we explain next, we profited from the last advances in object-oriented programming and image processing techniques, in addition to the constant progress in computing hardware performance and price reduction, to develop an inexpensive interface for the whole control room that uses touch screens and common PC/NT platforms instead of mice and workstations.

The interface plays a major role in this type of simulators and the configuration should allows many users can interact with the simulation in real time. As we mention above the interaction of the operators with the simulator is carried out through several touch screens displaying high fidelity of animated images of the elements in the control room during the operation (hard panels). The scale of the images is tried to be constant comparing with reality. In the prototype it was used a scale of 40-45%.

3.1 Application of Component Technology

The interaction of the operators with the simulator is carried out through several screens displaying high fidelity animated images of the elements in the control room. As we said above these screens are named soft panels.

The last advances in object oriented software development simplifies the fast construction of this kind of applications. We use Microsoft's ActiveX component technology to simulate the panels devices. With this kind of components it is possible to replicate easily and quickly the control room panels. This technology uses the standard Component Object Model (COM) to enable software components to interact with one another in a networked environment, regardless of the language in which they were created. Usually they are developed with Visual C++ to be inserted in Visual Basic forms. Looking for the best performance we use Visual C++ during the development of the whole application. Thinking in future improvements, an interesting feature of ActiveX technology is that it can be used to create applications to run on the Internet besides the desktop.

In our application every element in the control room becomes an ActiveX control (formerly known as an OLE control), i.e. an small in-process server that can be used in any OLE container. Thus, each control is a reusable software object that supports a wide variety of OLE functionality and can be easily customized to fit our simulation needs. Its appearance and behavior is defined by means of stock and custom properties and methods, similar in use and purpose, respectively, to member variables and member functions of a C++ class. The control container communicates with the object accessing to these properties and methods. The ActiveX control also sends messages, 'fires events' in the ActiveX jargon, to notify the container that something important has happened in the control, usually a change of state or an user action. With this approach we have achieved a solution that is economical both in cost and in effort in the construction of the soft panels.

The first step during the design of the simulator was the identification and classification of the current objects in the hard panels layouts. Basically, they can be of two types: switches and indicators. The first ones include all the objects able to receive input from the user (valves and pumps switches, buttons, keys, knobs, selectors, sliders...) while the others are solely used to display operational data. Restricted to the front panels of the control room we found more than five hundred objects of each type besides one thousand lamps and six hundred annunciators in the alarm panels that also have to be simulated. It is easy to understand the value of the reusability offered by this object oriented technology.

Then it was specified the functionality and properties of each object and they were classified again in smaller groups; for instance, there are black, green, red, blue and yellow valves switches, with different kinds of handles and with two, three or even four available positions which could either remain fixed or return to the central position after the user interaction. This task helped us to decide the number of distinct ActiveX controls to be used and their required properties that would serve to represent every control room component.

Due to the specific features of switches, we created for them a fully customized ActiveX control. Thereby, we could incorporate the needed functionality to allow them to run as independent client applications capable of communicating directly with OpenSim through the DDE server. The simulator signals corresponding to each one of the switch positions are defined as a control property. Its appearance, as happens with all the used controls, is specified by means of image properties. Besides a sound file can be associated with the object to provide the user with the appropriate feedback after the interaction.

The images are created from digital pictures of the real device treated as much as needed with image processing tools. With the advanced digital cameras and programs currently available, the obtained degree of physical fidelity is only limited by the effort devoted to this stage of the development process, in any event much lower than if we had have to draw the components with painting or design tools.

Based on the acquired experience, we created other customized ActiveX control to represent the lamps and all the switches with only two positions (mainly keys and pushbuttons). This simpler object, characterized by its two states, also includes the functionality to communicate directly with the simulator.

The rest of components (analog and digital indicators such us linear and angular gauges, odometers or LED's and some complex switches as selectors and knobs) were acquired from third party vendors. There are a lot of affordable ActiveX instrumentation libraries for sale and many of them include the kind of objects found in the control room of a nuclear plant. Since they are fully configurable and also permit to select an image as the object background is possible to represent with enormous fidelity almost any object. For instance, we have used components from Global Majic Software Inc. and Quinn-Curtis Inc. with good results. As we explain in the following section the container application must be responsible of the linkage of these third party objects to the simulator data.

3.2 Interactive Panels

As explained above, the ActiveX controls are incorporated into the windows or dialogs of control containers applications in order to be executed. Therefore every screen, or soft panel, that comprises the simulator interface is a dialog based ActiveX control container application. Besides holding the controls and showing them arranged as in the real panels, this application starts the execution of the DDE server that manages the communication of the controls with OpenSim, initializes the controls in the positions corresponding to the state of the sequence being simulated and updates the indicators with new data every time they are changed.

It must be kept in mind that while our customized components are directly linked to a simulator signal in the same way a real object is wired in the back side of the panel, the rest of components have to be associated to their corresponding OpenSim signal, or signals, in the container application. The index of the object in the dialog is used as a reference to identify the associated signal when the soft panel is refreshed. Thus, the container application must maintain its own conversation with the DDE server acting as a new OpenSim client.

The scale of the images in the interactive panels is tried to be constant comparing with reality. We use a scale of 40~45% and so this simulator will be a full scope half scale simulator with important physical space savings for accommodating the tool. This size is also intended to avoid the necessity of navigation through the panels or its elements, i.e., all the panels and its components will be visible while the simulator is running and accessible by a single touch on the screen without menus nor scrollbars.

Finally, to simplify even more the development process we have included a module in all the container applications that reads the numerous labels captions of the simulated panel from a text file and puts them on the corresponding tags already drawn in the soft panel. From this file the developer can also specify the font style and colors of the tag.

3.3 Alarms Panels

The same techniques have been applied to the development of alarms panels which reproduce the full logic and functionality of the real ones, including colors, flash frequency and sounds. The main difference with the rest of panels is that the pushbuttons that controls their operation (silence, acknowledge, reset and test functions) are not located in the same panel and so they do not have to be interactive.

Since there are only two types of distinct alarm panels we have developed just two container applications to represent the fifteen alarms panels found in the control room. Each annunciator is an ActiveX control which the container animates applying the panel functionality. In the same way we used with the labels of the interactive panels, a text file serves to specify the captions, font style and colors of the annunciators. Furthermore, the associated simulator signals and sound files are read from the file too, resulting in entirely configurable panels.

4 Simulator Configuration

As we said above the special features of the OpenSim NT™ environment facilitates the integration of new modules in the simulator configuration. This simulator extends a typical OpenSim project, as shown in figure 1, with the container applications that constitute the interactive and alarm panels and also with an extra module able to feed in real time the simulation data version of the updated plant Safety Parameter Display System [4] with the output of the sequence being simulated.

The SPDS module intends that the operators have, during the training sessions, the same information of the plant state that they use when working at the actual control room. The module can be launched and stopped from the instructor station with a single click of the mouse. It uses an especially developed DDE server to maintain updated, with the desired frequency, the more than six hundred signals currently required by the SPDS. The packets of simulated data are sent to the SPDS information system using FTP.

Figure 1 shows that, in our case, the system server and the simulator computer are the same machine, a PC with two Pentium II processors at 400 Mhz. The rest of computers do not demand particular features so inexpensive PCs of medium performance are enough.

With regard to the hardware responsible of the user interface we searched in the market many options: big-sized high resolution CRT monitors, LCD TFT flat displays or rear screen projectors for the visualization; and mice, remote controls or touch screens for the interaction. Eventually we have chosen 21" inches CRT monitors with a resolution of 1600x1200 pixels and resistive touch screens to implement the prototype presented in the following section. This decision seems to be, at the moment, the best solution keeping in mind both cost and performance. Smaller or less resolution displays are not useful to obtain clear images easily accessible by the user and other touch screens technologies do not offer the desired degree of accuracy. If available, these requirements applied to flat displays or projectors result in not affordable costs. We expect that in the near future technology advances will make these options more feasible. Then we could use them to show our already available soft panels without extra development effort.

Fig. 1. Simulator configuration

Each one of the soft panels runs on its own PC uniquely assigned. To simplify the initialization procedure the corresponding soft panel module is launched without any user action every time its PC is turned on. The instructor manages the execution of the simulator from the instructor station and the panels are updated automatically every time the initial conditions of the simulated sequence are changed. This feature eliminates the task of manual switch checking that must be accomplished in a hard panel simulator in those occasions.

5 Results and Conclusions

In order to test the users response to this kind of interface, a prototype of a full panel has been implemented. With this panel the operator is able to control the Emergency Core Cooling System (ECCS). It consists of eight touch screens displaying switches and indicators, two alarms panels and two SPDS monitors. The monitors are oriented in such a manner that reproduces the actual panel disposition. Figure 2 shows a composition image of four of the soft panels that compose the prototype.

The instructor and operators have tested the prototype using it to prepare and carry out the same kind of exercises they follow in the usual training sessions at the hard panel full scale simulator. In a short time the user gets familiarized with the simulator interface and interacts with the panels in a natural way. The results of this experience shows that this tool can fulfill many of the training requirements.

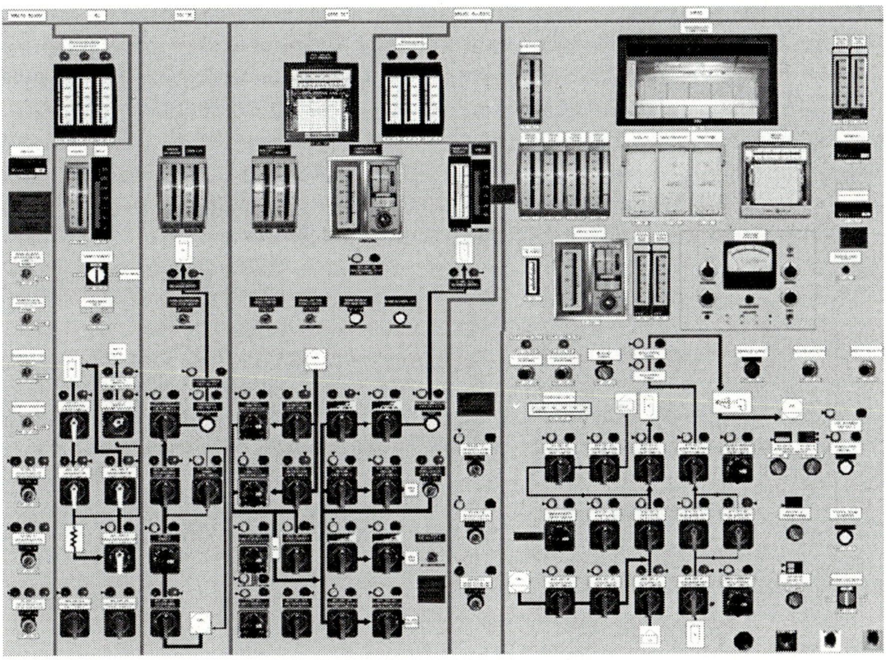

Fig. 2. Composition of four soft panels used in the prototype

We conclude that the component based soft panels interface approach represents an optimum solution in terms of performance, cost and effort in the development of full scope simulators. Apart from its suitability to be used as a training tool, this kind of interface increases the flexibility of the use of the simulator. For instance control room design changes can be proposed and tested in the simulator before its implantation, getting the previous approval of operators. Moreover, the possibility of distribution across the local area network simplifies its use by different departments. Finally, the open environment in which it is integrated guarantees the system extensibility.

The first step during the design was the identification and classification of the current objects in the layouts of the hard panels. Basically, they can be of two types: switches and indicators. After this classification it was specified the functionality and properties of each object. The development of the soft panels was performed using ActiveX components technology. Due the specific features of the switches, they were developed using Visual C++. The rest of the components were acquire from third party vendors.

With these components it is possible to reproduce easily and quickly the control panels of the control room. These include components such as gauges (linear and angular), sliders, selectors, knobs, leds, odometers, switches, alarms, etc. They can show images and sounds as resources of the simulated component and can be linked to a simulator signal in the same way a real object is wired in the back side of the panel. As an example, the Figure 1 shows a composition image of four soft panels. With this approach we have achieved a solution which is economical both in cost and in effort in the design of the soft panels.

The use of soft panels has increased the flexibility of the use of the simulator because it can be proposed design changes in the control room and tested with the simulator before its implantation getting the approval of operators. This can de done with lower maintenance costs. The next stage of the research will be the use of Java as a development language and the use of standard browsers such as Netscape or Internet Explorer for the displays of the soft panels. Moreover, the simulator executive could support distributed simulations of the systems over Internet.

References

1. Bustío, F., Corcuera, P., García, J., Mora, E.: Design and Implementation of Multimedia Environment for Simulation. Cybernetics and Systems, Taylor & Francis Publishers, Austria, Vol: 25, (1994) 63 - 71
2. Corcuera, P., Bustío, F., Mora, E.: Training Simulator for Garoña Nuclear Power Plant. Lectures Notes in Computer Science, Springer Verlag, Germany, Vol: 1030, (1996) 523 - 529
3. Ryan, J., Chan, S.: The trend towards Windows NT for use in simulations. 2º CSNI Specialist Meeting on Simulators and Plant Analyzers. Finland. (1997)
4. Corcuera, P., Mora, E., Zorrilla, M.: Distributed Graphic Visualization of Industrial Process Information in Real Time. 5th International Conference on Computer Aided Systems Theory and Technology (EUROCAST´97). Las Palmas de Gran Canaria. (1997)

Microscopic Randomness and "Fundamental Diagram" in the Traffic Flow Problem

H. Lehmann

GMD-FIRST
Rudower Chaussee 5, 12489 Berlin, Germany
heiko@first.gmd.de

Abstract. Starting from a continuous *follow-the-leader* model for individual vehicle motion a stochastic term is added in order to account for the elementary randomness of driving. In a suitably defined stationary state thus an equilibrium distribution of nearest-neighbour distances is spanned up which allows to understand the "fundamental diagram" of macroscopic traffic flow modelling as the ensemble average of the microscopic characteristic function. Estimates for the distribution width and a comparison with simulation results are given.

1 Introduction

Modelling of vehicular traffic flow by means of physical tools seems, on first glance, to follow a familiar pattern: microscopic deterministic rules for the motion of individual cars are complemented with an evolution picture of averaged macroscopic quantities, e.g. the probability distribution function $f(x, v, t)$ to find a car with velocity v at position x at any given time t. Prominent examples for the former are models on the basis of cellular automata (CA) [1] - [8] which - due to discretization and the replacement of Newtonian differential equations of motion by a set of rules - are extremely computer-friendly and can, thus, handle very large systems. Continuous microscopic models exist also; one of them will be the exemplary subject of more detailed argumentation further below.

Macroscopic modelling started with the observation by Lighthill and Whitham [9] that flowing traffic resembles flowing water and that hydrodynamics should provide a useful theoretical tool. Extensions of the theory were given for instance by Kühne [10,11], Kerner and Konhäuser [12,13] and Helbing [14]. The distribution function mentioned above as the center point of modelling was introduced by Prigogine and Herman [15] and carried further later on [16,17]. It should be mentioned that hydrodynamics and distribution function approach are closely connected: mean density $c(x, t)$, mean velocity $\bar{v}(x, t)$, mean velocity variance and so on are found as consecutive velocity moments $\int dv\, v^{\nu} f(x, v, t)$ of the distribution function.

Although the analogy to hydrodynamics seems to present a strong argument, a decisive qualification has to be made with respect to the numbers involved. In

F. Pichler, R. Moreno-Díaz, and P. Kopacek (Eds.): EUROCAST'99, LNCS 1798, pp. 343–356, 2000.

the case of an ordinary liquid or gas, the microscopic distribution function

$$\bar{N}(x,v,t) = \sum_i \delta(x - x_i(t))\,\delta(v - v_i(t)) \tag{1}$$

is averaged over a macroscopic number of individual particles i in order to arrive at $f(x,v,t)$ and yet the scales of the coarse-grained variables x and v are still very small in comparison with any structure of f, in other words, a "point" is well-defined. In traffic flow, on the other hand, the averaging ensemble may contain only some dozen individual cars and still spread over several hundred meters which is the same spatial scale as that of macroscopic structure evolution itself. As a consequence, a principal limit of resolution for macroscopic analysis arises, defining, thereby, also the adequate macroscopic time scale.

This blurring of scales also affects the nature of stochastic forces as will be dwelled upon in the following section. In a liquid or a gas, the stochastic picture of Einstein or Smoluchowski (see [18] for an overview) assumes a test particle to experience a large number of random, δ-like collisions during the propagation from one macroscopic "point" to another. The analysis of this kind of motion gave rise to an own branch of mathematics, i.e. the Ito and Stratonovich calculus. In the traffic flow problem, the elementary time step τ_1 is given by the driver's reaction time to sudden perturbations. Obviously, no faster processes can be imagined. This time step is, however, the same as the one that governs the deterministic microscopic motion itself. As a consequence, the conventional concept of very short, sudden acceleration (or deceleration) changes does not apply; in the overall motion the effects of deterministic and stochastic acceleration cannot be clearly separated anymore. Granting the drivers this kind of indecisiveness then leads to the assertion that for the individual vehicle a steady state $\dot{x}_i = const.$ can never be reached. It is only in terms of the averaged quantities (for example $f(x,v,t)$) and, consequently, on an appropriate time scale $\tau_2 \gg \tau_1$ that a steady state can be sensibly defined. Since such a steady state is the basis for the phenomenological relations of the macroscopic modelling school, τ_2 presents also the frame for the so-called "fundamental diagram" (FD). The FD relates - in a steady state - the mean density c of traffic flow to the mean speed $V_0(c)$ or the flux $Q_0(c) = cV_0(c)$ and has, obviously, to derive from the microscopic motions, interactions and, crucially, the "molecular chaos" of driving. It is the aim of this paper to clarify this relationship.

The following section is devoted to the design and study of an explicit stochastic microscopic model; section IV, then, discusses the ensuing implications for macroscopic relaxation equations.

2 Randomness in the Microscopic Motion: Model Build-Up

2.1 Optimal Velocity Model and Fundamental Diagram

It has been mentioned above that CA-based models hinge on a - sometimes brute force - discretization of phase space which makes them cumbersome to analyze (although studies of formalized models exist, such as [19]). On the other hand, CA models include the natural microscopic randomness of driving mentioned in the Introduction. Due to the discretization, however, this stochasticity produces unrealistically large fluctuations (a typical discretization of the velocity range would be into five steps; for the german autobahn maximum speed of 130 km/h this would entail a minimum of the random velocity changes of 26 km/h per simulation time step i.e. usually one second, resulting in near gravitational accelerations!). Thus, obviously, a need for an alternative approach to the randomness of driving is called for. It should be mentioned, though, that the CA stochasticity automatically acknowledges the other feature of microscopic fluctuativity discussed above, namely, to be of the same time scale as the deterministic motion itself which is given by the simulation time step.

As a continuous variant, the "Optimal Velocity Model" (OVM) [20,21] presents a set of coupled differential equations for the positions x_i of the ith vehicle:

$$\ddot{x}_i = \frac{1}{\tau_1}\left[V_B(\Delta x_i) - \dot{x}_i\right]$$
$$\Delta x_i = x_{i+1} - x_i \quad . \tag{2}$$

Here, τ_1 is the elementary time step mentioned above and found from experiments [20] to be of the order $\sim$ 0.5 *sec*. The crucial element of eqn. (2) is the function V_B, the so-called "optimal velocity" function. It gives the optimal speed a driver should assume in dependence on the distance to the car in front:

$$V_B(\Delta x_i) \;=\; v_0\left[\tanh\left(0.0860(\Delta x_i - 25)\right) + 0.97323\right] \tag{3}$$

(distances in meters). The action of V_B becomes clear immediately: zero distance dictates stand-still whereas for infinite distances Δx_i the maximum speed v_0 is assumed. It is intuitively clear that the degree of determinism presented by eqn. (2) is much too strong - it resembles a machine-like "cruise cotrol". In reality, drivers will rather fluctuate around the "optimal" velocity with the important boundary condition, however, to avoid crashes.

The function V_B is, furthermore, closely related to the "fundamental diagram" $V_0(c)$ of the macroscopic picture of traffic flow relating the mean density c of traffic to the mean speed, for instance in the parametrization of Kerner and Konhäuser [12]:

$$V_0 \;=\; V_K\left(c; v_0, c_{\max}\right) \;=$$
$$v_0\left(\left[1 + \exp\left[\left(\frac{c}{c_{\max}} - 0.25\right)/0.06\right]\right]^{-1} - 3.72\,10^{-6}\right) \tag{4}$$

(here, c in *vehicles* per *km* with the saturation density $c_{\max} \;=\; 200\,veh./km$ and v_0 the free maximum speed). V_0 acts in the same fashion as the optimal velocity function V_B of the OVM, the saturation density $c_{\max}$ corresponding to zero headway $\Delta x_i \;=\; 0$. However, it has to be emphasized again at this point that while eqn. (3) prescribes a pair interaction, eqn. (4) presents a collective phenomenon, i.e. the average over a number of pair interactions. Following the reasoning given above, this average must involve the time scale τ_2 and the microscopic randomness:

$$V_0 \sim \langle V_B \rangle_{\text{microscopic fluctuations}} \quad . \tag{5}$$

Relation (5) has the same character as, say, the relation between Lennard-Jones potential and Carnahan-Starling pressure in conventional statistical mechanics.

2.2 Random Acceleration

In order to account for the fluctuations in the drivers' behaviour, the usual procedure is to add a stochastic acceleration term on the rhs of eqn. (2). (It should be mentioned at this point that Kühne [11] proposed a stochastic acceleration in the equation for the macroscopic mean speed $\bar{v}(x,t)$. In the line of the argument developed above, however, the randomness of traffic flow has to be accounted for on the microscopic level; it is difficult to envisage concerted fluctuations of an entire coarse-graining ensemble of vehicles as is the basis for the macroscopic equations.) This term has to be designed, then, in such a way as to agree with the qualitative arguments given above: First, the minimum time scale on which the stochastic acceleration varies is given by the same τ_1 that governs the deterministic microscopic motion (2). Consequently, it cannot be a Langevin-type force but has to be an ordinary, continuous function of time. Second, the amplitude of the random acceleration has to scale in some form or other with the velocity itself in order to avoid crashes. Third, a time average on the macroscopic scale τ_2 equal to zero has to exist, thereby defining the steady state $\langle \dot{x}_i \rangle_{\tau_2} = const.$, the prerequisite for the existence of the fundamental diagram. These demands are met by the form:

$$\ddot{x}_i^{\text{random}} = \frac{\alpha}{\tau_1} V_B(\Delta x_i)\, \xi_i(t)$$
$$\xi_i(t) = C_\xi \sum_j^{N_\xi} \zeta_{ij} \sin\left[2\pi\omega_{ij}t + \phi_{ij}\right] \tag{6}$$

with random frequencies $\omega_{ij} \in [\omega_-, \omega_+]$, random phases $\phi_{ij} \in [0, 2\pi]$ and random amplitudes $\zeta_{ij} \in [0,1]$ for a number N_ξ of modes . (Note, that the negative range of amplitudes $[-1, 0]$ is subsumed into the random phases.) Since on the scale τ_2 the random forces must average to zero, the short-wave limit is given by $\omega_- = \tau_2^{-1}$, whereas the elementary time step presents the long-wave limit $\omega_+ = \tau_1^{-1}$. The numerical scaling factor α is to be determined from, desirably, experiments, or simulations.

The normalization of ξ_i is best connected with the average of its mean square on the time scale τ_2 as is customary in the context of stochastic problems (see [18]):

$$\langle \xi_i^2 \rangle_{\tau_2} = C_\xi^2 \, \frac{1}{2} \sum_j^{N_\xi} \langle \zeta_{ij} \rangle \tag{7}$$

such that for the ζ_{ij} being equidistributed on [0, 1] ξ_i is normalized to:

$$C_\xi = \sqrt{\frac{6}{N_\xi}} \tag{8}$$

Of course, any number of shapes for $\xi_i(t)$ can be imagined. The one given in (6) - (8) is the simplest one meeting the conditions set above. Let us define, then, the Random Optimal Velocity Model (ROVM):

$$\ddot{x}_i = \frac{1}{\tau_1}\left[V_B\left(\Delta x_i\right) - \dot{x}_i\right] + \frac{\alpha}{\tau_1} V_B\left(\Delta x_i\right) \xi_i(t) \quad . \tag{9}$$

A first qualitative point with respect to the behaviour of the ROVM can be made: The shape of the random function ξ_i is such that in situations with strong none-quilibrium, $V_B\ (\gg \vee \ll)\ \dot{x}_i$, the first term on the rhs of (9), i.e. the deterministic relaxation, dominates the motion. For $\dot{x}_i \sim V_B$, obviously, the stochastic acceleration does leading to a conventional multiplicative noise problem. Traffic flow can thus be thought of as a dynamics with controlled noise: there exists a band of velocity fluctuations in which stochastic motion is possible. Fluctuations leading out of this band will be damped out by a strong determinism (i.e. to assume the permissible maximum speed while avoiding crashes). The steady state of traffic flow then is, in this picture, characterized by stationary distributions of velocities, or, equivalently, nearest neighbour distances, within this velocity band. This distribution allows for evaluating averages as in (5). It should be commented that only on the time scale τ_2 a sufficient number of events occur in order to fill the distribution. In the following section, properties of this distribution are investigated.

3 Randomness in the Microscopic Motion: Ensemble Properties

As argued above, the continuum description of traffic flow necessarily involves the macroscopic time scale τ_2. The microscopic randomness - the "molecular chaos" of traffic flow, so to speak - as expressed by the random function $\xi_i(t)$ then, in local equilibrium, spans up the probability distribution of distances around some mean value $\bar{d}$, $P_d(d;\bar{d})$. In a steady state (as argued above, with respect to τ_2), P_d is stationary and allows to establish the connection between individual velocities and the fundamental diagram in the spirit of (5):

$$\begin{aligned} V_0(c) &= \int dd'\, V_B(d')\, P_d(d';\bar{d}) \\ c &= \frac{1}{\bar{d} + b} \end{aligned} \tag{10}$$

where b is an effective car length, typically chosen $5 \ldots 7.5\,m$. The following two paragraphs present discussions of P_d and the fundamental diagram.

3.1 Distance Variability

Analysis of the fluctuativity of the distance variables $d_i = x_{i+1} - x_i$ in a coarse-graining ensemble of vehicles can best be carried out starting from a rewritten ROVM:

$$\ddot{d}_i = \frac{1}{\tau_1}\left[V_B(d_{i+1}) - V_B(d_i) - \dot{d}_i\right] + \frac{\alpha}{\tau_1}\left(V_B(d_{i+1})\xi_{i+1} - V_B(d_i)\xi_i\right) \tag{11}$$

Assuming a traffic situation where distances in the ensemble vary only a little around some value $\bar{d}$, it is justified to linearize eqn. (11). Noting further that $\langle(\xi_{i+1} - \xi_i)^2\rangle = 2\langle\xi_i^2\rangle$ and assuming $d_{i+1} = \bar{d}$ (i.e. the frontrunner representing the ensemble average), one has for a small elongation Δd (vehicle index i dropped):

$$\Delta\ddot{d} = \frac{1}{\tau_1}\left[A_d\,\Delta d - \Delta\dot{d}\right] + \sqrt{2}\,\frac{\alpha}{\tau_1}V_B\bar{d}\,\xi(t) \tag{12}$$

with ξ constructed as in (6) and A_d shorthand for $\partial V_B/\partial d|_{\bar{d}}$. The solution of eqn. (12) can immediately be written down:

$$\Delta d \simeq 2\,\alpha\,\tau_1\,V_B(\bar{d})\,C_\xi \times \sum_j^{N_\xi} \zeta_j\left[\left(A_d\tau_1 - \Omega_j^2\right)^2 + \Omega_j^2\right]^{-1}$$
$$\left\{\Omega_j\cos[\Omega_j\eta + \phi_j] - \left(A_d\tau_1 - \Omega_j^2\right)\sin[\Omega_j\eta + \phi_j]\right\} \tag{13}$$

where

$$\Omega_j = 2\pi\omega_j\tau_1 \quad , \; \eta = t/\tau_1 \quad . \tag{14}$$

Integral to this solution is, of course, the mean speed $\langle\dot{x}_i\rangle_{\tau_2} = V_B(\bar{d})$ of the coarse-graining ensemble. The solution (13) displays further the property $\langle\Delta d\rangle_{\tau_2} = 0$, which is, however, a consequence of linearization. In a full analysis of the ROVM the mean value of Δd is itself expected to have a drift due to the asymmetry of $V_B(d)$.

Over long times τ_2, the microscopic events fill up $\Delta d(t)$ and thus the probability distribution $P_d(d;\,\bar{d})$. While it is impossible to extract the exact analytical shape of P_d from (13), a measure for its width can be found via the average of the mean square. As in the normalization of ξ_i (eqns. (7, 8)), this assigns every mode j a weight $\int \sin^2$ *or* $\cos^2 = 1/2$ and cancels cross terms whence eqn. (13) with $\langle\zeta_j^2\rangle = 1/3$ from the assumed equidistribution yields:

$$\langle(\Delta d)^2\rangle_{\tau_2} = 2\,\left(\alpha\,\tau_1\,V_B(\bar{d})\right)^2\,\frac{1}{N_\xi}\sum_j^{N_\xi}\left[\left(A_d\tau_1 - \Omega_j^2\right)^2 + \Omega_j^2\right]^{-1} \tag{15}$$

Assuming that each mode j will realize a frequency Ω_j with a certain probability (here for simplicity taken as $1/(\Omega_+ - \Omega_-)$) out of the interval $[\Omega_-,\,\Omega_+]$, the sum in (15) can be rewritten:

$$\langle(\Delta d)^2\rangle_{\tau_2} = \left(\alpha\,\tau_1\,V_B(\bar{d})\right)^2\int_{\Omega_-}^{\Omega_+} d\Omega'\,\frac{1}{\Omega_+ - \Omega_-}\left[\left(A_d\tau_1 - \Omega'^2\right)^2 + \Omega'^2\right]^{-1} \tag{16}$$

and evaluated to:

$$
\begin{aligned}
d \notin &\left[d^{(1)}(\tau_1), d^{(2)}(\tau_1)\right] \quad : \\
\langle(\Delta d)^2\rangle \;=\; &\frac{2\left(\alpha V_B(\bar{d})\tau_1\right)^2}{\Omega_+ \;-\; \Omega_-}\frac{1}{\sqrt{1-4A_d\tau_1}} \times \\
&\left\{ \frac{\arctan\left[\frac{z}{\sqrt{\frac{1}{2}-A_d\tau_1-\frac{\sqrt{1-4A_d\tau_1}}{2}}}\right]}{\sqrt{\frac{1}{2}-A_d\tau_1-\frac{\sqrt{1-4A_d\tau_1}}{2}}} - \right. \\
&\left. \frac{\arctan\left[\frac{z}{\sqrt{\frac{1}{2}-A_d\tau_1+\frac{\sqrt{1-4A_d\tau_1}}{2}}}\right]}{\sqrt{\frac{1}{2}-A_d\tau_1+\frac{\sqrt{1-4A_d\tau_1}}{2}}} - \right\}_{z=\Omega_-}^{z=\Omega_+}
\end{aligned}
\tag{17}
$$

and

$$
\begin{aligned}
d \in &\left[d^{(1)}(\tau_1), d^{(2)}(\tau_1)\right] \quad : \\
\langle(\Delta d)^2\rangle \;=\; &\frac{\left(\alpha V_B(\bar{d})\tau_1\right)^2}{\Omega_+ \;-\; \Omega_-}\frac{1}{4\left(A_d\tau_1\right)^{\frac{3}{2}}}\frac{1}{\sin\vartheta} \times \\
&\left\{ \sin\frac{\vartheta}{2}\,\ln\left[\frac{z^2+A_d\tau_1+2z\sqrt{A_d\tau_1}\cos\frac{\vartheta}{2}}{z^2+A_d\tau_1-2z\sqrt{A_d\tau_1}\cos\vartheta 2}\right] + \right. \\
&\left. 2\,\cos\frac{\vartheta}{2}\,\arctan\left[\frac{z^2 \;-\; A_d\tau_1}{2z\sqrt{A_d\tau_1}\sin\frac{\vartheta}{2}}\right]\right\}_{z=\Omega_-}^{z=\Omega_+}
\end{aligned}
\tag{18}
$$

where

$$\vartheta \;=\; \arccos\left[\frac{2A_d\tau_1 \;-\; 1}{2A_d\tau_1}\right] \quad . \tag{19}$$

The interval boundaries $d^{(1,2)}(\tau_1)$ are given by the roots of

$$4\left.\frac{\partial V_B}{\partial d}\right|_{d^{(1,2)}} - 1 \;=\; 0 \quad . \tag{20}$$

Only for $\tau_1 \;>\; \tau_c \;\simeq\; 0.18\,sec$ real solutions $d^{(1,2)}(\tau_1)$ and, correspondingly, the regime (18) exist such that τ_c with $d^{(1)}(\tau_c) \;=\; d^{(2)}(\tau_c) \;=\; 25\,m$ is established as a model-inherent critical value for the microscopic relaxation time. Fig. 3.1, however, shows the criticality of $\langle(\Delta d)^2\rangle$ with τ_1 to be merely mathematical, i.e. $\langle(\Delta d)^2\rangle$ to be continuous at the interval boundaries $d^{(1,2)}$. Nevertheless, it seems worth pointing out that the mathematical analysis of the ROVM yields a time scale which can, thus, be regarded as the analytical counterpart to the value $\tau_1 \simeq 0.5\,sec$ found from a fit of real-traffic data [20]. Furthermore, the dependence on the actual value of the individual microscopic relaxation time τ_1 is rather weak. This becomes plausible by comparison of the microscopic relaxation processes and the random wavelengths in $\xi(t)$, ω_{ij}^{-1}: reactions to the fluctuations of the car in front are faster or of the same temporal order as the fluctuations themselves such that in the coarse-graining average the τ_1-effect remains small.

Fig. 3.1, furthermore, clearly displays the effects of coupling the random acceleration $\xi(t)$ to the characteristic velocity function $V_B(d)$: for small values of d the fluctuations have to leave the traffic crash-free such that $\langle(\Delta d)^2\rangle \to 0$ in that limit. For large distances (or, see eq. (10), zero density and, thus, the interaction-free limit) $P_d(d;\, \bar{d} \to \infty) = P^{(0)}(d)$ becomes independent of the mean distance. Consequently, the corresponding velocity distribution does not depend on the density, either. In the modelling work of Prigogine and Herman [15] a strongly related single-particle property is introduced as the "distribution of desired speeds", representing differing intentions of drivers in non-interacting traffic. It should be pointed out, though, that, while similar in shape and effect, such a phenomenological distribution of deterministically set velocities is fundamentally different from one given rise to by microscopic fluctuations. The relation to the macroscopic traffic flow picture is discussed in more detail in sec. IV.

3.2 Velocity Fluctuations and the Fundamental Diagram

Within the assumption of local equilibrium the distribution of distances discussed in the foregoing paragraph is, in the spirit of eqn. (10), uniquely connected to that of the velocities such that the averaged steady-state velocity, i.e. the fundamental diagram $V_0(c)$, can be expressed as:

$$\bar{v} = V_0(c) = \int dd'\, V_B(d')\, P_d(d;\, \bar{d}) \quad . \tag{21}$$

Restricting the ROVM (9) to the assumption of instantaneous relaxation,

$$V_B\,(d_i) \simeq V_B(\bar{d}) \quad , \tag{22}$$

a first-order problem for the velocities $v_i = \dot{x}_i$ arises:

$$\dot{v}_i = \frac{1}{\tau_1}\left[V_B(\bar{d}) - v_i\right] + \alpha\, V_B(\bar{d})\, \xi_i(t) \tag{23}$$

This is the well studied Ornstein-Uhlenbeck process (if ξ_i is taken as customary white noise) for which the Fokker-Planck equation for the probability density $p(v,\, t)$ to realize a certain velocity v at time t reads as:

$$\begin{aligned} \partial_t\, p(w,t) &= \frac{1}{\tau_1}\, \partial_w\, (wp(w,t)) + \frac{(\alpha V_B(\bar{d}))^2}{2\tau_1}\, \partial^2_{w^2}\, p(w,t) \\ w &= v - V_B(\bar{d}) \quad . \end{aligned} \tag{24}$$

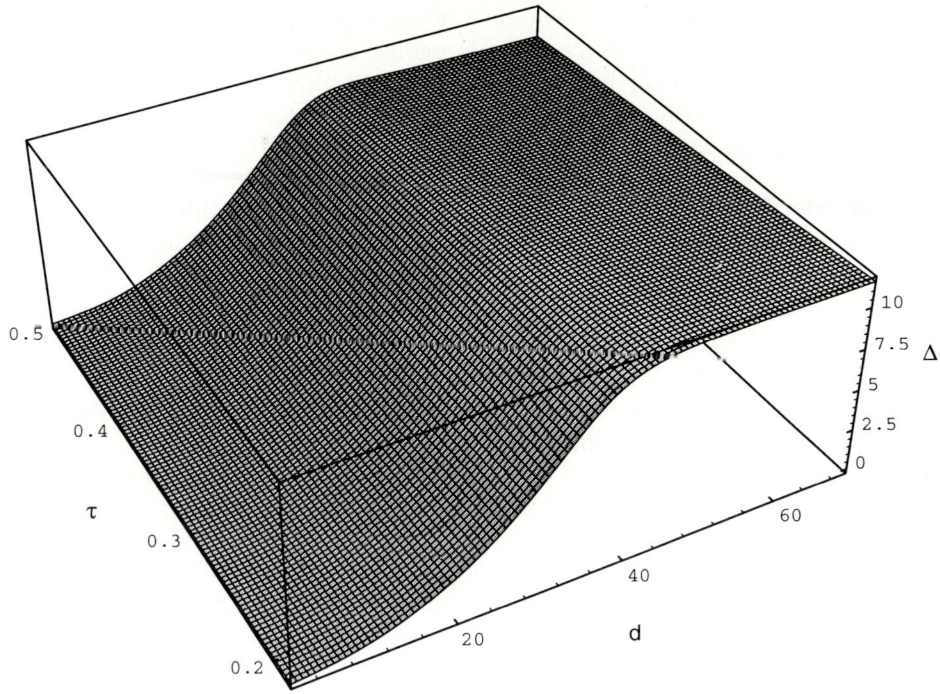

Fig. 1. distance variability $\Delta = \sqrt{\langle(\Delta d)^2\rangle}$ in m according to eqns. (17, 18) with numerical parameter $\alpha = 0.3$ and the frequency interval $\omega_- = 0.2\,sec^{-1}$, $\omega_+ = 2\,sec^{-1}$; the microscopic relaxation time τ ranges from the critical value τ_c to $\tau_1 = 0.5\,sec$ as taken from reference [20]; headways d in m.

The properties of this process are well known such that the results for velocity mean and variance can be written down immediately:

$$\langle v(t)\rangle = V_B(\bar{d}) + \left(v - V_B(\bar{d})\right)\, e^{-t/\tau_1}$$
$$\mathrm{var}\,\{v(t)\} = \frac{\left(\alpha V_B(\bar{d})\right)^2}{2}\,\left[1 - e^{-2t/\tau_1}\right] \tag{25}$$

The stationary solution $p_s(v)$: $\partial_t p_s(v,t) = 0$ presents, then, the Ornstein-Uhlenbeck estimate for the local equilibrium fluctuations of the velocities:

$$p_s(v;\bar{d}) \;=\; \sqrt{\pi}\,\alpha\, V_B(\bar{d})\, \exp\left[-\frac{(v - V_B(\bar{d}))^2}{(\alpha V_B(\bar{d}))^2}\right] \quad . \tag{26}$$

Of course, the Gaussian shape of solution (26) can only be a rather crude approximation since for certain velocities the probability has to be strictly zero in order to remain crash-free. This failure becomes decisive in the region near the permissible maximum speed v_0. Eqn. (21) clearly shows that in this limit the distances d may vary widely while the velocity distribution simultaneously narrows to a δ-shape. This effect is not contained in eqn. (26) which may serve, however, still as a first guess. The results of the foregoing paragraph, where the approximation (21) was taken further by the term $A_d\, d_i$, proved the τ_1-dependence of the distance distribution width $\sqrt{\langle(\Delta d)^2\rangle}$ (which is expected to arise in some scaled form in the velocity distribution function as well) to be, indeed, rather weak.

From the Ornstein-Uhlenbeck argument (26) and the result for $\sqrt{\langle(\Delta d)^2\rangle}$ it seems to be justified to guess the shape of the local equilibrium distance distribution P_d as:

$$P_d(d;\,\bar{d}) \;=\; \sqrt{2\pi\langle(\Delta d)^2\rangle}\,\exp\left[-\frac{(d-\bar{d})^2}{2\,\langle(\Delta d)^2\rangle}\right] \qquad (27)$$

whence it is possible, in the spirit of relation (10) to express the fundamental diagram in terms of microscopic features.

Fig. 3.2 shows the thus obtained $V_0(\bar{d})$ in comparison with the function $V_B(\bar{d})$ (which corresponds to the idealized state of "total cruise control") and results of a direct simulation for intermediate headways (for very small and very large headways the velocity fluctuations vanish as argued elsewhere). It is clearly recognizable that, due to the fluctuations, the average velocity of the coarse-graining ensemble lies below that of the corresponding single-particle optimal velocity V_B. Furthermore, the estimate (27) in eqn. (21) agrees remarkably well with the simulations of the full ROVM (for an ensemble of 5 vehicles the coupled eqns. (9) were solved with $N_\xi = 3$ random modes in the random functions $\xi_i(t)$, eqn. (6). As the elementary time $\tau_1 = \tau_c$ was used with the averaging interval $\tau_2 = 5\,sec$ and the spectrum of the random frequencies ω_{ij} the same as in fig. 3.1. In order to guarantee the τ_2-average as a true steady state, a non-fluctuating vehicle was placed in front of the averaging ensemble and the initial conditions set in accordance with the equilibrium solution $v = V_B(d)$.) In fig. 3.2 the velocity variance of the simulated ensemble is shown. Since, however, the microscopic velocity distribution function is not directly accessible, the shown quantity is the variance $\langle\left(V_B(d) - V_B\left(\bar{d}\right)\right)^2\rangle$. For large distances d the characteristic function V_B is constant whence the depicted variance tends to zero. Between this limit and the bound traffic limit $d \to 0$, in which the scaling of the fluctuations also dictates vanishing velocity variance, exists a transition regime with a variance maximum. This regime is important for the analysis of the critical properties of the ROVM. However, fig. 3.2 illustrates that, while qualitatively correct, the assumptions in deriving the distribution function P_d, eqn. (27), are not sufficient for an accurate analysis. Future work will have to be devoted to this problem.

4 Relaxation Regime in Macroscopic Dynamics

From the microscopic distribution function $\bar{N}(x,v,t)$, (1), the macroscopic one is obtained by an averaging procedure which involves, as reasoned above, the time scale τ_2:

$$f(x,v,t) \;=\; \langle\bar{N}(x,v,t)\rangle_{\tau_2} \quad . \qquad (28)$$

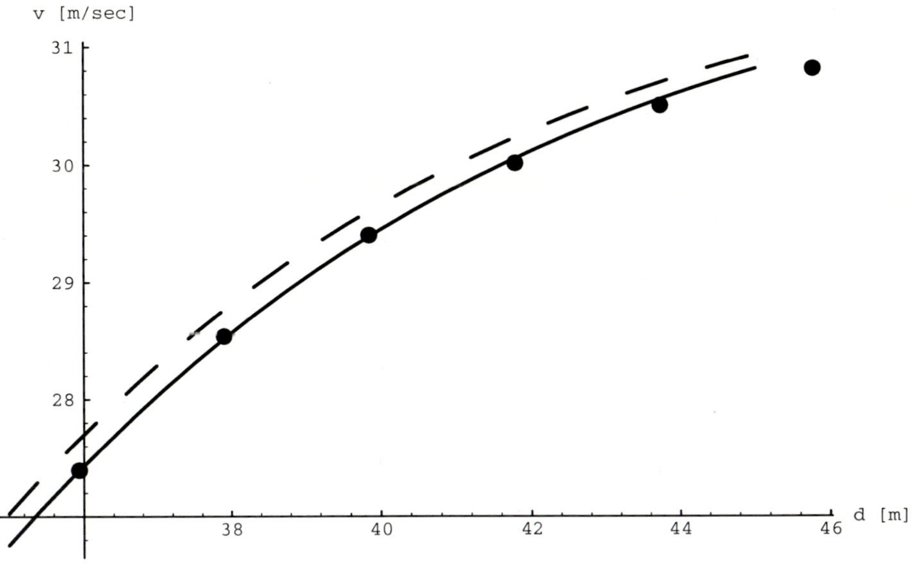

Fig. 2. average velocities in the ROVM for $\tau_1 = \tau_c$ (see eqn. (20)); solid line: fundamental diagram $V_0(\bar{d})$ according to eqns. (21) with etimate (27), dashed line: "total cruise-control" limit $V_B(\bar{d})$ as reference i.e. no microscopic randomness, points: simulations of the steady state average velocity (with $\tau_1 = \tau_c$ and $\tau_2 = 5\,sec$) of five coupled vehicles.

If the ROVM (9) is assumed to describe the true microscopics of traffic flow, the evolution equation for f is found from the total time derivative of $\bar{N}$:

$$\begin{aligned} \left\langle \frac{d\bar{N}}{dt} \right\rangle &= 0 \\ &= \left\langle \sum_i \dot{x}_i \; \delta'(x - x_i(t))\delta(v - \dot{x}_i\,(t)) + \ddot{x}_i \; \delta(x - x_i(t))\delta'(v - \dot{x}_i\,(t)) \right\rangle_{\tau_2} \end{aligned} \quad (29)$$

Inserting for $\ddot{x}_i$ the ROVM (9) establishes the connection between microscopic and macroscopic dynamics much in the same way as the relation between optimal velocity function $V_B(d)$ and fundamental diagram discussed above.

From the construction of f it is clear that τ_2 presents a natural limit for its resolution. Any perturbation on the scale τ_1 will need a number of pairwise interactions equal to the size of the averaging ensemble until it is filtered through to the coarse-grained variables. The second term on the rhs of eqn. (29) will thus constitute an evolution operator for the distribution function with τ_2 as the elementary time. This is a necessary consequence of the macroscopic description and should be pointed out here since the literature contains some confusing statements with this respect. Writing $\langle \ldots \rangle_{\tau_2} = 1/\tau_2 \int dt \ldots$, one can, with the assumption of local equilibrium (in the sense of the discussion around eqns. (13 -

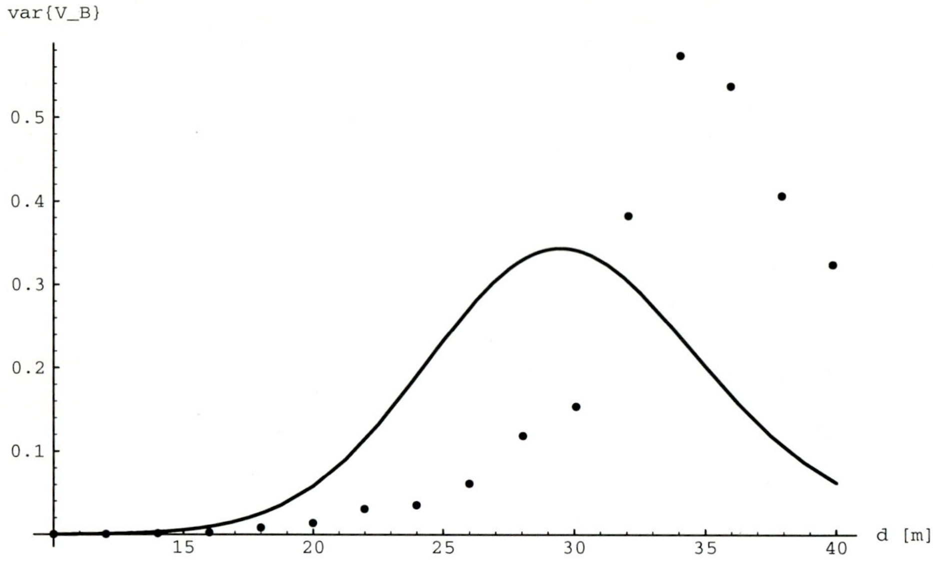

Fig. 3. steady state velocity variance $\mathrm{var}\{V_B\} = \langle (V_B(d) - V_B(\bar{d}))^2 \rangle$ with $P_d(d;\bar{d})$ according to (27) in comparison with the results of the simulation (points). While correct in shape, the estimate (27) is not sufficient to reproduce the measured variance. The transition regime from bound (small $\bar{d}$) to free traffic (large $\bar{d}$) with the variance maximum is important for the emergence of critical states in traffic flow.

16), i.e. $\dot{x}_i \neq const. \forall i$ but $\langle \dot{x}_i \rangle_{\tau_2} = const$) express the interaction term in (29):

$$\left\langle \frac{1}{\tau_1} \left[V_B(d_i) - \dot{x}_i \right] \delta(x - x_i(t)) \delta'(v - \dot{x}_i\,(t)) \right\rangle = \frac{1}{\tau_2} \frac{\partial}{\partial v} \left[(\langle V_B \rangle - v)\, f(x,v,t) \right] \quad , \tag{30}$$

resulting in the equation of motion for f:

$$\left[\partial_t + v \partial_x \right] f(x,v,t) = -\frac{1}{\tau_2} \partial_v \left[(V_0(c(x,t)) - v)\, f(x,v,t) \right] \quad . \tag{31}$$

Several comments should be made with respect to eqn. (31). First, the time scale $\tau_2 \sim 5 \ldots 7\,sec$ as found from measurements [14,10,12] has been justified by elementary coarse-graining considerations. Second, the form of the rhs of eqn. (31), although being different from the term $-(f - f_0)/\tau_2$ as given in [15] yields the same first two moment equations, i.e. the continuity and relaxation equation for the mean speed $\bar{v}$. Third, the microscopic randomness has been subsumed into the shape of the function $V_0(c)$. Of course, in strong non-equilibrium situation, local equilibrium will not be established and extra terms might emerge. It should be cautioned, though, that in such a case, the distribution function f with its problematic definition, (28), will not be a sensible entity of analysis anymore. Stochastic acceleration terms in equations for f or its moments seem, thus, ill motivated. Fourth, the interaction term of Prigogine and Herman [15] has to be commented on. From a gas kinetic analogue, those authors derived a "collision" term (which might sound rather provocative in the context of vehicular traffic). The notions of incoming and outgoing fluxes and probability of interaction become, while accurate for ordinary liquids or gases, however, problematic due to the small-sample size coarse-graining and the dimensionality of the problem. The divergences between Prigogine and Herman's model and eqn. (31) can best be highlighted by studying the velocity variance $\mathrm{var}\{v\} = \overline{(v - \bar{v})^2}$

$$[\partial_t + \bar{v}\partial_x]\, \mathrm{var}\{v\} + \frac{1}{c}\partial_x \left[c\overline{(v - \bar{v})^3} \right] = \left\{ \begin{array}{lcr} -2\mathrm{var}\{v\}/\tau_2 & : & \text{eqn. (31)} \\ -\left[\mathrm{var}\{v\} - \mathrm{var}^{(0)}\{v\}\right]/\tau_2 - (1-P)c\overline{(v - \bar{v})^3} & : & \text{ref. [15]} \end{array} \right. \tag{32}$$

The parameter P, the probability of passing, has to be set zero in order to compare the strictly one-lane OVM with the approach of Prigogine and Herman. While eqn. (31) leaves the interaction solely to an appropriately designed macroscopic characteristic function $V_0(c(x,t))$ (which can, as has been shown in the preceding sections, be referred to the microscopic interaction $V_B(d)$), the Boltzmann-like argument of Prigogine and Herman results in a coupling to the third moment of the velocity distribution as well as the necessity to define and determine the stationary distribution of desired velocities $f^{(0)}$. As has been pointed in the context of fig. 3.2, the velocity variance is a decisive quantity with respect to the criticality of the system. It can be expected, then, that the different forms in eqn. (32) will have consequences in the critical regime. Detailed studies have to be left, though, to future work.

5 Conclusions

The dichotomy between microscopic and macroscopic description of vehicular traffic is substantially different from the statistical physics of, say, ordinary liquids due to the temporal and spatial scales not being clearly separated. Viewing flowing traffic as a continuum therefore necessarily involves an averaging procedure on the scales of macroscopic structure evolution itself. Hence, the application of standard tools of statistical physics has to be taken *cum grano salis*. For example, the usual picture of a stochastic force as a series of δ-like interactions has to be modified to account for the comparable time scales of fluctuation and deterministic motion. With the due caution, however, modelling and analysis can be carried out.

The presented work had the aim to acknowledge the fluctuativity of the elementary driving process and to estimate its effects. Starting from the well-established "Optimal Velocity Model" (OVM) a stochastic acceleration was added to the equation of motion in order to account for the elementary randomness of driving. This acceleration was designed in such a way as to avoid crashes. Due to this random component, the individual cars can never reach a stationary state $\dot{x}_i = const.$. A stationary state of traffic can, therefore, only be defined if a time scale much larger than that of the OVM, $\tau_2 \gg \tau_1$, is introduced, and the corresponding averages are considered. (τ_2 is thus established as the natural time scale for the macroscopic equations.) The state $\langle \dot{x}_i \rangle_{\tau_2} = const$, then, involves a distribution of microscopic nearest-neighbour distances $P_d(d; \bar{d})$. With this distribution it is possible to connect the phenomenological fundamental diagram of macroscopic modelling to the optimal velocity function of the OVM/ROVM. A Gaussian form of P_d with an estimate for its width from a linearized form of the ROVM produced remarkable coincidence with direct simulations. Analysis of the nearest-neighbour distance variability in the ROVM, furthermore, revealed a model-inherent time constant $\tau_c \simeq 0.18\,sec$ as the analytical counter-part of the value $\tau_1 \simeq 0.5\,sec$ fitted from experiments. The attention of future work should be drawn to the study of flucutations in the critical regime where the nonlinear effects are expected to dominate.

References

1. K. Nagel and M. Schreckenberg, J. Phys. I (France) **2**, 2221 (1992)
2. N. Rajewsky and M. Schreckenberg, PHYSICA A **245**, 139 (1997)
3. O. Biham, A. Middleton and D. Levine, Phys. Rev A **46**, 6124 (1992)
4. T. Nagatani, PHYSICA A **246**, 460 (1997)
5. T. Nagatani, H. Emmerich and K. Naganishi, PHYSICA A **255**, 158 (1998)
6. D. A. Kurtze and D. Hong, Phys. Rev. E **52**, 218 (1995)
7. D. C. Hong, S. Yue, J. K. Rudra, M. Y. Choi and Y. W. Kim, Phys. Rev E **50**, 4123 (1994)
8. M. Gerwinski and J. Krug 188 (1999)
9. M. J. Lighthill and B. G. Whitham, Proc. Roy.Soc. London **A 229**, 317 (1955)
10. R. Kühne, Physik in unserer Zeit **15.3**, 84 (1984)
11. R. Kühne, Proc. 10th Int. Symp. on Transportation and Traffic Theory, Delft (1987)
12. B. S. Kerner and P. Konhäuser, Phys. Rev. E **50**, 54 (1994)
13. B. S. Kerner, S. L. Klenov and P. Konhäuser, Phys. Rev. E **56**, 4200 (1997)
14. D. Helbing, *Verkehrsdynamik* (in german), Springer, berlin, 1997
15. I. Prigogine and R. Herman, *Kinetic Theory of Vehicular Traffic*, American Elsevier, New York, 1971
16. H. Lehmann, Phys. Rev. E **54**, 6058, (1996)
17. C. Wagner, C. Hoffmann, R. Sollacher, J. Wagenhuber and B. Schürmann, Phys. Rev. E **54**, 5073 (1996)
18. C. W. Gardiner, *Handbook of Stochastic Methods*, Springer, Berlin (1997)
19. B. Derrida, E. Domany and D. Mukamel, J. Stat. Phys. **69**, 667 (1992)
20. M. Bando, K. Hasebe, A. Nakayama, A. Shibata and Y. Sugiyama, Phys. Rev. E **51**, 1035 (1995)
21. M. Bando, K. Hasebe, A. Nakayama, A. Shibata and Y. Sugiyama, Japan J. of Ind. and Appl. Math. **11**, 203 (1994)

Floating Car Data Analysis of Urban Road Networks

B. Kwella[1] and H. Lehmann[2]

[1] GMD-FIRST, Kekuléstr. 7, D-12489 Berlin, Germany
kwella@first.gmd.de
[2] gedas telematics, Berlin, Germany
Heiko.Lehmann@gedas-telematics.de

Abstract. A model for the analysis of *floating car data* (FCD) in urban road networks is proposed. In contrast to freeway traffic, structural features such as traffic lights are shown to dominate the dynamics - in a quasistationary picture - of such a network. Principles for obtaining a segment travel time depending on the mean vehicle density are discussed. Some remarks for the determination of this state variable are given, also on several segments.

1 Introduction

Modern traffic management systems are highly dependent on good knowledge of the current traffic loads in a route network. Therefore, the traffic has to be observed, and the measured traffic data is *online* communicated to a central unit. With additional information, extracted for example by police- and weather-reports, the incoming data is gathered in a database [14] and evaluated by the information system to provide a current picture of traffic state. The information on the state of the network and its future development can be used to control dynamic traffic guidance systems or directly for traffic light control.

Conventional vehicle-detecting systems work primarily with induction loops or infrared sensors. This leads to continous measurement of state variables at fixed points, like local traffic volumes q, which are defined as the number of vehicles per time interval passing a road-cross-section and local velocities v, the average speed of a number of cars at a fixed point.

Recent developments in the telematic sector have made it possible to equip more and more vehicles with traffic-telematic instruments, such as navigation support systems [3,15]. In dynamic route guidance drivers must be informed about the current traffic state of their route. Additionally, it is possible to use the telematic instruments as measuring devices. Thus, a new kind of data, called *floating car data* will be of increasing importance.

F. Pichler, R. Moreno-Díaz, and P. Kopacek (Eds.): EUROCAST'99, LNCS 1798, pp. 357–367, 2000.

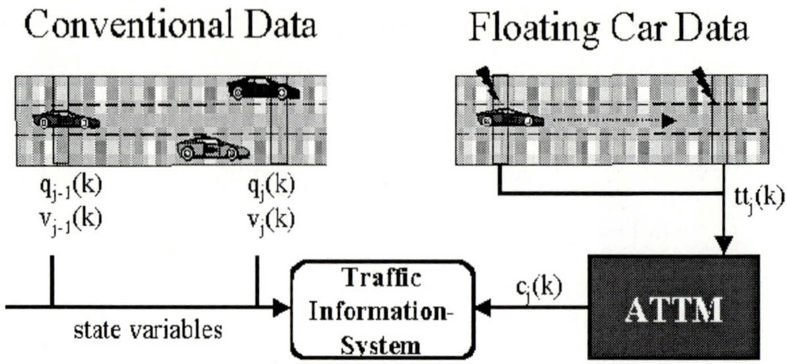

Fig. 1. Vehicle-Dectecting Technologies

In contrast to conventional data, *floating car data* is received by a fleet of individual cars. To reflect a general picture of the traffic situation, each of them is required to float with the traffic flow (hence the name of the data). Normally satellite-based navigation technologies, like the Global Positioning System (GPS), are used to determine the position of the test cars in the network and the current traffic information may be returned via GSM. These are the so-called *online* systems. To benefit from the structures in an urban area, the use of *offline* floating car systems seems more favourable. At certain fixed measuring points the cars equipped with tags are detected and their arrival time is transmitted *online.* They do not measure traffic densities or flow rates directly. It remains the task of a model to deduce these state variables from the given FCD.

In opposition to continuous measurements, *floating car data* are discrete events. When arriving they include information about the previous time period. FCD is time variant and locally distributed over the considered network. In addition to the obtained travel times, FCD also supply important information on the trajectories of the vehicles. This plays an important role in analyzing the so-called "origin destination matrix"(ODM) (see also [2,7,8]) and is not accessible for mere counting data.

In highway traffic the number of vehicles is maintained for long stretches of a road. Driving behaviour is described by the *Fundamental Diagram.* Assuming homogeneous conditions, simple inversion of the measured travel times yields the desired statements on the traffic situation. Other methods for analyzing traffic-data on motorways, for example with a Kalman-Filter, are described in [1].
In contrast, urban road networks are characterized by their structural features such as traffic lights, crossroads, number of lanes and so forth. This leads to a discontinuous traffic flow. Furthermore, the route between any two observation points will consist of several segments with highly fluctuating densities between them. Thus a more complex model is required for urban roads.

2 Aggregate Travel Time Model

The aim of the model is to deduce a state variable from a measured travel time t_{AB}. Therefore, the paths in the considered network are decomposed into certain segments, and for every segment a so-called "characteristic" $t_{\mathrm{i}}(c_{\mathrm{i}})$, which relates a travel time for this segment over a mean density, is determined.

$$t_{\mathrm{AB}} = \sum_{i=1}^{N} t_{\mathrm{i}}(c_{\mathrm{i}}) \tag{1}$$

The obtained segment-characteristics take account into the structural features of each segment.

For the development of the characteristic $t_{\mathrm{i}}(c_{\mathrm{i}})$ a mean density over a segment is assumed. This stationary request means that the passage over the segment has to be fast in comparison with a change of the overall traffic pattern. The requirement is fulfilled if, in segment i, the inflow q_{in} is equal or very similar to the outflow q_{out}.
Furthermore, it can not be the aim of the model to account for temporal structures finer than traffic lights periods. Since the test cars will be distributed evenly over time, the segment times experienced in reality present an average over the traffic light periods. Thus, the model works with mean densities and mean travel times. This leads to a description of traffic state in terms of aggregate variables.

The traffic in an urban network is dominated by waiting queues at crossroads and traffic-lights. So, a typical segment with length L of a route contains a queuing part L_{qu} and a free-flowing part $(L - L_{\mathrm{qu}})$.
The total time to pass such a segment is therefore a combination of the time to pass the free part and the queuing part, and, additionally a statistical waiting time at a red traffic-light, and it is called the *segment travel time*

$$t_{\mathrm{i}}(c_{\mathrm{i}}) = t_{\mathrm{free}} + t_{\mathrm{qu}} + t_{\mathrm{wait}} \tag{2}$$

The next paragraph deals with the development of this, therefore the indice $\{i\}$ for the segment itself is dropped whenever not necessary for clarity.

2.1 Segment Travel Times

Driving behaviour on freeway sections is described by the *Fundamental Diagram*, which relates traffic flow over mean density. This is also the basis of the driving velocity in the free flowing part of the segment. Various parametrizations exist, a summary of which can be found in [4]. Since the historical roots of the devising of the fundamental diagram lie in the study of freeway traffic, the applicability of the given forms to urban traffic has to be questioned. In [12] it was indeed shown that the figure of a speed-density relation in an urban context has a rather different shape. For a permitted maximum speed of 70 km/h and 3 lanes, for instance, it is given in [12] as:

$$v(c) = \begin{cases} 70 : 0 \leq c \leq 30 \\ -0.3c + 79 : 30 < c \leq 210 \\ -0.076c + 31.92 : 210 < c \leq 420 \end{cases} \tag{3}$$

where the velocities $v(c)$ are in km/h and the densities c are given in veh/km. The flow $q(c) = c * v(c)$ is shown in the figure 2.

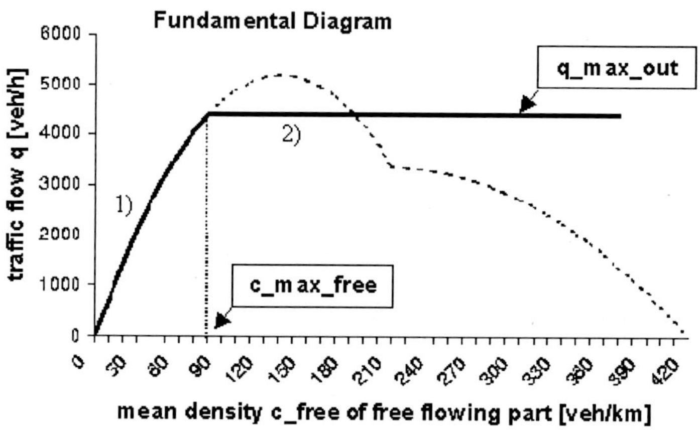

Fig. 2. Fundamental-Diagram for a 3-lanes segment with $v_{\text{free}} = 70\text{km/h}$

In an urban network the segments contain additionally a queuing part. Therefore, the mean density c of the entire segment is a linear combination of both parts. To determine the density, regard the characteristic size $q_{\text{max out}}$ of a segment, which can be defined for all segments with a traffic-light at their end points. This is the maximum mean exit rate of the segment, and it is given by the duration of the green phase T_{green} in relation to the total traffic light period T, the number of lanes n_{lanes} and an universal dissolution rate $1/q_{\text{diss}} = 1/1.8\ \text{sec}^{-1}$:

$$q_{\text{max out}} = \frac{T_{\text{green}}}{T} \frac{n_{\text{lanes}}}{q_{\text{diss}}} . \tag{4}$$

When the free flowing part is reaching the critical density $c_{\text{max free}}$ (see figure 2), the switching point of the model, called c_{green} is received:

$$\begin{aligned} c_{\text{green}} &:= c(c_{\text{max free}}) \\ &= c_{\text{max free}} + \frac{1}{L} \frac{T_{\text{green}} T_{\text{red}}}{1.8\,\text{sec}\, T} \left[n_{\text{lanes}} - 7.5\,\text{m} \times c_{\text{max free}} \right] . \end{aligned} \tag{5}$$

In this way two regions of flow can be distinguished:

$$1)\ c \leq c_{\text{green}}$$
$$2)\ c > c_{\text{green}}$$

Remember that in the stationary request the inflow equals the outflow and the definition of $q_{\text{max out}}$. For the region as indicated as 1) in figure 2, with flows below $q_{\text{max out}}$, the queue built up at the red phase is dissolved during the green phase. On the other hand, in the second regime (see 2) in figure 2) are always a number of cars, which cannot leave the segment during the first green phase.

Discussion is best started from the **critical case** $c = c_{\text{green}}$:

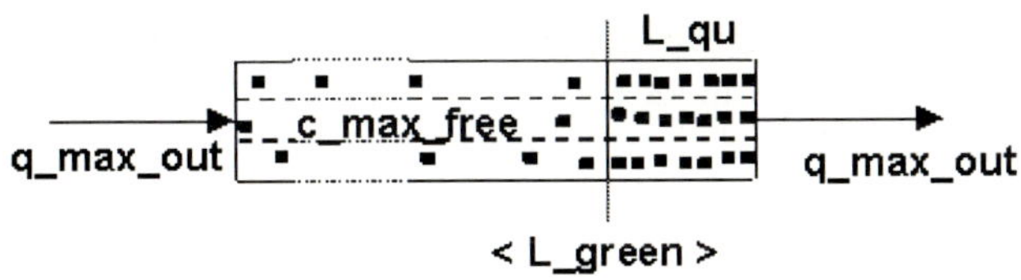

Fig. 3. Traffic situation for $c = c_{\text{green}}$

Here it is the first time that the inflow q_{in} equals the characteristic size $q_{\text{max out}}$, and the queue contains just the vehicle number

$$N_{\text{green}} = \frac{n_{\text{lanes}}}{1.8\text{sec}} T_{\text{green}} \ . \tag{6}$$

This queue is completely dissolved during a green phase. The length of this critical queue is given by:

$$\begin{aligned} L_{\text{green}} &= N_{\text{green}} \frac{7.5\,\text{m}}{n_{\text{lanes}}} \\ &= \frac{7.5\text{ m}}{1.8\text{ sec}} T_{\text{green}} \\ &= v_{\text{diss}}\, T_{\text{green}} \ . \end{aligned} \tag{7}$$

Note that, while the first line of eqn. (7) constructs L_{green} by assigning the N_{green} cars evenly to the n_{lanes} lanes, the last line gives it as the length driven by a floating car with the characteristic dissolution speed v_{diss} over the whole green phase.

The universal velocity $v_{\text{diss}} = 15$ km/h is a result of the effective car length $b = 7.5$ m combined with the universal dissolution rate q_{diss}. This velocity can be observed in a wide range of situations, like the dissolution of highway jams and waiting queues at urban crossroads, and can also be reproduced theoretically [5]. The length b is the key parameter in the modeling school based on cellular automata ([6,10,13] and a vast body of related and subsequent work).

On the other hand, the number of cars hitting the end of the queue during red phase is determined by the inflow rate $q_{\text{in}} = q_{\text{maxout}}$:

$$\begin{aligned} \langle L_{\text{green}} \rangle &= q_{\text{max out}} \, T_{\text{red}} \, \frac{7.5\,\text{m}}{n_{\text{lanes}}} \\ &= v_{\text{diss}} \, \frac{T_{\text{green}} T_{\text{red}}}{T} \, . \end{aligned} \tag{8}$$

Obviously, formulae (7) and (8) do not coincide. Since, however, the inflow rate is a quantity which is averaged over the whole traffic-lights period, equation (8) presents a mean value as has been indicated by the pointed brackets. Comparison of equations (7) and (8) then reveals that the effective time average for the queue build-up is the geometrical mean of both phases. For $T_{\text{green}} = T_{\text{red}} = T/2$ this mean is just $\langle L_{\text{green}} \rangle = 1/2 L_{\text{green}}$, whereas for asymmetric phases it is always smaller. The average length of the queue L_{qu} is then identified in the critical case as:

$$L_{\text{qu}} = \langle L_{\text{green}} \rangle \, . \tag{9}$$

Equations (7,8) furthermore reveal the remarkable fact that the eventual dissolution time t_{qu} for the critical queue contains only the characteristics of the traffic-lights. Since the queue is dissolved with v_{diss} one finds:

$$t_{\text{qu}} = \langle L_{\text{green}} \rangle / v_{\text{diss}} = \frac{T_{\text{green}} T_{\text{red}}}{T} \, . \tag{10}$$

Using the fundamental diagram it is implying, that the cars in the free-flowing part are driving at a velocity of $v_{\text{max free}}$ by a density of $c_{\text{max free}}$.
Thus, we can obtain this time as

$$t_{\text{free}} = (L - L_{\text{qu}}) / v_{\text{max free}} \, . \tag{11}$$

For **small densities** $c < c_{\text{green}}$, the in- and outflow of the segment is less than $q_{\text{max out}}$. This means that the average length L_{qu} of the queue is shorter than $\langle L_{\text{green}} \rangle$:

$$L_{\text{qu}} = q_{\text{in}} \, T_{\text{red}} \, \frac{7.5\text{m}}{n_{\text{lanes}}} \tag{12}$$

and the according dissolution time

$$t_{\text{qu}} = L_{\text{qu}} / v_{\text{diss}} = q_{\text{in}} \, T_{\text{red}} \, \frac{1.8\text{sec}}{n_{\text{lanes}}} \, . \tag{13}$$

In the free-flowing part the cars are driving with a mean speed of v_{free} by a density of c_{free} dependent on the inflow q_{in}. It leads to the time

$$t_{\text{free}} = (L - L_{\text{qu}}) / v_{\text{free}} \, . \tag{14}$$

It remains the second region with **high densities** $c > c_{\text{green}}$.
Here the average queue length L_{qu} is longer than $\langle L_{\text{green}} \rangle$. This means, a number of cars has to wait multiples of the total traffic light period T:

$$L_{\text{qu}} = \langle L_{\text{green}} \rangle + \langle L_{\text{ex}} \rangle \tag{15}$$

with a mean length

$$\langle L_{\text{ex}} \rangle = (c - c_{\text{green}}) L \frac{7.5\text{m}}{n_{\text{lanes}}}. \tag{16}$$

All cars in $L_{\text{qu}} - \langle L_{\text{green}} \rangle$ have to wait for an extra time, thus the total dissolution time assembles to

$$t_{\text{qu}} = \frac{L_{\text{qu}}}{v_{\text{diss}}} + \left(\frac{L_{\text{qu}}}{\langle L_{\text{green}} \rangle} - 1 \right) * T\,. \tag{17}$$

The velocity $v_{\text{max free}}$ over the mean density $c_{\text{max free}}$ for the free-flowing part is given as the result of the mean maximun outflow, it leads to:

$$t_{\text{free}} = (L - L_{\text{qu}}) / v_{\text{max free}}\,. \tag{18}$$

In contrast to the deterministically treated times t_{free} and t_{qu}, the time t_{wait} is a statistical value and the same for the two regions. The expectation value t_{wait} for the overall waiting time becomes the probability to hit a red traffic-light by the remaining waiting time for arriving at a time T' at the traffic light.

$$\begin{aligned} t_{\text{wait}} &= \frac{T_{\text{green}}}{T} * 0 + \frac{T_{\text{red}}}{T} (T_{\text{red}} - T') \\ &= \frac{1}{T} (T_{\text{red}}^2 - T_{\text{red}} T') = \frac{1}{T} \int_0^{T_{\text{red}}} (T_{\text{red}} - T') dT' \\ &= \frac{1}{2} \frac{T_{\text{red}}^2}{T}\,. \end{aligned}$$

Therefore, the sum of the three times define the travel time $t(c)$ (see eqn. (1)) over a mean density. We call this the *characteristic* of a segment, it is like a *Fundamental Diagram* for urban roads.
Receiving a measured travel time, it is now possible to invert the diagram at segment level to determine the traffic state variable.

The picture shows the model-equation for a real road segment with length = 1064 m, lanes = 3, total traffic light = 90 sec, green phase = 38 sec, red phase = 52 sec in comparison with travel times obtained by the mesoscopic simulation-tool DYNEMO [11].
In the low flow region cars can be found, that don't hit a red traffic-light. It is only the mere driving time. With approximately 70 km/h, they pass the segment in about 55 sec. Other cars have to wait at the red traffic light. The model shows an average.

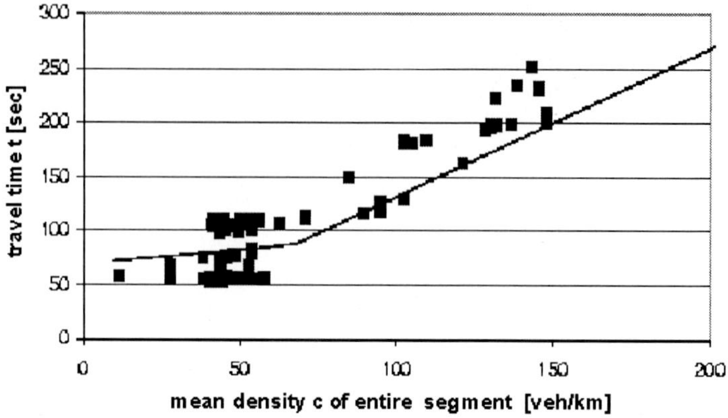

Fig. 4. Characteristic for a real road segment

It should be stressed that the model equations don't have any fit parameters, apart form the two universal numbers: The effective car length b and the dissolution rate q_{diss}. So, the obtained segment-characteristic shows a remarkable agreement with the simulation-results.

2.2 Segment Compilation

Having obtained an analytical form of the segment travel time as functions of the mean density, it is now possible to return to equation (1).
In the trivial case, where every segment is monitored by measuring points at its end, a picture of traffic state on the overall network can be immediately deduced.

To analyse a measured floating car travel time on certain segments, structural information and a-priori-knowledge will be used.

Figure 5 shows a typical situation, when receiving a measured travel time t_{AC} and another time t_{BC}. This leads to two equations with three unknown segment travel times:

$$t_{\text{AC}} = t_{\text{AX}}(c_{\text{AX}}) + t_{\text{XC}}(c_{\text{XC}})$$
$$t_{\text{BC}} = t_{\text{BX}}(c_{\text{BX}}) + t_{\text{XC}}(c_{\text{XC}})$$

The first information is obtained by comparison the measured times with a reference times

$$t_{\text{AC}}^{(0)} := L/v_{\text{free}} + t_{\text{wait}} + t_{\text{VEHspez}} \tag{19}$$

with v_{free} is the maximum permitted speed and t_{VEHspez} is an allowance dependent on the features of the measuring vehichle. If the comparison reveals that

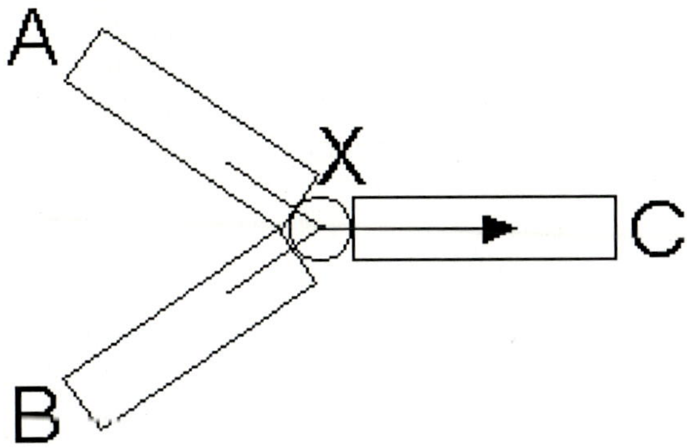

Fig. 5. Typical situation of overlapping paths at a crossroad

$t_{\mathrm{AC}} > t_{\mathrm{AC}}^{(0)}$ the floating car must have encountered an obstruction at some point of its journey, yet in which segment this occurred remains unclear.

If the considered segments have comparable structural features, the **similarity principle** is applicable, which estimates segment travel times by weighting with the corresponding reference times:

$$\text{for path AC: } t_{\mathrm{AX}} = t_{\mathrm{AX}}^{(0)} \frac{t_{\mathrm{AC}}}{t_{\mathrm{AC}}^{(0)}}, \; t_{\mathrm{XC}} = t_{\mathrm{XC}}^{(0)} \frac{t_{\mathrm{AC}}}{t_{\mathrm{AC}}^{(0)}} \tag{20}$$

and

$$\text{for path BC: } t_{\mathrm{BX}} = t_{\mathrm{BX}}^{(0)} \frac{t_{\mathrm{BC}}}{t_{\mathrm{BC}}^{(0)}}, \; t_{\mathrm{XC}} = t_{\mathrm{XC}}^{(0)} \frac{t_{\mathrm{BC}}}{t_{\mathrm{BC}}^{(0)}} \; . \tag{21}$$

Of course, the travel times for segment $\{XC\}$ in equations (20,21) will, in general, not agree. The solution nearest to the even distribution of delays, i.e. the one sought by the similarity principle is, then, given by constructing an appropriate mean value of the travel time t_{XC}. With just the arithmetic mean one arrives at the solution:

$$\begin{aligned} t_{\mathrm{XC}} &= \frac{1}{2} t_{\mathrm{XC}}^{(0)} \left[\frac{t_{\mathrm{AC}}}{t_{\mathrm{AC}}^{(0)}} + \frac{t_{\mathrm{BC}}}{t_{\mathrm{BC}}^{(0)}} \right] \\ t_{\mathrm{AX}} &= t_{\mathrm{AC}} - t_{\mathrm{XC}} \\ t_{\mathrm{BX}} &= t_{\mathrm{BC}} - t_{\mathrm{XC}} \end{aligned} \tag{22}$$

Inversion via the characteristic for the considered segment then yields the mean traffic densities.

Another idea is based on the assumption, that the traffic state in segment $\{XC\}$ is dependent on the previous states of the segments $\{AX\}$, $\{BX\}$. It leads to the **hereditary principle** and is in detail described in [9].

In combination with other structural information, such as the most probable distribution of traffic load on the net, travel time profiles or statistical splitting rates given by analyzing trajectories, these assumptions would be helpful in obtaining a picture of the traffic situation within highly branched road networks.

3 Conclusions

Floating car data present a modern way of analyzing traffic flow. But at first it must be guaranteed, that the general traffic situation is reflected by the measured travel times. Therefore, a sufficient number of vehicles is needed, which have to travel the path within a stationary situation. The received travel times have to be confirmed by an average. This request puts a constraint on the fleet of test cars and their distribution over the network.

Through recent developments in communication- and vehicle-detecting-technologies, like GPS and GSM, an increasing number of vehicles will be equipped with traffic telematic systems. In the future this vehicles then represent a measuring fleet, so that *floating car data* could cover space and time in a sufficient resolution.

This paper presents a model for urban road networks, which gives the mean vehicle density for a measured segment travel time. When assembling the segment travel times, extra reasoning is needed regarding the most probable distribution of traffic load on the net. In the future the calibration of the aggregate travel time model (ATTM) and the analysis of more complex structures will be continued. It is also planned to design an *online* simulation-tool based on FCD, which kindly note, that modern traffic information systems should be integrated tools, using a combination of different kinds of traffic data and additional information, such as knowledge of construction sites or cultural events, for obtaining an actual picture of the traffic situation in an urban network and its future development.

This project is supported by the administration of Berlin and is a part of BERTRAM, the BERlin TRAffic Management. The work is proceeding in cooperation with small middle enterprises SYSTRON and eloqu-metabasis, who makes real *floating car data* available.

References

1. G. Böker, *Traffic Control Systems on the Basis of Floating-Car-Data*, project report, Technical University Hamburg, 1999
2. J. Freund and T. Pöschel, PHYSICA A, **219**, 95 (1995)
3. gedas telematics GmbH, Berlin, private communication
4. D. Helbing, *Verkehrsdynamik*, Springer, Berlin, 1997
5. M. Herrmann and B.S. Kerner, PHYSICA A, **255**, 264 (1998)

6. S. Krauss, P. Wagner and C. Gawron, Phys. Rev. E **55**, 5597 (1997)
7. H. Lehmann, PHYSICA A, **230**, 202 (1996)
8. H. Lehmann, PHYSICA A, **247**, 405 (1997)
9. H. Lehmann, B. Kwella, *Traffic flow analysis in urban nets on the basis of floating car data*, preliminary project report, Berlin, 1999
10. K. Nagel and M. Schreckenberg, J. Phys. I (France), **2**, 2221 (1992)
11. ptv System Software und Consulting GmbH, Karlsruhe, also: Th. Schwerdtfeger, in: 9th Intl. Symp. on Transportation Traffic Theory, Proceedings, VNU Science Press, 1984
12. K. Putensen, *Ein makroskopisches Modell des Verkehrsablaufs auf Stadtstrassen und sein Anwendung in der Leittechnik*, Dissertation, Technical University Hamburg, 1994
13. N. Rajewsky and M. Schreckenberg, PHYSICA A, **245**, 139, (1997)
14. R.P. Schäfer, GMD-FIRST, Berlin, private communication
15. systron Systementwicklungsgesellschaft für angewandte Elektronik mbH, Berlin, private communication

Information Lost in the Hologram Subdividing Process

Grażyna Mulak[1], Leon Magiera[1], and Andrzej Mulak[2]

[1] Institute of Physics, Wrocław University of Technology, Wybrzeże Wyspiańskiego 27, 50-375 Wrocław, Poland
[2] Institute of Microsystem Techniques, Wrocław University of Technology, Wybrzeże Wyspiańskiego 27, 50-375 Wrocław, Poland

Abstract. The influence of the hologram division into parts on the height of the readout signal was considered. It was found that from the two waves onto which the signal can be divorced: the geometrical and the boundary wave, the latter is responsible for the transfer of the optical signal.

1 Introduction

Diffractive optical elements are inseparable connected with the contemporary communication techniques. They are suitable for realisation interconnections between the optoelectronic systems (microlenses and fibers), they permit data storage and package with the high densities, allow data processing, make logic operations. With the help of the optical element Fourier transform is achieved as quick as possible – with the light velocity. The light techniques on the one hand cause the increase of the speed of acting at full, limiting value, on the other hand lead to the miniaturization of the setup elements. The miniaturization is possible for the two reasons: at first – for the high density of information registration in media, in the second place – for the possibility of integration several functions into one component e.g. beam splitting, de ection and focusing. The holography, among the others, is the technique of producing diffractive optical elements. It is known, that information registered at the hologram plate may be read out from the piece of it. However, how far the hologram may be subdivided as far as recorded information will be possible to read out yet? The subdividing process, undoubtely irreversible, is associated with the entropy increasing. It is associated with the diminishing of the number of the interference structure lines contained in the element also. Then the resolving power of the element decreases in comparison with that one of the whole hologram, perceived as the diffraction grating.

2 Two – Dimensional Spatial Signal – the Distinction between the Geometrical and the Boundary Waves

The signal, which is obtained during the readout process, may be presented according Maggi–Rubinowicz concept [1], as a sum of two components: of the

F. Pichler, R. Moreno-Díaz, and P. Kopacek (Eds.): EUROCAST'99, LNCS 1798, pp. 368–373, 2000.

geometrical and of the boundary diffraction waves. This concept, originated from physical reasons, generalized by Miamoto and Wolf [2], and Rubinowicz also, is widely supported mathematicaly (stationary phase method). We show, that the lost of information caused by the hologram subdivision is directly related to the boundary wave amplitude. Interference of the object and the reference beam waves permits to record information about the object at the photographic plate. Through developing and bleaching process we obtain hologram. The illumination of the hologram with another wave permits us to readout information encoded in the system of interference fringes. The readout process is the diffraction process and then it may be described by the Kirchhoff–Fresnel integral

$$U(P) = k \int \int_S g(\xi, \eta) \exp(ikf(\xi, \eta)) \mathrm{d}\xi \, \mathrm{d}\eta \tag{1}$$

where:

P is the observation point,
$U(P)$ complex amplitude in the observation point,
$g(\xi, \eta)$ amplitude of the diffracted wave, immediately behind the hologram plate,
$f(\xi, \eta)$ phase function describing the phase of the wave perturbation coming from
(ξ, η) point at hologram to observation point,
$k = \frac{2\pi}{\lambda}$ wave number.

The integral is quickly oscillating function because of a great value of the wave number and then the integral (1) may be evaluated using the stationary phase method (**SPM**). The $U(P)$ according to **SPM** can be written as the sum of the geometrical ($\hat{U}_G$) and the boundary ($\hat{U}_B$) waves [1]

$$\hat{U}(P) = \hat{U}_G + \hat{U}_B \tag{2}$$

The hologram size, of necessity, must be limited. The influence of this limitation on the signal registered at P is represented by the boundary wave [3]

$$\hat{U}_B = \exp\left(-\frac{i\pi}{2}\right) \int_\Gamma g(\xi, \eta) \exp(ikf(\xi, \eta)) \frac{\partial f}{\partial n} |\nabla f|^{-2} \mathrm{d}\sigma \tag{3}$$

Where $\frac{\partial f}{\partial n}$ is the derivative of the phase function with respect to the outer normal of the boundary line.

3 Results

The idea of this paper is the following. Hologram of greater size may be divided into few number of the smaller holograms. The integrals boundary of the (3) type originating from the same element of the boundary of the two neighbouring subholograms are of the oposite sign according to opposite direction of the outer normal. Then, they cancel each other remaining the the contribution associated with the outer boundary only. As an example we consider axial hologram recorded and reconstructed with the $\lambda = 5 \cdot 10^{-5}$. The object – the lightening point located at the $z_0 = 1$ from the hologram plate represents signal. The reference wave is divergent spherical wave with the centre at $z_R = 1.2$ at the left from the holographic plate. The reconstructed wave is convergent to the $z_C = 1.1$ point on the right side of hologram. The results obtained for holograms of circular shape of various sizes are presented in [4].

As follows from these results – when the hologram radius increases, an amplitude of the signal tends to the geometrical wave amplitude, extremely high at geometrical focus. Similar results are obtained for the square hologram shape. Information flux (information entropy) defined as [5]

$$S = -\sum_i p_i \log(p_i) \tag{4}$$

in optics is transformed to the information content of the image (optical entropy) if an intensity distribution is interpreted as a density of probability

$$S(z) = \sum_i I_i(z) \log(I_i(z)) \tag{5}$$

where

z is the optical axis coordinate
$I_i(z)$ are the normalized different intensity levels at plane z

According to (5) – region of the focus is the region, where extreme of the entropy is expected. These results [4] give evidence that in the case circular symmetry the boundary wave causes the lost of information. The above results are more general and regards of the situation in which axial symmetry of the waves taking part in holographic imaging does not coincide with the boundary line shape. We consider the hologram of a square shape, cut off from the far, of axial region, where the density of the recorded interference fringes is high. The side of the square is equal to 800 λ and its centre distance from the axis is equal to 1. Then, remaining the centre position, the square side is diminished to 400 λ, 200 λ and 100 λ.

The respective amplitudes of the geometrical and the boundary waves and the amplitude of the total disturbance in these cases are presented at Fig. 1 and Fig. 2. The singularities of the geometrical and the boundary wave amplitudes (Fig. 1, 800 λ) corespond to geometrical and diffraction foci. The real amplitudes are finite, of course. The result of integration of (1) is presented at the bottom of this figure. From the comparison it follows that the maximal value of the amplitude corresponds to the boundary wave focus, the geometrical focus vanishes.

The information is degraded. Figs. 1 and 2 illustrate the sequence of further diminution. The signal decreases by reason of the decreasing amount of the energy which reach the hologram during the readout process.

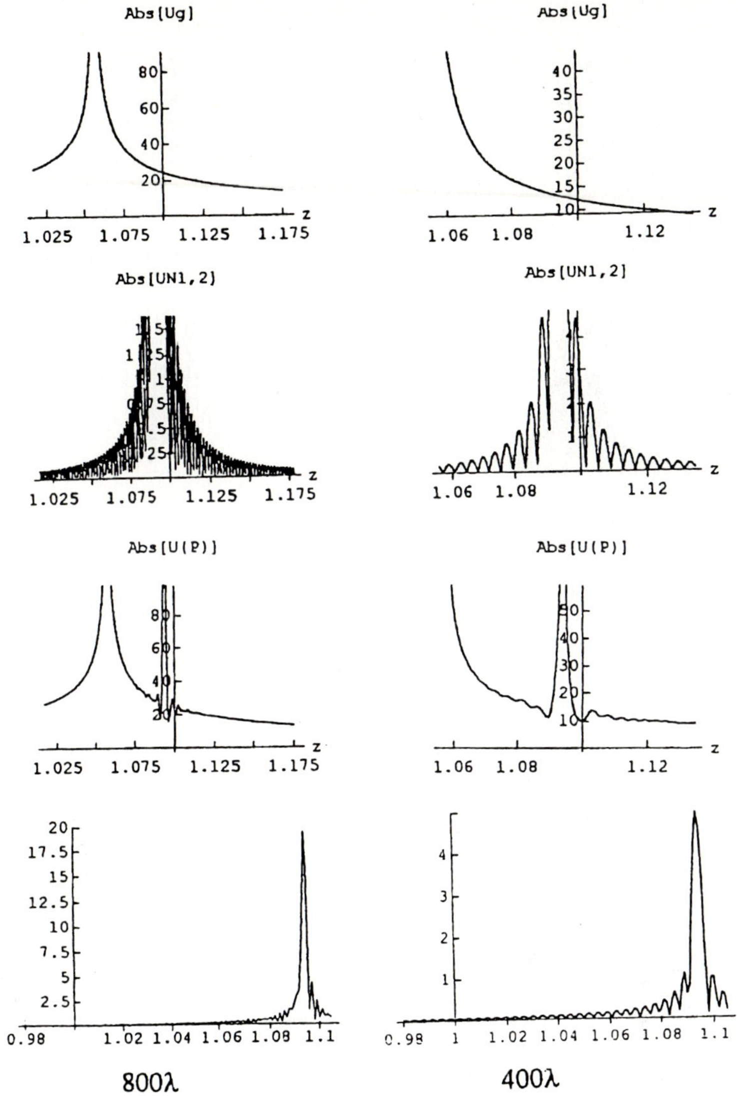

Fig. 1. The hologram of which the half sides are equal to 800 λ and 400 λ. From the top to the bottom are shown the distibutions of: – the geometrical wave amplitude, – the boundary wave amplitude evaluated as the contribution arising from the critical points at the edge, – the amplitude of the total disturbance evaluated by the stationary phase metod, – the ampiltude of the total distrubance obtained by numerical integration of Kirchhoff–Fresnel integral

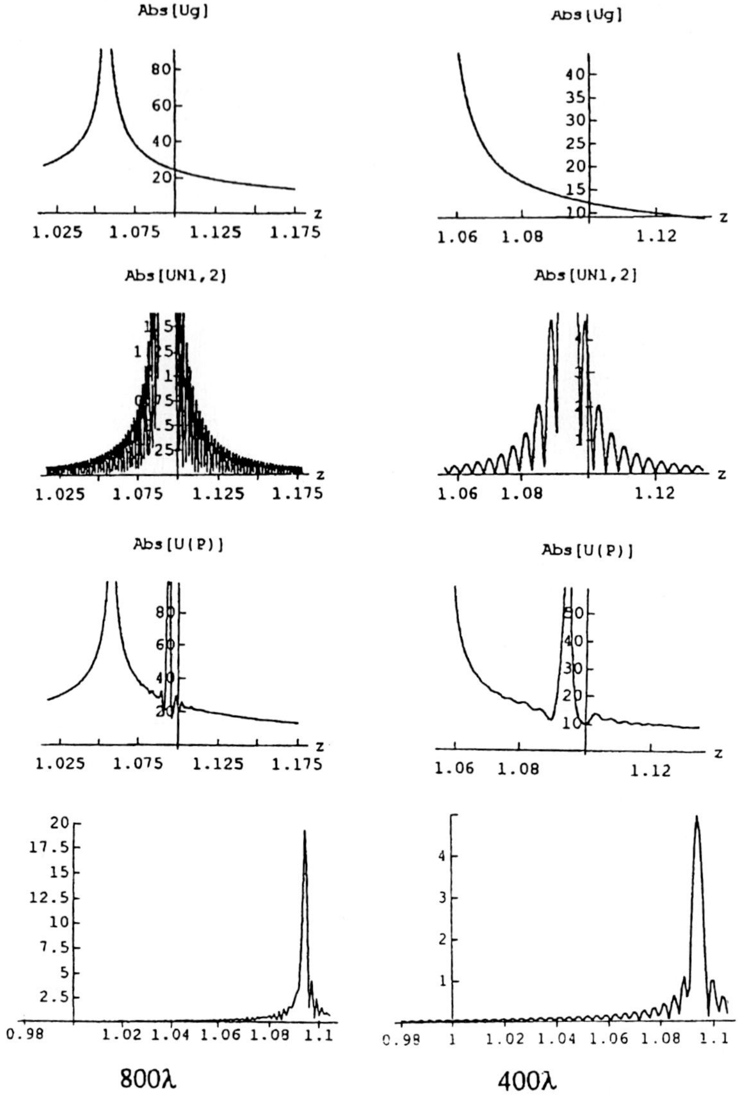

Fig. 2. The holograms of which the half sides are equal to 200 λ and 100 λ. All other descriptions as in figure 1

4 Conclusion

The limits of the hologram division, determining the height of the readout signal, are directly related to the boundary wave amplitude.

References

1. Rubinowicz A. Die Beugungswelle in der Kirchhoffschen Theorie der Beugung, Berlin, Springer Verlag, Warszawa, PWN (1966)
2. Miamoto K., Wolf E. Generalization of the Maggi–Rubinowicz Theory of the Diffraction Boundary Wave – Part I, J. Opt. Soc. Am. 52 (1962) pp. 615 625, Part II ibid., pp.626-637.
3. Fedoryuk M. V., Metod stacionarnoj fazy v mnogomernom slucae. Vklad ot granicy oblasti, Zh. v.m. i m.f. Vol.10 No2 (1970) pp. 287-299
4. Mulak G., Metoda stacjonarnej fazy w oszacowaniu amplitudy fali ugietej przez hologram, Prace Naukowe Instytutu Fizyki Politechniki Wroclawskiej 30, Seria: Monografie 19 (1991)
5. Marechal A., Francon M., Diffraction Structure des images, Paris 1960

5 SYSTEMS ENGINEERING AND SOFTWARE DEVELOPMENT

Electronic Performance Support Systems Challenges and Problems

Gerhard Chroust

Systems Engineering and Automation
Institute of System Sciences
Kepler University Linz
Altenbergerstr. 69, A-4040 Linz
Tel. +43-732-2468 866, Fax +43-732-2468-878
GC@sea.uni-linz.ac.at *

Abstract. After a short introduction to the concept of automation and computer based systems, chapter 2 presents the the concepts of Electronic Performance Support Systems (EPSS). Chapter 3 specifically discusses process-centered EPSS and their realization in software engineering and office automation. Different attitudes and requirements are pointed out. Chapter 4 argues that computer support makes it necessary to extend Scheer's Y (well-known in the CIM-area) to accomodate additional granularity. This discussion is carried over into chapter 5 where we map the relevant processes of software engineering and office automation to the processes identified in the extended Y from chapter 4.

1 Automation and Computer Based Systems

1.1 Automation

Automation has been defined (only as late as 1956!) [28] as the *replacement and enhancement of human work by computers* [28]. Machines, especially computers, have always acted as *amplifiers for human capabilities.* The current trend to ever increasing computational power at dramatically reduced cost and size will induce even more computer support. In many instances computers are *intelligence amplifiers*, analog to the power amplification qualities of classical machines.

1.2 Computer Based Systems (CBS)

The term EPSS has become a common name for ideas and environments already existent in the various domains to be discussed in this paper. In three domains of CBS (computer based systems) this trend is highly obvious. All three have an emphasis on processes and the computer support for their enactment.

* The research behind this paper was triggered and funded by Project P09372-PHY ('CAD für CIM') and P11824-TEC ('Autonomer reaktiver Roboterarm') of the Austrian Fond zur Förderung wissenschaftlicher Forschung

F. Pichler, R. Moreno-Díaz, and P. Kopacek (Eds.): EUROCAST'99, LNCS 1798, pp. 377–393, 2000.

Software Engineering
The main idea is to create a model of a system for information processing and to subsequently implement this system. The volatility and complexity of software engineering [12] makes it necessary to emphasise the adherence to a defined process to ensure quality [35].

Office automation Here the main objective is to organize a smooth and effective system to create and administer task in the office [40] [48] [58].

Flexible Manufacturing The major concern is the fully automatic and interruption-free running of a production cell to produce some goods [5] [42][43][44][6].

2 Electronic Performance Support Systems

For complex tasks it is not sufficient to amplify a single human capability, but one must take a systemic view and provide the user with an integrated environment which allows to solve the *complete task holistically.* We are challenged to build an environment for supporting humans in performing complex multi-step tasks. The term *Electronic Performance Support Systems (EPSS)* has been coined [31] for general supporting systems [13][7]. They are intended to *combine training, documentation, expert knowledge, feedback and obvious tools to provide technology users with the information they need, when they need it, where they need it, in the context of the situation they are in, without having to leave work to get it.*

One has to accept that the arrival of the new computer technology (especially the dramatic growth of computers with respect speed, power and memory at a dramatic reduction in cost) was the main driver for creating EPSS. Very often the scenario was of the type *'We have a solution - do you have a problem for it?.*

Nevertheless the overall requirements for such an environment is (Fig. 1):

- that it support humans in a 'natural, intuitive' way,
- is unobtrusive and
- becomes visible, only when it is needed or called upon.

Typically such an environment allows humans to perform six types of work (Fig. 1):

information access EPSS should provide the user with information, guidance and help: *"your unobtrusive assistant".*

creative work In all engineering disciplines creativity is the key to quality and cost-effectiveness, especially in the long run. The problem is that creativity needs a certain amount of freedom whereas quality (cf. ISO 9000, [39]) requires a certain adherence to predefined processes. One needs a delicate balance between prescribing a well defined process to be performed ('the human as a machine') and the freedom for engineers in their creative work ('the human as a free-wheeling spirit'). In the optimal case the EPSS provides to the engineer the much needed support for creativity: *"your helpful co-designer"*

calling of development tools This covers the 'mechanical, non-creative' part of every system development. Tools provide the necessary generative support by automatically deriving further intermediate products (e.g. executable code) from some previously specified docuements (e.g. low level design): *"your untiring ghost writer"*

process management This means to decide which tasks are still outstanding, which can be started and deciding which one to tackle next. It includes, in the extreme case, the detailed prescription of individual steps and their sequence in the form of a process model [15][26] : *"your unerring task manager"*

document management Storing, updating, classifying the various pieces of information arising in a project, including version and configuration management: *"your untiring registrar and archiver"*

communication with co-workers It is a fact that most work has to be done as a team effort. This means that an EPSS, even if conceptualized for the use by an individual, must enable a person to share his/her tasks, results and data with co-workers in order to perform all group related work (i.e. communication, coordination, consensus and collaboration) [21]:*"your effective team work organizer"*.

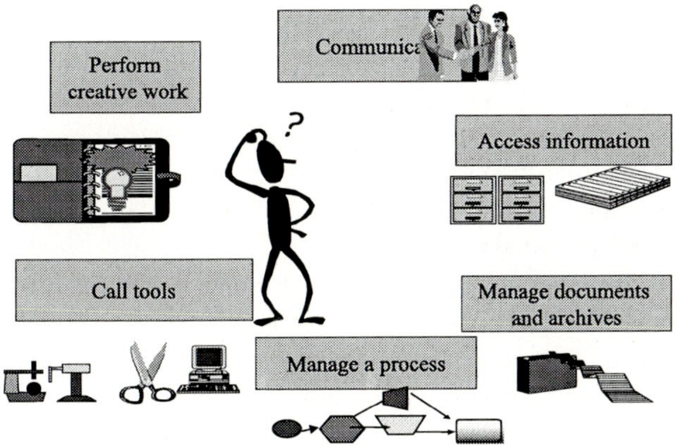

Fig. 1. Main Support of Electronic Performance Systems

2.1 Process-Centered and Non-process Centered EPSS

It has to be observed that in the three indicated areas the development/production process is complex and consists of many individual activities, which are intervowen and interdependent in many ways. That means that both the individual activities *and the process* are of importance. Errors in either one lead at least to suboptimal performance, in the worst case to unacceptable results. EPSS have two starting points:

bottom-up This is a natural approach in automation: gradually associating certain activities with mechanical tools and thus supporting/replacing human work. This leads to computer based systems of various form and complexity. In software engineering typically compilers were one of the earliest tools [49, 50]. In flexible manufacturing various (often still mechanical) tools were employed, and office automation used computing tools almost from its beginning (e.g. mechanical calculating machines [33]). One of the major problems of this 'bottom-up' approach is the exchange of data between the different tools, not only syntactically but also semantically. A problem which still has not been fully solved. For consistency reasons the best approach is to provide a common repository where all data are stored redundancy-free. Every tool takes its data from the repository and puts them there again. Obviously this means a considerable standardisation with respect to syntax and semantic. In the area of software engineering [22], [1] we therefore find EPSS in the form of a so-called *workbench* [15, 16]. We see several tools which share data via a common repository and we a see a process model where tool are attached to the various activities. For a full support of all software development activities both types of support are needed, although in many instances (typically the tool sets supporting Information Engineering (e.g. ADW, IEF etc.) do not have an explicit process model: It is tought by the tool provider and has to be internalized by the developers. A workbench contains an integrated set of tools, where the following aspects of integration are covered:

- **data integration** all data are freely shared by the various tools, the tools are able to interpret and use one anothers data
- **call-interface integration** all tools are called in a uniform way, consistent with the mental model of the user
- **philosophical integration** all tools obey to the same methodologial paradigm
- **functional coverage** the tools form an essentially non-overlapping set which cover in their totality all needed functions

Typical respresentatives are tool sets like ADW, IEF, INNOVATOR etc.

top-down The complexity of the complete production process, the interdependence of the various activities, and the number of people involved makes it necessary also to define and prescribe (to a certain amount) the sequence of these activities. In software engineering this is usually expressed in a so-called *process models*. One of the earliest was Boehm's waterfall model [9]. Typical examples are ADPS [2][3][14] [25] and MAESTRO II [46].
The process models V-Model [11], SSADM [29], and even ISO 12207 [36], when enacted, would also fall into this category. This has induced a considerable research and has gained considerable practical importance (cf. [30][57][11]), including standardisation efforts. A major problem is the danger of procrastination of the developers, over-bureaucratisation and a poor interface to tools (cf. ADPS [2][3][14][25]). In software engineering EPSS of this kind are often called 'process-centered Software Engineering Environment' [32]. For these EPSS the *process model*, usually derived from some develotoppment method, is central. The EPSS support the enactment of

such a model in the sense that for individual projects the process model is instantiated yielding an actual process.
From the very beginning of administration in the early forms of temple administrations [45][47] one needed on precise and clear prescription of proceeding - starting with religious ceremonies in pre-historical times. From these descriptions the processes to be performed have evolved and have instigated computer support *for the enactment of the prescribed process.*

By necessity EPSS need both types of computer support, but the emphasis is different. Fig. 2 shows - from the domain of software engineering - the main tasks of an EPSS.

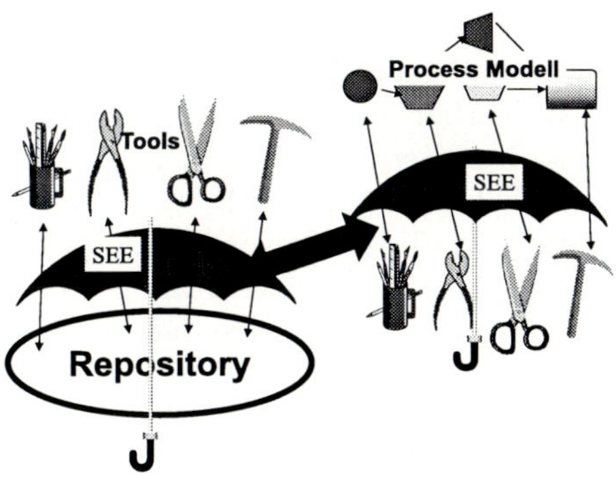

Fig. 2. Concept of a SEE

Above considerations can also be observed in the area of Office automation, where (cf. [48]) a distinction is made between 'workflow' and 'groupware'. Otto Petrovic defines: "Groupware is a jointly usable environment which supports teams to fulfill their common tasks". On the other hand we find: "Workflow is a system which completely defines, manages and executes workflow processes through the execution of software whose order of execution is driven by a computer representation of the workflow process logic" [59].

In the area of flexible manufacturing a distinction should be drawn between a manufacturing shop where several works produce some goods with the use of various tools and a flexible manufacturing cell where a precise control program controls all actions of tools, the intermediate storage and the robots [41, 44, 6].

3 Process-Centered EPSS

For the rest of this paper we will concentrate on process-centered EPSS. It turns out that the emphasis and the need for support and control is different in different domains.

As already stated in other publications, the consequence is that the basic structure of for such EPSS is the same, the emphasis and the specific design parameters differ, however [17].

3.1 Software Engineering

In software engineering (*"The establishment and use of sound engineering principles (methods) in order to obtain economically software that is reliable and works on real machines"* [4]) the main driving objectives are productivity and quality. In the beginning activities occurring late in the software production where subjected to automation in a recursive way: using software (compilers) to produce instruction for the computer. Thus that part of software which controls the computer hardware is *generated*, not hand-crafted by humans (cf. FORTRAN [37], PL/I etc. Gradually more and more tools covered more and more of the production process, including earlier and earlier phases of the production process: *CASE -computer aided software engineering.* Complete software engineering environments evolved [16][32] [34]. Initially there were more and more tools added to the support environment. With the growing complexity and expansion of the activities included in the development process the need for a process model became apparent (e.g. [11][14]). Various process models were proposed and even partially standardized (SSADM, ADPS; V-Modell, Merise, ISO 12207, etc.).

3.2 Office Automation

It is known that considerable productivity potential is still available via office automation. Large companies spent up to 50% on office costs, service oriented companies even up to 85% [55]. The current productivity increase in the office world is around 3 % which is not much better than the the learning curve. The current nature of office work does not allow too sophisticated tool to be employed - desk top publishing systems, calculators, automatic copying machines.

A major saving can be found in three areas:

- reducing transport delays by providing an *electronic act* which is moved in an network
- eliminating transfer costs due to change of medium (from paper to computer to paper etc.) by providing one electronic medium for every user
- eliminating the need for duplicating by providing *all* users with access to one and the same copy electronic copy.

Especially the electronic transport of documents induced the idea to provide an automatic routing, ie.e. a work flow model, embodying the rules prevalent in many government departments with their strict order of routing documents.

In this field we see less of the 'tool type' typically for software development (CASE-tools). The equivalent of this would be the so-called groupware tools which, however, do not so much impress by their tool capacity but rather by the chance that everybody can access all information available to everybody else, providing an environment for a group working together. In this area the emphasis is rather on defining the *flow of work - the workflow.*, When considering office automation, we should bear in mind, however, the great potential to create completely new solutions. These solutions are invariably process-centered EPSS.

3.3 Comparison of the Domains

In the previous sections we have discussed two important domains of CBS. We have also shown the dichotomy between process-centered and tool-centered EPSS. Table 1 illustrates some of these differences.

Table 1. Types of EPSS

	tool centered	process centered
software engineering	SEE	CASE-Tool
office automation	groupware system	workflow system

Despite the similarity of the underlying paradigm, these areas show considerable difference. We will discuss these differences from three viewpoints:

attitudes and basic requirements What do users of the process expect of their roles and the behaviour of the system?

analysis of key processes Based on an extension of Scheer's Y [52] the most important processes and their main component processes are identified. We will discuss the importance and the impact of these processes in the various domains. This will also give an indication of analogously processes in the other domain.

creativity Engineering is - at least it is claimed - a creative activity. But different amount of creativity and in different areas is needed in the different domains.

3.4 Attitudes and Basic Requirements

In previous publications major differences with respect to the users' attitude and expectations have been identified [19]. The main points of interest were:

- The user's role perception [8] in the process (from 'creative genius' to 'rule-obeying slave')
- The purpose of the process (from 'helping humans' to 'eliminating humans')
- The requirements on process documentation (from 'the result justifies the means' to 'the journey is the objective')
- Security requirements and back-up needs over system crashes (from 'let's just do it again' to 'it will never be the same again').
- control over process observation (from 'very strict' to 'very liberal').

- The amount of parallel work within a process respectively the amount of independent parallel processes (from 'numerous lean processes' to 'a few fat ones').
- The value and influence of creativity (from 'liberating from clerical drudgery' to 'stifling creativity').

Table 2 identifies some key differences.

Table 2. Attitudes towards EPSS

aspect	**software engineering**	**office automation**
user's perception	creative genius	law-abiding clerk
purpose of process	supporting human information processing	supporting human administration
creativity	fear of being oppressed	freed to think of the essential
process documentation	result justifies the means	the path is important
Security requirements and back-up needs	do not loose the creative contents	do not loose the information about taken steps going
control over process	liberal	very strict
parallelism of processes	a few large projects	many small processes

4 Analysis of Basic Processes: Refining Scheer's Y

The production of an industrial product is more than just an engineering effort. Numerous processes must assure that all 'accompanying processes' are also performed correctly. For the CIM-area Scheer's original Y (cf. Fig. 3, [52]) shows a simplified view of all departments involved. This 'Y' can be considered to be a general description for all types of 'production processes', be it for the production of material goods ('CIM') for the production of immaterial goods (e.g. software) or for the creation of immaterial decisions etc. (e.g. work flow). It identifies three major domains of Computer Integrated Manufacturing:

- a Production Planning and Control system (PPS),
- a Computer Aided Engineering System (CAE) and
- a Computer Aided Manufacturing System (CAM)

If we want quality processes, then the Deming-model is a good starting point. Mirroring Deming's improvement cycle means that we will have the four components as shown in Fig. 4.

If a process is enacted there are actually several separate but intertwined processes needed. If we follow W.E.Deming [27] there are basically four steps for quality improvement. If we apply this to the processes involved in production we have to

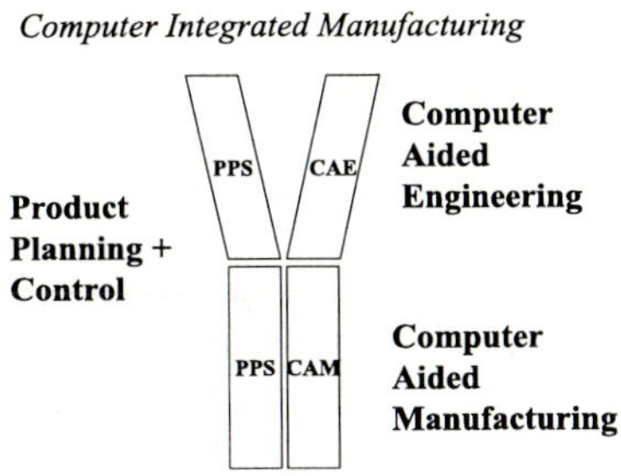

Fig. 3. Components of CIM (Scheer's Y)

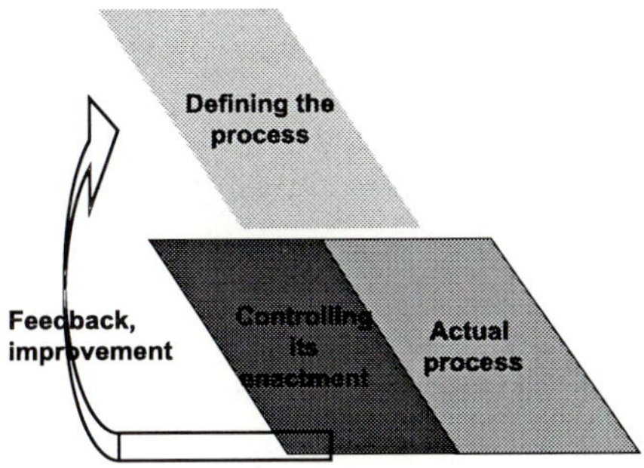

Fig. 4. Structure of a process

Plan We have to define the process we want to follow

Do We have to perform the process. This includes providing all the necessary materials and information for a successful enactment.

Check We have to control and check the enactment of the process. When the processes are enacted, their appropriate performance has to be controlled, introducing enactment control processes (cf. Fig. 6) and the various subprocesses of software engineering as described in ISO12207 [36] and ISO15504 [38]

Act This phase is concerned with feedback and prepares the process for the next iteration. For the time being we ignore this but we have to understand that is also an important part of the process, as especially the efforts in software engineering show [18][23][24].

For humans the distinctions is often not clear, large portions of the definition and steering process are often performed implicitly or even unconsciously. If such processes are subjected to automation also the accompanying processes have to be understood and enacted explicitly definition process for each of them, but this will not be further discussed.

Applying this idea to Scheer's Y we get a model of finer granularity. In the extended version we split the PPS process into two: A process concerned

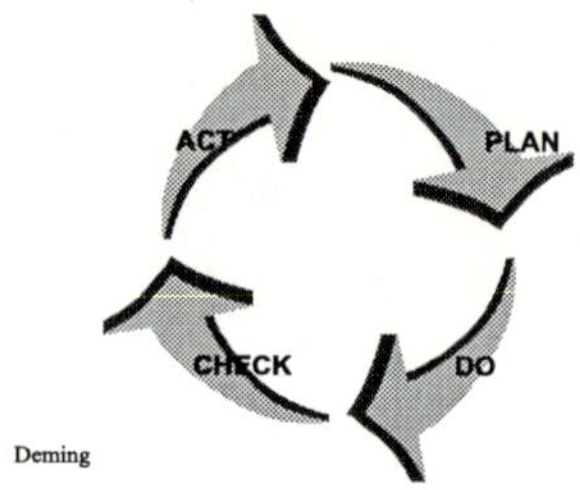

Fig. 5. Deming's improvement cycle

with establishing a market for the intended product by acquiring/identifying customers and a process for providing the necessary raw material to perform the actual engineering process. We will soon see that these two part of the process will have different importance in the discussed domains. Thus as Fig. 6 shows we should consider four essential processes contributing to the overall process. One process defines the 'product' to be produced, the other one is concerned with the demand for the product and as a consequence another process is concerned with providing the necessary material to produce the product, based on the information provided by the customer acquisition process.

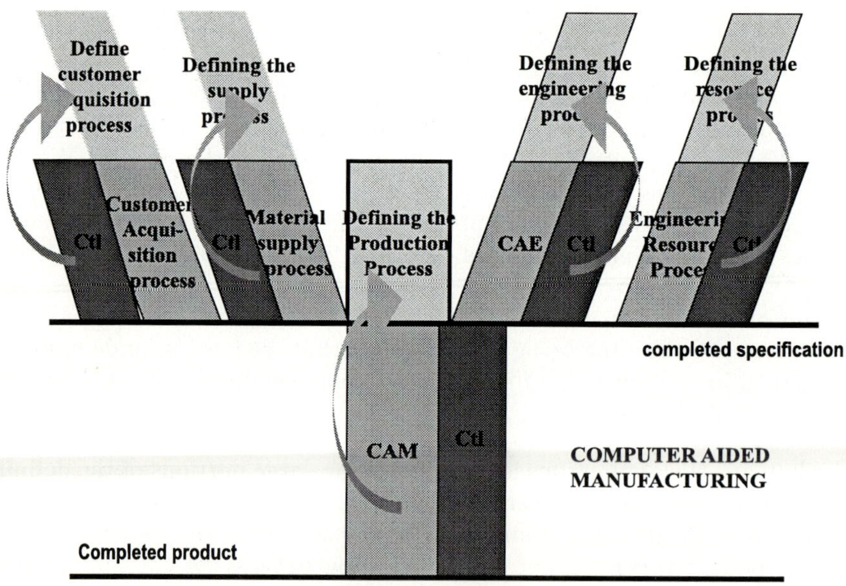

Fig. 6. Expanded Components of Scheer's Y

In the sequel we will discuss the importance and influence of these processes for the three domains identified above.

5 Matching Software and Office Automation to the Extended Y

5.1 Software Engineering

The key question is which parts of a software engineering process (e.g. [15][11] [10] are to be considered 'engineering' and which are to be considered 'production'. If we look at the individual steps of a software development process as most basically depicted in Boehm's famous 'Waterfall model', cf. [9]) one can argue convincingly that most of the software development process is *engineering* a system. Thus at least the early phases could be parallelled with the CAE-branch of the 'Y'.

Is there also a 'software production' in the sense of manufacturing? Let's start at the other end of the process. The preparation and distribution of the final copies of the developed software certainly is akin to 'production'. To answer that question we should investigate, what in the Y constitutes the boarder line between engineering and production: It is the complete construction documentation allowing a appropriately trained engineer to build the desired product any number of times. In software the cut-over point would probably a document at the level of low-level design, pseudo-code or equivalent. Based on it one has to produce an enactable product and to distribute it according to marketing advice etc.

5.2 Workflow

Given the definition of workflow, an EPSS supporting workflow would look like Fig. 7.

In the domain of office automation similar considerations as for software engineering are applicable. We could say that the main 'product' of a work flow operation is the successful performance of a business process. The 'product' of enacting some workflow procedure is some kind of document resulting from performing a business process. In many situations these are standardized routine processes like issuing a passport, certifying a state of affairs etc. In other cases more material has to be collected, like deciding on ordering some goods.

For the standard cases where an essentially automatic procedure is performed the enactment should be equated to the 'CAM' part of the 'Y'. In other cases, e.g. selecting tenders (cf. the software acquisition process described in ISO 12207 [36]) it is often at first necessary to define the documents to be used for this process, thus 'engineering' part of the end product. In this case a more or less large portion of the process would be associated with the CAE-branch of the Y.

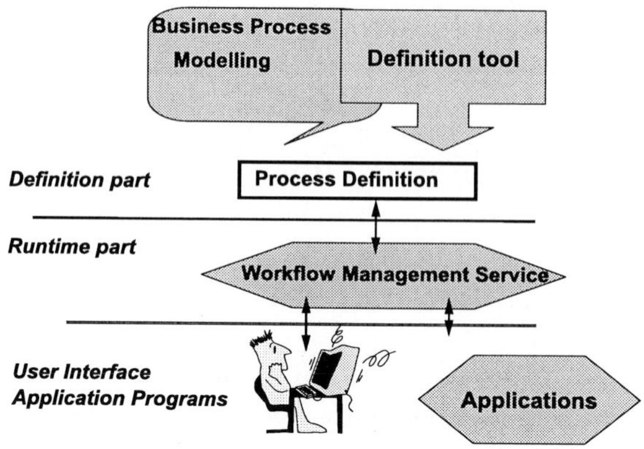

Fig. 7. Workflow Systems

Table 3. Subprocesses in PPS

	Supply Process (PPS)	
	Software Engineering	Workflow
Define	The supply process is rather simple, the major problem is finding good engineers. In the late phases of a software project, the supply challege is to find additional personnel for coding, for testing, correction and integration.	For Workflow the challenge is to express the desired business logic with an adequate representation enactable by the chosen workflow support system.
Enact. Ctl.	Given the simplicity of the supply process, control is not a major issues	Given the simplicity of the supply process, control is not a major issues
Enact.	routine management activity	routine management activity

5.3 Mapping Workflow, Software Engineering and the Extended Y

Tables 3, 4, and 5 comment the processes shown in Fig. 6.

While the original description of CIM by Scheer's famous 'Y' demonstrates the need for integration of production, it cannot be directly applied to software development or office automation. Both the original concept and our extensions restrict themselves to *process-centered* work. There is another large (and growing) domain where support systems typically help people perform their job *without* a predefined process. In the area of office automation this is covered by the concept of *groupware*. This dichotomy has to receive special attention. It seems that in software engineering it is rather apprarent by the gap between integrated CASE-tools and Process models [22].

Table 4. Subprocesses in CAE

	Engineering Process, CAE	
	Software Engineering	Workflow
Define	The choice of the 'correct' development process is of major importance [35][20][51] The 'correct' engineering process is still a major discussion point for theoreticians and practitioners. Methods and approaches still largely in flux. Often a different process is used for each project.	It seems that in the business world the options do not diverge considerably. There exist several well-established methods to express and record business processes [53][54]. Also the necessary components and their interplay are largely understood.
Enact. Ctl.	cost, deadlines, quality and development risk still largely out of control. Current attempts at capability assessments ([35][38] are still largely under discussion	Business has a tradition of controlling business processes, so theoretically there is not too much divergence. The challenge is to use EPSS to speed-up and optimize this control especially in view of economy, speed and globalisation, e.g. e-commerce.
Enact.	Due to the difficulties in definition and enactment control the actual engineering process is also wrought with problems. In many situations pseudo-creativity and ad-hocism win over orderly engineering principles.	The enactment has the challenge of using the chances modern information and communication technology

6 Conclusion

Software Engineering and Office Automation implement currently a very similar paradigm: Helping professionals to perform a better, more effective and more productive job by providing these professional with Electronic Performance Support Systems (EPSS). This similarity implies similar ways to implement an EPSS. There are, however, essential differences which warrant different designs and implementation, but still allow cross-fertilization of these domains. The paper tried to show that many of the concepts developed in one domain of the Information and Communication Technology-world carry - mutandis mutatis - to other areas.

In this paper extend the discussion of EPSS into various directions:

- Providing a more fine grained description of the individual processes (splitting them into 'definition', 'enactment' and 'enactment control'). *Retrospectively it seems useful to define also a supply process for engineering process, providing things like personnel, methods etc.*
- We have discussed the mapping of both software engineering and office automation onto the original concept (CIM). *It turned out that the interesting parts of software engineering cover essentially only the CAE-branch. For*

Table 5. Subprocesses in CAM

	Production Process (CAM)	
	Software Engineering	Workflow
Define	Most of the decisions are actually made in the engineering phase. It remains to define the sequence in which generation of code and testing is to be performed.	No large difference to defining the 'CAE'-process can be found.
Enact. Ctl.	Major problems result from errors in previous phases (Engineering process) which cause unexpected rework and design changes. The introduction of quality processes and the attempts to measure and improve the capability of a software producing organisation try to bring this phase under management control	Given an appropriate work flow environment most of the control is performed by the system. Introductory problems arise from a shifting of tasks (e.g. transport effort is greatly reduced, initial preparation of electronic documents needs additional work power [56]).
Enact.	To some extend automatic generation of the final product from design documents is available. The rest is mostly human work, thus error prone and not fully predictable.	not much difference to the Engineering process exists, since in the 'production phase' only less freedom of creating documents etc. can be noticeds.

workflow the mapping to either the CAE or the CAM branch depending on the flexibility/predictability of the process to be performed. But it seems that this is more a matter of definition than of inherent difference.

- We have also compared the attitude of team members when being supported by an EPSS. We have argued that - despite the overall similarity of the paradigm - there are significant differences in the domain of workflow and of software engineering.

The growing competition, globalisation and integration forces us not only provide computer support for all these activities via EPSS but also to consider merging and assimilating such environments for a seamless industrial operation. With this aim in mind this paper tries also tries to shed some light on the difference of automating software engineering versus office work.

References

1. ALTMANN, J. *Koooperative Softwareentwicklung* Reihe C, Technik u. Naturwissenschaften, Trauner, 1999 ISBN 3 85487 003 5.
2. BANDAT, K. *Process and Project Management in AD/Cycle* Proc. 31st GUIDE Spring Conference, Bordeaux, June, pp. 55–60.
3. BARTH, N. *Modellgetriebene Anwendungsentwicklung* IBM Nachrichten, 41:306:46–49.

4. BAUER, F.L. *Software Engineering* Information Processing, North Holland Publ. C, pp. 530.
5. BEDWORTH, D.D., M. HENDERSON , P. WOLFE *Computer Integrated Design and Manufacturing* McGraw-Hill Int. Edit. New York.
6. BENEDER, M. *Intelligent Work Cell Planning and Control, Prepr.* 4th Int. Workshop on Robotics, Alpe-Adria Region, Pörtschach, vol. 1, pp. 55–58.
7. BILL, D.T. *Transforming EPSS to Support Organisational Learning* WWW: http://www.centurion.sys.com/rtcl67.html, pp. 14.
8. BILL, D.T. *Contributing Influences on an Individual's Attidtude Towards a New Technology in the Workplace* WWW: http://www.centurion.sys.com/rtcl47.html, pp. 7.
9. BOEHM, B.W. *Software Engineering* IEEE Trans on Computers vol. C-25, no. 12, pp. 1226–1241.
10. BOEHM, B.W. *Understanding and Controlling Software Costs* Kugler H.J. (ed.): Information Processing, North Holland Publ.C, pp. 703–714.
11. BRÖHL, A.P. , W. E. DRÖSCHEL *Das V-Modell – Der Standard für die Softwareentwicklung mit Praxisleitfaden* Oldenbourg ISBN 3–486–22207–4.
12. BROOKS, F.P.JR. *No Silver Bullet – Essence and Accidents of Software Engineering* Kugler H.J. (ed.): Information Processing 86, IFIP Congress, pp. 1069–1076.
13. CCDB-VR2LINK *What's Performance Support?* WWW: http://www.vr2link.com/whatspss.html.
14. CHROUST, G., O. GSCHWANDTNER , D. MUTSCHMANN-SANCHEZ *Das Entwicklungssystem ADPS der IBM* Gutzwiller T., Österle H. (eds.): Anleitung zu einer praxisorientierten Software-Entwicklungsumgebung, Band 2 AIT Verlag München, pp. 123–148.
15. CHROUST, G. *Modelle der Software-Entwicklung – Aufbau und Interpretation von Vorgehensmodellen* Oldenbourg Verlag ISBN 3–486–21878–6.
16. CHROUST, G. *Software-Entwicklungsumgebungen – Synthese und Integration* Informatik-Spektrum, 15:7:165–174.
17. CHROUST, G. *Development Processes and their Enactment in Various Domains: A Summary and a Provocation* Cooke D., Hurley W.D., Mittermeir R., Rossak W. (eds.): Software Systems in Engineering at ETCE'95 Houston, ASME, New York vol. PD-67, pp. 183–189 ISBN 0–7918–1290–1.
18. CHROUST, G., (ed.) *Special Issue: ESPITI* Journal of System Architecture vol. 42 (1996) no. 8 ISSN 1383-7621.
19. CHROUST, G. , W. JACAK *Software Processes, Workflow and Work Cell Design* in: *Proc. EUROCAST 95, Innsbruck, Springer Lecture Notes on Computer Science 1996* ISBN 3-540-60748-X.
20. CHROUST, G. *What is a Software Process?* G. Chroust (ed.): Special Issue on ESPITI (European Software Process Improvement Training Initiative, Journal of Systems Architecture vol. 42(1996) no. 8, pp. 591–600 ISSN 1383-7621.
21. CHROUST, G. *Is there Groupware beyond E-Mail?* in: *Sugar, P., G. Chroust (eds.): CON'96: Future Information Technologies - 11th Austrian-Hungarian Informatics Conference*, pp. 137–145 Schriftenreihe der Österr. Computergesellschaft, No. 95, Oldenbourg 1996 ISBN 3-486-24059-5.
22. CHROUST, G. *Process Models, Integrated CASE-Tool Environments and Project Management - Birds of a feather or enemies for life?* in: *Montenegro S., Kneuper R., Müller-Luschnat G. (eds.): Vorgehensmodelle - Einführung, betrieblicher Einsatz, Werkzeug-Unterstützung und Migration - 4. Workshop 17-18 March, 1997, Berlin-Adlershof, GMD - Forschungszentrum Informationstechnik GmbH Berlin*, pp. 27–37 ISBN 3-88457-311X.

23. TEAM, SPIRE PROJECT *The SPIRE Handbook - Better, Faster, Cheaper - Software Development in Small Organisations*, chapter 10,11,12, pp. 131–200 Centre of Software Engineering Ltd, Dublin 9, Ireland ISBN 1-874303-02-9.
24. CHROUST, G. , P. GRÜNBACHER, (eds.) *Proceedings European Software Day, Milano, Sept. 1999* Österreichische Computer Gesellschaft Wien, 1999 ISBN 3-84403-199-8.
25. CORZILIUS, R., (ed.) *AD/Cycle – Ziele, Konzepte und Funktionen* Oldenbourg-Verlag München Wien.
26. DANGELMAIER, W. , W. FELSER *Vom Modell des Fertigungsprozesses zum Modell der Fertigungssteuerung* EMISA Forum 1995 no. 1, pp. 31–33.
27. DEMING, W. E. *Out of the Crisis* Massachusetts Inst. of Technology, 1986.
28. DIEBOLD, J. DOBERER, K.K. *Die automatische Fabrik - Ihre industriellen und sozialen Probleme* Nest Verlag, Frankfurt 1956.
29. DOWNS, E., P. CLARE , I. COE *Structured Systems Analysis and Design Method* Prentice Hall, Englewood Cliffs SSADM.
30. DOWSON, M., (ed.) *Iteration in the Software Process, 3rd International Software Process Workshop, Breckenridge Colorado* Nov. Participants' Proceedings.
31. FISCHER, O. , R. HORN *Electronic Performance Systems* Comm ACM vol. 40 (1997), no. 7, pp. 31–32.
32. GARG, P.K. , M. JAZAYERI, (eds.) *Process-Centered Software Engineering Environments* IEEE Computer Soc Press, USA ISBN 0-8186-7103-3.
33. GRÄF, M. (ED.) *350 Jahre Rechenmaschinen* Carl Hanser München.
34. HUENKE, H. (ED.) *Software Engineering Environments* Proceedings, Lahnstein, BRD, North Holland.
35. HUMPHREY, W.S. *Managing the Software Process* Addison-Wesley Reading Mass. 1989.
36. ISO, (ED.) *ISO-12207, Information Technology, Life Cycle Processes* Int. Org. for Standarization, ISO 12207.
37. ISO, (ED.) *ISO/IEC 1539:1997 Information technology – Programming languages – Fortran* Int. Org. for Standarization, ISO 1539, 1997.
38. (ISO) *ISO/IEC 15504: Software Process Assessment: Parts 1 - 9* Technical Report - Type 2, ISO/IEC, 1998 (approved for publication).
39. ISO, (ED.) *ISO 9000-1:1994 Quality management and quality assurance standards – Part 1: Guidelines for selection and use* Int. Org. for Standarization, ISO 9000, 1994.
40. JABLONSKI, S. *Workflow-Management-Systeme: Motivation, Modellierung, Architektur* Informatik Spektrum, 18:1:13–24.
41. JACAK, W. ROZENBLIT J. *Workcell Task Planning and Control* Cybernetics and Systems'92, ed. R. Trappl, World Scientific Press, Vienna, Singapore,, pp. 1479–1486 ISBN 981-02—X.
42. JACAK, W. *ICARS – Simulation-based CAPP/CAM System for Control Design in Manufacturing Automation* Wloka D. (ed.): The Handbook of Robot Simulation Systems John Wiley, London (in print).
43. JACAK, W. SIEROCKI, I. *Simulation-Based Intelligent Control Design in Manufacturing Automation* Crowley J., Dubrawski A. (eds): IRS-93, Intelligent Robotic Systems, Zakopane, Poland, July, pp. 81–90.
44. JACAK, W. *Hierarchical Control Systems in Manufacturing Automation* Proc. of 4th Conf. on Robotics, Wroclaw, Poland PWR-Press.
45. KLENGEL, H. *Handel und Händler im alten Orient* H. Böhlaus Nfg. Wien/Köln/Graz 1979.

46. MERBETH, G. *MAESTRO-II – das Integrierte CASE-System von Softlab* Balzert H. (ed.): CASE – Systeme und Werkzeuge 4. Auflage, B-I Wissenschaftsverlag, pp. 215–232.
47. NISSEN, H.J., P. DAMEROW , R. K. ENGLUND *Archaic Bookkeeping* The University of Chicago Press, 1993 ISBN 0-226-58659-6.
48. PETROVIC, O. *Workgroup Computing – Computergestützte Teamarbeit* Physica-Verlag 1993.
49. RUTISHAUSER, H. *Automatische Rechenplanfertigung bei programmgesteuerten Rechenmaschinen* Mitt Inst. f. Angew. Math., ETH Zuerich No. 3.
50. SAMMET, J.E. *Programming Languages: History and Fundamentals* Prentice-Hall Englewood Cliffs.
51. SCHÄFER, W. (ED.) *Software Process Technology – 4th European Workshop EWSPT'95 Noordwijkerhout, April* Springer Lecture Notes No. 913, Springer Berlin-Heidelberg ISBN 3–540–59205–9.
52. SCHEER, A.W. *CIM – Der computergesteuerte Industriebetrieb* Springer Berlin.
53. SCHEER, A.W. *ARIS - Modellierungsmethoden, Metamodelle, Anwendungen* Springer Berlin 1998 ISBN 3-540-64050-9.
54. SCHEER, A.W. *ARIS - Vom Geschäftsprozeß zum Anwendungssystem* Springer Berlin 1998 ISBN 3-540-63835-9.
55. SCHULTZ, (ED.) *Office Automation – Technologies for the s* Computer Technology Res. Corp., Charleston, SC.
56. SCHWAIGER, A. *Erfahrungen in der Wüstenrot Versicherungs-AG mit dem 'papierlosen' Büro* Techn. Report , IPO-Kompetenzzentrum, Kepler Univ. Linz, 3. Workshop, 19.1.1995.
57. TULLY, C. (ED.) *Representating and Enacting the Software Process* Proc. 4th Int. Software Process Workshop, May ACM Software Engineering Notes vol. 14, no. 4.
58. VON UTHMANN, C. , J. BERGERFURTH *Ein Metamodell als Grundlage zum Workflow-basierten Management von Produktionsprozessen* EMISA Forum 1999, no. 2, pp. 37–50.
59. WFMC, ED. *Glossary, – A Workflow Managment Coalition Specification* Workflow Management Coalition, Belgium Nov. 52 pp.

A Framework for the Elicitation, Evolution, and Traceability of System Requirements

Paul Grünbacher[1] and José Parets-Llorca[2]

[1] Systemtechnik und Automation, Johannes Kepler Universität Linz,
Altenbergerstr. 69, 4040 Linz, Austria,
pg@sea.uni-linz.ac.at

[2] Lenguajes y Sistemas Informaticos, Universidad de Granada,
Avda. Andalucia 38, 18071 Granada, Spain,
jparets@ugr.es

Abstract. In the software development process ideas and objectives are evolved into semiformal and formal representations allowing later execution of the system on a computer. Knowledge about the system is represented in various forms (e.g., natural language, UML models, source code, etc.). The evolutionary change of knowledge representations is a challenging issue. In this paper we propose a framework applicable to the elicitation and evolution of requirements in the software engineering process. We discuss architectural considerations for supporting traceability between artifacts of different CASE tools and present a partial implementation of the approach.

1 Introduction

In the software engineering process knowledge about the system to be developed is gathered and evolved using different representation mechanisms. Informal representations based on natural language are suited to involve non-technical stakeholders early in the life cycle. Semiformal representations which clearly establish syntactical rules and formal representations establishing syntactical and semantical rules are used mainly by technical stakeholders to design and develop the system (e.g., the UML class model, UML behavior model, a Java class). Our research interest is how knowledge can be represented and evolved integrating different representation mechanisms.

The inherent iterative and evolutionary characteristics of software development have led to the development of iterative and evolutionary life-cycles models [5,16]. However, modelling methods and tools supporting these processes often do not support evolution, i.e. they do not consider the inherent evolution of the produced models. As a result we are often lacking "connectors" between different methods and tools used in the life cycle.

The following scenario illustrates the discussed problems: A system analyst creates a UML Use Case Model, based on system requirements captured in natural language. The model contains instances of the UML artifacts *Actor* and *Use Case*, related by generalisations and associations. After finishing the Use

F. Pichler, R. Moreno-Díaz, and P. Kopacek (Eds.): EUROCAST'99, LNCS 1798, pp. 394–406, 2000.

Case Analysis a designer creates class models and state transition diagrams (STD) based on the Use Case Model. Some of the use cases will be evolved into class methods, which may be used as events in the STD. Later in the process the system analyst discovers that some assumptions in the use case model do not hold. So the designer has to create a new variation of the class model based on the new assumptions. In doing this, she has to take into account the dynamic, evolutionary constraints between artifacts of the Use Case model, the class model and the behaviour model (STD); i.e. a class should not be deleted if it corresponds to some use cases, the name of a method should not be changed if it participates in an event, ... The designer has to reconsider the models to produce a second version that considers the changes of the system analyst and maintains the coherence and integrity to her own models.

Although most CASE tools partially ensure the required integrity checks, a systematic approach does not exist and the lack of compatibility and communication between different methods and tools is evident. This becomes especially true if we scale this scenario to multiple technical and non-technical stakeholders using multiple representations. Especially in the early phases of software development a semantic distance between requirements and architectures exists. The semantic distance and mismatch between tools decreases in later phases of the life-cycle, probably due to the similarity of concepts used (e.g., Round-trip engineering between UML models and OO languages).

We selected the requirements engineering process to investigate the problems discussed above. The elicitation of requirements relies on successful involvement and interaction of multiple stakeholders and usually adopts natural language to represent knowledge. Evolution of informal knowledge on the one hand requires the ability to automatically transform representations into different representations. On the other hand we need to trace the path of evolution for certain semantic constructs.

This paper is structured in the following manner: Section 2 describes a case study we carried out to gain a better understanding of the challenges and issues. Section 3 presents a framework addressing the problems. Architectural considerations for interoperating software tools are proposed in Sections 4 and Section 5 demonstrated a partial implementation. We finish with conclusions and further work in Section 6.

2 WinWin and UML: A Case Study

In a previous experiment described in [14] we used the WinWin negotiation model [4,5] and the Unified Modelling Language (UML) [15] to capture, elicit and evolve requirements for a student registration system. Our process was organised in three steps carried out iteratively (see Fig. 1). Users and developers were involved in the negotiation stage. Requirements and System Engineers carried out the modelling stage.

"Negotiation" We captured and negotiated the requirements using the WinWin groupware tool involving success-critical stakeholders like customers, users, and developers.

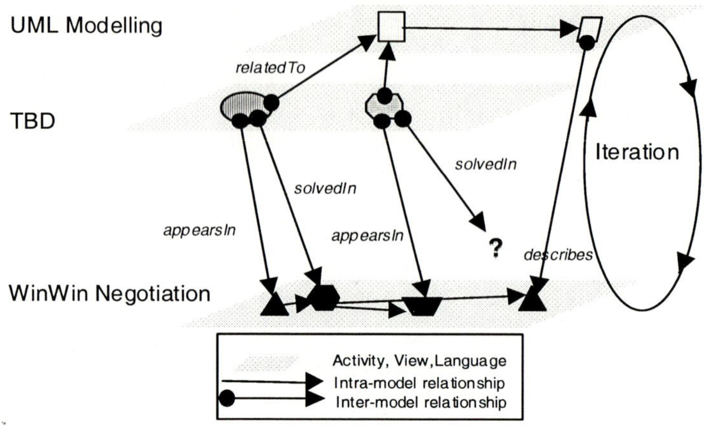

Fig. 1. Stages in the RE Process

"To-Be-Determined" After the initial negotiation process a Requirements Engineer analysed the negotiation results and identified open issues and problems captured in TBD artifacts (see Fig. 3).

"Modelling" The WinWin artifacts were then used together with the TBD artifacts to develop a model of the system using UML.

The major conclusions of this experiment can be summarised as follows:

Iteration and Evolution: Work products evolve based on previously achieved results. The requirements engineering process can be characterised as an iterative and evolutionary process [3,5,12,13]. While the WinWin system [5] improves the identification and negotiation of requirements, the transition to architecture and design (expressed in semiformal and formal representations) is still based on intuition and experience.

Importance of informal, "lightweight" approaches: In order to involve stakeholders inexperienced in system modelling and computer technology we have to adopt easy to use mechanisms to facilitate the interaction. The process should be performed intuitively benefiting from the semantic power of natural language. Crucial decisions in the process should be preserved to be available in further stages and projects.

"TBD" activity: Requirements captured in natural language can be evolved into a more formal representation through structuring and consolidation. This step also raises further issues and problems in the requirements and allows the establishment of useful links supporting traceability. Identified problems and links can be used in further refinements of the specification or in case of changes to the specification (design problems, maintenance, ...). Multiple views are required to involve technical and non-technical stakeholders and to re-negotiate the problems discovered by the requirements engineer during TBD to get a complete and consistent set of requirements.

3 Framework Characteristics

The previously described Requirements Engineering process involves multiple stakeholders, modelling languages, and tools. A framework useful to address evolution in the Requirements Engineering process should be based on such an iterative process and augmented with additional concepts supporting an operational management of evolution. In this section we describe the terms used throughout this paper using examples from the case study and show how our framework addresses static and evolutionary relationships as well as traceability and translatability.

3.1 Terms

Modelling language (representation language) Symbols, syntactical and semantical rules used for the elaboration of models. Examples: Unified Modelling Language, WinWin negotiation model

Model A representation of a system using a modelling language. Examples: UML class model for Project X, WinWin negotiations for Project X

Meta-model A model elaborated in a modelling language defining a set of meta-artifacts and -relationships. A meta-model defines the items available in a modelling language. Examples: WinWin negotiation model in UML, UML meta-model defined using UML

Meta-artifact A class of item used in a modelling language. Examples: "Use Case", "Actor", "WinCondition"

Artifact Instance of a meta-artifact. Examples: "Register for class"; "Student"; "Response time $<$ 10 seconds"

Notation A set of graphical symbols used to depict meta-artifacts and -relationships (e.g., arcs, icons, lines, ...). Examples: Rectangle to depict a UML class

View A subset of a model depicting a certain system aspect. Examples: UML Use Case Diagram "overall use case diagram"; WinWin Rationale View for project X

Tool Software capability supporting a notation and modelling language. Examples: UML CASE tool; WinWin Groupware Tool

(Intra-)model evolution Modifications of a system model using the same language (meta-model) resulting in a new version of a model. Example: Evolve domain class model version 1 to domain class model version 2. Typical model evolution operations are replace artifact, create new artifact version, delete artifact, or modify artifact.

Inter-model evolution Modifications requiring a change of the language (meta-model). Example: Use Case model developed based on WinWin agreements.

Intra-model relationships Relationships between the meta-artifacts of a modelling language. Examples: Issue $\rightarrow$ WinCondition, Actor $\rightarrow$ Class

Inter-model relationships Relationships between meta-artifacts used in different modelling languages (see Fig. 1). Example: WinCondition *describes* UseCase

We have to stress that the meta-level is always relative, and an infinite meta-meta-... may exist. We consider only two levels. Following the naming conventions described above we have to use the term meta-relationship instead of relationships, but we preferred to simplify the presentation.

Intra-model relationships are internal to a modelling language (tool). Inter-model relationships are partially language/tool independent and are our main interest. We distinguish static and evolutionary relationships.

3.2 Static Relationships

Static inter-model relationships link meta-artifacts of different modelling languages. As they are time-independent we describe them in a static meta-model (SMM) not considering evolution.

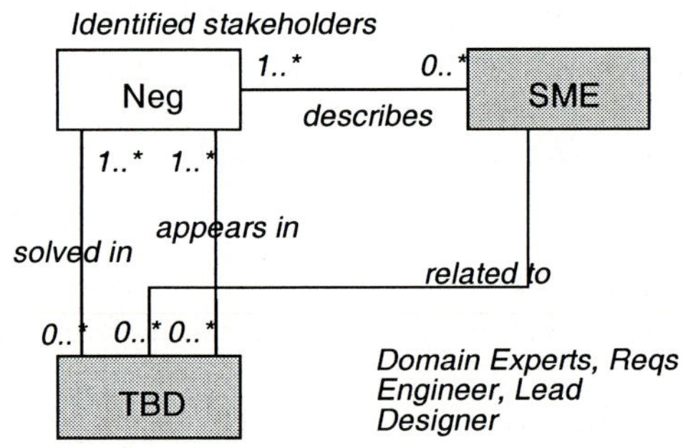

Fig. 2. Static top level meta-model

The UML Class Diagram in Fig. 2 shows main superclasses and static inter-model relationships. Each superclass corresponds to one stage of the case study process (Neg: WinWin Negotiation; TBD: To-Be-Determined; SME: UML Modelling). The class *Neg* represents the super class of the meta-artifacts used during the negotiation process. Each meta-artifact can have the following associations:

describes implies that a modelling artifact is explained by a negotiation artifact.
appears_in shows a problem detected in a negotiation artifact during the TBD stage
solved_in shows a problem detected in a negotiation artifact which has been solved in other negotiation artifact.
related_to links a problem to an SME artifact.

These relationships are established during the TBD activity to create links between artifacts of different languages/tools. During the experiment we identified major subclasses of TBD (Fig. 3):

Incompleteness indicates that insufficient information about a SME artifact has been specified during the negotiation process.

Conflict describes a constraint about a SME artifact existing between specifications made in different NEG artifacts.

Unsolved problem describes an open question about a SME artifact requiring further (re-)negotiation.

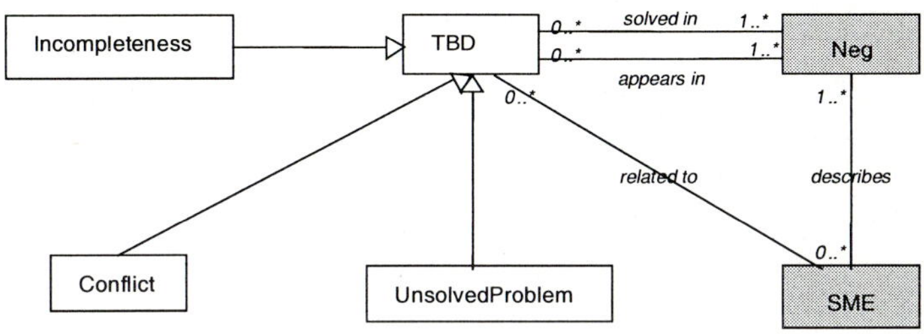

Fig. 3. TBD Meta-artifacts

Although we consider these classes as part of the SMM, as they do not impose restrictions on the evolution of the different artifacts in the different models they are very important in supporting the evolutionary process. Unsolved problems imply the need for further iterations in the Negotiation, TBD and Modelling activities. These artifacts are therefore evolutionary from the point of view of the process although not evolutionary from the point of view of the obtained products.

3.3 Evolutionary Relationships

The objective of the evolutionary meta-model (EMM) is to describe how meta-artifacts specified in the SMM can evolve depending on the inter-model and intra-model static relationships. Each meta-artifact is associated to a state transition diagram showing possible and allowed transitions depending on the state of the artifact in the course of development. We use UML State Diagrams to express the life cycle of these meta-artifacts as they allow us to express parallel states.

The behaviour of the meta-artifacts proposed in the SMM can be modelled in three parallel substates: NEG substate, SME substate and TBD substate. For each meta-artifact its corresponding substate is empty and has to be specialised at the subclass level.

An example for NEG class is shown in Fig. 4. The NEG class has the parallel substates SME and TBD. The concrete state of an artifact will depend on the established relationships in the SMM.

The NEG substate needs further specialisation in subclasses of NEG (see Fig. 5). Internal dynamic relationships are therefore defined inside and between

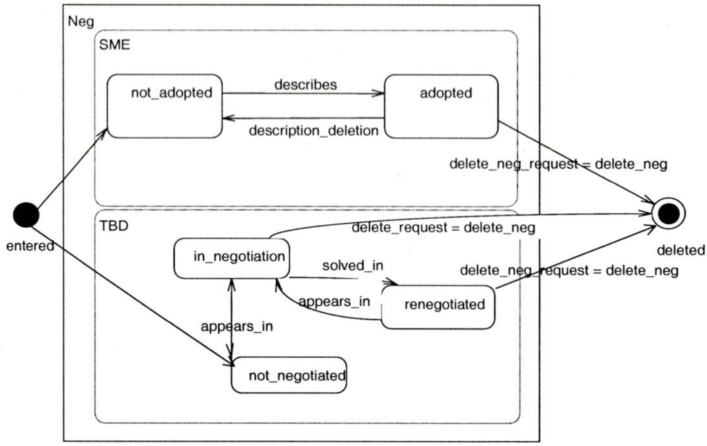

Fig. 4. Substates of Negotiation Meta-artifacts

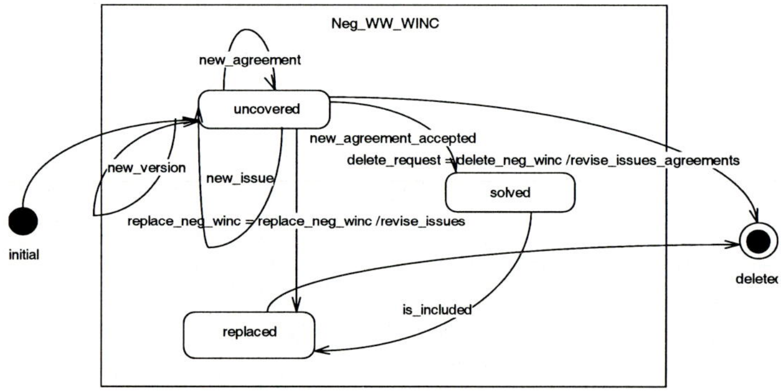

Fig. 5. States for Meta-artifact WinCondition

the internal artifacts of a tool, i.e. they are a intra-tool concern. External dynamic relationships are being modelled using the external static relationships, i.e. they have to be modelled when tools communicate. This absence of coupling between external and internal relationships, due to the hiding of the internal relationships, allows a separation of concerns and supports tool interoperability (see Section 4).

3.4 Traceability

Traceability refers to the ability to follow the evolution of an artifact in both a forward and backward direction and is important for intra-model evolution as well as inter-model evolution [7]. Traceability allows following an initial idea, through its development and specification, to its subsequent deployment and use, through periods of on-going refinement and iteration in any of these phases [9].

Table 1. Examples for Traceability for WinWin and UML with Traceability Attributes

WinWin artifacts	Accuracy	Completeness	UML artifact
WinCondition W17	Partially	Partially	Class "Student"
WinCondition W12	Largely	Fully	Use Case "Register for Class"
Option O4	Fully	Largely	Deployment Diagram "Deployment"

In our framework traceability is addressed by user established links describing relationships detected during the TBD stage (e.g., WinCondition → Class). The analyst is responsible to specify the links, there is no automatic conversion, however typical trace paths are known (e.g., WinCondition → Issue → Option → Agreement → Use Case → Class).

In the case study traceability has been addressed through the association *describes* between NEG and SME classes. We improved the semantic value of these simple associations by using further attributes to describe the links to specify the degree of "mapping". An example is given in Table 1 using the attributes completeness, accuracy, and a simple scale (fully, largely, partially, not).

3.5 Translatability

Translation, in a general sense, means the transformation of a model from one language to another. The notion of translatability implies the degree of achievable translation between the concepts of two languages. In our case, we reduce the concepts managed to the meta-artifacts used in the modelling language, so translatability can be defined as the degree of semantic equivalence between the meta-artifacts of two modelling languages. This is important for inter-model evolution. Beyond traceability translatability allows to establish more specific relationships that are be defined by the user as *Implemented_by, Included_in, Referenced_by, ...* also using scaled attributes. We need to establish a-priori mappings of semantic equivalence and specify links in advance to describe known possible mappings between meta-artifacts of modelling languages. Translatability can be achieved in two ways:

Manually During the TBD step specific qualifications of the describes relationships can be used to improve the semantic value of this relationship. This can be achieved using an ontology (thesaurus) allowing the definition of semantic equivalence between items. This step and the relationships established depend on the objectives of the modeller and of the tools involved. An ontology should be provided for each pair of tools/models in order to characterise the describe relationship. Examples are the qualifications *Included_in, Referenced_by, Is_Kind_Of, ...*

Automatically Later converters can be built based on the mapping for automatic conversion. Mapping between instances of tool meta-artifacts (ArtifactInstanceTool2 can be created based on ArtifactInstanceTool1). The conversion is not bi-directional and often partial. Examples are the qualifications

Table 2. Examples for Translatability for GroupSystems and UML with Attributes

Tool1	Meta-artifact1	Trust	Tool2	Meta-artifact2
GroupSystems	Use Case Analysis Level 2	Fully	UML	UseCase
GroupSystems	Included in	Fully	UML	Use Case
UML Tool	Actor	Largely	UML Tool	Class
GroupSystems	WinWinTree Level 2	Fully	UML	Class: stereotype WinCondition

Implemented_ by (UML Class → C++ class), *Extracted_from* (C++ class → UML class).
Both approaches can be mixed if the ontology defines the qualifications and the possibilities of automatic conversion. Table 2 shows some examples for translatability taken from the initial implementation of our approach relating artifacts in the GroupSystems negotiation system and artifacts of the UML (see Section 5).

3.6 Framework Generalisation

The framework previously described implies a negotiation activity/language, a system modelling activity/language and an intermediate step called TBD. Fig. 6 depicts a generalisation of this framework to any kind of tools and modelling language used during the software development process. In this generalisation the TBD step/tool is considered as another tool. Its aim is mainly to support the process of transition from one tool to another showing the general problems that should be further investigated. This role cannot be assigned to any usual CASE tool because they do not manage transitions. The main ideas in Fig. 6 are:

– The Meta-artifacts of different tools can be related by an association, which is qualified by a *LinkSpecifier.* This Specifier determines the traceability and translatability properties of the link. Artifacts of different modelling languages and tools can be traced and translated following the link specifiers.
– Each Artifact is simultaneously in different parallel states, belonging to the EMM of the association between Tool_MetaArtifacts, i.e. each pair of Tool_MetaArtifact will have an EMM determining the allowed states/transitions of its corresponding artifacts (instances).

Although this generalisation needs for further research is very interesting because simplifies the framework and includes the possibility of linking different tools by means of the establishment of the LinkSpecifier and the EMM. Especially, we have to test if the EMMs for each pair of meta-artifacts are always symmetric: *not_adopted/not_adopts; adopted/adopts.*

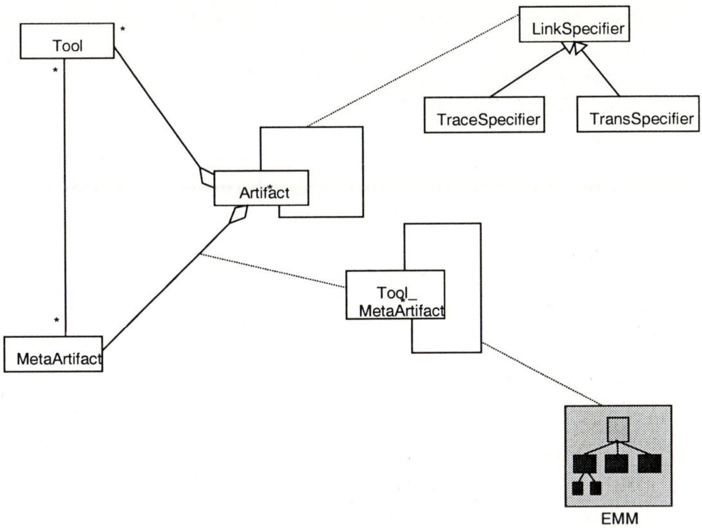

Fig. 6. General Traceability/Translatability Meta-Model

4 Architectural Considerations

Based on our proposed framework we discuss some basic interface requirements for tools to support synchronisation and maintaining consistency in evolutionary development. Existing approaches for method and tool integration use one the following approaches or a combination of them:

- Interchange formats like CDIF [6] or XMI [18]: While these approaches support the exchange of models between different tools there is no support for the evolution of meta-artifacts.
- Meta CASE technology [2,10,14] supports customisable meta-model and customisable notation to develop or enhance modelling approaches. Although the construction of new tools integrating different modelling languages is possible (cf. [14]), the problem is the difficult integration of existing tools.
- Repositories like Softlab's Enabler [17] or the Microsoft repository [11] provide a "database" for software engineering artifacts. Such environments support a customisable meta-model and provide multiple ways to access artifacts to provide basis for a tool suites.

There will be no universal semantics covering all combinations of tools. We prefer ad-hoc correlations between pairs of modelling languages and a general framework to cover further ad-hoc developments. Our approach towards realisation and implementation could be called "agent-based" supporting the following characteristics [1]:

- Connect independent tools through minimal interface and keep knowledge in the tools

- Build transducers and wrappers to tools ('agentification' [8])
- Adopt standard communication mechanisms (COM, KQML)
- Adopt standard interchange format (XML, KIF)
- Use facilitators for coordination and interaction of tools

Intra-model relationships and dynamics are private to each tool (see Fig. 5). External tools do not need to know internally handled relationships and states of an artifact. Each model/tool needs to know the use/realization of its artifacts (inter-state). Consequently each artifact has two parallel sub-states: The intra-state handled by its tool and the inter-state which should be public. In this way the required "API" can be kept small and writing tool wrappers is simplified.

What does this architecture imply for the design of the tools? They should be prepared to ask if one of its internal operations is allowed by an external future agent: e.g. to delete an artifact in WinWin the external agent should verify the external state of this artifact. Probably the kind of operations which require these queries are operations related with the identity of the artifacts (creation, deletion, renaming, ...) and general problems. Furthermore, the tools should be prepared to export its artifacts.

5 Implementation

We have partially implemented the concepts in order to:

- *Test and refine the framework:* Obviously, the framework should be instantiated for different representation instruments, taking into account the limited possibilities of translation between tools supporting different levels of granularity and, in most cases, different semantic constructs [14].
- *Explore architectures:* Evaluate state-of-the-art mechanisms (like COM and XML) to support tool interoperability as discussed in Section 4.
- *Explore traceability and translatability between informal and formal representations:* As pointed out before it is very difficult to establish semantic equivalencies due to the semantic gap between the meta-artifacts managed in the first stages of the software development. We believe that an adequate traceability mechanism is necessary to bridge these first phases. In further software development steps (design, implementation, test) more precise semantic equivalencies can be established defining new association between the involved meta-artifacts. For instance, between design and implementation meta-artifacts: *Implemented_by* and *Implements.*
- *Improve stakeholder interaction:* We replaced the WinWin Groupware tool used in the case study to explore the application of a Group Support Systems (GSS) to improve stakeholder involvement and interaction in Software Engineering and to provide support for WinWin related activities. The GSS Ventana GroupSystems supports group activities like idea generation, categorisation and prioritisation. This commercial package has been tailored to support WinWin related collaborative activities like gathering and elicitation of system requirements, categorisation and prioritisation of requirements, development of taxonomies, negotiation and conflict resolution following the WinWin approach, and the development of use cases.

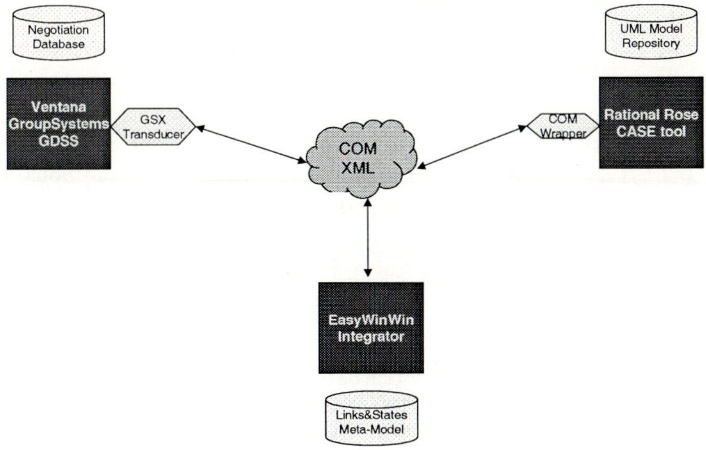

Fig. 7. Architecture

Fig. 7 shows the major elements the tool suite. The EasyWinWin integrator allows to establish traceability links manually and to initiate translations between WinWin artifacts in Ventana GroupSystems and UML artifacts in the Rational Rose CASE tool.

6 Conclusions

Given the available number of approaches and the differences in formality between modelling languages and tools we think that bridging gaps is often more important than inventing new approaches. The work shows that translatability, traceability and evolution using different modelling approaches is feasible. It is influenced by the degree of formality and the possibility of establishing semantic equivalence between modelling approaches. Semantic equivalence can be established in an intermediate step by an engineer familiar with the modelling approach and the domain using adequate computer-based tools.

We believe that frameworks integrating different ways of representing knowledge are useful in the system development process. Such bridges support an iterative and evolutionary development process, and help to integrate different stakeholders (users, developers, ...) by providing different views on the evolving model.

Acknowledgements. This research is supported by the Acciones Integradas program of cooperation between Austria (ACCINT 8/97) and Spain (HU97-0044).

References

1. Agent-based engineering Group: URL http://cdr.stanford.edu/ABE/
2. Alderson A.: Meta-CASE technology-Software Development Environments and CASE Technology, Lecture Notes in Computer Science 509, Springer, 1991.
3. Arthur L.J.: Rapid Evolutionary Development, John Wiley and Sons, Ltd. 1992.
4. Boehm B., In, H.: Identifying quality requirement conflicts. IEEE-SOFTWARE, 13(2):25ff., 1996.
5. Boehm B., Egyed A., Kwan J., Port D., Shah A., Madachy, R.: Using the WinWin Spiral Model: A Case Study, IEEE Computer, pp. 33–44, July 1998.
6. CDIF: CASE Data Interchange Format: EIA (Electronic Industries Association), URL: http://www.eigroup.org/cdif/
7. Dömges R., Pohl K.: Adapting Traceability Environments to Project Specific Needs, Communications of the ACM, Vol. 41/No. 12, pp. 54–62, 1998.
8. Genesereth M.R., Ketchpel S.P.: Software agents. Communications of the ACM, 37(7):48–53, 1994.
9. Gotel, C.: Contribution Structures for Requirements Traceability. Dissertation, University of London, Department of Computing, August 1995.
10. Grünbacher P.: Meta-CASE Based Realisation of Design Rationale Management Systems. In: Cooke, D. and Krämer, B., Sheu, P.Y., Tsai, J., and Mittermeier, R. (Eds.), IDPT 2nd International World Conference on Design and Process Technology, Austin, TX, pp. 445-451. SDPS, ISSN 1090-9389, 1996.
11. Microsoft Repository: URL http://msdn.microsoft.com/repository/
12. Parets J., Torres J.C., A Language for Describing Complex-Evolutive Software Systems, In: Computer Aided Systems Theory-EUROCAST'95. Pichler, F., Moreno-Diaz, R. (Eds.), Springer-Verlag, LNCS 1030, pp. 181–197, 1995.
13. Parets J., Torres J.C., Software Maintenance versus Software Evolution: An Approach to Software Systems Evolution. IEEE Conference and Workshop on Computer Based Systems (ECBS'96). Friedrichshafen, pp. 134–141, 1996.
14. Parets, J., Grünbacher, P., Capturing, Negotiating, and Evolving System Requirements: Bridging WinWin and the UML, In: Proc. EUROMICRO'99 Milan, vol. 2, pp. 252–259 September, 1999, ISBN 0-7695-0321-7.
15. Rumbaugh J., Jacobson I., Booch G., Unified Modeling, Language Reference Manual, ISBN: 0-201-30998-X, Addison Wesley, December 1997.
16. Rational Unified Process, URL http://www.rational.com/products/rup/
17. Enabler URL http://www.softlab.com
18. XML Meta data interchange URL http://www.omg.org/xml/

Development of a Precision Assembly System Using Selective Assembly and Micro Machining
(Evaluation of Combinatorial Optimization Method for Parts Matching)

Yasuhiro Yamada [1], Yoshiaki Komura [1], Junnosuke Mizutani [2], and Ikuo Tanabe [3]

[1] Dept. of Mechanical Engineering, Fukui University
9-1, Bunkyo 3-chome,Fukui-shi, Fukui 910-8507, Japan
Tel: +81-776-27-8545, FAX: +81-776-27-8539
{yyamada, komura}@mech.fukui-u.ac.jp
[2] Dept. of Electronic Control Engineering, Toyama National College of Maritime Technology
1-2, Ebie-Neriya Shinminato-shi, Toyama 933-0293, Japan
Tel: +81-766-86-5259, FAX: +81-766-86-5110
mizutani@toyama-cmt.ac.jp
[3] Central Machine Shop, Nagaoka University of Technology
1603-1, Kamitomioka,Nagaoka-shi, Niigata 940-2188, Japan
Tel: +81-25847-9862, FAX: +81-258-47-9770
tanabe@mech.nagaokaut.ac.jp

Abstract. In this paper we propose a new combinatorial optimization method for parts matching in a precision assembly system. This method combines selective assembly with micro machining so as to produce high precision assembled units. From the simulation results, it is presented that the system is effective in a small lot production, where higher assembly rates can not be attained with conventional methods, in such that a higher rate is obtained if micro machining is used alongside.

1 Introduction

With selective assembly based on the combinatorial optimization method, a combination of parts is optimized so that high quality assembled units are obtained in a lot unit at the maximum assembly rate while dimensions of each part are referred to appropriately[1-4]. The assembly method proposed in this study combines selective assembly based on a combinatorial optimization method with micro machining so as to produce high precision assembled units. This method is applied to the high precision assembly system where hole parts are combined with shaft parts. Further, micro machining used for the correction of the dimensions of the parts is taken into consideration during the optimization of the parts combination, and it is attempted to verify the performance by simulation.

F. Pichler, R. Moreno-Díaz, and P. Kopacek (Eds.): EUROCAST'99, LNCS 1798, pp. 407–413, 2000.

2 A Precision Assembly System Using Selective Assembly and Micro Machining

This system consists of (1) a diameter measuring instrument by which the hole diameter of the hole parts and the shaft diameter of the shaft parts D_{Hi}, D_{Sj} (mm) are measured, and stored together with part numbers i and j, (2) a computer through which the optimum combination of the parts is determined by way of the combinatorial optimization method while parts' data are referred to in a lot unit (lot size N), (3) a micro machining equipment by which the shaft diameters of the shaft parts at assembly are corrected so that the optimum combination may be obtained according to the instructions given by the computer, and (4) a selective assembly equipment by which combination target parts are selectively assembled according to the instructions given by the computer (Fig.1).

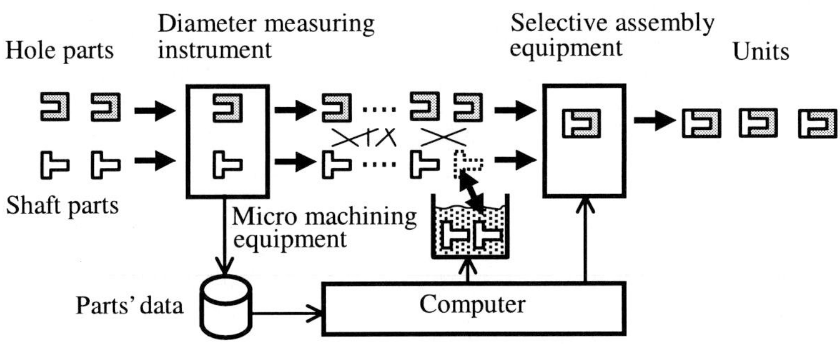

Fig.1. A precision assembly system using selective assembly and micro machining

3 Micro Machining

Micro machining used in this study is such that the parts to be machined are immersed into a tank of etching solution containing ceramic particles, and ultrasonic vibrations are applied for the sake of micro machining. This is to allow fine corrections of the dimensions so that the shaft diameters of the shaft parts may become suitable for the hole parts of objects to be combined while masking is provided for the plane not to be machined.

With this method, the dimension of machining δ• (μ•m) is proportional to the time of machining

t (min) and is expressed as follows using the machining rate a (•μm/min).

$$\delta = a \cdot t \tag{1}$$

The limit of machining dimensions by which the quality of the micro machined plane is ensured is referred to as δ•*, and machining dimensions are determined within the range of 0 <•δ ≤•δ*.

4 Combinatorial Optimization Method

Units are assembled by such a combination of the parts that at the time of assembly a clearance of $C_{ij}(=D_{Hi}-D_{Sj})$ meets with the following assembly rule.

$$C_L \leq C_{ij} \leq C_U \tag{2}$$

where C_L is lower limit of allowable clearance and C_U is upper limit of allowable clearance. Here we refer to N sets of D_{Hi} and D_{Sj}. When micro machining is applied to shaft parts, shaft diameter correction is possible up to $D_{Sj}-2\delta^*$ as far as D_{Sj} is concerned, and the assembly rule can be expressed as follows.

$$C_L - 2\delta^* \leq C_{ij} \leq C_U \tag{3}$$

In general, a variety of combinations of parts by which the number of assembled units reaches a maximum (maximum matching). Therefore, further considerations will be given to quality optimization of the assembled units. Difference of clearance C_{ij} and ideal clearance C_I (= $(C_L + C_U)/2$) (hereafter referred to as clearance difference) will be used as the index for quality of the assembled units. Among the clearance differences of the assembled units fulfilling the condition of maximum matching, there are some minimax matchings by which the maximum clearance difference becomes minimal. So, least square-sum minimax clearance is further obtained under the minimax condition and is referred to as least square-sum minimax matching.

As shown in Fig.2, two methods are available in which micro machining is taken into consideration for the optimization of the combinations.

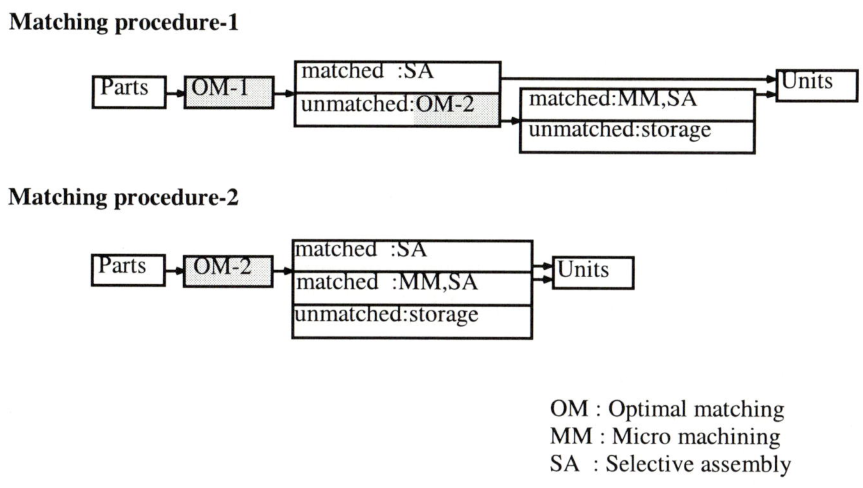

Fig. 2. Matching procedures

Optimal matching-1 and Optimal matching-2 are, respectively, defined as follows:

(1) Optimal matching-1

$$\text{Minimize } (\max_{CL \le Cij \le CU} L_{ij})^2 \cdot \sum_i \sum_{CL \le Cij \le CU} \Box_{i,j} + N \cdot \{ (C_U - C_L) / 2\}^2 \cdot (N - \sum \sum_{CL \le Cij \le CU} \Box_{ij}) + \sum_i \sum L_{ij}^2 \cdot \Box_{ij}$$

subject to

$$L_{ij} = \begin{cases} C_{ij} - C_I & \Box\ C_L \le C_{ij} \le C_U \\ N^{1/2} \cdot (C_U - C_L)/2 & \Box\ \text{otherwise} \end{cases}$$

where

$$\sum_i \Box_{ij} = 1 \ (i = 1, \cdots, N), \ \sum_j \Box_{ij} = 1 \ (j = 1, \cdots, N), \ \Box_{ij} = \{0,1\} \ (i, j = 1, \cdots, N) \tag{4}$$

(2) Optimal matching-2

$$\text{Minimize } (\max_{CL \Box 2\delta^* \le Cij \le CU} L_{ij})^2 \cdot \sum \sum_{CL \Box 2\delta^* \le Cij \le CU} \Box_{ij} + \Box \cdot \{ (C_U - C_L)/2 + 2\delta^* \}^2 \cdot (\Box - \sum \sum_{CL \Box 2\delta^* \le Cij \le CU} \Box_{ij}) + \sum_i \sum_j L_{ij}^2 \cdot \Box_{ij}$$

subject to

$$L_{ij} = \begin{cases} C_{ij} - C_I & \Box C_L - 2\delta^* \le C_{ij} \le C_U \\ \Box^{1/2} \cdot \{(C_U - C_L)/2 + 2\delta^*\} & \Box \text{otherwise} \end{cases}$$

where

$$\sum_i \Box_{ij} = 1 \ (i = 1, \cdots, N), \ \sum_j \Box_{ij} = 1 \ (j = 1, \cdots, N), \ \Box_{ij} = \{0,1\} \ (i, j = 1, \cdots, N) \tag{5}$$

In the case of optimal matching-2, micro machining dimension δ_j to be applied to the shaft part j is obtained from the following equation. δ_j is designed so that the clearance after the machining may not exceed the ideal clearance.

$$\delta_j \begin{cases} (C_I \Box C_{ij})/2, & C_I - 2\delta^* \le C_{ij} \Box C_L \\ \delta^* & , C_L - 2\delta^* \le C_{ij} < C_I - 2\delta^* \end{cases} \tag{6}$$

The number of assembled units m is given by

$$m = \begin{cases} \sum_i \sum_{j, \, CL \le Cij \le CU} \Box_{ij} & \Box\text{Optimal matching-1}, \\ \sum \sum_{CL \Box 2\delta^* \le Cij \le CU} \Box_{ij} & \Box\text{Optimal matching-2}. \end{cases} \tag{7}$$

Using m, the assembly rate$\Box$ is given by

$$p = m / N. \tag{8}$$

5 Simulation

We examine a numerical example with the parameters given in Table 1. The distributions of D_H and D_S are, respectively, assumed to follow normal distributions. Performance of the two matching procedures are compared for lot sizes between 5 and 100 in the Monte Carlo simulation (the number of simulations is 1000 in each lot size). The result of the example simulation is illustrated in Fig.3 to Fig.5.

Table 1. An example problem

D_H	$N(8.4990, 0.0015^2)$ mm
D_S	$N(8.4945, 0.0015^2)$ mm
C_L	4.0 μm
C_U	5.0 μm
a	6.0 μm/min
δ^*	10.0 μm

It is noticed from Fig. 3 that as for the mean assembly rate, the larger the lot size, the more combinatorial matchings become available. When both selective assembly and micro machining are used, higher mean assembly rates are attained even when the lot sizes are small as shown. Effectiveness of this system has thus been revealed.

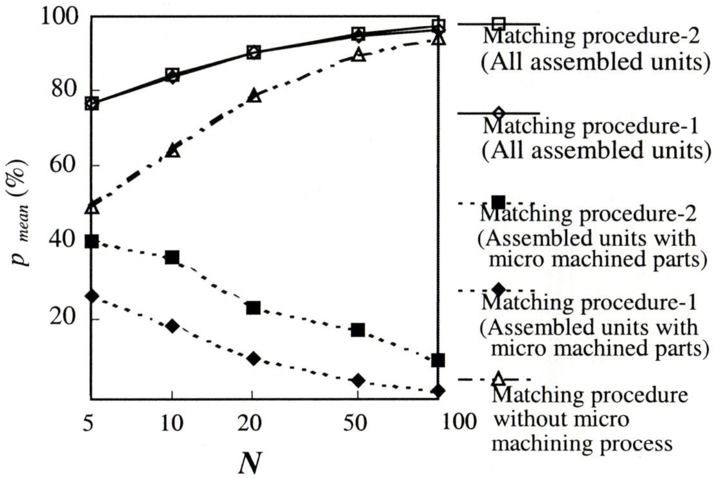

Fig. 3. Relationship between p_{mean} and N

In Fig.4, many assembled units with a clearance close to the ideal clearance are obtained within the assembly rule. The frequencies of the assembled units being micro machined are concentrated at the center of the assembly rule.

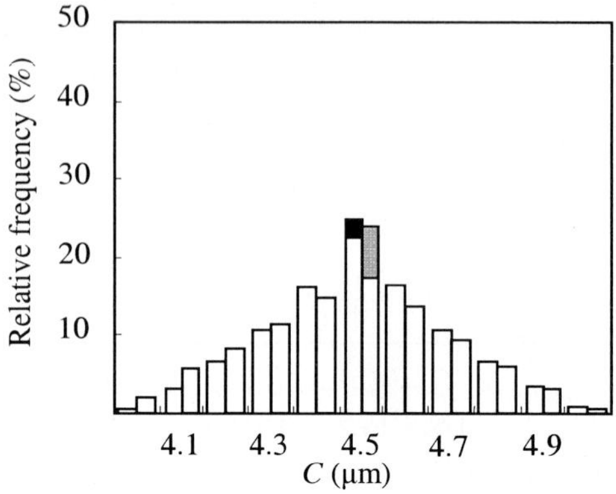

	Matching procedure-1	Matching procedure-2
Units with micro machined parts	■	■
Units without micro machined parts	□	▩

Fig.4. Distribution of C (N =100)

In Fig. 5, the maximum micro machining dimension by matching procedure-2 is 3.5μm which is almost 1/3 of the 10.0μm of matching procedure-1. The micro machining used in this study has the distinctive features that the machining of all shaft parts subject to the machining can be started at once and that the machining time is determined depending upon the maximum machining dimension. Short machining time is preferable from the viewpoint of a reduction of scattering of the quality of the micro machined plane. It can be said from the above that matching procedure-2 is superior to matching procedure-1.

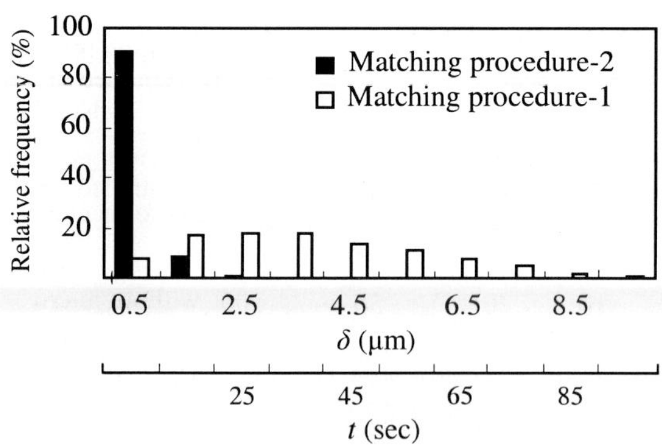

Fig. 5. Distribution of δ and t (N=100)

6 Conclusions

A high precision assembly system based on the combination of selective assembly and micro machining through a combinatorial optimization method is proposed. The quality of the assembled units is optimized while preserving the maximum assembly rate taking micro machining into account. The following findings are obtained by simulating combinatorial optimization methods of parts.

(1) This system is effective in a small lot production, where higher assembly rates can not be attained with conventional methods, in such that a higher rate is obtained if micro machining is used alongside.

(2) With the combinatorial optimization method for parts where micro machining is taken into consideration in advance, maximum micro machining dimension and maximum machining time of the parts can be greatly reduced and the number of assembled units close to the ideal state is increased.

References

[1] Yamada,Y. and Kobayashi,T. (1993) : Combinatorial Optimization of Selective Assembly - Minimum Variance Criterion-. J. Jpn. Soc. Quality Control,23(4),409-415,(in Japanese).

[2] Yamada,Y. and Kobayashi,T. (1994) : Combinatorial Optimization of Selective Assembly - A Case Study of the Automobile Engine-. J. Oper. Res. Soc. Jpn., 39,556-560,(in Japanese).

[3] Yamada,Y.(1994) : Combinatorial Optimization of Selective Assembly in Continuous Production.JSME,60(573),1877-1881,(in Japanese).

[4] Yamada,Y.,Tanabe,I. And Takada,K. (1996) : Behavior Analysis of Automatic Assembly System with Selective Assembly Optimization Function Journal of Advanced Automation Technology,8(1),36-41.

Computer Aided Planning System of a Flexible Microrobot-Based Microassembly Station

S. Fatikow, J. Seyfried, and A. Faizullin

University of Karlsruhe, Institute for Process Control and Robotics
Kaiserstr. 12 (40.28), 76128 Karlsruhe, Germany
{fatikow, seyfried, faizulli} @ira.uka.de

Abstract. The assembly of complex microsystems consisting of several single components (i.e. hybrid microsystems) is a task which has to be solved to make mass production of microsystems possible. Therefore, it is necessary to introduce flexible, highly precise and fast microassembly methods. In this paper, the control system of a microrobot-based microassembly desktop station that has been developed at the University of Karlsruhe, will be presented from the lower to the planning levels. This comprises vision-based closed-loop control, user interfaces, a re-configurable computer-array, execution planning and assembly planning algorithms tailored to the needs of the microassembly station.

1 Introduction

The slow adoption of microsystem technology (MST) by industry has made it clear that many problems exist in mass-producing microsystems. Most batch processes are rarely applicable for the production of complex microsystems which consist of microcomponents made of different materials and which are manufactured using different techniques. This means that individual components must be very exactly assembled in one or more steps to form the desired microsystem. The microassembly systems which exist today are usually only fit for a specific task and depend on the manual agility of an operator or consist of microassembly robots with conventional drives which are extremely expensive and are due to mechanical wear and frequent maintenance. With increasing workplace miniaturization, however, it becomes more and more difficult to use conventional manipulation robots for assembling microsystems. The manipulation accuracy is mechanically limited for conventional robots, since disturbing influences which

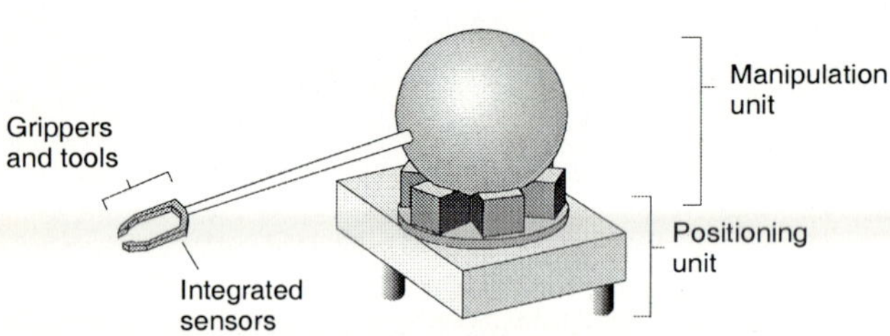

Fig. 1. Mobile micro-manipulation robots

F. Pichler, R. Moreno-Díaz, and P. Kopacek (Eds.): EUROCAST'99, LNCS 1798, pp. 414–434, 2000.

are often negligible in the macro-world, such as fabrication defects, friction, thermal expansion or computational errors, play an important part in the microworld. The next point is that the positioning accuracy and the tolerances of the microcomponents lie in the nanometer range, a few orders of magnitude smaller than in conventional assembly. These accuracy requirements can be obtained with microrobots having highly precise direct drives utilizing MST and advanced closed-loop control. A microrobot-based desktop station may offer the desired features and versatility.

2 Concept of a Microassembly Desktop Station

The concept of an automated microassembly desktop station, in which microassembly tasks can be executed using a closed-loop control approach, has been developed at our Institute. Each robot currently used in this station is equipped with a piezo-electrically driven base platform and a powerful manipulation unit which can be furnished with different tools, Fig. 1. The robots are mechatronic systems based on microsystem technology principles; they were presented in [1]. The most difficult task in the development of such a multi-robot station is the development of a suitable control and planning system, which should be scaleable and take into account the special requirements of microassembly.

2.1 Microrobots

The concept of mobile micromanipulation robots is shown in Figure 1. In order to manipulate an object, the robot travels to the working space using its positioning unit, and picks up the part by positioning the gripper using the manipulation unit. A gripped part can then be positioned using the manipulation unit or the positioning unit if it is to be transferred to a remote location.

Both the manipulation and the positioning unit employ a microstep sequence for micropositioning; the robot has three piezo-legs which can be bent in any direction, allowing a precise positioning in the range of 3 μm with a motion resolution of 10 nm [1]. If a bigger distance is to be bridged, the legs are bent to the maximum stroke and then, a "step" is performed. By repeating this sequence with a frequency of up to 10 kHz, speeds of up to 5 cm/s can be obtained (depending on the robot prototype,

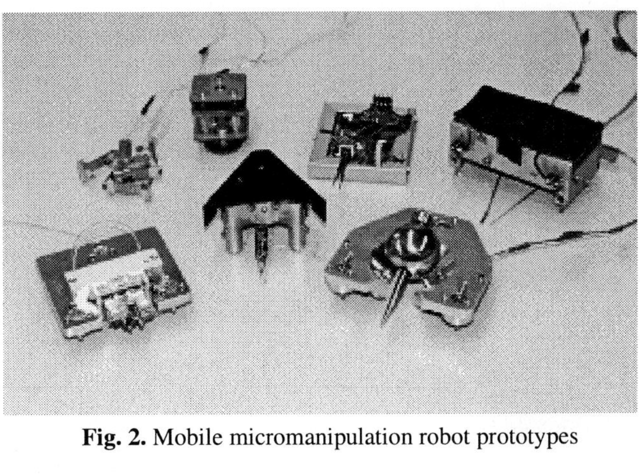

Fig. 2. Mobile micromanipulation robot prototypes

the base material and other factors of the surroundings, which has to be taken care of by the control system [2]).

Figure 2 shows several implemented prototypes. The two prototypes in the front row, RobotMan and Miniman III (front left and front right) have been developed for the use with an optical microscope and a scanning electron microscope (SEM) applications, respectively. They were both designed for the special microscope's requirements.

2.2 Hardware of the Station

Figure 3 gives an overview over the station's components [3]. The robots and the station's peripheral devices (an XY-stage, control of microscope parameters, lighting parameters, etc.) are controlled by a central hybrid parallel computer array. The visual sensors (CCD cameras) of the station are connected to a frame grabber with a vision system which detects the robots, the robots' grippers and the microobjects under the microscope. This setup of a local and a global sensor system is commonly used for micromanipulation stations [4], [5].

Figure 4 presents the architecture of the computer system. The parallel control computer consists of two types of modules, namely Intel Pentium-based PC104 modules and micro-controller modules using the Siemens C167. The latter are used for high speed I/O tasks (e.g. up to 1,092,000 analog values per second for the control of the positioning platform), while the Pentium modules are responsible for vision, closed loop control [6] and communication tasks. The different types of modules are connected by dual ported RAMs (DPRs), one of which connects two computer modules and offers a communication means of high bandwidth (64 Mbit/s) and low response time (360 ns). This is the base for any real-time control

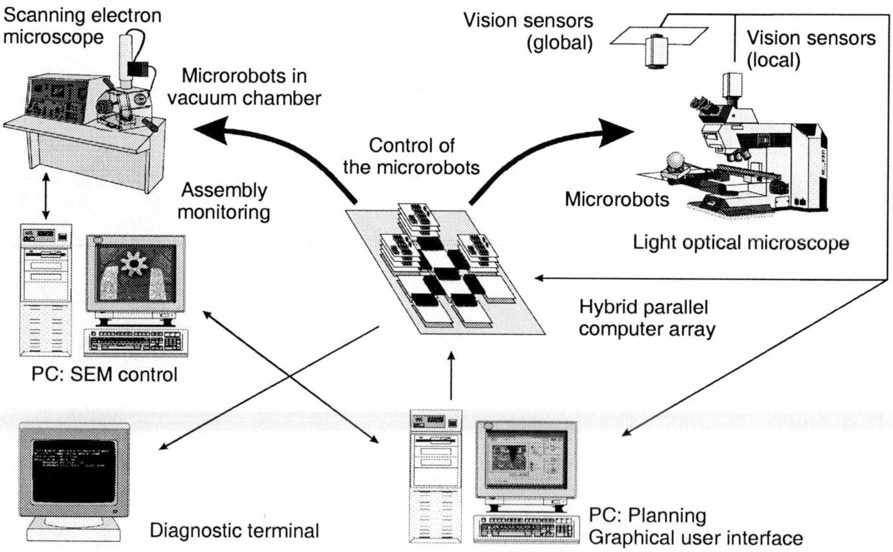

Fig. 3. Components of the desktop station

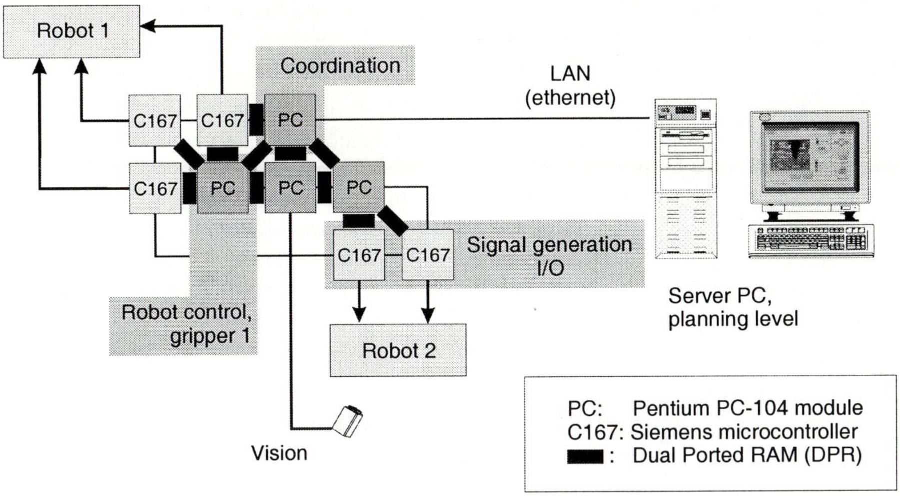

Fig. 4. The control computer array

used in the station [7]. The architecture of the computer array makes it easily scalable. A new robot can be integrated into the system by adding the appropriate micro-controller modules and a PC module and connecting them via DPR. The array nodes have detection algorithms which identify their surroundings at startup. By propagating the local maps from the PC modules to the server, the complete topology of the computer array can be determined automatically after a change. By displaying the map graphically, the user can easily assign modules to robots and other tasks.

Linux is used as the operating system for the control computer. To meet the real-time requirements of robot control, Real-time Linux was chosen [8], which offers a real-time scheduler that runs the conventional Linux Kernel as one task and allows Real-time tasks which cannot be interrupted to run with a given priority.

3 Comparison of Micro- and Macroassembly

Microassembly differs greatly from conventional assembly in some aspects. These differences are caused by the small dimensions of the parts to be assembled. The biggest problems arise from surface forces. These become dominant for small parts (e.g. cubes with approximately 1 mm^3) because when miniaturizing an object, the surface decreases by the second power, while the volume (i.e., the object's mass) decreases by the third.

This gives rise to problems due to

- electrostatic forces (electrostatically charged objects or base materials
- surface tension forces (humidity of the surroundings)
- Van der Waals forces for very small objects

The effects of the surroundings can be avoided using a suitable air-conditioning or clean-room environment, but the other effects are inherent to the manipulation of microscopic parts. Figure 5 illustrates the relationship between part size and adhesive forces. Below a certain "critical mass[1]" m_c of the parts, adhesive forces play a dominant role and make manipulations unpredictable if no actions are taken to cope with these problems.

Figure 6 shows an example of an object (a microgearwheel with a diameter of less than 1 mm) sticking to the gripper of a robot. First, the object is grasped (Figure 6, top), then the gripper was opened, but the object stuck to the left gripper of the robot (Figure 6, bottom). The cross-hair in the picture marks the original center of the gearwheel to illustrate the sticking effect. Also, the reverse effect can be observed: parts "jumping" from the gripper due to electrostatic charge.

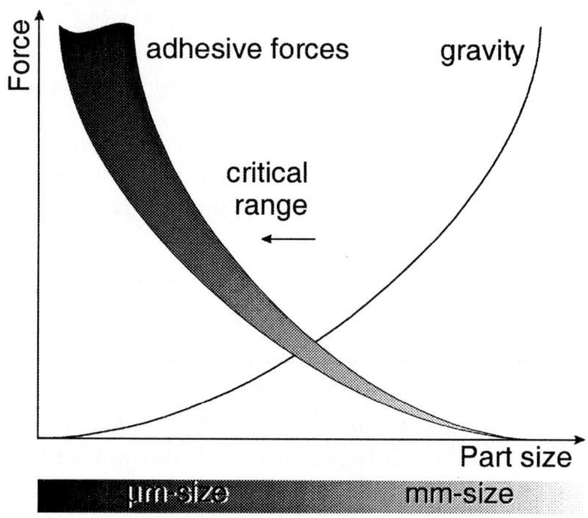

Fig. 5. Adhesive forces caused by small dimensions

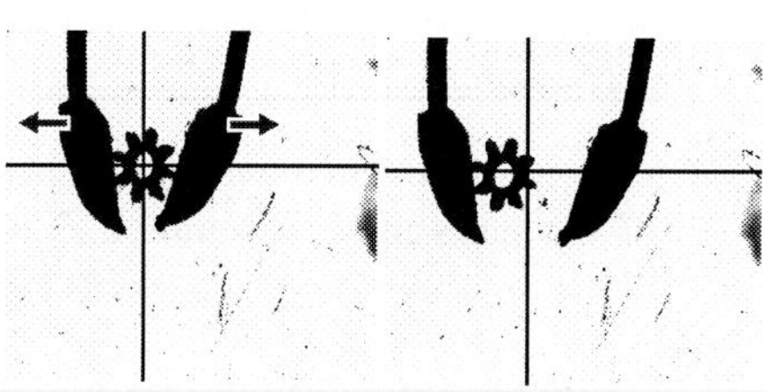

Fig. 6. Object sticking to gripper

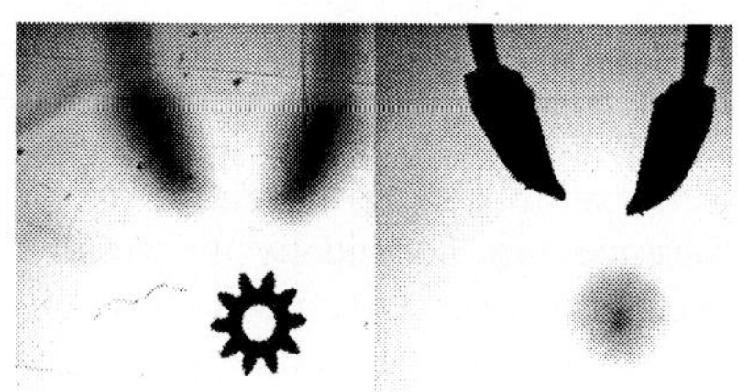

Fig. 7. Limited depth of field

[1] in this case, in the microscopic sense

One possible approach to avoid sticking objects is to involve a second robot which is equipped with a "helping hand", which consists of a simple needle-shaped gripper tip to brush off the object, minimizing the contact faces by the small dimensions of the needle. Even with this approach, it is quite obvious that microassembly demands for sensor surveillance – not only for assembly operations, but also for operations as simple as grasping or dropping a part. Micro-effects should also be taken care of already at the planning level[2]. Unfortunately, vision in microassembly also suffers from the small dimensions of the parts. In the case of an optical microscope, the main problems are a small depth of field and view. Figure 7 shows a microobject and the gripper which is a few mm above the object. Depending on the focal plane of the microscope, either the gripper or the object is in focus. The picture shown is approximately 4 mm wide, which gives an idea who hard it is to keep track of all the objects located in different parts feeding devices on the working space of the robots, which can extend to several dm^2. Also, this makes it harder to perform tasks by several robots cooperatively.

The limited field of view is also a problem in SEM applications, while SEMs offer an excellent depth of field. On the other hand, electron microscopy imposes several restrictions on the possible processes like vacuum compatibility, problems due to the electron beam or problems due to non-conductive materials which pose a challenge to the electron optics.

The problem of the limited field of view can be solved by the use of a second CCD camera ("global camera" in Figure 3). This makes it possible to track several robots traveling on the platform while a simultaneous micromanipulation under the microscope is possible. We are also evaluating this concept for scanning electron microscopy [9]. The other problems of vision in microassembly have been discussed in [10]. It is obvious that this additional vision sensor – along with multi-robot operations and precautions to compensate for micro-effects – makes the control architecture more complex. In the next section, we will present how we addressed these topics.

4 Control System Architecture

Figure 8 shows an overview over the station's software architecture. In order to compensate the problems of microassembly, a special assembly planning module has been developed [11], [12], [13] based on [14]. This module takes into account the given resources of the station (e.g. different sensors which insure controllability of assembly steps in different directions of observation). In the decomposition phase, the assembly plan is allocated to robots according to their operational capabilities.

In order to make multi-robot operations possible and allow the user to add and remove robots easily from the system, a distributed approach was chosen which relies on a central supervision and execution unit which provides the local robot control objects with the desired motion sequences. The communication framework

---

[2] i.e. by always mounting small objects onto a bigger and therefore heavier sub-assembly which is not affected by micro-effects. This will be discussed in Section 5.

is based on CORBA which is extended to make dynamic real-time execution of tasks possible. In this Section, we will describe this architecture in detail starting with the mapping of physical objects in the station to logical objects in the following sub-section.

4.1 Modeling of the Station's Components

To be able to scale the system as easily as possible, the physical objects involved in the microassembly station were mapped to C++ objects by a permutation mapping. Thereby, all real robots RR_i, $i \in \{1,...,n\}$ are represented by a robot object RO_i, $i \in \{1,...,n\}$, derived from the class `robot` *CR*. The sensor objects are divided into the two classes local sensors and global sensors and are represented by the objects G_i, $i \in \{1,...,n_g\}$ and L_i, $i \in \{1,...,n_l\}$ of the classes *CG* and *CL*, respectively. Objects to be assembled are also represented by (micro-) objects O^j_i, $i \in \{1,...,o\}$, $j \in \{1,...,k\}$ of the class CO^j, which play an important role in the manipulation process, as we will see. The class identifier *j* denotes which type the physical object belongs to, i.e. gearwheel, axle, etc. Furthermore, a class representing the XY-stage, *CT* exists. As it is common practice in object-oriented programming, these objects are derived from abstract base classes *ACi*, $i \in \{R,G,L,O,T\}$, which contain methods common to all objects of a given type, e.g. the robot base class can be given motion commands, can report its current position and orientation, etc.

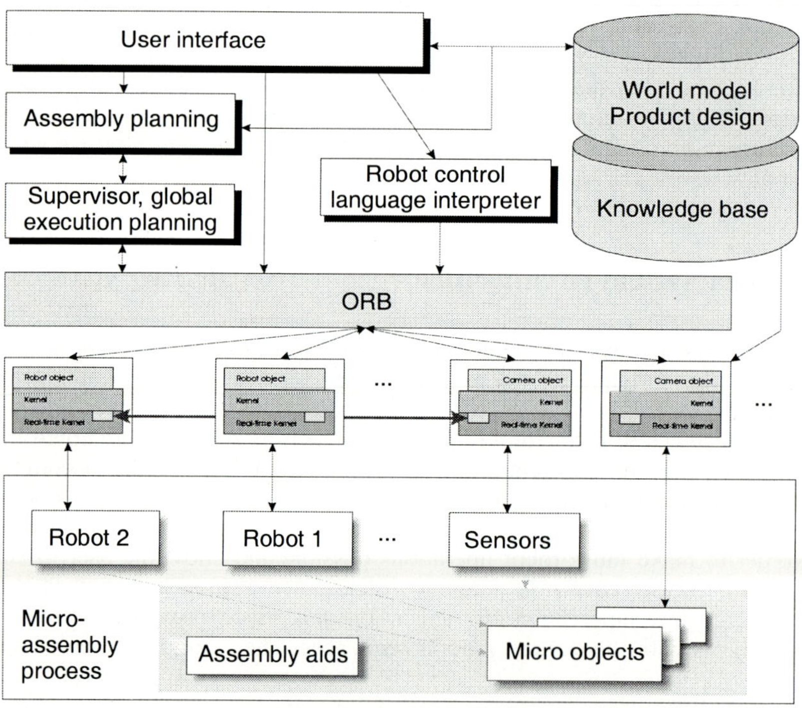

Fig. 8. Software architecture

To keep track of the processes in the station and to avoid deadlocks and resource collisions, all aforementioned objects have an internal state which can also be queried by a supervisor and is set by different actions. The objects' states are set by actions which are executed in the station, e.g. "grip object", "drop object", "goto position", etc. To guarantee a deadlock-free execution, all state transitions were modeled using finite state machines and the possible object transitions were checked using Petri-nets. As an example, we will discuss the most complex object, the robot object. Figure 9 shows the possible states of the robot class. The transitions *Ra...Rk* correspond to special actions that may be taken in the station, Table 1.

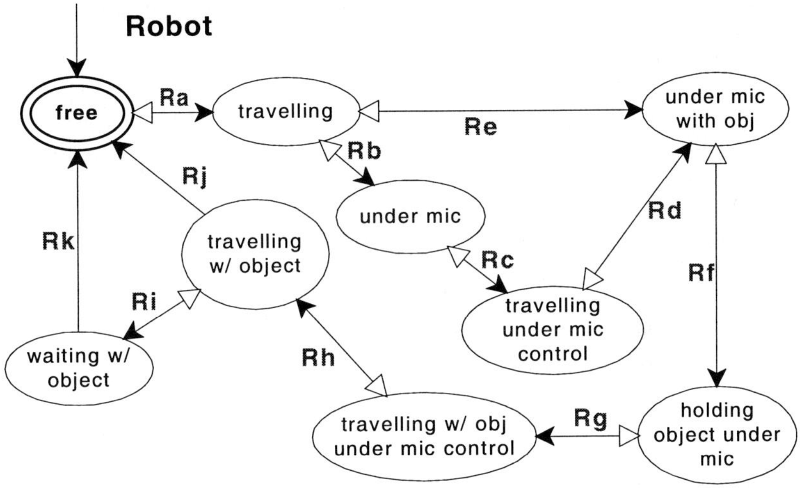

Fig. 9. State chart of the robot class

Table 1 State transitions for the robot class

Action	Obj.	Local	Global	Remarks
Ra			Ga, Gb	Motion using global camera
Rb		Le	-Ga, -Gb	Mic position, local camera
Rc	Oc			moving under mic. control
Rd			-Ga, -Gb	Mic pos. local cam w/ obj
Re		Lb		Mic pos. local cam w/ obj
Rf		Lb		Gripping obj
Rg				moving under mic control
Rh		Lc	Ga, Gb	leaving mic pos
Ri			-Ga, -Gb	yielding
Rj	Of			Dropping object?!
Rk	Of			Dropping object?!

The columns "Obj", "Local" and "Remote" denote which state changes in other classes are caused by the actions. E.g. recognizing the robot's gripper under the microscope means performing vision with the local camera, therefore this action triggers and requires the *Le* transition of the `Local camera` object. Note that transitions *Rj* and *Rk* are "dangerous" and should be avoided because dropping an object without sensor surveillance leads to unpredictable results (sticking to the gripper in the worst case, which cannot quite be called "dropping" the object), as we have seen in Section 3. These transitions are modeled for the sake of completeness, but they are never requested by assembly or execution planning.

After having designed the state transitions for all classes *{CR,CG,CL,CO}*, the possible actions which can take place in the station were identified (events which trigger a state change in one or several objects). To prove that this system is able to perform the desired processes and to prove the absence of inherent dead-lock situations, we mapped the state changes onto a petri-net. In this net, transitions correspond to actions in the station during an assembly process, and places to states of the objects of the control system. The net for a single-robot system consists of 22 places and 16 transitions. The reachability tree proved that this net is ordinary, conservative, bound, safe and life. For the assembly process, this means that no resource collisions or multiple assignments may occur and no deadlock-situations can be reached accidentally. The analysis of a two-robot scenario led to a net with 31 places and 33 transitions with the same properties.

Figure 10 shows a result of the simulation of the net: all possible processes in the station were calculated; depicted is a pick-and-place operation under microscope control. Here, the robot first travels to the microscope using the global camera (T2), reaches the microscope (T7), then, the XY-stage and the robot are moved simultaneously (T9) to bring the microobject into the field of view of the microscope, next, the object is gripped (T10) and finally dropped under the microscope.

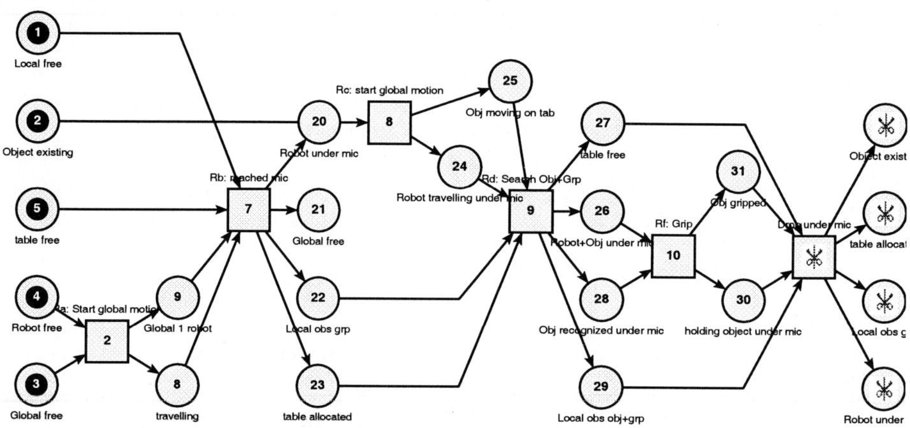

Fig. 10. Simulation result of the petri-net model: pick and place under sensor surveillance

4.2 Negotiated Real-time Connections

CORBA provides an excellent framework for scheduling and coordinating the execution process in the station, as it has already been implemented in higher-level coordination of manufacturing systems [15], [16], but in order to be able to control the robot system in real-time, we propose an extension to the CORBA mechanisms.

Normally, components are managed by the ORB of the CORBA system which dispatches the RPC-based requests from clients to servers (Fig. 8). To make these requests real-time capable, two approaches can be taken: either, the recently proposed Real-time CORBA [17] could be employed and the communications channel could be changed to a real-time channel (which Ethernet is actually not). This could be done by employing the TCP/IP stack for the DPRs which we have developed and therefore using a real-time medium for communications (the DPRs). This is based on a kernel module which makes the Dual ported RAM cards behave like a conventional network card. Two problems remain unsolved using this approach: first, there is no implementation of the real-time CORBA available today, and second, neither the objects *RO, G, L, O* and *CT* nor kernel module for network communications are real-time modules by default. Implementing the complete objects as real-time modules would mean a big overhead since large portions of these objects are not subject to real-time constraints.

A second way to guarantee real-time behavior of the control system is presented in the following. Suppose two objects which require a real-time communication channel, e.g. a robot object *R* and a camera object *G*. To avoid having to implement the whole objects as real-time kernel modules (which would imply that many non-critical functions are executed in real-time), the time-critical procedures can be identified and located in a real-time module. When the robot object is to perform a task which requires real-time communications with the camera object, it establishes a real-time communication channel by sending the camera object a handle to the DPR memory used for real-time communication.

Figure 11 gives an overview over this kind of negotiated real-time communication (NRC). The fact that the scheduling of assembly tasks in the station is collision-free as shown in Section 4.1 grants that no dead-lock situations can occur during negotiated real-time operation. Note that only a small part of the robot

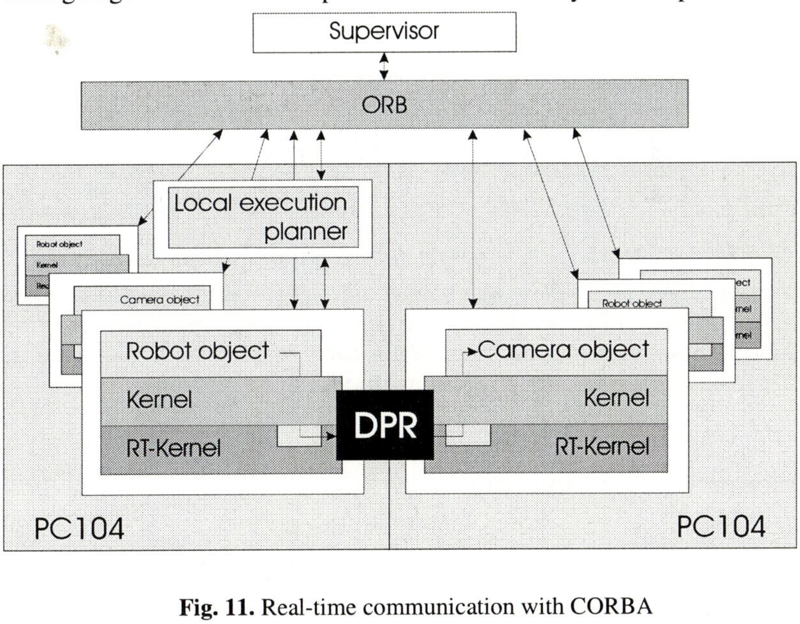

Fig. 11. Real-time communication with CORBA

and camera object is a real-time module. Thereby, only time-critical tasks run in the real-time kernel, which improves the overall response times of the computer module for non-real-time tasks. The control algorithms are designed in such a way that interrupts from the upper control levels still can be executed by cyclically yielding control from the real-time module. Figure 12 shows an example how NRC is established and how objects change their states according to commands from the supervisor module. The task consists of detecting an object on the XY-stage and moving a robot to that position. First, The microobject is told to locate itself on the base. Therefore, it allocates the XY-stage (table) and the local (microscope) camera. When the local camera acknowledges this request (acknowledge messages are not shown for the sake of clarity), DPR handles are exchanged and NRC is established. The next task which is dispatched by the supervisor is "GotoMic" which tells a robot to move to the microscope position. The robot first uses the global camera to reach the microscope position and next the local camera to be able to pick up the part. When observing the robot and the microobject (i.e. an assembly process), the camera object is connected to *RO* <u>and</u> the O_i via NRC. This is necessary since both the robot's gripper and the object have to be detected in order to manipulate the object.

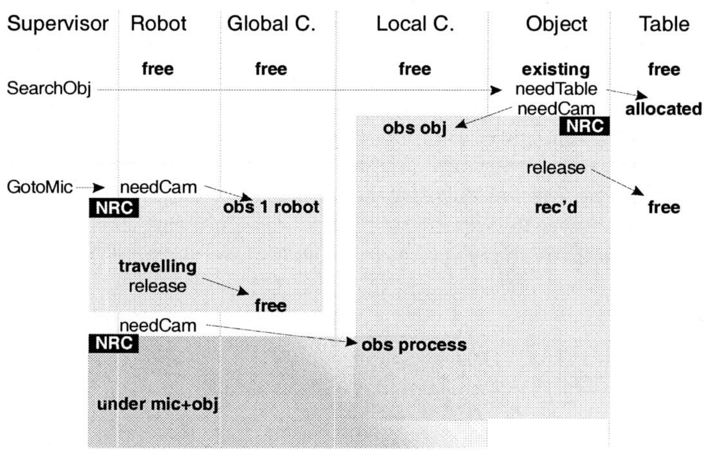

Fig. 12. Communication to establish NRC (simplified)

The microobject O_i contains special methods specific to the physical microobject at hand. Different objects require different grasping strategies and mounting techniques. These properties are specific for each object class CO^j derived from the object base class for a specific physical microobject.

5 Microassembly Planning

As mentioned in Section 3, efforts should be taken to avoid or minimize micro-effects on every level of the planning and execution system. After discussing how to cope with micro-effects by taking them into account when designing the control system, we will now analyze how they can be taken care of already at the planning level.

Considering the scenario in Figure 6, it is clear that mounting a part p_i of mass m_i, $m_i<m_c$ onto a small subassembly $p_s=p_j+p_{j+1}+\ldots+p_k$ with $m_s<m_c$ and $m_i+m_s<m_c$ leads to the same problem. Performing such a subassembly can be avoided already at the planning level if all parts with this interfering "micro-property" are marked accordingly in the parts database and the assembly planner optimizes the assembly process in such a way that offending subassembly steps are avoided. A special case would be a complete assembly p with

$$p=\sum_{i=1}^{n} p_i \text{ and } \sum_{i=1}^{n} m_i < m_c$$

which can be the case for small microsystems. Here, the assembly planner indicates that using an assembly aid is necessary on which the assembly takes place and which has a mass $m>>m_c$. Apart from this, the assembly planning must also guarantee visibility and feasibility of all assembly steps. Therefore, the assembly planning module presented in [9] based on [14] was extended to account for micro-effects as mentioned above.

The assembly planning system offers a intuitive user interface which integrates into the user interface of the whole robot system. The parts specifications are entered using a CSG (constructive solid geometry) system, Figure 13. The planning module generates an assembly plan based on the parts specifications and decomposes the plan according to the number of robots and their capabilities concerning the necessary operations for the assembly, Figure 14 [12].

If an operation fails due to a robot which is not responding or a faulty manipulation, the incident is reported to the supervisor which analyzes the fault and either requests manual assistance or re-planning from the planning module. In case of a necessary re-planning, the system is capable to perform the re-planning online if the number of parts involved in the subassembly is sufficiently small[3]. The

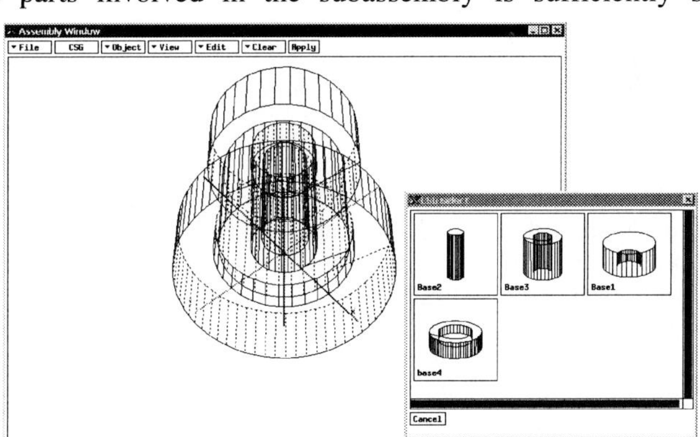

Fig. 13. Assembly planning: CSG model

[3] the system is able to perform re-planning for 7 parts in 0.3 seconds on a Intel PII-400 system [13].

decomposition of very large assemblies into several sub-assemblies is essential since assembly planning is NP-complete [18], [19], and normally, the user is able to easily identify subassemblies which are suitable for separate planning.

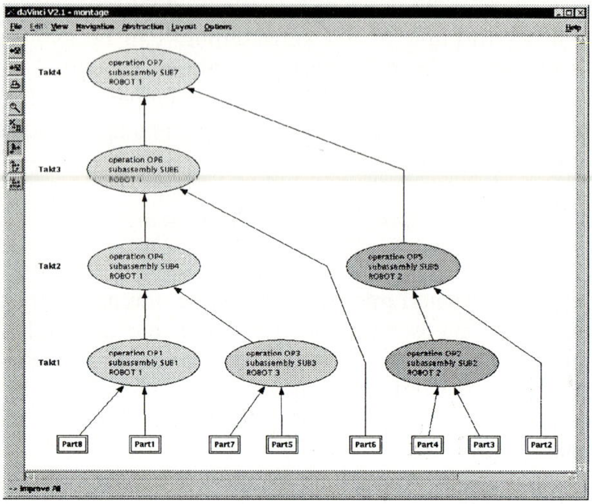

Fig. 14. Assembly plan decomposed for three robots

5.1 Experimental Results

The motion performance of a robot of type t is described in the planning system in a rather simple way by a 6-tuple

$$\boldsymbol{R}_t = \{d_{+x}, d_{-x}, d_{+y}, d_{-y}, d_{+z}, d_{-z}\}, \quad t=1,..,T, \tag{1}$$

where T is the number of a type of the manipulator; $d_p=1$, if the manipulator of type t is able to move the gripped part in the direction p, otherwise $d_p=0$.

In [11, 12], complete algorithms for the generation of feasible assembly sequences and for selecting the best assembly plan were introduced.

The planning system developed has been tested by an example of an automatic assembly planning of the worldwide smallest micromotor, which is made by Faulhaber, Germany (Figure 15) [20].

For testing, two groups of robots were supposed to be employed. The first group includes $\boldsymbol{R}_1 = \boldsymbol{R}_2 = \{0,0,1,0,0,0\}$ and the second group includes $\boldsymbol{R}_3 = \boldsymbol{R}_4 = \{0,0,0,1,0,0\}$ (1). The product reasoning and assembly planning was implemented on an Intel Pentium II-400 PC under Linux 2.0.36. The base language for programming was C.

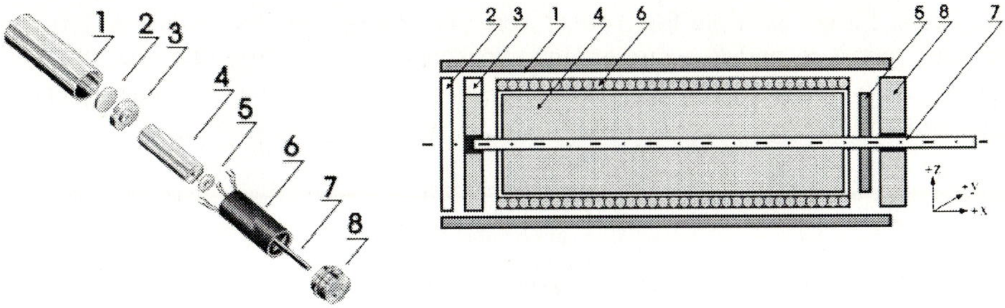

(a) (b)

Fig. 15. Assembly example: (a) initial configuration and (b) final configuration

Example 1. The best assembly plan is decomposed for three robots (Figure 16). In this example, three robots were working to perform the task without errors (Table 2).

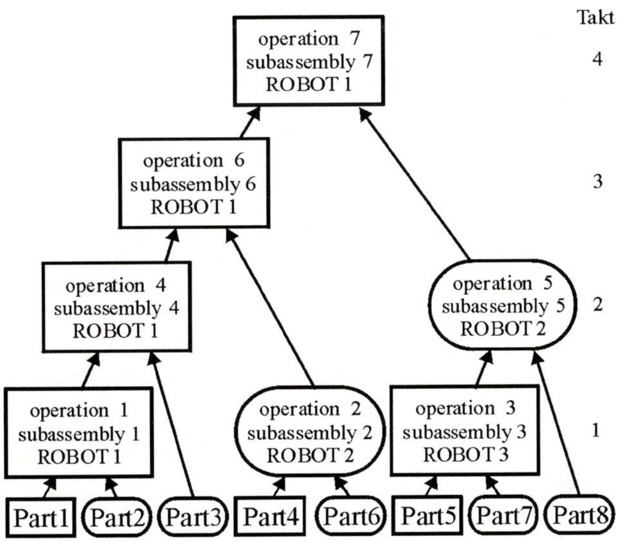

Figure 16 The best assembly plan.

Table 2 The assembly protocol.

Tact	Robot	Operation	Gripped part	Base part	Status
1	1	1	part 2	part 1	yes
	2	2	part 6	part 4	yes
	3	3	part 7	part 5	yes
2	1	4	part 3	sub 1	yes
	2	5	part 8	sub 3	yes
3	1	6	sub 2	sub 4	yes
4	1	7	sub 5	sub 6	yes

Example 2. In this example four robots were performing the task (Figure 17). In the first tact, robots $\boldsymbol{R}_1$ and $\boldsymbol{R}_2$ were assumed broken (Table 3). From then on, the only working robots were $\boldsymbol{R}_3$ and $\boldsymbol{R}_4$. Table 4 shows the time needed for planning or for re-planning.

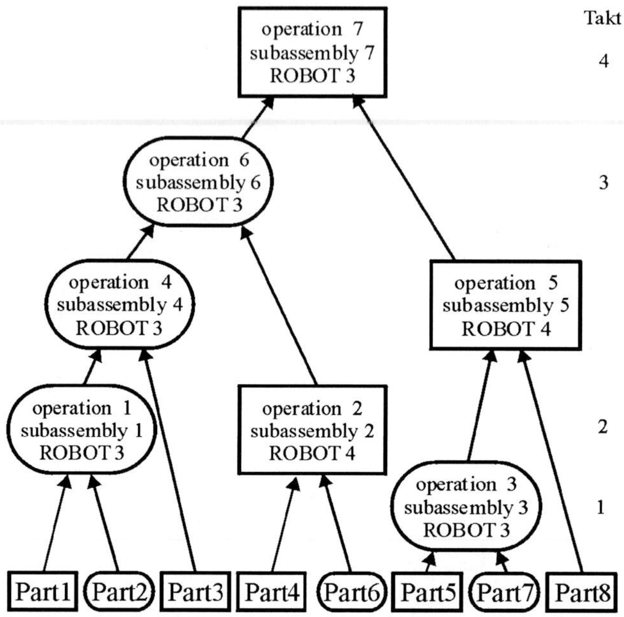

Fig. 17. The best assembly plan with re-planning and re-decomposition.

Table 3 The assembly protocol with re-planning and re-decomposition.

Tact	Robot	Operation	Gripped part	Base part	Status
1	1	1	part 2	part 1	no
	2	2	part 6	part 4	no
	3	3	part 7	part 5	yes
Re-decomposition					
2	3	1	part 2	part 1	yes
	4	2	part 6	part 4	yes
Re-planning					
3	3	4	sub 1	part 3	yes
	4	5	sub 3	part 8	yes
4	3	6	sub 4	sub 2	yes
5	3	7	sub 6	sub 5	yes

Table 4 Time spent for planning (re-planning).

Number of parts	Number of sequences	Time, sec
8	192,512	3.677321
7	12,288	0.232050
6	1,280	0.023115
5	192	0.003416
4	24	0.001002
3	4	0.000704
2	2	0.000659

6 Execution Planner

The motion control of the robot is based on its geometric description. The aim is to control the robot movements in a such way that, first, the working part of the robot endeffector (its tip, as a rule) is moved from the initial (actual) point to the aspired end-point and, second, the certain orientation of the robot in the final state is achieved. The requirements for the robot movement depend on the task to be performed. One can distinguish between a transportation task (coarse motion) and a micromanipulation task (fine motion). In the first case, the robot has to transport microobjects over a relatively long distance so that the highest motion speed and an optimal trajectory must be provided. In the second case the robot manipulates with microobjects under a microscope; it means that the endeffector tip must stay in the view field of the microscope during the robot motion.

In Figure 18 the rectangular coordinate system XOY describes the robot position. The polar coordinate system $x_i o_i \varphi_i$, i=1,...,3 describes the step direction of the leg i in regard to the robot platform. The coordinates (X_A, Y_A) of the endeffector tip and angle α_A between the axis OX and the symmetry axis of the robot passing through the back leg (X_{O2}, Y_{O2}) serve as the sensor information from a CCD-camera for a closed-loop control of the robot. This information fully determines the position and orientation of the microrobot in the initial state. The geometric model of the microrobot platform is described by the following constant values:

$$\begin{cases} \phi = \angle O_1AO_2 = \angle O_2AO_3 \\ l_1 = |AO_1| = |AO_3| \\ l_2 = |O_1O_3| \\ l_3 = |AO_2| \end{cases}$$

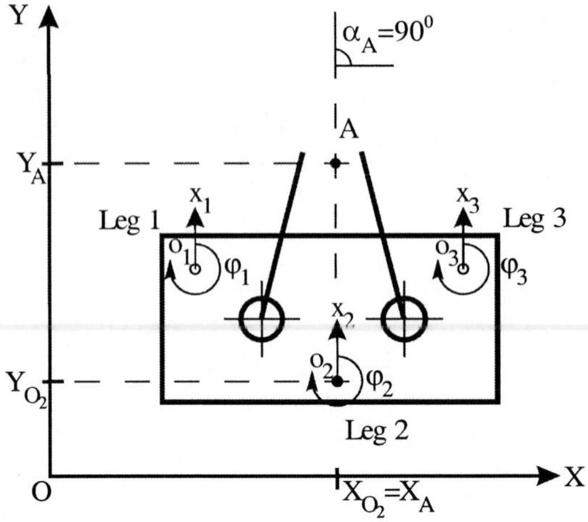

Fig. 18 The base orientation of the microrobot

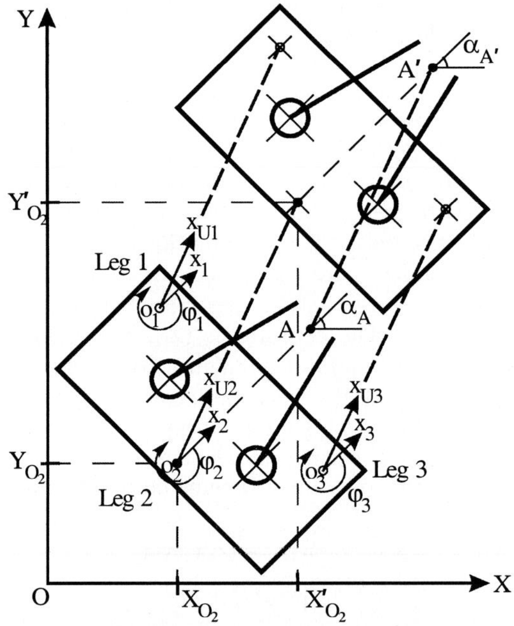

Fig. 19 Linear motion of the robot

The base orientation of the platform corresponds to the platform position at angle α_A=90°. So, a task of motion control of the microrobot can be formulated in the following way. In the coordinate system XOY, the initial parameter set $\{(X_{O2}, Y_{O2}), \alpha_A\}$ and the final parameter set $\{(X'_{O2}, Y'_{O2}), \alpha'_A\}$ are given. It is necessary to calculate the motion vector values $U_i = (x_{Ui}, \varphi_{Ui})$, i=1,...,3 in correspondence to the coordinate system $x_i o_i \varphi_i$.

6.1 Linear Motion

To simplify the description, suppose $\alpha_A=\alpha_B$. For performing linear motion from point A to point B (Figure 19) motion vector $U_i = (x_{Ui}, \varphi_{Ui})$ must be simultaneously applied to all three piezolegs. Coordinates (x_{Ui}, φ_{Ui}) of vector U_i can be calculated in the following way:

$$\begin{cases} x_{U1} : x_{U2} : x_{U3} = 1:1:1 \\ \varphi_{U1} = \varphi_{U2} = \varphi_{U3} \end{cases} \tag{2}$$

6.2 Rotation

Motion vectors Ui for all three piezolegs, which are necessary for rotation around the point C, are shown in Figure 20. The vectors $Ui = (x_{Ui}, \varphi_{Ui})$ are also tangential to the rotation circumference. The motion vectors Ui are calculated as follows:

$$\begin{cases} x_{U1} : x_{U2} : x_{U3} = |Co_1| : |Co_2| : |Co_3| \\ \varphi_i = 90 + \alpha - \beta_i, \quad \text{where } i = 1,\ldots,3 \\ \beta_i = \operatorname{arctg}\left(\dfrac{|Y_C - Y_{Oi}|}{|X_C - X_{Oi}|}\right) \end{cases} \tag{3}$$

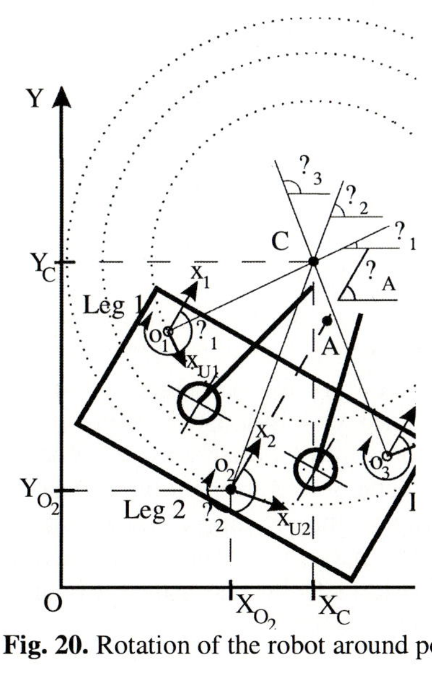

Fig. 20. Rotation of the robot around point C

6.3 Advanced Motion

This motion consists of a (simultaneous) superposition of a rotation motion at an angle θ and a linear motion at a distance L. One advantage of this method is a possibility to move along an optimal trajectory in minimal time. Its disadvantage is that the actual direction of the linear robot motion is determined by its current orientation and must be continuously corrected (Figure 21).

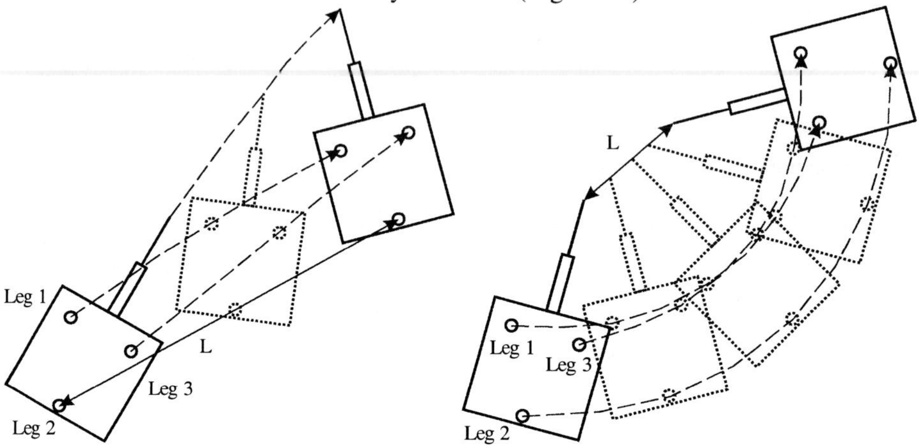

Fig. 21. Robot motion during the coordinated motion
a) coarse motion: rotation center – leg 2; b) fine motion: rotation center – endeffector

To determine the motion vectors $U_i = (x_{Ui}, \varphi_{Ui})$, i=1,...,3 an interaction of both a rotating and a translational component of platform movement must be considered. Vector U_i for piezoleg i is calculated as a sum of motion vectors U_i^{trans} and U_i^{rotat} :

$$U_i = U_i^{trans} + U_i^{rotat} \tag{4}$$

7 Conclusion

In this paper, we have presented a microassembly system based on mobile microrobots. It incorporates several vision sensors to perform robot motions and micromanipulations under sensor control either on an XY-stage under an optical microscope or inside the vacuum chamber of a scanning electron microscope.

Manipulation of objects smaller than 1 mm requires special methods due to micro-effects like adhesion of the parts to the gripper. These effects make vision systems indispensable; they also have to be taken into account both at the control and already at the planning level.

The micromanipulation station uses a multi-robot approach to speed up assembly by parallel operations. This also makes special operations possible which cope with micro-effects, like a “helping hand”.

The control system of the presented station is tailored to the requirements of microassembly and is highly scaleable due to a transparent mapping of physical components in the station to logical objects. Thereby, adding a new component like a robot is very easy. Moreover, the representation of physical microobjects by logical objects grants optimal adaptability to new types of objects.

8 Acknowledgements

This paper is based on research done at the Institute for Process Control and Robotics (Head - Prof. Dr.-Ing. H. Woern). The research work has been supported by the German federal ministry for education, science, research and technology (Grant No. 16SV600/0), and the European Commission (Esprit Project “MINIMAN”, Grant No. 33915).

References

[1] Fatikow S, Magnussen B. and Rembold U. (1995): A Piezoelectric Mobile Robot for Handling of Microobjects, Proc. of the Int. Symp. on Microsystems, Intelligent Materials and Robot, Sendai

[2] Santa K., Fatikow S. and Felso G. (1999): Control of microassembly-robots by using fuzzy-Logic and neural networks, Journal Computers in Industry, Vol. 39, 1999, Elsevier Science, Netherlands

[3] Fatikow S. (1996): An Automated Micromanipulation Desktop-Station Based on Mobile Piezo-electric Microrobots, Proc. of the SPIE. Symp. on Intelligent Systems & Advanced Manufacturing, 2906, Boston

[4] Allegro S. and Jacot J. (1997): Automated Microassembly by Means of a Micromanipulator and External Sensors, Proc. of the Int. Conf. Microrobotics and Micromanipulation, SPIE ’97, Vol. 3202, Pittsburgh, USA

[5] Allegro S. (1997): Use of a Leica DM RXA Microscope as Optical Sensor for Automated Microassembly, Scientific and Technical Information Vol. XI, No. 5

[6] Santa, K., Seyfried, J., Fatikow, S., and Munassypov, R. (1997): Control of a three-leg piezoelectric microrobot with two friction-driven manipulators, Proc. of Micromechanics Europe, Southampton

[7] Woern H., Seyfried J., Fatikow S. and Santa K. (1998): Information Processing in a Flexible Robot-Based Microassembly Station, Proc. of the Int. Symp. on Information Control Problems in Manufacturing, Nancy

[8] RTL: RT-Linux as an Embedded Operating System, Jerry Epplin, Embedded Systems Programming, October 1997

[9] Fatikow S., Seyfried J., Fahlbusch St., Buerkle A. and Schmoeckel F. (1999): A Flexible Microrobot-based Microassembly Station, Journal of Intelligent & Robotic Systems, in print

[10] Buerkle A. and Fatikow S. (1999): Computer Vision Based Controll System of a Piezoelectric Microrobot, Proc. of the Intl. Conf. on Computational Intelligence for Modelling, Control and Automation, Vienna, Austria

[11] Fatikow S., Munassypov R. and Rembold U. (1998): Assembly Planning and Plan Decomposition in an Automated Microrobot-Based Microassembly Desktop Station", Journal of Intelligent Manufacturing, No. 9, Chapman & Hall, London.

[12] Woern H., Seyfried J., Fatikow S. and Faizullin A. (1999): Assembly Planning in a Flexible Micro-assembly Station, Proc. of the International Workshop on Intelligent Manufacturing Systems, Leuven, Belgium

[13] Fatikow S., Faizullin A. and Seyfried J. (1999): Computer Aided Planning System of a Flexible Microrobot-based Microassembly Station"; Proc. of the 7th International Workshop on Computer Aided Design Theory and Technology, Wien, Austria

[14] Homem de Mello L.S and Sanderson A.C (1990): AND/OR Graph representation of assembly plans, IEEE Transactions on Robotics and Automation 6(2)

[15] Pancerella C. and Whiteside R. (1996): Using CORBA to integrate manufacturing cells to a virtual enterprise, SPIE Vol. 2913, Proc. of Plug and Play Software for Agile Manufacturing, Boston, USA

[16] Whiteside R., Pancerella C. and Klevgard P. (1997): A CORBA-Based Manufacturing Environment, Proc. of the Hawaii Intl. Conf. on System Sciences, Maui, Hawaii

[17] Real-time Corba, OMG TC Document orbos/99-02-12, March 1, 1999, http://www.omg.org/cgi-bin/doc?orbos/99-02-12

[18] Kavraki, L. and Kolountzakis, M. (1995): Partitioning a Planar Assembly into two Connected Parts is NP-complete, Information Processing Letters, vol. 55

[19] Kavraki, L., Latombe, J.-C., and Wilson, R. (1993): On the Complexity of Assembly Partitioning, Information Processing Letters, vol. 48

[20] Dr. Fritz Faulhaber GmbH Co. KG, http://www.faulhaber.de/dseiten/mikrom.htm

A Formalisation of the Evolution of Software Systems[1]

Juan Jesús Torres Carbonell[1] and José Parets-Llorca[2]

[1]Secretaría General de Comunicaciones. Ministerio de Fomento.
Plaza de Cibeles, s/n. 28071 Madrid.
jj.torres@sgc.mfom.es
[2]Dpto. Lenguajes y Sistemas Informáticos. E.T.S. Ingeniería Informática
Avda. Andalucía, 38, 18071 Granada.
jparets@ugr.es

Abstract. One of the most interesting and probably more difficult challenges in developing Software Systems is the modelling of their evolutionary capacity, that is to say, to gather the possibility that the Software Systems will go ahead in the future with the necessary changes to adapt to the environment using a different and new functionality. Modelling this evolution requires to have into account what kind of changes and modifications could follow and support a Software System during its life and also during its development. The evolutionary characteristics of a Software System can be approached by abstract evolutionary models, which can be further formalised. This formalisation makes operational the abstract evolutionary models and allows a kind of representation of the evolutionary process that could support the specification and mapping into concrete specification and implementation tools. These tools further allow us to obtain concrete and functional Software Systems.

1 Introduction

Software Systems change and suffer modifications imposed by the modeller (developer) during their development process. This is not the unique evolution that Software Systems must support. They must evolve later along their functioning life, because they are part of Information Systems and the interaction with the environment concern them in a way that requires to perform changes in the structure or functionality in order to continue being useful. That is to say, they ought to assimilate the modifications suffered by the Information System, as well as the new requirements not expressed before.

With these evolutionary necessities in mind, we can think over two forms of evolution of Software Systems. The modeller is the most important element in both because he is responsible of the design and modelling process and the capacity of the Software System to evolve when immersed in the Information System:

1) The modeller driven evolution: implies a direct intervention of the modeller changing the Software System.

[1] This research is supported by a R+D project of the Spanish CICYT (TIC97-0593-C05-04).

F. Pichler, R. Moreno-Díaz, and P. Kopacek (Eds.): EUROCAST'99, LNCS 1798, pp. 435–449, 2000.

2) 7The self-evolution of the Software System depending on certain mechanism defined by the modeller.

In a previous work [10], we presented a biological approachto the modelling of the evolution of Software Systems, using an interdisciplinary point of view within the System Theory framework [4] and having in mind the existence of some degree of isomorphism [2], that allows us the use of models from different disciplines. Later we proposed [11] a general framework of the modelling of the evolution process of Software Systems in which a three layerstructure was presented:

1) Abstract level: Mechanisms and models of software evolution related to the activity of the modelling (development) systems.
2) Formalisation level: Formal models of evolution which try to make operational the previous mechanisms and models.
3) Specification level: The previous formal models should be mapped into concrete representation and implementation tools, which allow us to obtain functional Software Systems.

In this paper the development of the formalisation level will be presented using the following organisation:

Firstly we introduce the elements of the structure of a Software System that will be needed for further explanations. Secondly, the mechanisms of evolution are presented, followed by the enumeration of the abstract models of evolution. Later, a detailed explanation of the formalisation of the models is developed. Finally, the possible applications of the formal models will be outlined.

2 Structure of a Software System

For us, following Parets [6] and Parets et al. [7] [9], a Software System (SS) is a set of processors that interact between them and with the environment, in such a way that the whole Software System could be viewed as a processor from the functional point of view. Furthermore, the functionality of a Software System as an object is reached by the processors that belong to it, through the activation of theirs actions. Parets [6] proposes a representation of Software Systems using concepts from the General System Theory [4]: processors, environments, systems, actions, events and decisions. This structure use a historical representation of the functioning (System Functional History) and a historical representation of the structure (System Structural History). Parets introduces the concept of Software Metasystem (MS): a System that allows the interaction between the Software System and the Development System. The elements of this basic structure are the followings (Figure 1):

1. *Action Interface (AI)*: Using this interface messages and functional actions pass to and from de System (the SS and the MS). This is the way to communicate with the System's *functional* environment.
2. *Evolution Interface (EI)*: Through this interface structural actions and messages related to them pass to and from the System. Actions and messages come from the Metasystem to the Software System, and from the Modeller to the Metasystem (the *development* environment).
3. *Processing Structure*: Set of processors that work within the system, performing actions.

4. *System Functional History (SFH)*: Memory of the functional actions, represented by functional events, performed by the System or its processors.
5. *System Structural History (SSH)*: Memory of the structural actions developed over the system, represented by structural events.
6. *System Decisional History (SSH)*: Memory of the decisional actions developed over the system, represented by decisional events.
7. *Decision Subsystem (DS)*: Processors that take decisions about the actions performed by the System
8. *Genetic Subsystem (GS)*: Processor that take decision about the evolutionary actions.

The actions are related to events that symbolise them. These events are symbols of the execution of actions and are recorded in the history of the System. Anaya et al. [1] proposed that the functionality of a System is determined by:

1. Spontaneous actions performed by the System's processors when the established conditions are satisfied according to the Decision Subsystem.
2. Execution of actions required by the environment through the Action Interface.
3. Execution of actions started by 1) and 2).

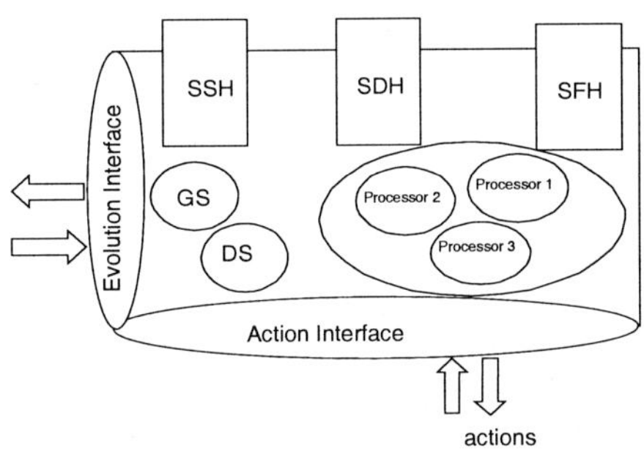

Fig. 1. Elements of the basic structure of a Software System

Functional Structure The functional structure of a System, common to the Software System and the Metasystem, isthe following:

Definition 1 (System Structure). The System Structure is defined as SS =(PS, ES, M) where PS is theprocessing structure, ES is the entry structure and M is the system memory, s.t. (such that):

System Structure: (PS, ES, M)
PS = {P | P is a processor for the System}
P = { AS | AS is an action in the System}
ES = {AS | AS is an action of the System Interface}

This definition allows that the System behaves as a processor, and so it can replace a processor. We could say that the ES of a System is a processor.

AS = (f_{ACT}, Conditions)

f_{ACT}: Activity Function.

Conditions: established over the activity function.

M = memory of the System: (SFH, SSH, SDH)

SFH = sequence of stimulus happened over the ES and events of the PS. Represents the sequence of actions performed through time, i.e. the *functional state* of the System.

SSH = sequence of structural events that correspond to functional events of the Metasystem. Represent the *structural state* of the System.

HDS = sequence of decision events previous, during and at the end of the actions. Represent the *decisional state* of the System.

Following Parets [5], we think that a Metasystem, which is part of the development system, exists. Its main function is to perform changes in the structure of the Software System. The Software System and the Metasystem are isomorphic, that is to say, they have both the same structure, and the Functional History of the Metasystem records its functional actions. This implies that the Structural History of the System is a subset of the Functional History of the Metasystem.

Both Software System and Metasystem can perform several adaptations following the same models. In the rest of the paper we will use the term *System* to include Software Systems and Metasystems, unless we make an explicit distinction.

3 Mechanisms of Evolution

Mechanisms are abstract schemes of co-ordinated activities of evolution performed by the different components of a System. This activities produce a modification of the System. Mechanisms represent the ways used by a System to change itself in order to evolve

We propose two kinds of mechanisms borrowed from biology: *Heredity mechanisms* and *Adaptation mechanisms.*

Due to its high level of abstraction, both of them are, as general concepts, independent of any structure, although they will have a concrete representation depending on the structure. The mechanisms are also independent of the evolution model used.

In certain way, we must accept that the modeller act as Meta-Mechanism that is capable of conducting the role of the mechanisms.

3.1 Adaptation

Adaptation is based in the necessity of accommodation, learning, mutation and differentiation according to the requirements of the environment. This is the general way used to modify its structure (assimilation, adaptation by mutation/differentiation) or to modify the use of its structure (accommodation, adaptation by accommodation/learning).

Adaptation by Accomodation/Learning. We understand the concept of accommodation as the possibility of modifying the interrelations of the System processors without modification of their composition [5]. It is a process that does not obey pre-programmed rules, in such a way that there is no uniform answer.

This kind of adaptation occurs in a favourable *functional* environment, in which the System is able to act using the actions implemented in its action interface. We accept the possibility that a learning process that helps the System to reach the better answer exists. This implies that a System can adapt to the environment, learning the better way of using its own structure without changing it.

Adaptation by Mutation/Differentiation. This is a more radical process than Accommodation/Learning, and it will imply structural changes, producing new possibilities of accommodation. This kind of Adaptation occurs within a resisting environment.

The structural change requires the intervention of the Metasystem, because Systems in general and Software Systems in particular are not selfending (autofinalitarios, buscarlo en bibliografía). If the affected is the Metasystem, the changes require the intervention of the Modeller (as Meta-Metasystem).

3.2 Heredity

The heredity mechanisms is the general way used to produce *descendant* Software Systems that inherits the adaptations performed by their parent(s). Some inference engine, with decision capacities and present in the system, decides which will be the structure of the next generation of descendant Software Systems. This decision will take into account the structural properties with adaptive survival value present in the historical memory of the system.

We have to remark that not exclusively the initial structures are inherited. The modifications performed as a consequence of adaptations by mutation/differentiation, and the capacity to go ahead with adaptations by accomodation/learning are also inherited. This implies that structural modifications which involves to the way used to adapt by accommodation learning can be introduced during the evolution process.

4 Models of Evolution

The second element of the Abstract Level are the Models of Evolution. This Models are symbolic representations of the possible ways of introducing changes in the Software System by the interaction of the Adaptation mechanism, the Heredity mechanism and the Modeller. In considering these models we have into account the abstract structure of interaction proposed in [6]:

a) The Modeller which decides what changes have to be carried out in the Software System.

b) The Metasystem which symbolises the modeller actions.

c) The Software System which *suffers* changes.

Several evolutionary models were presented in [11], all of them supported by the Adaptation and Heredity mechanisms:

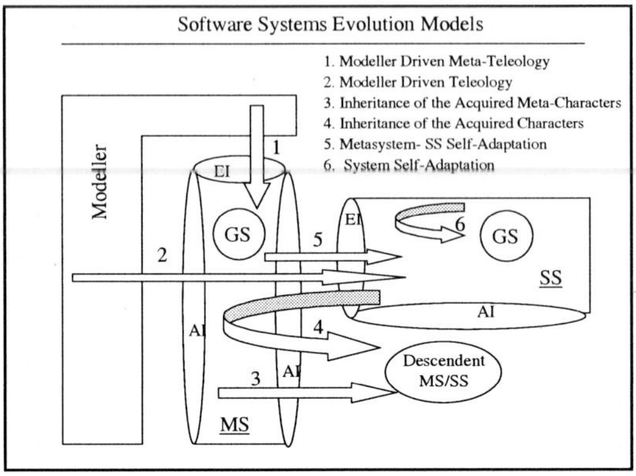

Fig. 2. Relation between Modeller, Metasystem, Software System and Models of Evolution

The Models proposed are:

Modeller Driven Meta-Teleology. The Modeller carries out modifications of the structure of the Metasystem (Figure 2, arrow 1). This Model applies the Mechanism of Adaptation by Mutation/Differentiation.

In this case, the activity of the Modeller is performed using actions of the Evolution Interface of the Metasystem.

This modifications could be stored in the descendant Metasystems. The modeller can change the viewpoint and discernment of the MS and therefore change the criterions used to decide about adaptation and inheritance.

Modeller Driven Teleology. The Modeller through the Action Interface of the Metasystem produces changes in the Software System (Figure 2, arrow 2). This Model applies the Mechanism of Adaptation by Mutation/Differentiation.

This model, as the former, reflects the direct intervention of the Modeller in the evolution process, through the Action Interface of the Metasystem and the Evolution Interface of the Software System.

This changes also are supposed adaptive and potentially inheritable, then they could be incorporated to descendant Software Systems.

Inheritance of the Acquired Meta-Characters. A new Descendant Metasystem, which could inherit the new meta-characters acquired by the old MetaSystem during its evolution, is produced (Figure 2, arrow 3). This Model applies the Mechanism of Heredity.

The initiative to produce a new system belongs to the Modeller, even in the case when the Genetic Subsystem of the Metasystem take part in the process.

Inheritance of the Acquired Characters. A new Software System is produced, which could inherit the new characters acquired by the old Software System during its evolution (Figure 2, arrow 4). This Model also applies the Mechanism of Heredity.

In this case the initiative to produce a new system begins in the Modeller, like in the last model, and the Genetic Subsystem of the Metasystem also takes part. The Genetic Subsystem of the Software System is not involved in this type of changes, because the aim of this process is the production of a new entity.

Metasystem-SS Self-Adaptation. The Metasystem interacts with the Software System in order to produce structural changes of the S.S. without intervention of the modeller (Figure 2, arrow 4). This Model applies the Mechanism of Adaptation by Mutation/Differentiation.

The decision about the adaptation is taken by the Genetic Subsystem of the Metasystem

System Self-Adaptation. The Software System (or Metasystem) performs an adaptive process without intervention of the Modeller and the Metasystem (Figure 2, arrow 6). This Model applies the Mechanism of Adaptation by Accomodation/Learning and does not produces structural modifications.

This model was not included in former proposal [10], [11]. We think that it must be included to gather the possibility of adaptation of a SS by itself , with no structural change,.

In this case the decision is taken by the Genetic Subsystem of the Software System, because there is no structural modification

5 Formal Models

In order to make operational the previous mechanisms and models, we could try to find concrete representations in programming languages or in software development tools. This approach, very practical, has the problem of saving the distances between the previous concepts, proposed in an abstract level, and the concepts used in programming languages and tools. We consider that an intermediate step, consisting of the formalisation of the previous models in order to specify evolutionary structures, can be very fruitful and translatable to other representations.

After studying different formal tools, usually applied in computer science, as Petri Nets and state transition diagrams, we considered that the formalism proposed by Holland [3] have the components that we need. We have to remark that the Formal Models are independent of the specification further developed from them.

The formalisation is established according to *adaptive plans*, *operators*, *structures* and *adaptation or fitness functions* of the Models of Evolution. The main objective is to explain the Models of Evolution as adaptive plans that apply operators to structures of the Software System or MetaSystem. These adaptive plans evaluate functions about the performed adaptations trying to establish its viability and interest.

Figure 3 present the elements of this formalisation:

- AI — Action Interfaces of the Software System and Metasystem.
- EI — Evolution Interface of the Software System and Metasystem.
- GS — Genetic Subsystem of the Software System and Metasystem.
- A_k — Structure Ak of the MS.
- A_n — New structure of the MS after the application of operators O_n.
- A_i — Structure Ai of the SS.
- A_j — New structure of the SS after the application of operators O_j.
- I_e — Inputs from the environment.
- I_m — Inputs from the Modeller.
- $F(A_i)$ — Function fitness of the SS with structure A_i.
- P_i — Adaptive plan which implies the application of the operators O_i.
- $F(A_k)$ — Function fitness of the MS with structure A_k .
- O_i — Operators that do not modify the structure A_i of the SS .
- P_j — Adaptive plan elaborated by the MS, that implies the application of O_j.
- O_j — Operators that modify the structure A_i of the SS, yielding the new structure A_j.
- P_k — Adaptive plan elaborated by the MS, that implies the application of the operators O_k.
- O_k — Operators that do not modify the structure A_k of the MS.
- P_n — Adpative plan elaborated by the MS, that implies the application of the operators O_n.
- O_n — Operators that modify the structure A_k of the MS, yielding the new structure A_n.

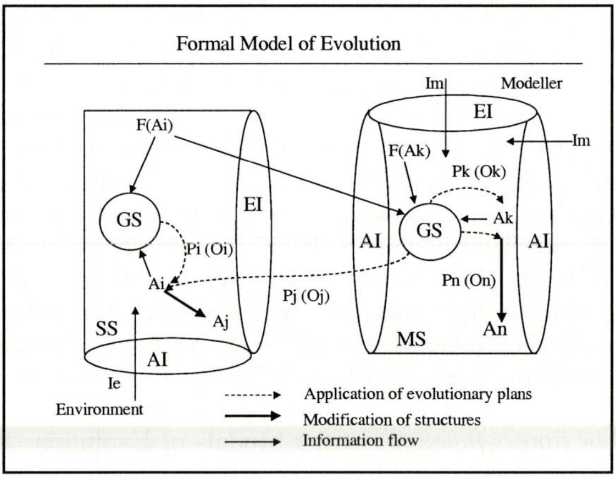

Fig. 3. Relations between the elements of the formalisation, Modeller, environment, Software System and Software Metasystem

In the next paragraphs we present the formalisation of each evolutionary model.

5.1 Modeller Driven MetaTeleology

Definition 2 (Modeler Driven MetaTeleology). Let a Software MetaSystem ,MS=($PS|_{MS}$, $ES|_{MS}$, $M|_{MS}$) . An Evolutionary process by Modeller Driven MetaTeleology, MDMT, is a function defined over the Software MetaSystem, MS, s.t.:

$$\text{Source (MS)} \xrightarrow{\text{MDMT}} \text{Target (MS)}$$

If and only if MDMT use the Mechanism of Adaptation by Mutation/Differentiation, and one of the following conditions is satisfied:

i) $\exists\, AS_1 \in EI|_{MS} \subset ES|_{MS}$, s.t.,

$$\text{source }(PS|_{MS}) \xrightarrow{AS_1} \text{target}(PS|_{MS}) \Rightarrow \text{Target }(PS|_{MS}) \neq \text{Source }(PS|_{MS})$$

and then Source (MS) $\neq$ Target (MS)

ii) $\exists\, AS_1 \in EI|_{MS} \subset ES|_{MS}$, s.t.,

$$\text{source }(ES|_{MS}) \xrightarrow{AS_1} \text{target}(ES|_{MS}) \Rightarrow \text{Target }(ES|_{MS}) \neq \text{Source }(ES|_{MS})$$

and then Source (MS) $\neq$ Target (MS)

That is to say :

There is a structural change of the Software MetaSystem through its Evolution Interface. This structural change can affect the Processing Structure (i) or the Entry Structure (ii).

Then, we have the following function:

$$\text{MDMT} : I_m \times F(A_k) \times A_k \longrightarrow P_n = \{O_n \mid O_n = AS \in EI|_{MS} \subset ES|_{MS}\}$$

That is to say, an evolutionary process by Modeller Driven MetaTeleology is a function of the input from the modeller, I_m , the function fitness of the Software MetaSystem, $F(A_k)$, and the actual structure of the Software MetaSystem, A_k. As consequence an operator, O_n, that modifies the structure, A_n, is applied.

5.2 Modeller DrivenTeleology

Definition 3 (Modeller Driven Teleology). Let a Software System, SS=($PS|_{SS}$, $ES|_{SS}$, $M|_{SS}$), and a Software Metasystem, MS=($PS|_{MS}$, $ES|_{MS}$, $M|_{MS}$). An Evolutionary process by Modeller Driven Teleology, MDT, is a function defined over the Software System, SS, s.t.:

$$\text{Source (SS)} \xrightarrow{\text{MDT}} \text{Target (SS)}$$

If and only if MDT use the Mechanism of Adaptation by Mutation/Differentiation, and one of the following conditions is satisfied:

(i) $\exists\, AS_1 \in AI|_{MS} \subset ES|_{MS}$ and $\exists\, AS_2 \in EI|_{SS} \subset ES|_{MS}$, s.t.,

$$\text{source }(PS|_{SS}) \xrightarrow{AS_1 \bullet AS_2} \text{target}(PS|_{SS}) \Rightarrow \text{Target }(PS|_{SS}) \neq \text{Source }(PS|_{SS})$$

and then Source (SS) $\neq$ Target (SS)

(ii) $\exists\, AS_1 \in AI|_{MS} \subset ES|_{MS}$ and $\exists\, AS_2 \in EI|_{SS} \subset ES|_{SS}$, s.t.,

$$\text{source}(ES|_{SS}) \xrightarrow{AS_1 \bullet AS_2} \text{target}(ES|_{SS}) \Rightarrow \text{Target}(ES|_{SS}) \neq \text{Source}(ES|_{SS})$$

and then Source (SS) $\neq$ Target (SS)

That is to say:

There is a structural change of the Software System through the Action Interface of the Metasystem Software and the Evolution Interface of the Software System. This structural change can affects the Processing Structure (i) or the Entry Structure (ii).

So, we have the following function:

$$MDT : (I_e \times F(A_i) \times A_i) \bullet (I_m \times F(A_k) \times A_k) \longrightarrow$$

$$P_j = \{O_j \mid O_j = AS_1 \bullet AS_2, \text{ s.t. } (AS_1 \in IA|_{MS} \subset ES|_{MS}) \text{ and } (AS_2 \in EI|_{SS} \subset ES|_{SS})\}$$

That is to say, an evolutionary process by Modeller Driven Teleology is a function of the input from the environment, I_e , the function fitness of the Software System, $F(A_i)$, the actual structure of the Software System, A_i; and could take into account the input from the modeller, I_m , the function fitness of the MetaSystem, $F(A_k)$, and the actual structure of the Metasystem, A_k. As consequence an operator, O_j, that modifies the structure, A_j, is applied.

5.3 Inheritance of the Acquired MetaCharacters

Definición 4 (Inheritance of the Acquired MetaCharacteres). Let a Software Metasystem, $MS=(PS|_{MS}, ES|_{MS}, M|_{MS})$. An Inheritance of Acquired MetaCharacters, IAMC, is a funtion defined over the Software MetaSystem, MS, s.t.:

$$\text{Source}(MS) \xrightarrow{IAMC} \text{Target}(MS)$$

If and only if IAMC use the Mechanism of Heredity, and then the following condition is satisfied:

$$\exists AS \in EI|_{MS} \subset ES|_{MS}, \text{ s.t., source}(MS) \xrightarrow{AS} \text{target}(MS) \Rightarrow$$
$$\{\text{Target}(PS|_{MS})=\text{Source}(PS|_{MS}) \text{ and}$$
$$\text{Target}(ES|_{SS})=\text{Source}(ES|_{SS}) \text{ and}$$
$$\text{Target}(M|_{SS})=\varnothing\}$$

That is to say:

There is an action of the Evolution Interface of the Metasystem that produce a new Software Metasystem with the same structure than the old one, and whose Memory is empty.

So, we have the following function:

$$IAMC: I_m \times F(A_k) \times A_k \longrightarrow P_k'=\{O_k' \mid O_k' = AS \in EI|_{MS} \subset ES|_{MS}\}$$

That is to say, an evolutionary process by Inheritance of Acquired MetaCharacteres is a function of the input from the modeller, I_m , the function fitness of the Software MetaSystem, $F(A_k)$, and the actual structure of the Software MetaSystem, A_k. As consequence an operator, O_k', that produces a new Software Metasystem, A_k', is applied.

5.4 Inheritance of the Acquired Characters

Definición 5 (Inheritance of the Acquired Characteres). Let a Software System $SS=(PS|_{SS}, ES|_{SS}, M|_{SS})$, and a Software Metasystem $MS=(PS|_{MS}, ES|_{MS}, M|_{MS})$. An Inheritance of Acquired Characters, IAC, is a funtion defined over the Software System, SS, s.t.:

$$\text{Source (SS)} \xrightarrow{\text{IAC}} \text{Target (SS)}$$

If and only if IAC use the Mechanism of Heredity, and then the following condition is satisfied:

$$\exists AS \in AI|_{MS} \subset ES|_{MS}\text{, s.t., source (SS)} \xrightarrow{AS} \text{target(SS)} \Rightarrow$$
$$\{\text{Target}(PS|_{SS})=\text{Source }(PS|_{SS}) \text{ and}$$
$$\text{Target}(ES|_{SS})=\text{Source}(ES|_{SS}) \text{ and}$$
$$\text{Target}(M|_{SS})=\varnothing\}$$

That is to say:

There is an action of the Action Interface of the Metasystem that produce a new Software System with the same structure than the old one, and whose Memory is empty.

So, we have the following function:

$$IAC : I_m \text{ x } F(A_i) \text{ x } A_i \longrightarrow P_i\text{'}= \{O_i\text{'} \mid O_i\text{'}=AS\in AI|_{MS} \subset ES|_{MS}\}$$

That is to say, an evolutionary process by Inheritance of Acquired Characteres is a function of the input from the modeller, I_m , the function fitness of the Software System, $F(A_i)$, and the actual structure of the Software System, A_i. As consequence an operator, O_i', that produces a new Software System, A_i', is applied.

5.5 Metasystem-Software System SelfAdaptation

Definición 6 (Metasystem-Software System SelfAdaptation). Let a Software System $SS=(PS|_{SS}, ES|_{SS}, M|_{SS})$, and a Software Metasystem $MS=(PS|_{MS}, ES|_{MS}, M|_{MS})$. An evolutionary proces by Metasystem-Software System SelfAdaptation, MSSS, is a function defined over the Software System, SS, s.t.;

$$\text{Source (SS)} \xrightarrow{\text{MSSS}} \text{Target (SS)}$$

If and only if MSSS use the Mechanism of Adaptation by Mutation/Differentiation and one of the following conditions is satisfied:

(i) $\exists AS_1 \in AI|_{MS} \subset ES|_{MS}$ and $\exists AS_2 \in EI|_{SS} \subset ES|_{SS}$, s.t.,

$$\text{source }(PS|_{SS}) \xrightarrow{AS_1 \bullet AS_2} \text{target}(PS|_{SS}) \Rightarrow \text{Target }(PS|_{SS}) \neq \text{Source }(PS|_{SS})$$

and then Source (SS) $\neq$ Target (SS)

(ii) $\exists AS_1 \in AI|_{MS} \subset ES|_{MS}$ and $\exists AS_2 \in EI|_{SS} \subset ES|_{SS}$, s.t.,

$$\text{source }(ES|_{SS}) \xrightarrow{AS_1 \bullet AS_2} \text{target}(ES|_{SS}) \Rightarrow \text{Target }(ES|_{SS}) \neq \text{Source }(ES|_{SS})$$

and then Source (SS) $\neq$ Target (SS)

That is to say:

There is a structural change of the Software System through the Action Interface of the Metasystem Software and the Evolution Interface of the Software System. This structural change can affects the Processing Structure or the Entry Structure. The difference between this model and Modeller Driven Teleology is that in MetaSystem Software SystemSelfAdaptation there is not intervention of the Modeller.

So, we have the following function:

$$\text{MSSS}: (I_e \times F(A_i) \times A_i) \bullet (F(A_k) \times A_k) \longrightarrow P_i = \{O_i \mid O_i = AS_1 \bullet AS_2, \text{ s.t. } (AS_1 \in IA|_{MS} \subset ES|_{MS}) \text{ and } (AS_2 \in EI|_{SS} \subset ES|_{SS},)\}$$

That is to say, an evolutionary process by MetaSystem-Software System SelfAdaptation is a function of the input from the environment, I_e , the function fitness of the Software System, $F(A_i)$, the actual structure of the Software System, A_i; and could take into account the function fitness of the MetaSystem, $F(A_k)$, and the actual structure of the Metasystem, A_k. As consequence an operator, O_j, that modifies the structure, A_j, is applied without intervention of the Modeller.

5.6 System SelfAdaptation

This model of evolution can be used by the Software System and the Metasystem. So, in the formalisation definiton we will refer to System.

Definición 7 (Software System SelfAdaptation). Let a System S=(PS, ES, M). An evolutionary process by System SelfAdaptation, SSA, is a function defined over the System, S, s.t.;

$$\text{Source}(S) \xrightarrow{\text{SSA}} \text{Target}(S)$$

If and only if SSA use the Mechanism of Adaptation by Accomodation/Learning and one of the following conditions is satisfied:

i) $\exists AS \in PS, \text{ s.t., source}(M) \xrightarrow{\text{SSA}} \text{target}(M) \Rightarrow \text{Target}(M) \neq \text{Source}(M)$

ii) $\text{Target}(f_{ACT}|_{PS}) \neq \text{Source}(f_{ACT}|_{PS})$

iii) $\text{Target}(f_{ACT}|_{ES}) \neq \text{Source}(f_{ACT}|_{ES})$

That is to say:

i) There is a change in the Memory of the System caused by an action of the Processing Structure of that system that affects the further reasoning process and so it is an adaptation.

ii) There is a change in the function of activity, that is, a change in the way used by the System to do the action

iii) There is a Change in the Action of the System's Interface.

So, we have the following functions:

Applied to a Software System:

$$\text{SSSA}: (I_e \times F(A_i) \times A_i) \longrightarrow P_i = \{O_i \mid O_i = AS, \text{ s.t. } AS_1 \in PS|_{SS}\}$$

Applied to a Software Metasystem:

$$\text{MSSA}: (I_m \times F(A_k) \times A_k) \longrightarrow P_k = \{O_k \mid O_k = AS, \text{ s.t. } AS \in PS|_{MS}\}$$

That is to say, an evolutionary process by System SelfAdaptation can take one of the following faces:

1. When performed by a Software System, it is a function of the input from the environment, I_e , the function fitness of the Software System, $F(A_i)$ and the actual structure of the Software System, A_i. As consequence an operator, O_i, that does not modify the structure, A_i, is applied.
2. When performed by a Software MetaSystem, it is a function of the input from the modeller, I_m (but not the decision), the function fitness of the Software MetaSystem, $F(A_k)$ and the actual structure of the Software MetaSystem, A_k. As consequence an operator, O_k, that does not modify the structure, A_k, is applied.

5.7 Overview of the Evolutionary Process

The proposed formalisation establishes the existence of different structures during the life of a System These structures are produced by the application of two kinds of operators

1. Operators which, modify it (i.e., O_j, O_n),
2. Operators which impliesdifferent "uses" or states of the.System. These operators do not modify the structure, but the way the System uses it (i.e., O_i, O_k).

This implies that a System could present a structure and use at a concrete moment (instant). This allows to conceive the evolutionary process as follows (Figure 4):

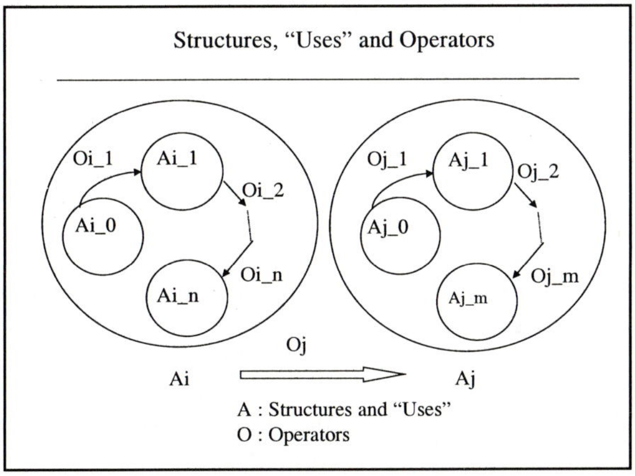

Fig. 4. Transformation between structures and uses

1. Within a structure (i.e., A_i) the different ways of use of that structure represent different states or "uses" (i.e., A_{i_0}, …, A_{i_n}).
2. The structure and the "use" identify the present System's situation, which can be call STAGE of evolution.
3. Moving from the state A_{i_n-1} to the state A_{i_n} requires the application of the operator O_{i_n} that does not modify the structure.
4. Moving from the structure A_i to the structure A_j requires the application of the operator O_j.
5. Given a System an evolutionary process with no structural modification can exists, just moving from one state to another within the current structure.

6. When a new System is produced by cloning or development ex-novo, it is thrown in the state ("use") A_{0_0}.
7. The state A_{0_0} represents the initial definition of the structure. This structure could be the structure initially defined by the modeller or that of the System from which is produced .
8. Given a structure, whatever use or state, the System can move to another System, that is to say, from any state within a structure the System can move to another state that belongs to a different structure.
9. When the structure of a System is modified, for example from A_i to A_j, the new state reached is A_{j_0} which is the first state of the structure A_j. There is *no a-priori* defined characteristics about the conditions that state A_{j_0} must accomplish, because the following modification at each instant is unknown
10. The number of reachable states and structures is, *a priori*, finite.

6 Utility of the Formalised Models and Further Research

The previous formalisation allows the manipulation of the models of evolution. The modelling of the evolutionary characteristics of Software Systems requires a double perspective: Firstly, from the point of view of the selfsame modelling process followed by the modeller, and, secondly, from the point of view of the future adaptation of a functional Software System during its life.

The modelling process. Applying the formalised framework two important aspects must be considered:

a) *The inclusion in a modelling methodology*In this case the modeller should take into account the characteristics of the future SS in the developed models. This requires an adaptation of the methodology in order to provide the necessary diagrams, actions, or schemes. At present we are working on the possibility of adaptation of UML [12] in order to include the evolutionary process following the pattern established by the proposed.formalisation
b) *The adaptation of CASE tools.* These tools ought to offer the possibility of mapping the evolutionary characteristics into the developed SS . In order to do that the firs step is the specification of the formalised models. The framework of levels proposed in [11] allows the use of different Knowledge Representation Tools to implement the formal models. During the selection of the Knwoledge Representation Tool, two important aspects have to be considered: the K.R.T. will determine the instance of Software System that finally will be generated; and that reasoning about evolution requires reasoning about the time.

An example of this kind of tools is being implemented by our group under the name of HEDES Software Specification, Design and Evolution Tool), using temporal logic and object-oriented programming as specification tools [8].

The life of the Software System. This point of view must paid attention to the future necessities of the Software System, concerning evolutionary characteristics and mainly the possibility of further adaptation. The necessity of adaptation and evolution can appear in a variety of systems that must adapt (accomodate) themselves to different conditions of hardware or environment pressure whilst they are funtioning. An interesting example of this kind of systems are autonomous agents that must perform an activity in not well known environmentes.

These capacities of evolution and adaptation of S.S. can be modelled using the formalisation proposed and the abstract models to establish what kind of evolution is appropriate. We are researching on this possibility in the HEDES project.

References

1. Anaya,A. ; Rodríguez,M.J.; Parets,J.. (1996). 'Representation and Management of Memory and Decision in Evolving Software Systems' in: 'Computer Aided Systems Theory - EUROCAST'97'. Pichler,F. ; Moreno-Días,R. (Eds.). Springer-Verlag. LNCS 1333. p. 71- 82.
2. Bertalanffy, L.Von (1968). 'General System Theory'. New York. George Braziller, Inc. (fr. tran. 'Théorie générale des systèmes. Paris. Dunod. 1973).
3. Holland, J.H. (1992). 'Adaptation in Natural and Artificial Systems'.The MIT Press. Cambridge. Massachusetts.
4. Le Moigne, J.L. (1977, 1983, 1990). 'La théorie du système général. Théorie de la modélisation'. Paris. Presse Universitaires de France.
5. Le Moigne, J.L. (1990). 'La modélisation des systèmes complexes'. PARIS. Dunod.
6. Parets, J. (1995). 'Reflexiones sobre el proceso de concepción de sistemas complejos. MEDES: un método de especificación, desarrolo y evolución de sistemas software'. (Ph.D. Thesis). Granada. Dpto. de Lenguajes y Sistemas Informáticos. Universidad de Granada.
7. Parets, J. ; Anaya,A. ; Rodríguez,M.J. ; Paderewski,P. (1994). 'A Representation of Software Systems Evolution Based on the Theory of the General System' in: 'Computer Aided Systems Theory - EUROCAST'93'. Pichler,F. ; Moreno-Díaz,R. (Eds.). Springer-Verlag. LNCS 763. 96- 109.
8. Parets, J.; Rodríguez, M.J.; Paderewsky, P.; Anaya, A. (1994). 'HEDES: A System Theory based tool to support evolutionary Software Systems". EUROCAST'99. 69-71. Viena. September 1999.
9. Parets,J.; Torres,J.C (1996) "Software Maintenance versus Software Evolution: An Approach to Software Systems Evolution". IEEE Conference and Workshop on Computer Based Systems (ECBS'96). Friedrichafen. March 1996. pp. 134-141.
10. Torres-Carbonell, J.J.; Parets-Llorca J.(1996) 'Biological Evolutive Models Applied to the Evolution of Software Systems'. Procc. of the Third European Congress on Systems Science. Rome 1-4 Oct. 1996. Pp 705-709.
11. Torres-Carbonell, J.J.; Parets-Llorca J.(1999). "Modelling the evolution of Software Systems: A General Framework". 4th Europenan Congress on System Sciences. Septiembre 1999. Valencia-Ibiza.
12. Unified Modelling Language. http://www.rational.com/uml. Booch, G.; Jacobson, I.; Rumbaugh, J., et al. (1997). "The Unified Modeling Language for Object-Oriented Development v 1.1, UML Summary, UML Notation Guide, UML Semantics". Rational Software Corporation. September 1997.

HEDES: A System Theory Based Tool to Support Evolutionary Software Systems[1]

M.J. Rodríguez, J. Parets, P. Paderewski, A. Anaya, and M.V. Hurtado

Departamento de Lenguajes y Sistemas Informáticos
Universidad de Granada. ETS Ingeniería Informática.
Avda. Andalucía, 38. 18071 –GRANADA (Spain)
Phone: +34 58 243 180. Fax: +34 58 243 179
mjfortiz@ugr.es

Abstract. Software systems evolve over time. Traditional software development methods and tools support partial aspects of this evolving process. Over the last few years we have researched into theoretical evolution search models which might be applied to the development of software systems. The Theory of Systems and biology have interesting views on the evolution process, very different from genetic algorithms, which may aid in the development of software systems and CASE tools. Our aim in this paper is to present a first version of a tool (HEDES) which includes these models and implements them in an object-oriented language (VisualWorks 3.0), using first-order temporal logic as support. In addition, some lessons learnt in the development of these complex and changing tools will be outlined, especially the importance of an iterative lifecycle in object-oriented development, a high degree of cohesion of the development team and the need for flexible and rapid ways of communicating new ideas using discussion meetings, Internet facilities and development support tools.

1 Introduction

Software development is a process which more and more is classified as incremental and iterative in the literature. Classic life cycles such as the waterfall model have given way to other life cycles which try to reflect this iterativity (prototyping, spiral, fountain,...). Moreover, these life cycles consider that software validation must directly involve the users by means of executable prototypes.

Our outlook, in this sense, is that software development methods and CASE tools which support them have to offer a large amount of help during the prototyping process and during transformations of the elaborate software. In order to offer this support, it would appear to be necessary to have software evolution models which might be transferred to specific tools.

For several years now, we have been carrying out research into this type of models and we have attempted to obtain operational representations. Some results from this conceptual and theoretical work have appeared in other works [16],[17],[14],[15],

[1] This research is supported by an R+D project of the Spanish CICYT (TIC97-0593-C05-04) a subproject of the MENHIR project (TIC97-0593-C05-01)

F. Pichler, R. Moreno-Díaz, and P. Kopacek (Eds.): EUROCAST'99, LNCS 1798, pp. 450–464, 2000.

[1], [2]. In this paper, we present the design and implementation of a prototype of a tool (called HEDES) which collects the results from these previous works.
In order to make the explanation understandable, we summarise our view on the evolutionary problem in the first and second sections. Then we present the design of the class structure and we briefly comment on the most important groups of classes (subsystems). Finally, we offer comments about the implementation details, and the characteristics of the life cycle used during the development of the tool are outlined.

2 Software Development: An Evolutionary Process

Since Software Engineering came into being in 1968, different authors have commented on the importance of the software development process. Most handbooks on this discipline insist on its importance and the study of the 'software process' is an active branch with strong industrial projection [10]. The main argument provided by these authors is that the quality of the product obtained (the software) depends on the quality of its development process.
[6] offer a framework which structures the points of view under which the development process can be studied. This framework, which we have reinterpreted in Figure 1, provides an interesting integrated *process-product* model. The Software Production Process is responsible for obtaining the product. The Software Process Support includes the necessary elements to support the production process (technology and production models).The Metaprocess will be responsible for evolving the previous processes. The central idea in this framework is the existence of close interdependency between the three processes and the products elaborated, furthermore its evolutionary nature, mainly conducted by the Metaprocess .

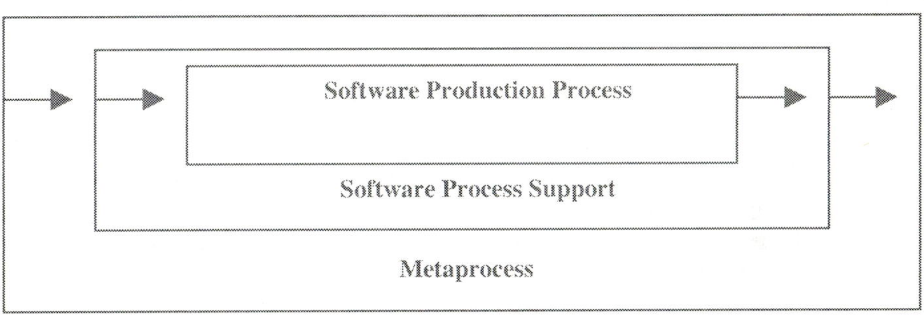

Fig.1. Software Development Procces

In parallel to this emphasis on the process, some methods and methodological tools have been developed. The main outcome of this tendency is the progressive systematization of Requirements Engineering and Design processes by means of languages and automatic tools, which offer support, in Conradi's terminology, to the Software Production Process.
However, when most of the methods and tools are examined in detail, a great abandonment of their application process is observed [16]. Each one of them has a list of models to be produced and a series of detailed steps to be applied, but very few

emphasise the fact that the models elaborated should be modifiable and navigable. The tools which implement the methods try to mitigate, generally in a heuristic way, some of these problems without offering specific schemas for transformation and navigability. Only partial contributions to these problems are offered by concepts handled in certain fields of Requirements Engineering: traceability, reusability of specifications, etc.

From our point of view, a wider and more integrated vision of the software production process is needed. This view should take into account a principal and well known argument: the Software Systems change. This change is usually produced by a change in the environment in which the Software must work and the change is effectively produced by a Development System, i.e., a group of people who are involved in producing and maintaining this software. Obviously, notions such as maintenance, reutilization, transformation of class structures, schemas evolution, etc. imply important contributions to this approach, but they are partial approximations. An abstract view of the changing process allows the development of models which integrate a great number of these concepts.

Although this view has been presented in our aforementioned works, we will try to summarise it from an applied perspective. As a starting point we should outline one important premise: we do not find global evolution models in the Software Engineering field, however, System Theory and biology have been tackling evolution problems for a long time now.

The contributions from Le Moigne [11] on System Theory are especially useful. This author indicates the possibility of identification of *no self-evolutionary* systems which are unable to evolve by themselves and, on the other hand, *self-evolutionary* systems which have the capacity of producing their own evolution. Software Systems belong to the first group and the Development Systems belong to the second. To a certain extent, the latter are intelligent systems. Le Moigne proposes a canonical representation based on three identifiable subsystems useful in modelling *no self-evolutionary systems*:

- The Operation Subsystem which executes actions in the system.
- The Information Subsystem which has an informational representation of the actions executed and the decisions taken in the system. The Information Subsystem handles the system memory.
- The Decision Subsystem which decides the actions to be executed depending on the prior system memory state.

On the other hand, biology has developed two main models for the evolutionary process:

1. The natural selection model (Darwin) and mutation: only the best adapted mutations survive.
2. The adaptive or environment pressure model (Lamarck): the individuals which can adapt in an environment incorporate these adaptations into the genetic material and transmit them to their descendants.

Darwin's model has been used as an analogy in genetic algorithms, but it is not useful in modelling software evolution process, because evolution is directed by the Development System. However, although the Lamarkian model is nowadays not admitted for biological systems, it is a very interesting approach in considering the evolution of software and allows an easy incorporation of the adaptations into the environment, .

We have incorporated Le Moigne's Theory and the adaptive model in a method called MEDES, the main objective of which is to experiment on the evolution field. The most relevant characteristics of this experimental method are described in the next section.

3 MEDES: A Model for Software Evolution

The canonical structure proposed by Le Moigne for *noself-evolutionary* system allows us to have a basic structure for a software system, consisting of three main subsystems. In addition, software evolution can be defined by considering this structure:

Definition 1.- The evolution of a Software System is a transformation of its structure (action patterns, decision patterns and memorisation) over time. These transformations will imply changes in the functionality of the system.
Because software evolution is not a spontaneous process and is performed by the Development System, we can offer a second definition which involves it:
Definition 2.- The evolution of a Software System is a transformation of the structure (action patterns, decision patterns and memorisation) over time, produced by the Development System. The evolution of the Software System is the main functional capacity of the Development System.
Because the Development System will evolve the Software System and it cannot evolve by itself, Lamark's analogy allows us to conceive the Development System as the environment which will produce pressures on the software. This evolved software will keep the most interesting modifications executed by the Development System in its structure.
A practical view of this perspective may be observed in Figure 2. The Development System will transform the Software System. In order to execute this task it must have a tool, that we call Metasystem, which helps in the process. The Metasystem works like a CASE tool that supports the evolutionary process. We can distinguish five forms of interaction depending on the importance of the Metasystem and the possibilities of inheriting (in biological sense) certain acquired characters by the System and the Metasystem [22].

1. Evolution of the Software System carried out by the Development System: it is the usual form of evolution in which the development team transforms the Software System helped by a tool.
2. Evolution of the Metasystem carried out by the Development System: the Development System evolves the Metasystem.
3. Evolution of the System carried out by the Metasystem: the Metasystem could be provided with certain learning capabilities and could modify the System as a result of changes in the environment of use for the software system.
4. Inheritance of acquired Metacharacter: reuse of the most interesting Metasystem characteristics.
5. Inheritance of acquired characters: reuse of the most interesting System characteristics.

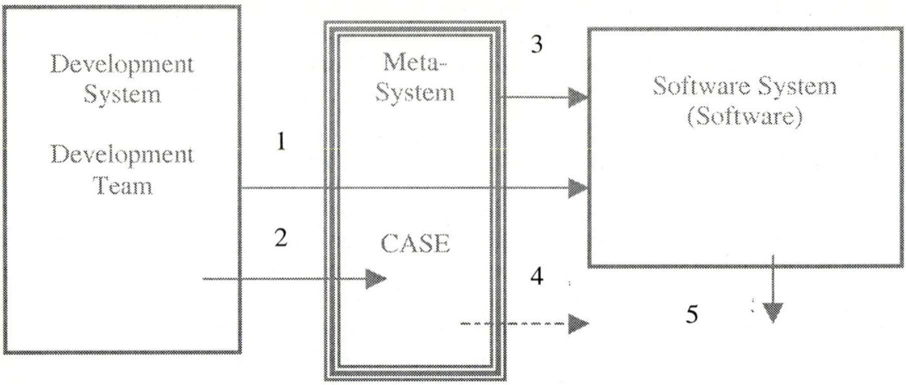

Fig.2. Operational Approach to Software Evolution

Obviously, the evolution definitions as well as the previous model of the evolutionary process are excessively abstract and to work with operational representations that can be implemented it is necessary, in general, to find representations of all three structural elements in a system, which is the main objective, i.e.:

1) The action Subsystem: the concept processor which carries out actions allows us to represent this subsystem. For establishing an understandable analogy, each one of our processors is similar to an active agent.
2) The decision Subsystem: a distributed decision strategy linked to the processor actions in the form of pre -conditions, during -conditions and post -conditions is used. Decisions are expressed in temporal first order [8].
3) The memorisation Subsystem: it is constituted by a history of the action results which the processors execute. These results are called Events and the memory of Events is called System Functional History.

A detailed description of the exact structure of these subsystems may be found in [16]. The most important result of this previous theoretical work derives from the possibility of representing and implementing Software Systems and the Metasystem using the same concepts, i.e., the system operation (functionality) and its evolution are homogeneous. In the following section we present the current state of the tool (HEDES) which implements these three subsystems and forms 1 and 2 of the aforementioned evolution.

4 HEDES Architectural Design

4.1 General Structure

HEDES is, mainly, the tool which implements the concepts developed in MEDES. Hence, the different concepts are mapped into classes in a nearly univocal way. These classes will allow both the software system operation and its subsequent evolution. Figure 3 shows the class structure designed and further implemented in Smalltalk .

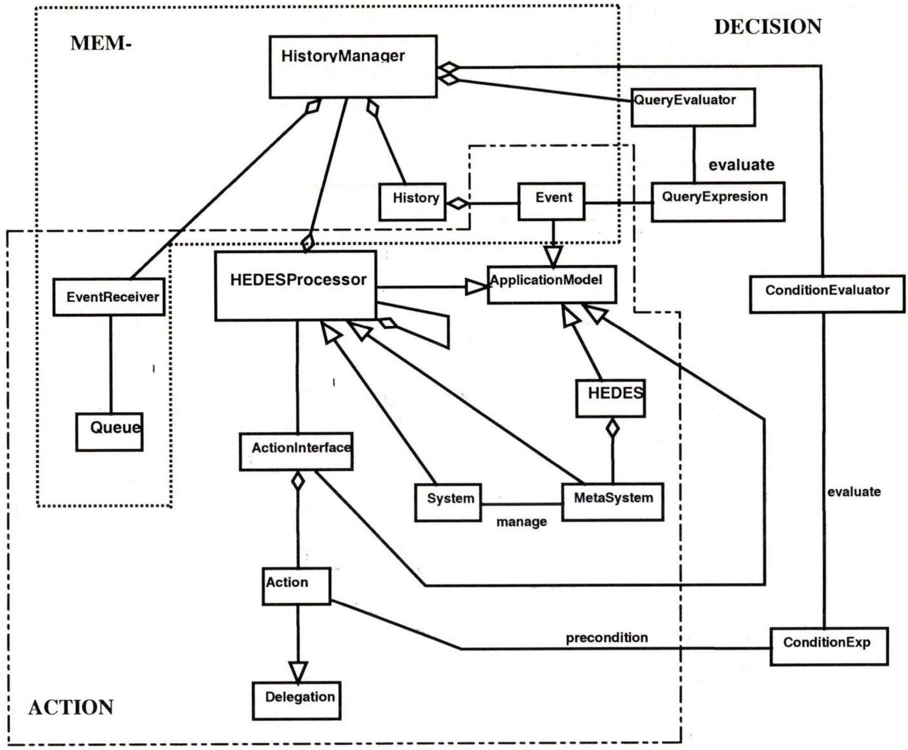

Fig.3. HEDES class structure

For each one of the three subsystems considered in the previous section, we have designed several interrelated classes. We emphasise the Action Subsystem, because it constitutes the central part of the prototype .
The kernel of the Action Subsystem is the HEDES Processor class which allows the structuring of the Systems, the Metasystem and the processors that formthem. The activity of these processors is regulated by the presence of a history (History Manager), belonging to the Memorisation Subsystem, and the possibility of executing actions (Action) depending on the evaluation of preconditions (ConditionExp) and queies (QueryExp) over the history. This evaluation depends on the Decision Subsystem. The control of activation depends on an event manager (EventReceiver) which works linked to the history. This class plays a double role: it stores the events, as part of the Memorisation Subsystem, and activates the processors, as part of the Action Subsystem. In order to systematise the explanation, we comment on each one of these subsystems separately ending with a section devoted to the evolution of the systems.

4.2 Memorisation Subsystem: History and History Manager

The kernel of the System memory is constituted by the events that have been produced inside it. These events (Event class), together with the instant in which they

were generated, are stored in the System memory (History class) by an events manager (EventReceiver). All these classes are co-ordinated by a manager of the history (History Manager). Figure 4 depicts the interactions between these components.
At present the Events are forms with possible queries (QueryExp) associated to the fields. A more detailed description of history structure and the additional information about the events may be found in [2].

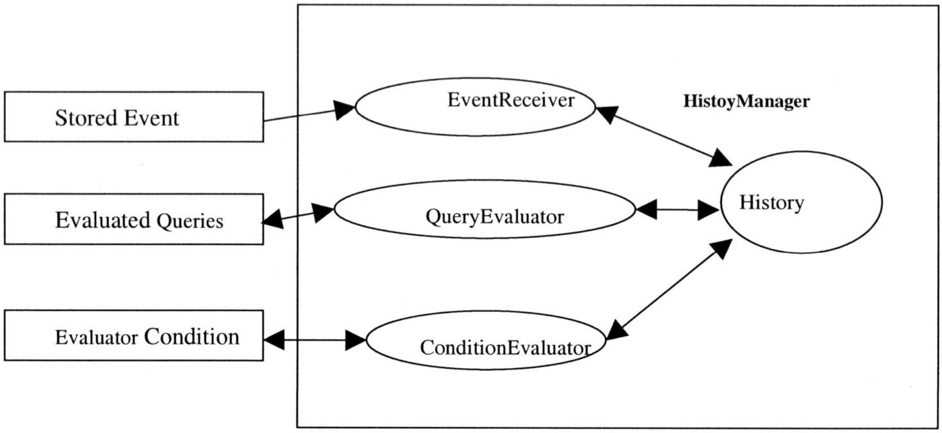

Fig. 4. History Manager activity

4.3 Action Subsystem

4.3.1 Structure of the System and Metasystem

In HEDES, a System and a Metasystem are special types of processors which, are themselves made up by processors. Therefore, the classes that implement them are subclasses of HEDES Processor. A mechanism of aggregation is used to form the different elements for a processor.
A System is created by a Metasystem and implements a prototype of an application that is directly executable by the user. All systems are instances of the System class. The design of the Metasystem determines a specific form for creating systems. It provides internal rules to determine the coherence and consistence of the systems developed. These rules are invariants of the structure of a system and are formally described as preconditions for the processor actions of the Metasystem [19]. Each particular Metasystem is an instance of the Metasystem class. Since a Metasystem has the same structure as a System, and it is a specialised system, it should be a subclass of the System. However, due to the fact that the Metasystem allows the systems to be created and evolved , the egg and the chicken problem arises. In order to solve this, we have taken the decision to implement the Metasystem before the System, creating it as a subclass of the HEDES Processor, and, fixing the invariants for its structure inside it.
In this version, only one of the component mechanisms, aggregation, has been implemented. The interaction between systems, through their interfaces, will be one of

the goals of the second version. With this version, the previous problem will also be solved by means of direct interaction between a Metasystem and its systems.

4.3.2 The Mechanism for Controlling Activation

When an action is carried out, a new event is generated and stored in the history by the Memorisation Subsystem. Each action includes references to other actions in its precondition. If the actions in a precondition were carried out, and new events were obtained, the precondition could hold and the action with that precondition would be activated.

The activation control for the processors in a specific system is carried out by its EventReceiver, one of the components of the History Manager. It is responsible for informing the processors when a new event has been recorded in the history. Each processor could activate one or more of their actions.

To put this functionality into practice, the EventReceiver needs to maintain specific information which relates the events, the processors and the actions of the processors. The actions that are part of each action precondition in each processor allow us to determine which processor may be activated when an event is recorded, giving it the name of the actions, which could potentially begin to be carried out. Therefore, the processors try to prove whether their actions can be carried out or not when a possibility of success exists, instead of attempting it continuously.

Since the systems evolve, new processors can be created or deleted or they can modify their behaviour by adding or dropping their actions. Also the action preconditions and the actions attributes can be updated. These modifications are reported to the EventReceiver dynamically in order to update its information about the system relationships and to maintain the consistency with the changes carried out in it.

4.3.3 Delegated Actions Versus Inherence

Actions are carried out by processors and their execution results are events. A similarity relationship between a processor and an object could be established. The actions of a processor correspond to the object protocol, and, the events correspond to the methods. Following the same philosophy, a class of objects would be equivalent to a processor class. This will imply that all processors in the same class will have the same behaviour.

But, previous studies [20],[21] show that the mechanisms of cloning and delegation, characteristics of object-oriented languages without classes, are able to simulate inheritance and they are also more expressive and dynamic than the traditional mechanism for inheritance.

Due to the fact that our main objective is the dynamic prototyping of the system represented, the delegation mechanism is used instead of inheritance. This forces the delegation chains to change at runtime and, at the same time, the characteristics of the actions of a processor.In our case, this implies that the pre-, during-, and post- conditions, and the events associated to the execution of an action may be delegated by a processor to another processor. In order to provide this dynamicity for the activity of the processors it is only necessary that the Action class uses the delegation mechanism.

4.4 Decision Subsystem: Mechanism for Activating Actions

Each action of a processor has an associated pre-condition (object of the ConditionExp class). An action is only carried out if its precondition holds. The preconditions are evaluated over the actions that have been carried out previously, i.e., over the events stored in the history managed by the History Manager. Thus, when an action is going to be carried out by a processor, it sends a request to its History Manager which queries its history to evaluate the pre-condition and answers whether or not the action can be carried out. The responsible for evaluating the pre-condition is an object of the ConditionEvaluator class and it is a component of the History Manager. From a formula or expression which constitutes the condition to be evaluated, the ConditionEvaluator creates atomic formulas which are evaluated more easily over the history. The ConditionEvaluator combines the partial results of the atomic formula evaluation and gives the processor a *true* or *false* answer.

It might be necessary to make a query on the possible values that the attributes of an action may have in order to carry it out. The user can give these values, or, otherwise, they can be obtained as the result of a query on the events of other actions stored in the history. The QueryEvaluator is responsible for evaluating these queries (objects from the QueryExp class), which itself is a component of the History Manager. An event will have some QueryExp linked, as much as one per attribute. The functioning of the QueryEvaluator is similar to the functioning of the ConditionEvaluator.

4.5 Evolution: A Form of Functioning

In our software model three aspects are considered: functioning, structure and evolution. These three aspects interact and their relationships are established during the life of the software and while it is functioning. From the moment when changes are produced in the structure of a software system, its functioning changes, i.e., it is evolving. Changes of structure should be indicated in the same way in which the carrying out of an action is required without having to stop the functioning of the system. This homogeneity, shown in [19] allows the evolutionary transformations of a system to be executed in the same way as its functionality, i.e., using the elements implemented and described beforehand. In fact, the evolution is carried out by the Metasystem using a special system history (System Structural History) in which the events stored are events of structural transformation. Its homogeneity with the Functional History allows the HistoryManager class to be used in order to manage both functioning actions and evolutionary actions. The only difference between both histories is the origin of the stimulus needed to carry out an action: the Metasystem in the evolution case, and the user in the case of normal system functioning.

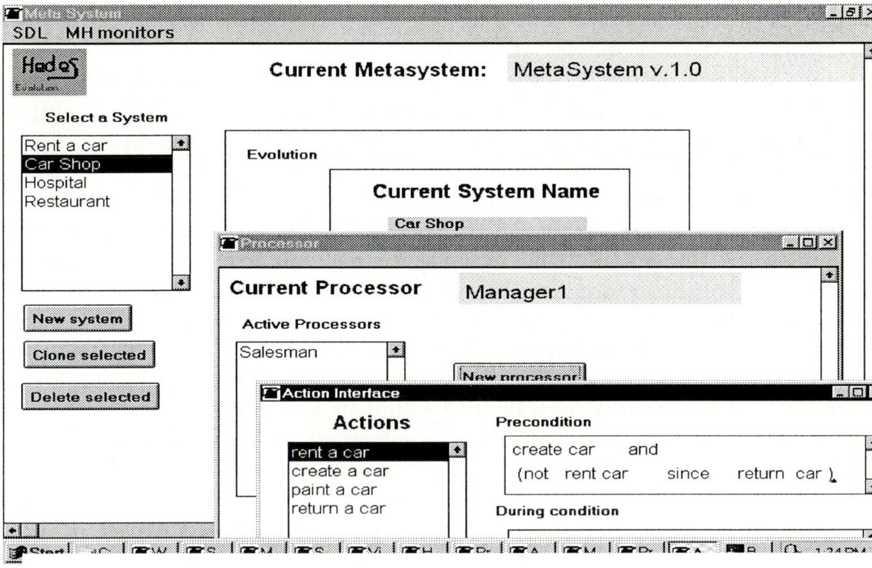

Fig. 5. User interface of the HEDES tool

5 Prototyping in Smalltalk (VisualWorks®[2])

HEDES implementation has been carried out in the object oriented environment VisualWorks 3.0. It is a visual programming environment fully implemented in Smalltalk. The choice of this language is based, besides the group's experience with group programming on it, on reasons of a conceptual nature:

- It is a reflexive language implemented on itself in which, thereby, the manipulation of the classes and their modification at execution time are carried out in the same way as the manipulation of the instances of the classes. In fact, both classes and metaclasses are objects. Due to the requirements of homogeneity for the structure and functioning, the HEDES tool has to be created with a language, which provides instructions to carry out the typical functions of an application, and to change the structure of the application itself and the programming environment at execution time. Therefore, reflexive programming languages are necessary to do this.
- It has an object-oriented visual environment with a client-server architecture. Its interaction model is MVC (*Model-View-Controller*) and it is well implemented and defined. It also has well-defined interfaces with relational and object-oriented databases.
- Due to its flexibility, the mechanism of delegation, much more dynamic than the inherence, has been implemented in it.
- It provides tools for constructing compilers which have facilitated the implementation of the temporal logic evaluator.

[2] VisualWorks 3.0 is a trademark of ObjectShare, Inc.

Other more object-oriented languages have been evaluated (Eiffel, C++ and Java) and discarded, at least in this phase of the project, basically because they do not allow full management of classes as objects.

The implementation details of the prototype are not especially relevant because the architectural design has been translated into the language without difficulty. Figure 5 shows a detail of the user interface of the HEDES tool. The provision of new forms of Metasystem and the user interfaces designer (Canvas) used to design the events stand out in this figure.

One important element is the implementation of the cloning and delegation mechanism in this language. Smalltalk is a language with simple inheritance, but, the accessibility of the mechanism for messages, allows its modification and the addition of the possibility of execution of delegated methods [20]. From the different delegation mechanisms available [7] we have chosen the delegation by object in which an object delegates, by defect, all its messages to another object. This mechanism allows the mechanism of delegation by message to be emulated. The characteristics that allow the resolution of the delegated methods are implemented in the Delegation class, from which the Action class inherits.

6 An Evolutionary Life Cycle for an Evolutionary Tool

The tool, as it has been developed, supports the evolution form 1 considered in Section 3. Our activity as a development team corresponds to the evolution form 2, i.e., the evolution of the Metasystem. Our work consists of the elaboration of versions of the Metasystem with progressive functionality. The design and implementation of a tool are not easy when the implementation details are not known beforehand, and, when the work is executed by a group of people in which each person needs the development works from the others. We decided, due to this co-ordination difficulty, to establish an evolutionary life cycle to carry out the development work in an incremental form, with the necessities being negotiated amongst the members of the team. Periodic meetings, and the sending of e-mails in the event of emergencies, are used for negotiating. The schema of the work followed is a modification of the classical approach in version controls on object orientation [3]:

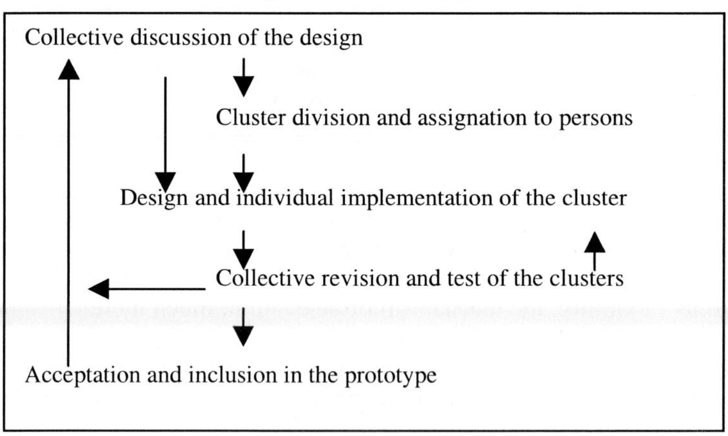

Fig. 6. Schema of work

In order to co-ordinate the development work, HEDES was divided into clusters [13]. The tool used to manage the configuration is SourceSafe[3], which allows us to control the software and the documentation generated. As usual in this kind of tools, two dimensions are considered:

- HEDES components: the storage and access control of the components of the work team are controlled. In the Smalltalk language, our case, each cluster corresponds with a category of classes and there is only one person responsible for each cluster.
- Versions of a cluster: each person can obtain new versions of its clusters and can control their changes.

Each group member has reading access to the clusters of the other members. Therefore, the work carried out by others can be observed and local copies from it can be made. Copies allow the execution and test of the own classes which need methods implemented in different clusters. Each test of the prototype implies a revision from which a new version is obtained (evolutionary prototyping). Sometimes, the revision of a cluster determines that the clusters with which it is related also have to be revised. The person responsible for a cluster is the only person who can change it, and sometimes, as in the previous case, because a third person required it.

There is a version of the work in which the clusters already tested and accepted are introduced, which means that the prototype always has an acceptable consistence. Obviously, the management of this consistence is the responsibility of a team member.

7 Related Works

In the evolution software field, the closest works are those carried out by the authors who work on object orientation. Gibbs [9] interpret software evolution as the changes carried out in the class definitions. In Casais [5] the evolution of a class structure is considered, taking into account the operations to modify that structure, the restrictions for the modifications, and, the effects of these modifications in the class structure and in the existent instances of these classes. In the work developed by Lieberherr [12] in the context of the DEMETER project, the notion of an adaptive program is introduced, providing an instantiable propagation pattern for different structurations of classes. It uses heterogeneous graphs for the representation of the structure of class relationships, in which, the nodes are the different kinds of classes and the arcs are the different kinds of relations between them. The program adaptability to the structure comes from the instantiation of the propagation directives over the structure, which, are finally translated in running time on the structure of the graph.

In the context of the MENHIR project , there are aspects of MEDES, that also exist in other methods such as OASIS [4] and TESORO [23]. OASIS, in an non-explicit form, it incorporates the action and decision subsystem.., Evolutionary aspects are now being considered, adding the Metaclass concept [18]. TESORO has an object model with events which describe their behaviour using transition rules for the changes of state. The objects work in a concurrent way and interact between one another. One of its more interesting elements is the use of restrictions on the objects, which specify the state of the objects, their behaviour depending on the state

---

[3] SourceSafe is a registered trademark of Microsoft, Inc.

However none of these authors or methods adopt a global model for evolution which allows us to consider the evolution in the software production process, therefore, they do not explicitly identify the subsystems that have been considered in this work. Besides, none of them take into account the evolutionary possibility of the tools for producing software (our Metasystem).

8 Conclusions and Future Work

The design and implementation of a first prototype of a CASE tool to develop evolutionary software systems have been presented. This first approximation includes fundamental concepts of the General System Theory of Le Moigne and an analogy to the Lamarkian model of evolution. Our main objective with this prototype was to show the utility of this tool. Some important conclusions may be taken from this work:

- System Theory provides a good support in the development of this kind of tool. We show that the concepts are used not only at a theoretical level, but also as practical support in the architecture and design of the tool.

-The notions of adaptation and inheritance, especially from a Lamarkian point of view, are very fruitful in modeling evolution [22]

The lessons learnt in constructing this tool, which supports the evolution of Software Systems, are being used in supporting its OWN evolution. In this sense, we are observing that sound System Theory concepts provide us the migration to the next versions of the tool with minimum development effort. In the near future, our work will be centred on a progressive refinement of the prototype with the aim of achieving a tool that may be useful for carrying out the prototyping of information systems. The main improvements are:

1. Action Subsystem:
   - The control of activation will be improved, by eliminating more information on the preconditions for the actions of the processors.
   - The possibility of designing events will be amplified.
   - Direct communication between processors by means of their interface will be added.
   - The structure for the Metasystem will be redesigned.

2. Decision Subsystem:
   - Improvements in the conditions and queries associated with the events will be included.
   - The evaluation of the conditions will be optimised.

3. Memorisation subsystem:
   - Mechanism to delete and optimise the memory will be added.
4. Evolution:
   - The inclusion of the kinds of evolution not considered till now (3, 4 and 5 from Figure 2) will be studied.
5. Optimisation:
   - The possibility of defining transformations from the history to state variables will be considered.

References

[1] Anaya, A., Rodríguez, M.J., Paderewski, P., Parets, P.: Time in the evolution and functionality of information systems. Jornadas de trabajo en Ingeniería del Software. Seville (1996) 161-170

[2] Anaya, A., Rodríguez, M.J., Parets, J.: Representation and management of memory and decision in evolving software systems. In: F.Pichler, R. Moreno Díaz (eds.): Computer Aided Systems Theory- EUROCAST'97 Lecture notes in Computer Science 1333. Springer-Verlag, Berlin (1997) 71-82

[3] Björnerstedt, A., Hulten,C.: Version Control in an Object-Oriented Architecture. In: W. Kim, Lochovsky (eds.). Object Oriented Concepts, Databases and Applications.. Adisson-Wesley.(1989) 451-485

[4] Carsí, J.A., Ramos, I., Bonet, B., Jaén, F.J., Penadés, M.C.: Implementación de la Metaclase de un entorno de prototipación automática basado en OASIS. Proyecto PASO PC-310. (1996)

[5] Casais, E.: Managing Class Evolution In Object-Oriented Systems. In: *'Object Management'.* Tsichritzis, D. GENEVE. Centre Universitaire d'Informatique(1990) 133-195

[6] Conradi,R., Fernström,C., Fuggetta, A.: A Conceptual Framework for Evolving Software Processes. Vol 18 num 4.ACM SIGSOFT SEN (1993) 26-34

[7] Dony,C., Malefant,J., Cointe, P.: Prototype based languages: From a new taxonomy of constructive proposals and their validation. Proc. ACM OOPSLA'92. Vol 27 num 12. ACM SIGPLAN Notices(1992) 201-217

[8] Gabbay,D.M., Reynolds, M.: Towards a computational treatment of time. In: D.M.Gabbay, C.J.Hogger: Handbook of logic in Artificial Intelligence and Logic Programming. Vol. 4: Epistemic and Temporal Reasoning. Oxford Science Publications. Oxford: Clarendon press (1995) 351-431

[9] Gibbs, S., Tsichritzis, D., Casais, E., Nierstrasz, O.: Class Management for Software Communities. *CACM 33*.9. (1990) 90-103

[10] Grünbacher, P.: Facilitating requirements engineering through computer supported cooperative work. Proc. IDIMT96, (1996)

[11] Le Moigne, J.L.: La Théorie du systéme général. Théorie de la modelisation. Paris: Presses Universitaires de France (1990)

[12] Lieberherr, K.J.: Adaptive Object-Oriented Software. The Demeter Method with Propagation Patterns, BOSTON: PWS Pub.Co. (1995)

[13] Meyer, B.: Object-Oriented software construction. Prentice-Hall. 2d. edn. (1997).

[14] Parets, J., Torres, J.C.: A Language for Describing Complex-Evolutive Software System. In: F.Pichler, R. Moreno Díaz (eds.): Computer Aided Systems Theory- EUROCAST'95. Lecture notes in Computer Science 1030. Springer-Verlag, Berlin (1996) 181-199

[15] Parets, J., Torres, J.C.: Software Maintenance versus Software Evolution: An Approach to Software System Evolution. IEEE Conference and Workshop an Computer Based Systems. Friedrichafen. March (1996), 134-141

[16] Parets, J.: Reflexiones sobre el proceso de concepción de sistemas complejos: MEDES: un método de especificación, desarrollo y evolución de sistemas software. Doctoral Thesis. Universidad de Granada (1995)

[17] Parets,J., Anaya, A., Rodríguez, M.J., Parewski, P.: A Representation of Software Systems Evolution Based on the Theory of the General System. In: F.Pichler, R. Moreno Díaz (eds.): Computer Aided Systems Theory- EUROCAST'93. Lecture notes in Computer Science 763. Springer-Verlag, Berlin (1994) 96-109

[18] Ramos, I., Pelechano, V., Penadés, M.C., Bonet, B., Canós, J.H., Pastor,O. Análisis y diseño orientado a objetos de un entorno de prototipación automática basado en OASIS. Proyecto PASO PC-310. (1995)

[19] Rodriguez, M.J., Parets, J.: Cambio en la estructura de decisión en un sistema evolutivo. II Jornadas de Ingeniería del Software. JIS'97, San Sebastian (Spain) (1997) 259-274

[20] Sánchez, G., Parets, J.: Delegation versus Inheritance: a mixed approach in Smalltalk. Grupo GEDES. Working paper. (1998).

[21] Sánchez, G., Parets, j.: Mecanismos alternativos en lenguajes orientados a objetos. NOVATICA, 144, (1995) 51-58

[22] Torres, J. J., Parets, J.: Biological Evolutive Models Applied to the Evolution of Software Systems. Third European Congress on Systems Science, Rome, October (1996) 705-710

[23] Torres, J.: Especificaciones orientadas a objetos basadas en restricciones, prototipado en un lenguaje orientado a objetos. Doctoral Thesis, Seville University, (1997).

Vertical Partitioning Algorithms in Distributed Databases

M. E. Zorrilla, E. Mora, P. Corcuera, and J. Fernández

Department of Applied Mathematics and Computer Sciences.
University of Cantabria.
Avda. De los Castros s/n 39005 Santander. SPAIN
Tno: 433-42-201363
Fax: 34-42-201829
mezorr@ macc.unican.es

Abstract. Data distribution is a crucial problem affecting the cost and efficient use of these systems. The problem is further exacerbated by the lack of methods and support tools for the design of distributed databases. This paper outlines some of the main techniques currently used for data distribution, such as vertical partitioning and replication. Two vertical fragmentation methods are described, the classic NAVATHE method and the newer FURD method, as well as the two proposed in this paper, the FURD-FDEZ and the FURD WITH REPLICATION methods.

1 Introduction

Up to the present date, most information systems have been located in centralised systems. Now, however, communication networks and particularly Internet offer new alternatives which allow us to obtain the data wherever it is handled.
Data distribution is a crucial problem affecting the cost and efficient use of these systems. The problem is further exacerbated by the lack of methods and support tools for the design of distributed databases.

This paper outlines some of the main techniques currently used for data distribution, such as vertical partitioning and replication. Two vertical fragmentation methods are described, the classic NAVATHE method and the newer FURD method, as well as the two proposed in this paper, the FURD-FDEZ and the FURD WITH REPLICATION methods.

Vertical fragmentation is not only of interest within the context of distributed databases, but also in the field of centralised databases. In the latter case, the main aim is to detect the most active queries in order to allocate the relevant data in faster memory subsystems, thus reducing the number of page accesses.

It is worth pointing out that the vertical fragmentation of a relation R produces new relations, R1, R2,...Rn, which contain a subset of the attributes of R amongst which the primary key must be found. The main objective of this operation is to divide the

F. Pichler, R. Moreno-Díaz, and P. Kopacek (Eds.): EUROCAST'99, LNCS 1798, pp. 465–474, 2000.

relation into a set of smaller relations so that most user applications execute processes which affect attributes of only one of these fragments.

2 Navathe Method [5]

The aim of the first method studied, the Navathe partitioning method, is to regroup the columns of a table in the sites where they are most frequently requested in order to minimise the number of joining operations. The following stages are required for fragmentation using the Navathe method:

Define the Attribute Usage Matrix.

Build the Attribute Affinity Matrix (clustering algorithm).

Divide the Attribute Affinity Matrix (partitioning algorithm).

The first stage involves the identification of the processes to be executed in the distributed database. With these, a two dimensional matrix is built for each of the tables at the database, in which the attributes of the table are arranged horizontally to the matrix and the processes which have access to this table are arranged vertically. This matrix will contain a 1 if the attribute is requested by the query and a zero if not. Once the Attribute Usage Matrix has been defined, the frequency accesses must be assigned, that is, the number of times each process is requested from each site.

In the second stage, the Attribute Affinity Matrix is built. This matrix shows the set of attributes which are most likely to be requested at the same time. It is obtained using expression 1:

$$aff(A_i, A_j) = \sum_{\forall S_l} ref_l(q_k) acc_l(q_k) \tag{1}$$

where, $ref_l(q_k)$ is the number of accesses to attributes (A_i, A_j) for each execution of application q_k at site S_l and $acc_l(q_k)$ is the application access frecuency sites.

Then, taking as input the initial Attribute Affinity Matrix, and permuting rows and columns, the Clustered Affinity Matrix is obtained, using the Bond Energy Algorithm. This permutation is carried out in such a way that the overall Affinity Measurement (AM) is maximised, through expression 2:

$$AM = \sum_{i=1}^{n} \sum_{j=1}^{n} aff(A_i, A_j)[aff(A_i, A_{j-1}) + aff(A_i, A_{j+1})] \quad where, \tag{2}$$

$$aff(A_0, A_j) = aff(A_i, A_0) = aff(A_{n+1}, A_j) = aff(A_i, A_{n+1}) = 0$$

Next, in order to group those attributes whose affinity is maximum, the contribution defined by expression 3 is used :

$$cont(A_i, A_j, A_k) = 2bond(A_i, A_k) + 2bond(A_k, A_j) - 2bond(A_i, A_j) \qquad where, \tag{3}$$

$$bond(A_i, A_y) = \sum_{z=1}^{n} aff(A_z, A_x) aff(A_z, A_y)$$

As the cluster matrix obtained can have more than two fragments, it is necessary to resort to a recursive algorithm which finds the sets of attributes which are accessed solely, or mainly, by various sets of processes. This makes up the third and final stage of the Navathe method.

3 Furd Method [6]

The FURD (Fragmentacion, Ubicación y Reubicación de Datos) is a method based on the minimisation of a single objective function which enables the attributes of relations to be allocated and/or reallocated in the nodes of the network, thus reducing communication costs. That is, it not only takes into account the frequency of access but also the means of communication linking the sites. In contrast with the traditional Navathe approach which consists of two stages, fragmentation and allocation, this method performs the two operations at the same time.

The objective function is a form of cost assessment and consists of two terms:

1. In the first term, the communication costs generated by the transmission of the data required to satisfy the queries from each of the nodes are modeled.

2. In the second term, the communication costs generated by the migration of data from one node to another in a change of design are modeled.

$$min \qquad z = \sum_k \sum_j f_{kj} \sum_m \sum_t q_{km} l_{km} c_{jt} x_{mt} + \sum_m \sum_j \sum_k a_{mj} c_{jt} d_m x_{mt} \tag{4}$$

where,

f_{kj} = frequency of accesses of query k from node j.

q_{km} = parameter indicating with 1 if query k uses attribute m, and with 0 otherwise.

l_{km} = number of communication packages required to transport the attribute m required for the query $k = p_m s_k / PA$ [1]

where, p_m is the size in bytes of attribute m;

s_k is the selectivity of the query and

PA is the size in bytes of the communication packages;

c_{jt} = cost of communication between node j and node t

x_{mt} = decision variable equal to 1 if the attribute m is stored in node t and 0 otherwise.

a_{mj} = parameter which indicates with 1 if the attribute m is currently stored in node j.

[1] S_k and PA take the unit value in order to compare them with Navathe method.

d_m = number of communication packages needed to change the allocation of attribute m to another node.

In this algorithm the primary key of the relation to be fragmented is not considered. This will be added in each fragment when the process is finished in order to maintain integrity and to make possible the reconstruction. Replicated attributes will not be considered either, which means that each of the attributes of the relation is stored in a single node, so that the following is fulfilled:

$$\sum_t x_{mt} = 1 \tag{5}$$

Moreover, each attribute must be stored in a node which executes at least one query which requests it, so that

$$\sum_k q_{km} \varphi_{kt} \geq x_m \tag{6}$$

where, φ_{kt} is equals 1 if $f_{kt} > 0$, ó 0 if $f_{kt} = 0$

In J. Pérez et all. (6) a comparative analysis is made of results of the Navathe and FURD methods under certain conditions, their conclusion being that both methods offer identical fragmentation schemas. It is also observed that the FURD method simplifies the calculation and performs the fragmentation and the location at the same time.

The FURD method also offers the possibility of evaluating the cost of the migration of fragments from site to site in order to adapt to the changes in the transaction patterns of the above design. Since it is considered that a change in the system design must be a decision of the system administrator, only the first term has been considered here.

4 Furd-Fernández Method

In the case of vertical partitioning to distribute the columns of a table in different sites, each fragment must include the primary key of the original table. Since neither the Navathe nor the FURD method contemplate the influence of the size of the index on the fragmantation, the FURZ-FDEZ method is now proposed, providing an algorithm which takes into account this situation. This method is based on the FURD method and makes use of the same initial data.

Why evaluate the index cost?. The index is made up of one or more attributes and as such, has an influence on the communication cost since it is composed of a number of bytes and, as with any attribute, it depends on the network costs, on the frequencies of the queries to each field and on the union with other fields of the same process.

Algorithm:

The method aims to offer an improvement on the FURD method. The process is the same, but, over the cost of each field in each combination of sites obtained with FURD, this method adds the calculated index cost, bearing in mind that if there is another field situated at the same site, the index cost to be added will be half, and if there is a third, a third and so on successively.

In order to avoid a dependence on the order of selection of the fields in the process, a combination of sites is chosen and the alllocation of each one is varied field by field in each site, and if the cost obtained is low, the rest of the assigned fields are again varied.

Inclusion of the Index Cost

For the calculation of the index cost, the attributes of each query and their frequency are taken into account.

On initiating the calculation, there are no clusters with other fields (these are just beginning) and, thus, the first result will be as if the first field were the only one in a specific fragment. In this case, the additional cost of the index is evaluated according to:

$$\text{ACUA} = \text{Additional\_cost\_unique\_attribute} = \text{cx} * (\text{ti} / \text{tx}) \tag{7}$$

where,

c_x = FURD cost of field x

t_i = index size

t_x = size of field x

Next, the following steps must be taken for each field at each site:

1. Calculate the added cost if the field were the only one in the fragment, using expression 7.
2. Calculate the importance, I, in so many to one, of each attribute in each site using the query/attribute table, according to expression 8:

$$I = pa / pt \tag{8}$$

where pa is the FURD cost, for a specific node and query, that is:

$$pa = \sum_{w} f_{kl} c_{lw} \tag{9}$$

and pt is the cost in the rest of the sites:

$$pt = \sum_{k} \sum_{w} f_{kl} c_{lw} q_{km} \tag{10}$$

3. Calculate the number of coincidences between the query/attribute table and the allocations table obtained up to the present moment.
4. Obtain the additional cost of the index using expression (11):

$$\text{Additional\_index\_cost} = \text{ACUA} * \left(\frac{\text{I}}{\text{sum of coincidences}}\right) \tag{11}$$

The cost matrix obtained after carrying out this process will offer the vertical fragmentation schema, taking into account the index.

5 Método Furd With Replication

The methods outlined above, the Navathe, FURD and FURD-FDEZ methods, are based on the vertical fragmentation and the allocation of each attribute in a single site - that in which the cost in resources is lowest. This section analyses ways of optimising the cost by allocating each attribute to one, to several or to all of the sites competing for the attributes.

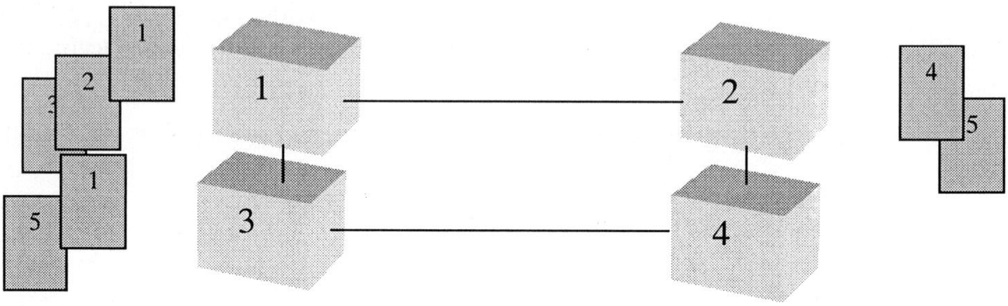

Fig. 1. A replicated enviroment

For this new method, it is taken into account that:

1. Every time a field has to be updated, this must be performed in all existing copies of the field.
2. If the queries are reading queries, they will be brought from the site which has the copy requiring the lowest cost, unless it is found in the site where the query is made.

The parameter **remote writing accesses** (A) is defined, indicating the value , in so many to one, of the updating processes. If the parameter tends towards zero (most of the queries are reading queries) there will be a tendency towards full replication, while if it tends towards one (most of the processes are writing processes) there will tend to be one attribute in each site.

At the same time, the matrix M is defined, that is a squared matrix of a range equal to the number of sites, such that; Mii is the cost an attribute has when it is at site i and due only to its own trajectory, and Mij is the cost an attribute has after being brought from site j to node i.

The value of Mij is found using the following expression for each attribute m of the table:

$$M_{ij} = c_{ij} \sum_{k} f_{kj} q_{km} l_{km} \tag{12}$$

where:

f_{kj} = frequency of access of query k from node j.

q_{km} = parameter indicating with 1 if query k uses attribute m, and with 0 otherwise.

l_{km} = number of communication packages required to transport the attribute m required by query k = $p_m s_k/PA$
where,

p_m is the size in bytes of attribute m;

s_k is the selectivity of the query and;

PA is the size in bytes of the communication packages;

c_{ij}= is the cost of communication between node j and node i

The algorithm consists of the following steps which are to be carried out with each of the attributes of the table:

1. A function is created which obtains all the possible combinations of the attributes in the sites (if there are four sites, the function would go from combination 0001 to 1111).
2. The costs for each field are found in each of the combinations (sites(i)), so that:

$$\text{Total\_cost} = \sum \cos t \quad i$$

2.1. If the node i to be analysed has the field which is to be evaluated, then the cost is obtained using expression 13:

$$\text{Cost i} = \text{Mii} + [\text{A} * \sum(M_{ij})\ ] \tag{13}$$

2.2. If the field to be evaluated is not found in the node under analysis, then the cost is obtained using expression 14:

$$\text{Cost i} = (M_{ij}) \tag{14}$$

where function P is a function which obtains the possible elements (Mij, selects the lowest and multiplies the rest by remote_writing_accesses. If there are no elements, its value is zero). The value of j is that of the nodes which the copy has.

3. The cost is checked. If it is the lowest, the combination is kept, and we return to point 1.

When the algorithm is finished, the schema of the fragmentation with replication of the attributes which form the relation will be obtianed.

6 DDB Design Simulation Tool

An application has been developed in Visual Basic for the use of these two variants of the FURD method and itself. The application takes as input data the information on the structure of the table to be partitioned, system costs and frequency of queries in order to facilitate the distribution of its columns, according to the chosen algorithm, with the minimum cost.

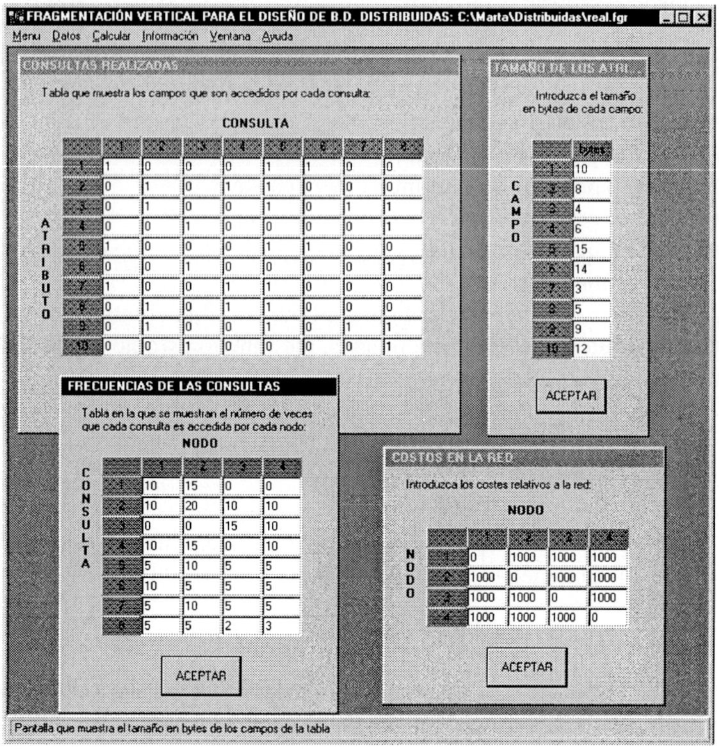

Fig. 2. . Initial conditions data

Figure 2 below shows the data input screens. With the calculate option, the various methods are accessed and the results are shown as in Figure 3 and 4. This tool is available in Spanish.

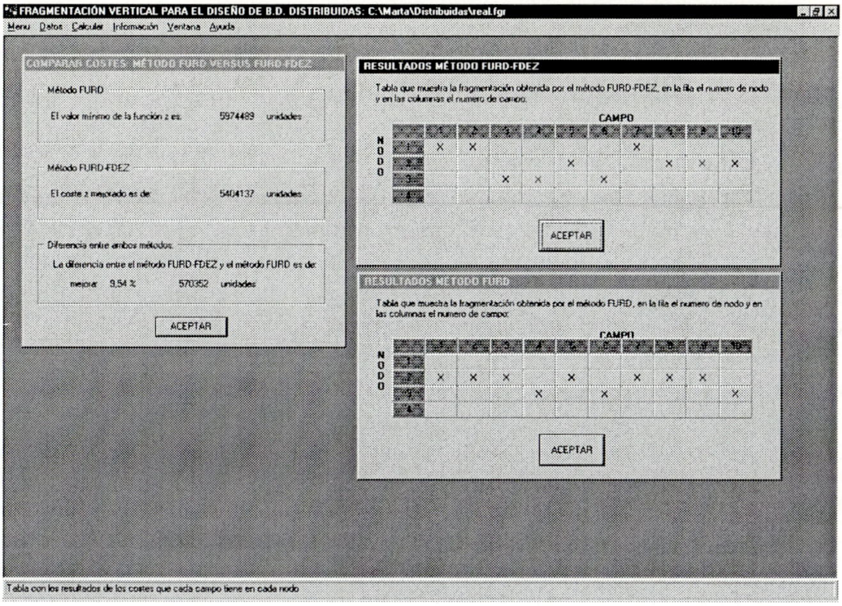

Fig. 3. Furd vs Furd-Fdez with index size of 6

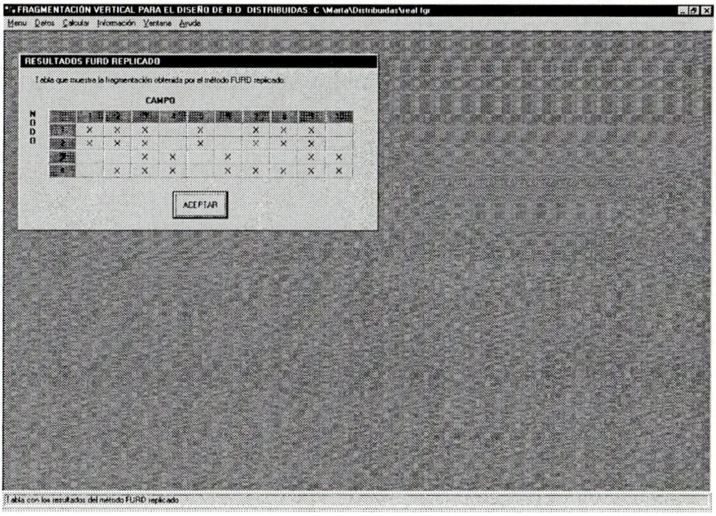

Fig. 4. Furd with replication with 15 % of written access

7 Conclusions

The aim of this paper is to contribute towards improving distributed database design and, in particular, the study of the vertical fragmentation of relations and the replication of columns.

An analysis of the two existing methods, the Navathe and the FURD methods, has been made. Two improvements on the above are proposed: the FURZ-FDEZ method which incorporates the incidence of the index cost of the fragmentation and the FURD method with Replication which offers, depending on the percentage of accesses with updating, the replication schema with the lowest cost.

Finally, an application has been developed in Visual Basic for the use of these two variants of the FURD method.

References

1. Apers, P.M.G. Data Allocation in Distributed Database Systems. ACM Trans. On Database Systems, vol.13 nº3, (1988) pp.263-304.
2. Fernández, J. Desarrollo de una aplicación para el diseño de bases de datos distribuidas: definición de nuevos algoritmos de fragmentación vertical. MSc. Thesis. University of Cantabria. February. (1998).
3. Navathe, S., Ceri, S., Wiederhold, G. , Dou, J. Vertical Partitioning Algorithms for Database Design. ACM Trans. On Database Systems, vol.9 nº4 (1984) pp.680-710.
4. Navathe, S., Muthuraj, R., Chakravarthy, S. A Formal Approach to the Vertical Partitioning Problem in Distributed Database Design. Proc. IEEE, ISBN:081863330 1,1993.
5. Ozu, M., Valduriez, P. Principles of Distributed Database Systems. Englewood Cliffs, N.J. Prentice-Hall. (1991).
6. Pérez, J.,.Pazos, R., Rodriguez, G. Fragmentación, Ubicación y Reubicación Dinámica de Datos en Bases de Datos Distribuidas. II Jornadas de Investigación y Docencia de Base de Datos. Madrid, (1997).

Decision Based Adaptive Model for Managing Software Development Projects

Manfred Mauerkirchner

Polytechnic University of Upper Austria, A-4272 Hagenberg, Austria,
manfred.mauerkirchner@fhs-hagenberg.ac.at,
http://www.fhs-hagenberg.ac.at

Abstract. The article is about the management of an important part of the development process of a software system: the phase of incremental implementation. Implementation is based on the design of the system architecture and its components, its main task is the production of modules and their integration.
Project management has to plan and control the flow of project activities. Besides logical dependencies between them it has to take into consideration particularly the aspects of time and availability of qualified resources. In addition, each calculated plan depends upon several restricting constraints, which are influenced by real time decisions of the environmental system.
Therefore a model of such a supporting management system is dynamic and non deterministic in its nature, it needs the ability of adapting itsself due to external decisions.

1 Introduction

Traditionally, software development is described by different process oriented models, i.e. classic sequential software-life-cycle-model, waterfall-model, spiral-model, prototyping-model, object-oriented-model and mixture forms. They are characterized by using repeating phases, which build the so-called *software life cycle*. Unavoidable tasks in this area of professional software production are analysis, design and implementation[2]. Based on the design of component and logical architecture of the software system, the following chapters focus on the phase of *incremental implementation*, i.e. the realization of parts of programming code and their integration.

First, the result of the design process is a set of interrelated functions working together with one common objective, which build the logical description of the software system. The description methods range from simple unstructured documents until complete repositories, which often enable automatic code generation too. Secondly, a decomposition of the problem into physical parts, so-called components, is a necessary step. Using the bottom-up method the implementation phase starts with the production of programming code of elementary predesigned modules, followed by their integration into more comprehensive modules, etc. It is very important to carry out a review of the complete design of the

F. Pichler, R. Moreno-Díaz, and P. Kopacek (Eds.): EUROCAST'99, LNCS 1798, pp. 475–485, 2000.

software system after the completion of each single software module; in that way the tasks of design and implementation influence other (incremental design and implementation). Nowadays, the items analysis and design are substituted by the term *modelling*. A modern but already well known graphic oriented description language, which supports this kind of modelling and which is independent of specific process models, is UML (Unified Modelling Language) [4].

Modelling software development, several specific difficulties have to be taken into consideration. Most involved processes are creative in their nature and executed by human resources. With regard to the need of integration of permanent modifications, performed while the real software project is running, the underlying project structure has to be dynamic, or the entire system is non deterministic[7]. Besides, from the viewpoint of the project manager a software development project is represented by a set of manageable units called *project activities* and a set of human resources called *resource pool*. The main target of every managing instance, i.e. every project management system (PMS), is the support of planning and controlling the project flow. Nowadays time-to-market is the primary goal concerning software production[3] [12]. Planning is not only restricted to correspond with the logical project flow, but also has to assign optimal resources in such a way that the overall project time gets minimized with addional constraints concerning budget[5]. Control has to ensure the agreement of the real project with the current project plan. If deviations occur, several external decisions have to be made and, based on them, a new planning is necessary, followed by the integration of the new created plan into the running system[9]. To be successful in this respect the underlying model has to be "decision based" and "adaptive" (see topic of the article).

1.1 System Levels

Fig.1 is depicting the different levels of the involved systems and the central role of the so-called *project plan*: The DMS (Decision Making System) belongs to a higher level than the PMS (Project Management System) and RPS (Real Project System), the object "project plan" figures as synchronizing object for all systems. Vertices describe communication channels between the systems, where channel 1 stands for monitor information of current activities generated by the PMS and channel 2 is for reports coming from the RPS. It is the responsibility of the project manager (DMS) to observe deviations between planned and current project flow and to make decisions. These decisions are brought via updating interface into the running PMS, which is symbolized by channel 3.

Therefore the typical tasks of all systems can be summarized as followed:

- PMS: Import of modifications, export of informations concerning running activities, calculation of optimal project plans
- RPS: Execution of all planned activities, generation of reports
- DMS: Comparison of PMS informations and RPS reports, generation of structured decisions (in the case of deviations)

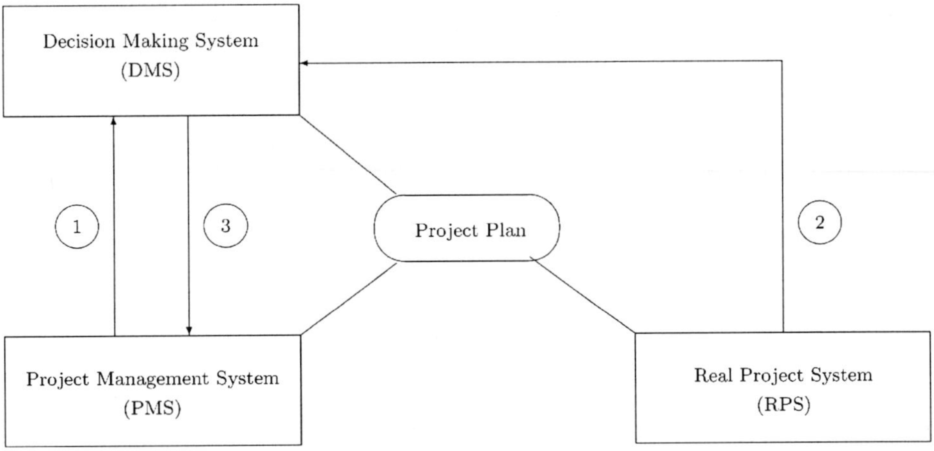

Fig. 1. Systems Communication

1.2 Decision Based Synchronization

Systems are synchronous if they work like described in the current project plan. Disturbances can be

- time deviations of project activities
- changes within the resource pool (see 2.1)
- changes in decomposition graph (see 2.1)
- not accomplished quality criteria

It is the most challenging task for the members of the DMS (i.e. project manager, quality manager, etc.) to observe the current project flow permanently and to discover differences between reality and project plan as fast as possible. In the case of a noticed deviation, a decision which intervention type should be used for correction is immediately necessary. Dependent on the above listed items we can distinguish between

- simple intervention (correction of remaining activity time, modification of resource status) and
- aggregate intervention (modification of system granularity, realization of repeating loops).

Figure 2 illustrates the roles of the different involved systems and their project plan from the beginning of a project. All kinds of interventions need an input at time t_{fix} using the update interface of the PMS monitor (IRQ-1). The asynchronous phase up to that point is called *reaction time interval* T_{react} and can only be influenced by the decision system. At any time, therefore also at t_{fix}, the PMS is able to calculate the *remaining time interval* T_{plan} until the next relevant change of system state takes place (must be a start or a termination of a

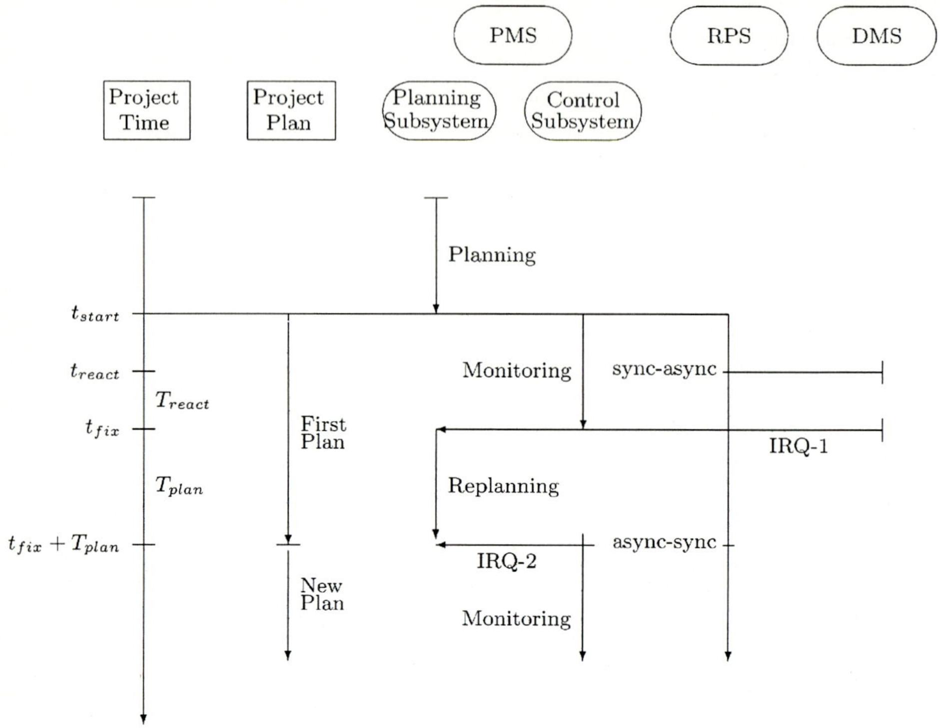

Fig. 2. Systems Synchronization

project activity). This time interval is available to start the planning subsystem at time t_{fix} for the purpose of replanning (creation of a new optimized project plan of all not yet successful terminated project activities). It is clear that the efficiency of the optimization (quality of new plan) depends on

- the length of the available time interval,
- the complexity of the remaining project structure (activities and resources),
- the effectivity of the planning subsystem.

Nevertheless, in any case new specific project plans are created and the best one gets integrated at time $t_{fix} + T_{plan}$(IRQ-2). This new project plan object implies that all other systems (DMS, RPS) have to adapt their organizations. The sum of both time intervals (reaction time T_{react} + planning time T_{plan}) defines the *asynchronous phase* of the involved systems (like exhibited in fig.2). Together with the documentation of the intervention decisions the currently adapted project plan is a perfect detailed description of the project history.

2 Project Management System

2.1 Basic Notions

To describe the logical dependencies of a software system a project structure is represented by a directed non circular graph with so-called *project subtasks* as nodes, where each subtask is a comprehensive container of an arbitrary sequence of project activities. Activities are the elementary building units of our model and characterized by the assignability of an unique type of human skill, by an externally generated time and a (optional) boundary concerning costs. Mapping the software design into a work breakdown structure, one has to observe that at one extreme a project activity can be started only in the case of successful termination of its predecessor activity (concerning the same subtask), and at the other that the first activity of a new subtask will only be able to begin if all predecessor subtasks of the graph are finished successfully[8].

Resources are restricted to human resources because this type is the most important in software development projects. Each resource of an existing resource pool is characterized by a discrete set of *resource steps* to enable the specification of variable costs and skills of each person. An amount belongs to each step of a person to specify the costs per hour, and addional values (so-called *capacities*) to define the personnel skills for performing the designated project activities. Furthermore, a database for storage of relationships concerning cooperation between all human resources is defined, which is necessary to build optimal resource groups for each single project activity[8]. With respect to observed deviations between the planned project flow and its realization and due to modification decisions of the project manager, the database gets updated automatically.

2.2 Formalism of Base Model

To be successful in building a model of a project management system, which also represents the dynamic behavior, some further common basic requirements have to be accomplished:

- integration of time
- ability for simulation

In our case the Discrete Event System Specification (DEVS) formalism is used to describe each project subtask and each human resource[13].

$$DEVS_m = (X_m, S_m, Y_m, \delta_m^{int}, \delta_m^{ext}, \lambda_m, \tau_m), m \in M$$
$$DEVS_r = (X_r, S_r, Y_r, \delta_r^{int}, \delta_r^{ext}, \lambda_r, \tau_r), r \in R$$

where M is the set of project-subtasks and R the set of human resources. X is a set, the names of external event types, S the set of the specific sequential

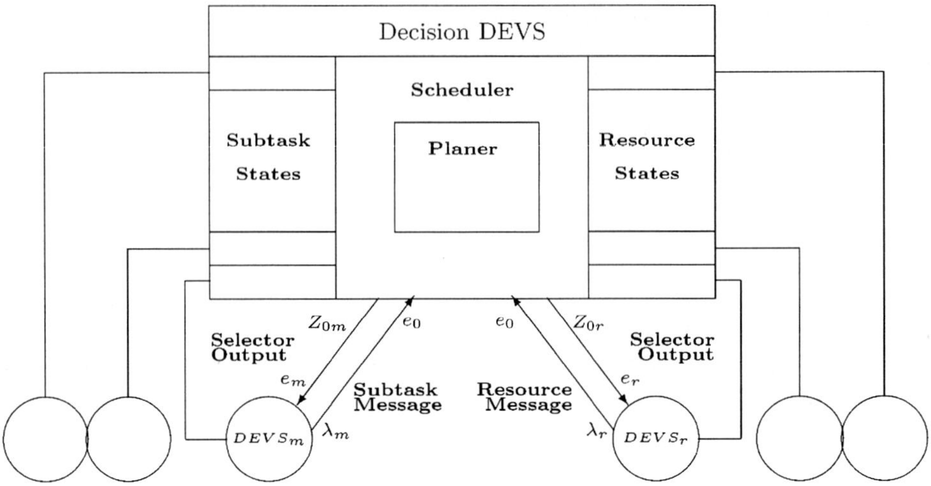

Fig. 3. DEVS Network

states, Y the output set, δ^{int} the internal transition function, δ^{ext} the external transition function, λ the specific output function and τ the so-called time advance function[8].

Direct communication between all these atomic DEVS systems (subtask DEVS: $DEVS_m$, resource DEVS: $DEVS_r$) is not possible, it gets managed by an addional supervising DEVS system, called decision DEVS: $DEVS_0$, which has to plan, to schedule and also to master all activities.

$$DEVS_0 = (X_0, S_0, Y_0, \delta_0^{int}, \delta_0^{ext}, \lambda_0, \tau_0, \{Z_{0,i} \mid i \in M \cup R\})$$
$$Z_{0,i} \;:\; S_0 \rightarrow X_i, i \in M \cup R$$

$Z_{0,i}$ describes the 0-to-i output translation and is called selector.

With respect to the recently defined logical dependencies, all above mentioned DEVS systems (DEVS components) are linked to a DEVS net, called project DEVS, which is another very complex DEVS system[13]. The result of coupling DEVS components is

$$DEVS_N = (X_N, S_N, Y_N, \delta_N^{int}, \delta_N^{ext}, \lambda_N, \tau_N)$$

where, for instance, $S_N = \times S_i, i \epsilon M \cup R \cup \{0\}$ (for a detailed description see [10]). Based on that formalism, the following example illustrates the typical usage of the DEVS net (see fig.3):

A certain activity is ready to start and sends a specific request (subtask message: $\lambda_m \equiv e_0$) to the decision system, which has to find an optimal group of resources, by proving the dynamic states of all resources of the pool (selector output: $Z_{0,r} \equiv e_r$, resource message: $\lambda_r \equiv e_0$). If no qualified resources are available, a so-called *passive waiting time* has to be realized, if not, two other subcases are possible. First, if at least one member of the assigned optimal group is still busy for another project activity, a so-called *active waiting time* can be calculated and realized. Secondly, if all group members are in state free, the requesting activity can be started immediately (selector output: $Z_{0,m} = e_m$).

2.3 Subsystems

The entire Project Management System (PMS) can be broken down into two subsystems, both using the same base formalism (DEVS network $DEVS_N$): the off-line runable *Planning Subsystem* (see fig.2, fig.4, fig.5) and the real time *Control Subsystem* (see fig.2, fig.5).

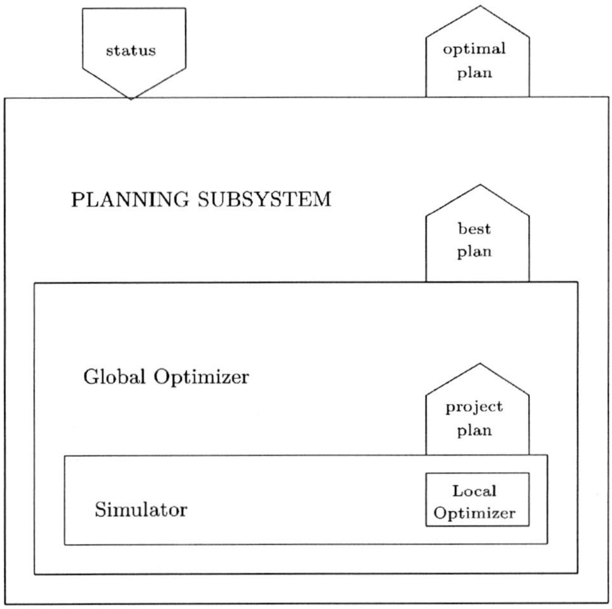

Fig. 4. Optimization

It is the primary target of the planning system to produce project plans whenever needed and as efficient as possible. In order to be successful (and with respect to the objective function, which is multiobjective and nonanalytical), the method of simulation has to be used in general; particularly evolutionary computation based methods (i.e. genetic algorithms [1]) for local optimization and numeric non gradient methods for global optimization get used (see also fig.4). The control system has to carry out four important tasks (fig.5):

- emulation of the project flow
- display of relevant real time informations
- integration of external (DMS generated) interventions
- supervising of the entire PMS

Based on the current project plan and the specific $DEVS_N$, the DEVS formalism is also used to realize the real time *emulation* mechanism (Event Base Coordination DEVS), which enable the so-called *monitor display* of project flow informations (EBC messages, see fig.5). For a more detailed explanation see[9].

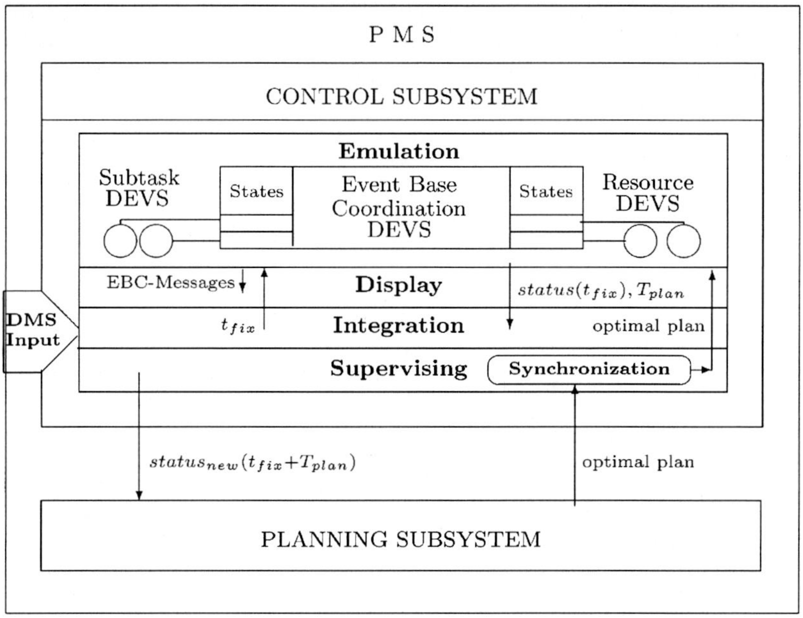

Fig. 5. PMS

Two types of real time information are shown in table 1 (concerning subtasks) and table 2 (concerning resources). Line 1 and line 2 of table 1 describe project activities in passive waiting states (no qualified resources available), line 3 and 5 are examples for busy subtasks (characterized by an explicit resource set and a calculated execution time interval) and line 4 stands for an activity in an active waiting state (resource set and calculated waiting time available). From the viewpoint of the project management line 3 and line 5 are implicit definitions of so-called *milestones*[3]. Whenever necessary, remaining time informations, corresponding to the actual time, can be obtained from column 8 of table 1. It is clear that this kind of information can be generated only for subtasks in states like presented in lines 3,4,5.

Lines 1,4,5,6 of table 2 are self illustrating, line 2 and line 3 need some additional explanation. Line 2 means that resource 2 is actually busy for activity

Table 1. Subtask Information

Subtask Number	Node Number	Activity Number	Subtask State	Resource Set	Time Execute	Time Wait	Time Remain
4	5	3	ready	-	-	-	-
3	4	1	ready	-	-	-	-
5	6	2	busy	1,2	3.4	-	-
7	10	1	ready	2,3	-	2.1	-
6	9	2	busy	4	6.3	-	-

2 of subtask 5 but already reserved for subtask 7, line 3 describes a prereservation of resource 3 for subtask 7 (therefore resource 3 is not available for other project activities).

Table 2. Resource Information

Resource Number	Resource State-1	Subtask Number	Activity Number	Resource State-2	Subtask Number	Activity Number
1	busy	5	2	-	-	-
2	res	7	1	busy	5	2
3	res	7	1	free	-	-
4	busy	6	2	-	-	-
5	free	-	-	-	-	-
6	zero	-	-	-	-	-

Besides, the monitor window, based on table 1 and 2 is used to realize an input (update) interface for simple interventions by the DMS. Permissible modifications are for instance:

- increasing of the remaining time of a busy activity (lines 3,5 - column 8 - table 1)
- decreasing of the remaining time of a busy activity (lines 3,5 - column 8 - table 1)
- state transition of a resource from zero to free (line 6 - column 2 - table 2)
- state transition of a resource to zero, independent of the current state (lines 1,2,3,4,5 - column 2 - table 2)

If a resource group is involved, time modification automatically leads to corrections within the cooperation database (see 2.1). It is an essential feature of the control subsystem to establish a link to the remaining times of all other busy activities if you select the second modification (i.e. decreasing of a remaining time). It is clear that each correction of such a remaining time must be a positive number, and that the duration T_{plan} can be defined as minimum of all calculated remaining time intervals (lines 3,4,5 of table 1).

On the one hand, the input of time t_{fix} marks up the start of the *replanning phase* of the PMS (duration T_{plan}), but on the other hand it is the beginning of the so-called *integration phase* of the control subsystem (see fig.5). First, an immediate storage of the entire running emulation state into a structure named $status(t_{fix})$ is performed (the emulation keeps running obviously). Secondly, a new system state - $status_{new}(t_{fix} + T_{plan})$, which depends on the future time $t_{fix} + T_{plan}$ and all forced modifications, has to be calculated and stored. Finally, a transmission of the stored and already modified structure takes place; the remaining project flow has to be replanned under the new circumstances by the planning subsystem. Because both subsystems (planning system and control system) are built on the same base model, each stored system state of the emulator is usable to initialize a (new) planning session. Transmission, start and termination of the planning subsystem are managed by the *supervising* instance of the control subsystem (IRQs of fig.2, fig.5). At time $t_{fix} + T_{plan}$ the hitherto best project plan (called *optimal plan*) has to be taken over into the original running emulation system substituting the old project plan object. All involved systems are now synchronized based on DMS decisions; the underlying model is adapted.

3 Conclusion

From the system point of view, each model of a software development process, but rather the task of managing it, is a living system or complex adaptive system (CAS)[6]. Adaption of our system (PMS) is implied by interrupts of a decision making environmental system (DMS). Our PMS provides an interactive interface to support such intervenient decisions:

– permanent display of relevant real time informations
– transparent input of structured modifications

It is the main responsibility of the DMS to determine time and type of interventions (availability of resources, degree of accomplishment of quality criteria and milestones are crucial aspects). Automation of these tasks in addition will lead to a more comprehensive model or higher level system: a *self adapting model* for managing software development projects.

Apart from a few exceptions, which do not restrict generality, the presented theoretical concept is already implemented as a prototype (the very complex user interface for aggregate interventions is not realized yet, see 1.2). To satisfy all kinds of requirements (i.e. support of bit operations in genetic algorithms, modern graphical user interface), the application framework MS-VisualC++ is used[11]. Because multitasking is an unavoidable precondition for our PMS, the operating system MS-WindowsNT was taken as a basis.

References

1. P.J.Angeline, (1995): Adaptive and Self-Adaptive Evolutionary Computation, in *Computational Intelligence: A Dynamic System Perspective, IEEE Press*
2. B.S.Blanchard and W.J.Fabrycky, (1990): Systems Engineering and Analysis, *Verlag Prentice Hall*
3. M.Burghardt, (1993): Projektmanagement: Leitfaden fuer die Planung, Ueberwachung und Steuerung von Entwicklungsprojekten, zweite ueberarbeitete Auflage, *Verlag Siemens AG, Berlin und Muenchen*
4. R.Burkhardt, (1997): UML-Unified Modelling Language, *Addison-Wesley Longman Verlag, Bonn*
5. P.Cabalar, R.P.Otero, M.Carbacos and A.Barreiro, (1997): Introducing Planning in Discrete Event Systems, in *Lecture Notes in Computer Science - Proceedings EUROCAST'97, Las Palmas de Gran Canaria, Spain*
6. J.Casti, (1996): Would-Be Worlds: How Simulation Is Changing the Frontiers of Science, *John Wiley & Sons*
7. B.Curtis, M.Kellner and J.Over (1992): Process Modelling, in *Communications of the ACM, Vol.35(9)*
8. M.Mauerkirchner, (1997): Event Based Modelling and Control of Software Processes, in *Engineering of Computer-Based Systems - Proceeding ECBS'97 Monterey, California*
9. M.Mauerkirchner, (1997): Dynamic Discrete Model of Simulation Based Planning and Control System for Software Project Development, *PhD, Johannes Kepler University, Linz, Austria*
10. M.Mauerkirchner, (1997): Event Based Simulation of Software Development Project Planning, in *Lecture Notes in Computer Science - Proceedings EUROCAST'97 Las Palmas de Gran Canaria, Spain*
11. M.Mauerkirchner, (1997): Dokumentation der Implementierung von PMS, *Linz, Austria*
12. A.Pagnoni, (1990): Project Engineering, *Springer Verlag*
13. B.P.Zeigler, (1984): Multifacetted Modelling and Discrete Event Simulation, *Academic Press, San Diego*

A Fractal Software Complexity Metric Analyser

Vili Podgorelec, Peter Kokol, and Milan Zorman

Laboratory for System Design, University of Maribor - FERI
Smetanova 17, SI-2000 Maribor, Slovenia
{Vili.Podgorelec,Kokol,Milan.Zorman}@uni-mb.si

Abstract. As we try to fulfill the requirements upon quality of the software products we cannot avoid the use of the complexity metrics. There is a lot of different metrics and also there are hundreds of tools for analyzing the software with some of those metrics. However, since all tools available are concentrated only on some specific programming metrics, for a comprehensive analysis one has to use a lot of different tools. We wanted to derive an environment that would include all of the mostly used metrics, and since we are also developing new metrics ourselves, we developed a tool called Software Complexity Analyzer, that beside classical metrics incorporates also more universal fractal metrics.

1 Introduction: Fractal Complexity Measure

Software complexity is aimed to objectively associate a number with a program, based on the degree of presence or absence of certain characteristics of software. It is assumed that software complexity is related with such features of software like number of errors left in the software, effort to design, test or maintain a software product, development time, maintenance cost, etc. The main weaknesses of the traditional software complexity metrics are:

- language dependency
- form dependency
- the output of a traditional complexity metric is a number, usually without any "physical" meaning and unit.

The majority of experts agree that complexity is one of the most relevant characteristics of computer programs. For example Brooks states that computer software is the most complex entity among human made artifacts [Broo87]. But what is complexity and how can we measure it? There are two possible, not completely distinct, viewpoints:

- the classical computational complexity [Coh86, Weg95] and
- recent "science of complexity" [Pin88, Mor95].

F. Pichler, R. Moreno-Díaz, and P. Kopacek (Eds.): EUROCAST'99, LNCS 1798, pp. 486–493, 2000.

The first viewpoint is well know and researched. Thereafter it is much more interesting to concentrate on the second view and the aim of this section is to present some ideas and results concerning it.

1.1 Complexity

According to Morowitz [Mor95] the complex systems share certain features like having a large number of elements, possessing high dimensionality and representing an extended space of possibilities. Such systems are hierarchies consisting of different levels each having its own principles, laws and structures. The most powerful approach for studying such systems was reductionism, the attempt to understand each level in terms of the next lower level. The big weakness of the reductionistic approach is that it is unable to explain how properties at one level emerge from the next lower level and how to understand the emergence at all. The problem is not the poverty of prediction, but its richness. The operations applied to the entities of one level generate so enormous many possibilities at the next level that it is very difficult to make any conclusions. This demands radical pruning and the main task of complexity as a discipline is to find out the common features of pruning (or more generally selection) algorithms across hierarchical levels and diverse subject matters.

1.2 Quantitative Properties of Complexity

Many different quantities have been proposed as measures of complexity. Gell-Mann [Gell95] suggests they have to be many different measures to capture all our intuitive ideas about what is meant by complexity. Some of the quantities are computational complexity, information content, algorithmic information content, the length of a concise description of a set of the entity's regularities [Gell95], logical depth [Gell95], etc., (in contemplating various phenomena we frequently have to distinguish between effective complexity and logical depth - for example some very complex behavior patterns can be generated from very simple formulas like Mandelbrot's fractal set, energy levels of atomic nuclei, the unified quantum theory, etc.- that means that they have little effective complexity and great logical depth). A more concrete measure of complexity, based on the generalization of the entropy, is correlation [Schen93], which can be relatively easy to calculate for a special kind of systems, namely the systems which can be represented as strings of symbols.

1.3 Complexity and Computer Programs

Computer programs, including popular information systems, usually consist of (or at least they should) number of entities like subroutines, modules, functions, etc., on different hierarchical levels. Concerning "laws of software engineering" or the concepts of programming languages [Watt90] the emergent characteristics of above entities must be very different from the emergent characteristics of the program as the whole. Indeed, programming techniques as stepwise refinement, top-down design, bottom up design or more modern object oriented programming are only meaningful if different hierarchical levels of a program have distinguishable characteristics.

Computer programs are conventionaly analyzed using the computational complexity or measured using complexity metrics [Cont86, Fent91, Schne94]. Another way to asses complexity is, for example, to use Fractal Metrics [Kok94]. But as we can see from above we can regard computer programs from the viewpoint of "complexity as a discipline" and according to that apply various possible complexity measures. The fact that a computer program is a string of symbols, introduces an elegant method to asses the complexity - namely to calculate long range correlations between symbols, an approach which has been successfully used in the DNA decoding [Buld94] and on human writings [Schen93].

1.4 Long Range Power Law Correlations

Long range power law correlations (LRC) have been discovered in a wide variety of systems. Recognizing a LRC is very important for understanding the system's behavior, since we can quantify it with a critical exponent. Quantification of this kind of scaling behavior for apparently unrelated systems allows us to recognize similarities between different systems, leading to underlying unifications. For example, the recent research has shown that DNA sequences and human writings can be analysed using very similar techniques, and we are interested if alike approach can be applied to computer software, too.

In order to analyze the long range correlations in a string of symbols we must first map the string into a random walk model [Buld94]. The advantage of this method over more traditional power spectrum or direct correlation of string's correlation is that it yields high quality scaling data [Schen93].

1.5 Fractal Software Complexity Measure: α metric

Alpha metric [Kok98b, Kok99] is based on the long range correlation calculation. In this paper we will use the so-called CHAR method described by Kokol [Kok98b]. A character is taken to be the basic symbol of a computer program. Each character is then transformed into a six bit long binary representation according to a fixed **code table** (i.e. we have 64 different codes for the six bit representation – in our case we assigned 56 codes for the letters and the remaining codes for special symbols like period, comma, mathematical operators, etc) is used. The obtained binary string is then transformed into a two dimensional Brownian walk model (Brownian walk in the text which follows) using each bit as a one move - the 0 as a step down and the 1 as a step up.

An important statistical quantity characterising any walk is the root of mean square fluctuation F about the average of the displacement. In a two-dimensional Brownian walk model the F is defined as:

$$F^2(l) \equiv \overline{[\Delta y(l,l_0)]^2} - \left[\overline{\Delta y(l,l_0)}\right]^2$$

where

$$\Delta y(l,l_0) \equiv y(l_0 + l) - y(l_0)$$

l is the distance between two points of the walk on the X axis

l_0 is the initial position (beginning point) on the X axis where the calculation of *F(l)* for one pass starts

y is the position of the walk – the distance between the initial position and the current position on Y axis

and the bars indicate the average over all positions l_0.

The *F(l)* can distinguish between two possible types of behaviour:

- if the string sequence is uncorrelated (normal random walk) or there are local correlations extending up to a characteristic range i.e. Markov chains or symbolic sequences generated by regular grammars, then

$$F(l) \approx l^{0.5}$$

- if there is no characteristic length and the correlations are “infinite” then the scaling property of *F(l)* is described by a power law

$$F(l) \approx l^{\alpha} \text{and } \alpha \neq 0.5.$$

The power law is most easily recognised if we plot *F(l)* and *l* on a double logarithmic scale. If a power law describes the scaling property then the resulting curve is linear and the slope of the curve represents α. In the case that there are long range correlations in the program analysed, α should not be equal to 0.5.

2 Software Complexity Analyzer

To confirm our theoretical findings about new fractal metrics, to compare them with the conventional software metrics and of course to be able to analyze large quantities of commercial programs we developed a program tool called Software Complexity Analyzer (Figure 1). We wanted to derive an environment that would include all of the most often used conventional complexity metrics together with our new, more general fractal metric.

Our interest is primarily focused into analyzing source code of computer programs. Since we want the tool to be as programming language independent as possible, it has to provide a transparent way to handle different programming languages. In this manner every programming language is described with a set of attributes needed for analysis. Once the set of needed attributes is described, all kinds of analyses can be performed upon source code of a specific programming language. We have already described some general purpose programming languages (Java, C, C++, Pascal, Fortran), others can be added through an intuitive user interface. For the analyzing purposes we also added the possibility of generating syntactically and semantically correct random programs (GRP).

Sometimes it is useful not only to analyze the source code but also to analyze some other program representation, for example object code or a compiled program. Therefore it is possible to analyze binary data and in this way compare the growth of complexity through the whole life cycle of a computer program.

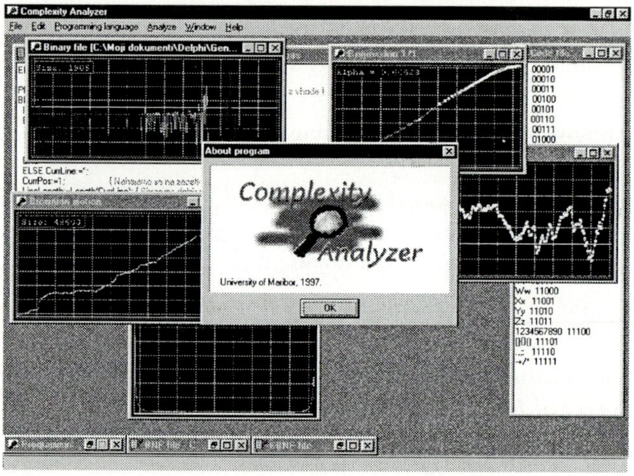

Fig 1. Working environment of Software Complexity Analyzer.

In order to analyze data with α-metric they have to be transformed into Brownian motion, a sequence of 0 and 1. For binary data representation we just have to select which bits will be considered. For source code of computer programs, we have more options: char method requires coding table, where each character is described with a binary sequence, categories method requires each language category to have its own binary sequence, and operator-operand method requires such a sequence for operators/operands. From a Brownian motion model a regression curve can be calculated, from which we can obtain α coefficient. Regression curve, Fourier transform, α local slopes and similar interesting data can also be presented in visual form as graphs.

To avoid repeating the same procedures over and over again, and to avoid analyzing once already analyzed programs, simple database has been added that can store all interesting data about a computer program, like general information, obtained results from analysis, etc. and also programs, that are being analyzed. Such database shows its real value especially when performing some larger analyzing experiments. Beside the calculations and analysis that are provided by Software Complexity Analyzer, we would also like to perform some less used statistical analysis or include obtained results in reports. Therefore we can export data to some general purpose (like MS Excel) or more specialized programs (like SPSS, Statistic or Chaos Analyzer).

3 Application of the Tool and Conclusion

Among others applications the tool as been used to analyze the NIAM formal specification conceptual schemes and sub – schemes. All sub - schemes have been ranked according to their assessed α – metric values. The sub – scheme's ranks for each conceptual scheme have been then averaged. According to these average ranks the conceptual schemes have been ordered. Opposite to our expectation the intuitive order and the calculated order didn't match. Indeed the statistical analysis using Spearman correlation showed statistically insignificant (Spearman R = -0.28) reciprocal relation between α – metric complexity and intuitive complexity. Contrary, the internal ranking of sub – schemes within each conceptual scheme was in match with intuitive order both individually and in average. In fact the majority of internal and intuitive orders within conceptual schemes where statistically significantly correlated (on the p = 0.05 level). This last observation revealed the hypothesis that there is something wrong with the intuitive order of conceptual schemes. A closer examination exposed the following three assumptions about the complexity of sub – schemas:

1. conceptual schemes with ***equality relation*** have the highest ranks and are thereafter the most complex;
2. conceptual schemes with exactly one ***totality relation*** have the lowest ranks and are as a consequence less complex, in fact it seems that exactly one totality relation per scheme reduces the complexity.
3. Total functions (relation in the middle) have highest middle ranks and thereafter "high – middle" complexities.
4. All other relations contribute by approximately one quarter of the equality relation.

According to above we constructed a simple ***ranking function*** f_R and used it produce order generated (Table 2). We can see that the ranks generated by the α – metric and the ranks generated by the ranking function f_R are in close match. Indeed, the Spearman's correlation coefficient is 0.88, meaning that the correlation between these two orders is statistically significant (p = 0.000) which proves the assumptions 1 to 4 introduced before.

The results obtained by using the tool can help us a lot by gaining insight into the complexity of the formal specification. As a consequence we can use them as a guideline for designing less complex specifications which are then first easier to understand by end users (making them more valid) and second easier to implement.

Table 1. Ranking of NIAM's schemes according to ranking function.

SCHEMA	Rank defined by α – metric	Rank defined by ranking function
1	6	20
2	10	13
3	10	10
4	20	16
5	10	6
6	20	19
7	16	17
8	23	22
9	20	21
10	16	18
11	14	15
12	2	1
13	9	12
14	10	8
15	16	14
16	26	26
17	19	23
18	6	7
19	5	9
20	25	24
21	1	4
22	23	25
23	14	11
24	2	3
25	4	5
26	6	2

References

[Abr96] Abrial, J.R.: The B–Book: Assigning Programs to meanings, Cambridge University Press (1996)

[Broo87] Brooks, P.F.: No silver bullet: essence and accidents of software engineering, IEEE Computer, 20(4) 10-19 (1987)

[Buld94] Buldyrev, S.V. et al.: Fractals in Biology and Medicine: From DNA to the Heartbeat, Fractals in Science (Eds. Bunde, A, Havlin, S), Springer Verlag (1994)

[Coh86] Cohen, B, Harwood, W.T., Jackson, M.I.: The specification of complex systems, Addison Wesley (1986)

[Cont86] Conte S.D., Dunsmore, H.F., Shen, V.Y.: Software engineering metrics and models, Benjamin/Cummings, Menlo Park (1986)

[Fent91] Fenton, N.E.: Software Metrics: A Rigorous Approach, Chapman & Hall (1991)

[Gell95] Gell-Mann, M.: What is complexity, Complexity 1(1) 16-19 (1995)

[Kok94] Kokol, P.: Searching For Fractal Structure in Computer Programs, SIGPLAN 29(1) (1994)

[Kok97] Kokol, P, Brest, J, umer, V.: Long-range correlations in computer programs, Cybernetics and systems 28(1) 43-57 (1997)

[Kok98a] Kokol, P., Brest, J.: Fractal structure of random programs, SIGPLAN notices 33(6) 33-38 (1998)

[Kok98b] Kokol, P., Podgorelec, V., Brest, J.: A wishful complexity metric, In Proceedings of FESMA (Eds: Combes H et al). Technologish Institut 235–246 (1998)

[Kok99] Kokol, P., Podgorelec, V., Zorman, M., Pighin, M.: Alpha - a generic software complexity metric, Project control for software quality (Eds: Rob J. Kusters et al.), Maastricht : Shaker Publishing BV 397-405 (1999)

[Mor95] Morowitz, H.: The Emergence of Complexity, Complexity 1(1) 4 (1995)

[Pin88] Pines, D. (Ed.): Emerging syntheses in science, Addison Wesley (1988)

[Schen93] Schenkel, A., Zhang, J., Zhang, Y.: Long range correlations in human writings, Fractals 1(1) 47-55 (1993)

[Schne94] Schneidewind, N.F.: Methodology for Validating Software Metrics, IEEE Trans Soft Eng 18(5) 410-422 (1994)

[Spi94] Spivey, J.M.: La notation Z, Traduction de M. Lemoine, Paris, Masson (1994)

[Watt90] Watt, D.A.: Programming Language Concepts and Paradigms, Prentice Hall (1990)

[Weg95] Wegner, P., Israel, M (Eds.): Symposium on Computational Complexity and the Nature of Computer Science, Computing Surveys 27(1) 5-62 (1995)

6 ARTIFICIAL INTELLIGENT SYSTEMS AND CONTROL

Systems Approach to Attention Mechanisms in the Visual Pathway

Roberto Moreno-Díaz jr.[1,2], Juan Carlos Quevedo-Losada[1], and Alexis Quesada-Arencibia[2]

[1]Instituto Universitario de Ciencias y Tecnologías Cibernéticas, Univ. de Las Palmas de Gran Canaria, Edificio de Informática y Matemáticas, Campus de Tafira, E-35017 Las Palmas, Spain
{rmorenoj,jcquevedo}@dis.ulpgc.es
[2]Departamento de Informática y Matemáticas, Univ. de Las Palmas de Gran Canaria, Edificio de Informática y Matemáticas, Campus de Tafira, E-35017 Las Palmas, Spain
alex@ciicc.ulpgc.es

Abstract. Presynaptic Inhibition (PI) basically consists of the strong suppression of a neuron's response before the stimulus reaches the synaptic terminals mediated by a second, inhibitory, neuron. It has a long lasting effect, greatly potentiated by the action of anaesthetics, that has been observed in motorneurons and in several other places of nervous systems, mainly in sensory processing. In this paper we will focus on several different ways of modelling the effect of Presynaptic Inhibition(PI) in the visual pathway as well as the different artificial counterparts derived from such modelling, mainly in two directions: the possibility of computing invariant representations against general changes in illumination of the input image impinging the retina (which is equivalent to a low-level non linear information processing filter) and the role of PI as selector of sets of stimulae that have to be derived to higher brain areas, which, in turn, is equivalent to a "higher-level filter" of information, in the sense of "filtering" the possible semantic content of the information that is allowed to reach later stages of processing.

1 Lettvin's Divisional Inhibition. Invariant Computation Using PI-Like Mechanisms

One of the first known formalisms intended to describe the effect of presynaptic inhibition is due to Lettvin [1], who named it Linear Divisional Inhibition. Lettvin suggested that inhibition may cause a change in membrane permeability at the point of inhibition equivalent to a change in electrical conductivity. Such a change will act as an electric shunt, and it follows that if E is the excitation on a fibre that receives divisional inhibition I, the resultant activity, A, is:

$$A = \frac{E}{1 + I/I_0} \quad . \tag{1}$$

where I_0 is a constant. For $I \gg I_0$, $A = I_0 E/I$.

F. Pichler, R. Moreno-Díaz, and P. Kopacek (Eds.): EUROCAST'99, LNCS 1798, pp. 497–505, 2000.

This mechanism was used by Moreno-Díaz [2] as part of the operations carried out by a frog retinal group two ganglion cell model to account for the temporal behaviour of the cells , where two kinds of presynaptic inhibition were assumed: linear divisional and nonlinear.

In its simplest form, the same mechanism can be used to build a neuron-like network to compute invariances against global changes in its input. This would be a desirable goal of the usual pre-processing of an image, the resulting image being sent to a higher level stage to be analyzed. Some preprocessing characteristics have been described to be present in retinal computation, but no known mechanism, besides adaptation, have been described to obtain a representation which is invariant against global illumination changes. In the model that follows, parallelism is a need for the system to work properly. We will assume that processors (ganglion cells) are arranged in layers, that some kind of computation is done by every cell on their receptive fields and that the output of the layer is a transformation of the original data. There will also be a plexiform layer where the inhibition between cells take place [3] (Fig. 1).

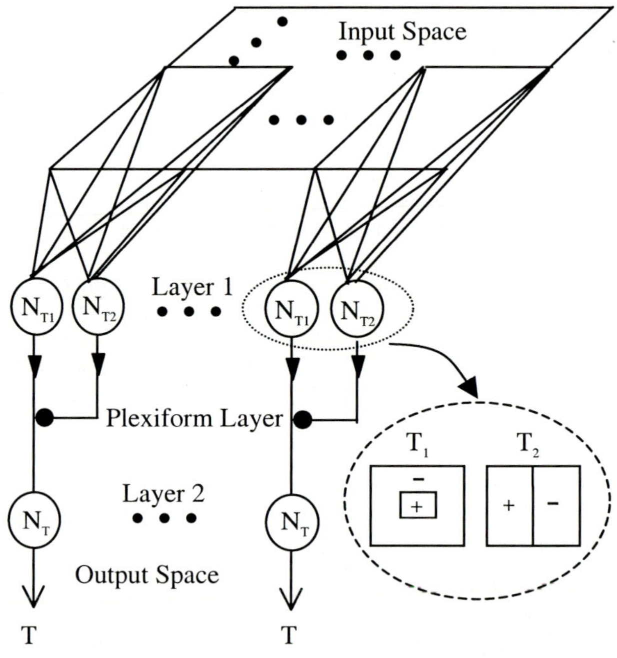

Fig. 1. A parallel processing layer of units performing a model of PI activity

Let then I(x',y',t) be a representation of the original input image impinging on the fotorreceptor layer and T the output of our system as measured in the axons of the ganglion cells. The effect of the presynaptic inhibition will be defined in two parts: let be T_1(I(x',y',t)) and T_2(I(x',y',t)) two linear transformations whose kernels are respectively W_1(x,y,x',y',t) and W_2(x,y,x',y',t) such that:

$$\int_c W_i(x, y, x', y', t)dx'dy' = 0 \ . \tag{2}$$

where c is the domain of each receptive field. This accounts for the fact that the known receptive fields structures in retinal ganglion cells have mutually cancelling regions, e.g., inhibitory and excitatory regions that cancel each other (center-surround or bar-type regions). This is a basic requisite in the description that follows.

The effect of the transformations T_1 and T_2 on the input image is expressed as usual:

$$T_i(I(x', y', t)) = \int_c I(x', y', t)W_i(x, y, x', y', t)dx'dy' \ . \tag{3}$$

and the presynaptic inhibition effect would then be expressed as the ratio: $T=T_1/T_2$.

In these conditions is easy to prove that a change of I(x,y,t) like: I(x,y,t)=KI(x,y,t)+R(t), where k is a constant and R(t) a function of time representing a global change on light intensity over the retina, would not affect the calculation of T. Thus, when $T_2 >> T_1$

$$T' = \frac{T_1(I'(\overline{X'}, t))}{T_2(I'(\overline{X'}, t))} = \frac{\int_c kI(\overline{X'}, t)W_1(\overline{X}, \overline{X'}, t)dx'dy' + \int_c R(t)W_1(\overline{X}, \overline{X'}, t)dx'dy'}{\int_c kI(\overline{X'}, t)W_2(\overline{X}, \overline{X'}, t)dx'dy' + \int_c R(t)W_2(\overline{X}, \overline{X'}, t)dx'dy'} =$$

$$= \frac{k\int_c I(\overline{X'}, t)W_1(\overline{X}, \overline{X'}, t)dx'dy' + R(t)\int_c W_1(\overline{X}, \overline{X'}, t)dx'dy'}{k\int_c I(\overline{X'}, t)W_2(\overline{X}, \overline{X'}, t)dx'dy' + R(t)\int_c W_2(\overline{X}, \overline{X'}, t)dx'dy'} = \frac{kT_1}{kT_2} = T \ . \tag{4}$$

where

$$\overline{X'} = (x', y') \ and \ \overline{X} = (x, y) \ . \tag{5}$$

A plausible place in the nervous system to locate this kind of invariant computation is the LGN. The output of information from retina through the optic nerve follows three channels towards higher brain areas, being the geniculo-cortical pathway the one receiving more fibres. In the LGN a topographical representation of the whole retina can be found. The cells in the LGN are arranged in six perfectly defined layers and attending to the size of the cells these layers could be divided into two groups: the magnocellular layer and the parvocellular layer [4]. The magnocellular layer is formed by big cells working on the illumination characteristics of the input image and the parvocellular layer includes smaller neurons involved in color coding.

Our mechanism can be assumed to work in the magnocellular layer. The transformations T_1 and T_2 are computed by the ganglion cells and the result reaches the LGN via the optic nerve. The cells at the magnocellular layer of the LGN would be the units that compute the invariant representation T using presynaptic inhibition (see Fig. 1). The simplicity of the figure mimics the simplicity found in the physiology of LGN where each cell receives only a few input lines from retina including inhibitory effects. Thus, one of the possible outputs of the LGN would be an invariant representation of the light pattern already coded by ganglion cells.

2 Model Refining. Non Linear Divisional Inhibition

Despite the above model is quite useful when specific illumination conditions are present, there are certain formal objections that make it unsatisfactory, both from a formal as well as an applied point of view. First, we can start taking into account the idealized performance of PI (Fig. 2) [5].

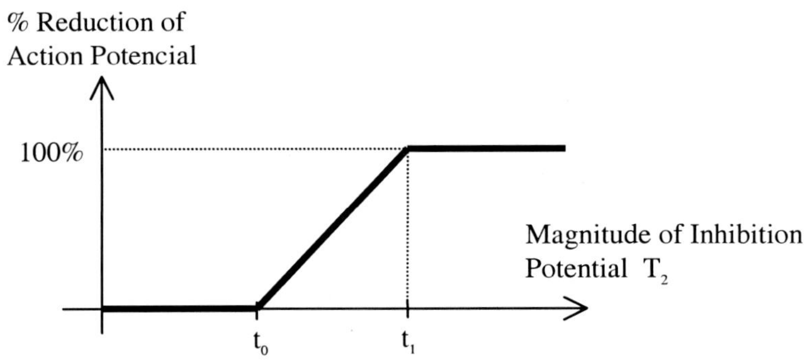

Fig. 2. Idealized performance of PI

As we can observe in the graphic, when T_2 is smaller than a certain threshold (t_0) the Presynaptic Inhibition mechanism does not reduce the action potential at all. From the threshold t_0 to t_1 PI presents a linear performance, and for values of T_2 grater than this threshold PI reduces the action potential to zero. Thus the idealized performance of T can be expressed as follow:

$$T = f(T_1,T_2) = \begin{cases} T_1 & if\ T_2 \le t_0 \\ f'(T_1,T_2) & if\ t_0 < T_2 < t_1 \\ 0 & if\ T_2 \ge t_1 \end{cases} . \tag{6}$$

Using the original analysis of Lettvin, already mentioned at the beginning of this paper, we obtain the following expression:

$$T = \frac{T_1}{1+\frac{T_2}{I_0}} . \tag{7}$$

to model Presynaptic Inhibition, where I_0 is a constant. This comes from considering the action of PI as an electric shunt.

It has been used a linear function in the denominator. However, the function that better fits the ideal performance has a shape shown in Fig. 3:

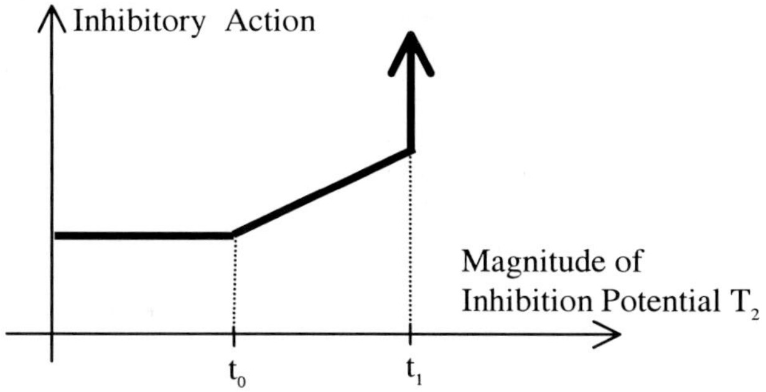

Fig. 3. Better-fit representation of PI function

$$f(T_2) = \begin{cases} 1 & if\ T_2 \le t_0 \\ aT_2 + b & if\ t_0 < T_2 < t_1 \\ \infty & if\ T_2 \ge t_1 \end{cases} . \quad \textbf{(8)}$$

Now we show a comparative graphic where we can observe the functions mentioned above and the idealized function in Fig. 4, and at the same time we present another non-linear function that fits better into the idealized function (we use an exponential function):

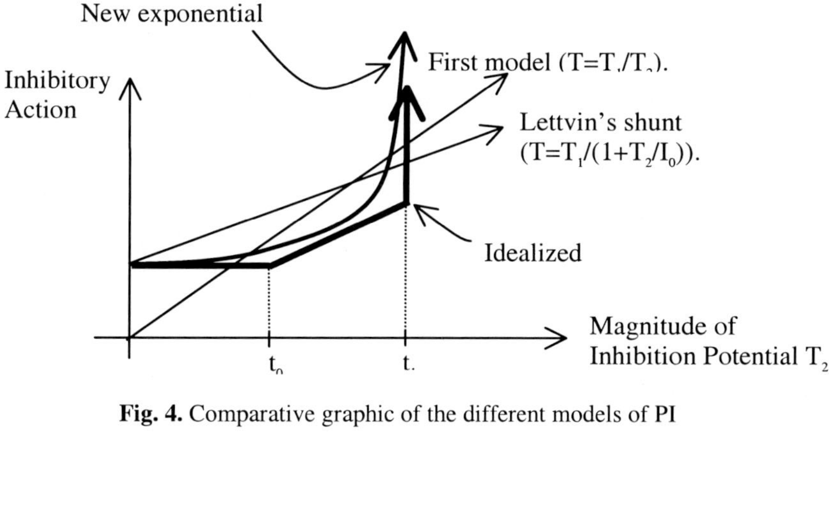

Fig. 4. Comparative graphic of the different models of PI

The new model is an exponential function as follows:

$$f(T_2) = \frac{T_2}{a} * e^{\frac{T_2}{b}} + 1 \quad \textbf{(9)}$$

$$T = \frac{T_1}{\frac{T_2}{a} * e^{\frac{T_2}{b}} + 1} \quad \textbf{(10)}$$

where a and b are constants that should be obtained depending on t_0 and t_1 and trying to make the best fit with the idealized function. This model is similar to Schypperheyn's model, who named it Non-Linear Divisional Inhibition [6].

To illustrate this point, we developed an example where invariances against global uniform illumination changes can be observed (Fig. 5). In Fig. 5a we have the original image, which is a 132x102 pixel black and white image with 256 grey levels. In Fig. 5b we present the result of performing the operation indicated in expression (10), where we used Newton Filters [7], both to calculate T_1 as well as T_2. In Fig. 5c we can observe the original image (I(x',y',t)) transformed as follows:

$$I(x', y', t) + R(t) = I(x', y', t) + 200, \textit{ that is, } R(t) = 200 \quad \textbf{(11)}$$

Finally, in Fig. 5d we show the result of performing the same operation as in the previous case, and we can verify that this image is equal to the image of Fig. 5b, confirming the theoretic results explained in the first point of the paper. Besides, the improvements to which this new exponential model contributes can be seen graphically on Fig. 4, since it fits better into the idealized function. We can see this in the implementation of the examples where if we do not use this model we would have problems in the limits, that is to say when T_2<=t_0 and T_2>=t_1. In this case the other models differ strongly. The implementation of this examples has been carried out by means of a program developed in the programming language Borland Delphi Professional 3.0 for Windows 95.

3 A Possible Mechanism to Control Visual Attention and Information Flow via Pi

The Theory of Vision's goal is, basically, to build a theoretical and practical framework to explain the visual function in live beings and its possible artificial counterparts. The sense of vision is, in most of species, the one that processes the biggest amount of information coming from the outside, being the most important in guiding its behaviour. The ability to discriminate certain parameters and locate what part of all that information is significant is crucial for both natural and artificial complex systems [8].

In a previous paper [9] a theoretical construct called Directed Foveal Transform, TFD, was presented as the first step in modelling the attentional mechanisms ruling the visual processes of vertebrates. The kernel of the TFD is a variation of the moving average that presents the highest acuity on a particular area in the retina (called fovea). Originally this area of better resolution expanded from the center of the image. In the TFD, the dominia on which the best resolution is placed can be located around any point in the image and at the same time any foveal size can be especified. It is also possible to define, on the parafoveal zone, transformations where the completeness is not a crucial factor. Thus, we see the goal of defining this transformations in different image areas as performing an "interest attractor". On the parafoveal zones information is extracted but there is a loss of resolution, but once the event or characteristic is discriminated, the fovea can be placed over it in order to perform a more detailed operation. There is a basic idea underlying it: the economy of computations: no known visual system has large fovei (at least they do not cover more than 10% of the total system), otherwise the total information to process and transmit would be too high to cope with, needing then more complex and bigger nervous tracts and much more connections in intermediate stages of the visual pathway. Thus, the goal could be to achieve certain "economical balance" between structural complexity and detailed information to be transmitted.

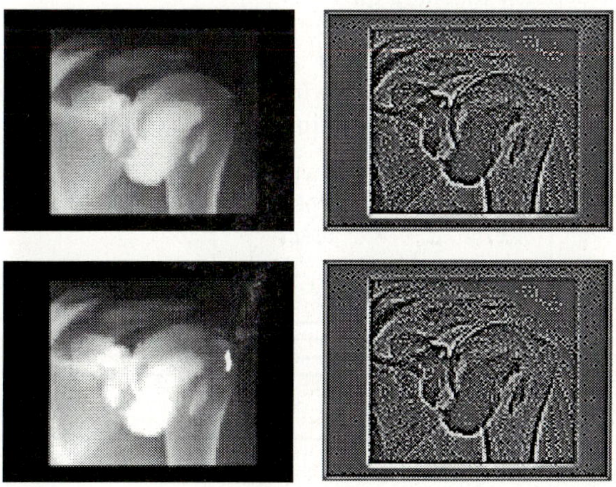

Fig. 5. A X-ray image of the shoulder. An example of invariance computation; From left to right and from top to bottom we have figures 5a, 5b, 5c and 5d: original image, processed image, displaced illumination image and processed image after displacement

A step forward in the simulation of information flowing and attention mechanisms in the visual pathway would be to combine the above mentioned concept with the action of the presynaptic inhibition. The goal is then to highlight some feature or object in the image, blurring (or loosing resolution) the rest of it since we assume that it does not contain anything relevant. In the original TFD formulation, a three-step procedure was designed to locate a visual event on the input image and concentrate the highest amount of computation on it:

1. First, it is necessary to locate the center of luminance of the region/object. The coordinates of the center of luminance will act as the coordinates for the center of the fovea.

2. Second, the outermost point of the object is located and the distance from the center of the fovea is calculated.
3. We use this data in order to calculate the size of the fovea to be used.

In order to include PI, a feedback line from the cortex is necessary to allow the information flow within the topographical representation of the retina that is present in the lateral geniculate nucleus [10]. In Fig. 6 we can see the proposed architecture. Thus, the mechanism will perform as follows:

1. A first retinal description in terms of contrast, edges, color and movement parameters, present in the axons of retinal ganglion cells, reaches the LGN. These parameter calculations are already studied in previous reports [10,11,12].
2. This information is transformed in a second, higher-semantic content representation in terms of location of center of gravity of moving objects, high luminance or color- components areas and sent to the first layers of visual cortex.
3. The decision of selecting the geographic position of the object of interest is made in the cortex, that sends a feed back signal to the LGN, and via PI controls what information is effectively reaching the LGN.

This mechanism could be proposed as a "first-order" attention focusing mechanism. Note that it has nothing to do with "willingness" or purpose of behaviour, it is closer to an automatic response depending on outside stimulae.

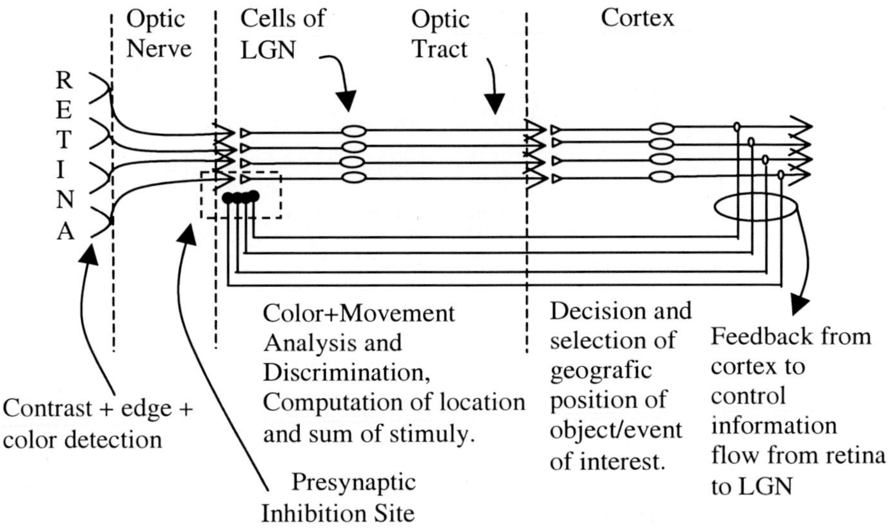

Fig. 6. Representation of the feedback lines coming from the cortex into the LGN including the PI mechanism

4 Conclusions

Although more information on the real connectivity of PI in LGN is needed, two immediate operations can be thought of being performed by that kind of mechanism: invariant computation and control of information flow. Mechanisms for obtaining invariant representations of features in the first steps of information processing are important in Nervous Systems since they provide the basis of reliable pattern and movement discrimination. In order to do this, it is plausible to expect that the visual cortex works, at least to some extent in the first steps, with a low-level invariant version of the input image. On the other hand, there is a need of controlling the total amount of information reaching effectively the brain, and this can be achieved, according to the connectivity of the first stages of the visual pathway, by means of two structures: first the existence of a foveal area in retina in which the resolution is maximal, and second, a control of the information flow in the LGN. In both, invariance computing and information control, the mechanism of presynaptic inhibition may play a crucial role.

Regarding the artificial counterpart and applicability of the ideas presented before, the mechanism of PI, implemented in a parallel fashion, is suitable of being used in artificial perceptual systems, mainly in image processing, to obtain image representations invariant against general changes in illumination.

References

1. Lettvin, J. Y.: Form-function relations in neurons, Research Lab. of Electronics, MIT Quarterly Progress Report (1962) 333-335
2. Moreno-Díaz, R.: An analytical model of the Group 2 ganglion cell in the frog's retina, Instrumentation Lab., MIT Quarterly Progress Report (1965) 1-34
3. Muñoz-Blanco, J. A.: Jerarquización de estructuras de nivel bajo y medio para reconocimiento visual, PhD Dissertation, University of Las Palmas de Gran Canaria (1987)
4. Truex, R. C., Carpenter, M. B.: Human Neuroanatomy, Williams and Wilkins, Baltimore, USA (1969)
5. Graham, B., Redman, S.: A Simulation of Action Potentials in Synaptic Boutons During Presynaptic Inhibition, Journal of Neurophysiology, Vol 71, No 2 (1994)
6. Hagiwara, S., Tasaki, I.,: A Studdy of the Mechanism of Impulse Tranmission Across the Giant Synapse of the Squid, Journal of Neurophysiology, Vol 143 (1958)
7. Moreno-Díaz jr., R., Computación Paralela y Distribuida: relación estructura-función en retinas, PhD Dissertation, University of Las Palmas de Gran Canaria (1993)
8. Kandel, E. R.,: Processing of form and movement in the visual system, Sensory Systems of the Brain: Sensation and Perception, Chapter 29, Part V (1990)
9. Quevedo-Losada, J. C., Bolívar-Toledo, O., Moreno-Díaz jr, R.: Image Transforms based on Retinal Concepts, R. Trappl Ed. Cybernetics and Systems 98, University of Vienna, Austria (1998) 312-316
10. Moreno-Díaz jr., R.: On structure, function and time in retinae, R. Moreno-Díaz and Mira-Mira Eds. Brain Processes, Theories and Models, The MIT Press, Cambridge, Mass, USA (1996) 430-435
11. Alemán-Flores, M., Leibovic, K. N., Moreno-Díaz jr., R.: A computational model for visual size, location and movement, R. Moreno-Díaz and F. Pichler Eds. Computer Aided Systems Theory, Springer Lecture Notes in Computer Science, Vol. 1333 (1997) 406-419
12. Moreno-Díaz, R., Alemán-Flores, M., Moreno-Díaz jr, R.:A bio-inspired method for visual movement discrimination, R. Trappl Ed. Cybernetics and Systems 98, University of Vienna, Austria (1998) 307-311

On Completness in Early Vision from Systems Theory

O. Bolívar-Toledo[1], J.A. Muñoz Blanco[1], S. Candela Solá[1], and R .Moreno-Díaz[2]

[1]Departamento de Informática y Sistemas.Universidad de Las Palmas de Gran Canaria. Campus de Tafira, 35017 Las Palmas. Spain
(obolivar,jamunoz,scandela)@dis.ulpgc.es
[2]Instituto Universiatario de Ciencia y Tecnología Cibernética. Universidad de Las Palmas de Gran Canaria. Campus de Tafira, 35017 Las Palmas. Spain
moreno@ciicc.ulpgc.es

Abstract.This paper shows the interaction among the algebraic-analytical data fields theory, the complete descriptions that may be truncated for practical visual tasks and the different level of processing in the early visual pathway, in order to establish a measure of the representation of similarity between images. Our work is based on the approach of modeling the representational capabilities of systems of receptive fields found in early mammalian vision. It is well known that a representation scheme with a metric in the space of representation induce, in a natural way, a similarity measure between images. We propose in this paper to compare different representation schemes, based in previous theorems about completnes on data fields, according to the induced similarity measure.

1 Introduction

At all levels of the visual system, complex objects seems to be codified by the activities of population, networks or cells, and the representation of a particular object can be widely distributed through one or more visual areas. Many of the low-level areas of the mammalian visual system are retinotopically organized, that is to say, with an organization or projection that preserve, in certain degree, the topography of the retina. A unit that it is part of this retinotopics map, normally respond in a selective manner to an stimuli localized in a place of the visual field called the Receptive Field.

The receptive field of a neuron anywhere in the visual pathway is defined as that portion of the visual field whose estimulation affects the response of the neuron. Computational models of perception usually assume that the neuron performs a spatial integration over its receptive field, and that its output activity is a (possibly nonlinear) function of

$$\iint_{RF} K(x,y)I(x,y)dxdy \tag{1}$$

where I(x, y) is the input, and K(x, y) is a weighting kernel that describes the relative contribution of different locations within the receptive field to the output.

F. Pichler, R. Moreno-Díaz, and P. Kopacek (Eds.): EUROCAST'99, LNCS 1798, pp. 506–513, 2000.

Also, it is well known that nervous systems exhibit a remarkable convergence and divergence of signal traffic [1]. In this sense information can be processed and conserved in detail through overlapping receptive field, being the variation of the degree of overlapping closely related to the degree of divergence in the system. This suggest a need for preserving information while it is passed from one level to another.

The retina of the vertebrates can be considered as a system that processes by layers the visual information that comes in from the outer world; the information progresses longitudinally as well as transversally involving all of retinal neurons from photoreceptors to the ganglion cells. As a first stage in that progresses, the retinal image is sampling by a dense mosaic of fotoreceptors cells which through a transduction mechanism, provide the electrical signal which is later and gradually processed at the different stages of the visual pathway.

Cell recordings in the retina of cats and monkeys have uncovered two classes of retinal ganglion cells whose axons form the optic nerve fibers along which visual information is carried to the lateral geniculate nucleus before reaching visual cortex. These two types, termed X and Y cells both have receptive fields with an antagonistic centre-surround organization, whose shape may be modeled as the difference of two gaussians (DOG).

The axons of the ganglions cells form the optic nerve, which send the outputs to the lateral geniculate body, which constitute the first visual signal stop previous to reach visual cortex. It is well known that centre-surround circularly symmetric receptive fields, can simulate the representation at lateral geniculate body level.

There are neurophisyological evidence that most of the neurons of visual cortex shows a diferenciated behaviour characterized by the responses to a particular stimulus, like edges, borders or lines and that this behaviour depends in a great way of orientation. So different orientations has to be considered to model the orientation selectivity. The x/y asymmetry was inspired by the shape of the receptive fields of the simple cells in the primary visual cortex of mammals, and made the units orientation selective.

2 Complete Description of a Data Field

As well as neurophysiologists worry about what natural neuron receptive fields are and what they are doing, persons working in Artificial Vision worry about complete descriptions and representations of images in order to approach nature. A description is said to be complete, at a given level of processing, if it contains all necessary data and data properties to achieve some goal. From an analytical point of view, a complete description needs the preservation of the number of degrees of freedom or the number of independent properties of the visual field.

From a more generic and abstract point of view, we have consider a Data Field as the data available to feed receptive fields. To sample or to carry out a Data Field partition means to compute over sensorial data in parallel with some overlapping degree. In this way, a unidimensional data field of length N and resolution R is an ordered set of N places, such that a number, complex or real, with resolution R can be assigned to each place.

From a data field D, addressable by index i, a *Complete Transformation* is a rule to construct a second field D' addressable by index j, from which field D can be recovered.

$$D(N,R) \Leftrightarrow D'(N',R') \tag{2}$$

A more realistic approximation take us to introduce the concept of "extended degrees of freedom" of a data field transformation which involves not only the number of places N , but the resolution of them R. Completness requires as necessary condition that: $N \cdot R = N' \cdot R'$. Therefore, a possible temptation could be to assign to the places i, a greater resolution capacity such that if R'>R then N'<N. This only would be possible through an additional structure of assignation of meanings, which suppose a jump of level. And viceversa, if in a natural or artificial process we find a situation such that R'>R and consequently N>N', it must exist an additional structure in order to establish the necessary coding at a superior level.

This considerations have an interesting conceptual incidence in current research of "retinal coding", on what is known as "multiple meanings" in retinal processing for the vertebrate retinae. Thus it seems that retinal cells compress the original numbers of places in the retina at the expense of an increase of the resolution in tectum (frog) or cortex (cat), through a mechanism of coding-decoding of higher level present in said tectum or cortex; so that the extended number of degrees of freedom is mantained.

The Receptive Fields System probably form the more relevant computational mechanism employed in the biological information process, and particularly in vision. The fact that different areas of a receptive field can contribute in a different way to the activity of the unit, in accordance with the profile or weight function of the Receptive Field, take us to consider two completely different aspects included in it which are: the function and the structure [2]. This is related with concepts concerning completnees of a visual transformation with respect to the receptive field, where the data is selected and to the function performed by the transform. This treatment emphasizes the separation of receptive field and function in visual processing and in general in parallel computation.

These approach gave rise to a theory, with their respective theorems, lemmas and demonstrations, about Algebraic-Analytic Transforms in Data Fields [3]. The main objective of this theory was the design of Complete Transforms, both from the structural (data field partition carried out) and the functional (set of functions applied) point of view. One interesting analytic result was that the partition carried out over the data field, per se, could generate a complete description, with the same properties usually required to justify the functionals, that is to say, completeness, capacity to decrease the degree of freedom and security with respect to escotomas.

In this context we have introduced the concept of Progressive Resolution Transform, in which it was realized that no matter what function it is performed, a vast class of partition could be complete. That is, given an image (a data field in fact, because the image fills a data field in memory), then one can, and it is convenient, to separate receptive fields from functions, so that a data field partition, by itself, can be complete. The following theorem folds:

For a data field of N degrees of freedom (pixels), given a dimension d<N for receptive fields of computing units, there exists a complete PRT, which is formed by

all independent receptive fields of dimension d plus d-1 receptive fields which must be unitary.

Next figure is an illustration of a PRT foveal transform.

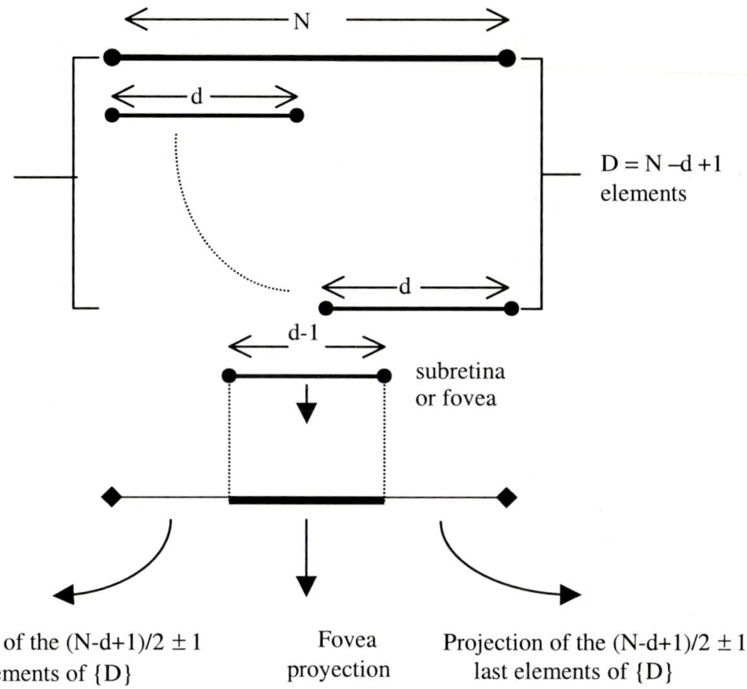

Fig. 1. Illustration of a PRT fovel transform

The next logical problem was one of trade off, which is probably what happens in the retina, since the resolution of fibers is not necessarily preserved, and there is a multiple meaning coding, which in our terminology mean that each fiber of the optic nerve send information pertaining to more than one operation on the receptive field. There is a second theorem for that and it goes as follows:

For a data field of N degrees of freedom, and an arbitrary partition giving L independent receptive fields such that M=N/L is an integer, then, the computation of M functional coefficients linearly independent in each partition provides for a complete description of the data field.

In a more general way, we can conclude that in Artificial Vision and probably in natural vision too, it exists a basic trade off between functional and data field partitions to maintain the degree of freedom imposed by the objectives of the problem.

Nature does not follow theorems exactly, but they provide us with a clear way to approach nature. On any case, the parallel computing structures which result, are both illuminating to understand natural systems and to build artificial ones.

3 Representation Spaces

In Artificial Vision and in Image Processing also, it is frequently required to deal with descriptors proceeding from transformed spaces. The typical way to face the problem consists of spreading the spatial image in terms of a new set of coefficients. The justification of this expansion in visual recognition and image processing is based on a conjecture [4], which has not been yet reduce to formal theorems and therefore has no theoretical sustent.

As we have indicated previously on the objective of the representation of data fields consists of obtaining alternative descriptions of the fields in order to generate a new pattern, according to certain selection rules. The methods are in essence those of a selection of coefficients in a complete representation.

There are several methods for extracting features (principal components analysis, projection pursuit...). In this work we chose to explore a considerably simpler method of feature extraction, based on the notion of localized receptive field, borrowed from neurobiology of vision. So, we can consider the representation of the visual appearance of an object by a pattern of receptive field activities, or partition/functional, as a method of characteristic extraction.

It is well known that a representation scheme with a metric in the space of representation induce, in a natural way, a similarity measure between images. We propose in this paper to compare different representation schemes according to the induced similarity measure, based in the above theorems about completnees on data fields.

In the generation of the representation space, we may consider different factors like the complexity of the data fields, the extension and overlapping rate of the receptive fields, the nature and number of descriptors in each one of them, the resolution, etc. In this work we have considered the representational capabilities of systems of receptive fields found in early mammalian vision [5], under the assumption that the successive stages of processing remap the retinal representation space in a manner that make similar stimuli closer to each other, and dissimilar stimuli farther apart [6]. Concretely, we treat the visual pathway through three stages:

1. Pixel based receptive fields, simulating the representation at photoreceptor level.
2. Centre-surround circularly symmetric receptive fields, simulating the representation at lateral geniculate body level.
3. Receptive fields based on oriented difference of Gaussians or (DOG), simulating the representation at primary visual area level.

Dimensionality reduction must take place before classification is attempted. An intuitive description of the goal of the dimensionality reduction is the extraction of low-dimensional features from the original representation space. Our program reduces the dimensionality of the input images by converting them into vectors of RF responses. For the purpose of pattern classification it is important to concentrate on dimensionality reduction methods that allow discrimination between class, rather than faithful representation of the data.

Using for classification the vector of activities of the receptive fields, chosen according to certain criterium, which may be random or not, instead of the original image, effectively reduced the dimensionality of the input to a level which may be accepted provide a good performance.

Indeed, the vectors of RF activities proved to be adequate for representing images for recognition, although it would be impossible to recover from it the original structure of the image.

4 Computational Applications

As a data field we have consider as set of images composed by 256*256 images faces of four individuals (figure 2). The different images of each face varied in viewing position, illumination direction and facial expression.

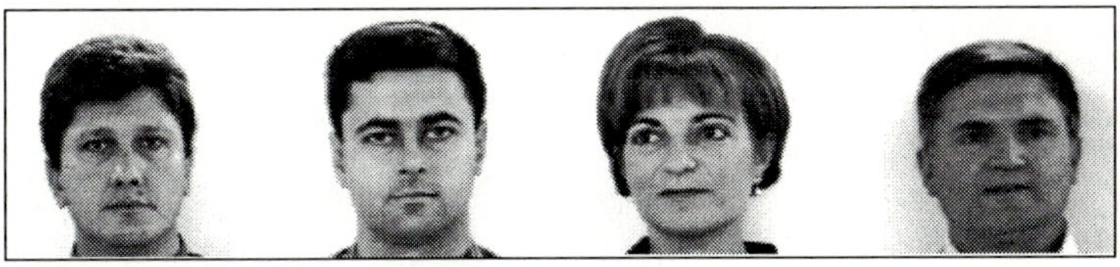

Fig. 2. Set of images

As we have indicated previously, we treat the visual pathway through three stages corresponding at the successive levels of processing in the visual pathway.

The simulation at a fotorreceptor level correspond to a resampling of the input image with different degrees of resolution, so we could applied any pure resolution transform. To simulate the second stage corresponding to the lateral geniculate body level, we consider centre-surround circularly symmetric receptive fields. In this case, there are enough neurophisiologycal evidence that the psychophysical channels correspond closely to the LOG operators, so we consider as a functional kernels

$$\nabla^2 G(x, y) = \frac{1}{2\pi\sigma^2}\left(2 - \left(\frac{x^2 + y^2}{\sigma^2}\right)\right)e^{-\frac{x^2 + y^2}{2\sigma^2}} \tag{3}$$

Receptive fields based on oriented difference of Gaussians or (DOG), simulating the representation at primary visual area level. In this situation we use functionals kernels like

$$M(x,y) = e^{-\frac{y^2}{2\sigma_v^2}} \left(\frac{x^2}{\sigma^2} - 1\right) e^{-\frac{x^2}{2\sigma^2}} \tag{4}$$

being the horizontal cross section of the directional operator a one-dimensional second derivative of a gaussian and its vertical cross a Gaussian.

In each case we applied an algebraic analytical transform to the set of proposed images, by selecting a set of receptive fields and functionals, according to the theorem proposed about complete representations.

As a final result we can observe in figure 3 the similarity measure induced by the each one of the above representations: The clustering in the retinal or pixel representation space is clearly inferior to the clustering in the LGN and VI. The performance of the LGN representation is nearly identical to that of the linear cortex model.

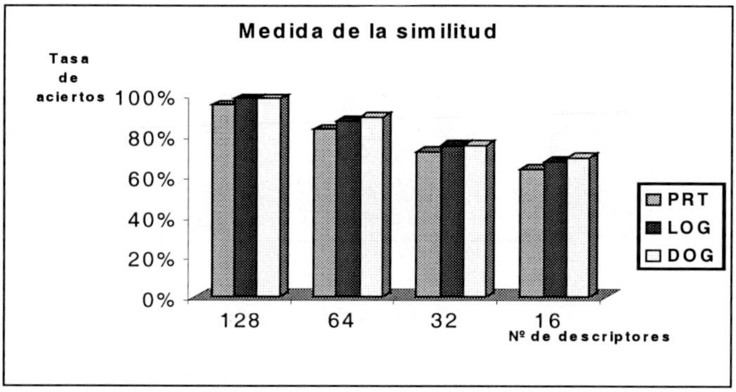

Fig. 3. Similarity measure

5 Conclusions

We present a theoretical analysis and computational experiments that compare the visual representations through the successive levels of the visual pathway, from the retina to the linear cortical units.

In spite of receptive fields compute local properties, these must be related in context to the whole image on the retina, in fact the visual cortex contains a topological projection of the retina. For example, in the visual cortex, locally computed variables such as edge orientation, spatial frequency and binocular disparity are embedded in a map of the retinal image transmitted via the lateral geniculate nucleus. Thus we have serial as well as parallel processing. This suggests a need for preserving information while it is passed from one level to another.

Additional motivation for RF-based representation is provided by recent work on the modeling of human performance in a variety of visuals tasks commonly referred to as *hiperacuity*. (In these tasks, the human visual system exhibits spatial resolution that is far below photoreceptor spacing, presumably due to the integration of information over extended regions of the retina)

References

1. Leibovic K.N.: Science of Vision. Springer Verlag, Berlin Heidelberg New York (1992).
2. Moreno Díaz, R. jr., Bolivar O.: Preservation of Information in Retinal Systems: Completness, Structure and Function. Cybernetics and Systems Research. (1994)World-Pub.730-736. Edited by R.Trappl. Viena.
3. Bolivar O., Candela,S., Moreno-Díaz, R: Complete Transforms and their Incidence in Artifitial Perception Systems Theory. Lecture Notes in Computer Science, vol 585. Springer Verlag, Berlin Heidelberg New York (1991) 514-525.
4. Muñoz, J.A., Bolivar,O., Candela,S., Moreno-Díaz, R.: Sobre la Generación de nuevos Espacios de Representación en Visión Artificial. Revista de la Academia Canaria de Ciencias. Vol 3. Nº 2 (1991) 105-118.
5. Edelman, S.: Representing Three-dimensional Objects by Sets of Activities of Receptive Fields. Biological Cybernetics, vol 70 (1993), 37-45.
6. Weiss Y., Edelman S.: Representation of Similarity as a Goal of Early Visual Processing. Computation in Neural Systems vol 6 (1995) 19-41.

McCulloch Program II in Artificial Systems and Lastres Theorem

E. Rovaris[1], F. Eugenio[1], and R. Moreno-Díaz[2]

[1]Escuela de Ingeniería de Telecomunicaci6n,Universidad de Las Palmas.
[2]Facultad de Informática. Universidad de Las Palmas (Gran Canaria).

Abstract. In this article an original formulation of McCulloch´s Artificial Program II is presented. A general methodology in order to build layered distributed granular computing machines is obtained. We apply this Program II to the classification stage of a pattern recognition system leading to Lastres Approach.

1 Introduction

Neural Network Theory formally begins with [1] McCulloch and Pitts´ works in 1943. When considering the interrelations between natural and artificial neural nets the subjects may be grouped into two clearly different frames, referred to as McCulloch Program-I and McCulloch Program-II. Program-I aim was to find out biological counterparts of logic machines. It is over in such a way that the level of logic machines ends in automata theory.

Program-II is much more concrete being a synthesis program which can be formulated for natural systems as: "Given a natural information processing system formed by sensors, effectors and decission making processes, the aim is to find out a neuron-like net, reliable and secure, meeting that function".

In this paper we reformulate Program-II for artificial systems and then we focus towards a procedure for the granular decomposition of a vision system and the classification stage. This constitutes a concretion of Program-II for recognition and classification artificial systems which led to Lastres Theorem.

2 Reformulating Program-II

Reformulation of Program-II can be done in the following terms:

a) Identification of a subsystem (perceptual, decission or action) choosing a descriptive level and defining its properties. For example, it could be a subsystem belonging to an artificial vision system implemented on a Von Neumann computer.
b) Synthesis of a parallel distributed neuron-like net, cooperative secure and reliable, duplicating the system with a layered structure of computational units.

F. Pichler, R. Moreno-Díaz, and P. Kopacek (Eds.): EUROCAST'99, LNCS 1798, pp. 514–520, 2000.

For a well defined problem (id est, a pattern recognition system), Program-II is applied to each of the functional stages, thus obtaining the neuron-like network. Computational units can be dessignated as artificial neurons.

A simplified version of the network structure corresponding to a generalized robotic situation is as follows [2,3]:·

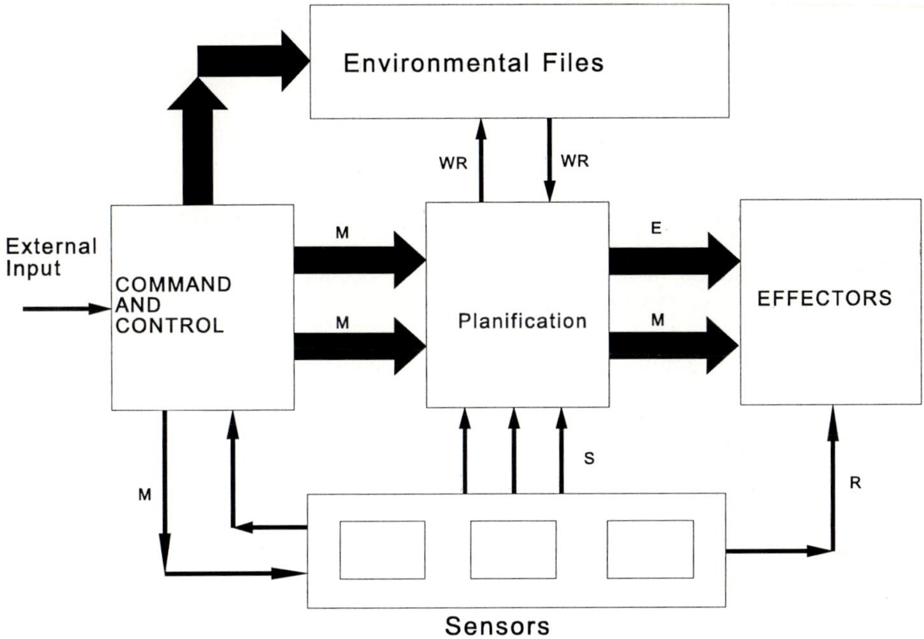

The whole system has a set of exclusive modes of *behaviour,* each mode determines the overall task being performed. Selection of a particular mode is carried out by means of Command and Control Subsystem based upon the sensorial information and the system state. This decision is modulated by the external input (in practice, the operator console).

Information concerning the selected mode M is sent to sensors which must be tuned in order to optimize data acquisition relative to action mode. That information is also sent to the subsystem named Planning in the Environmental Representation.

Finally, the selected mode run and control the process of objective setting, planification and execution considering sensorial data S processed at a high level. Updated environmental representations are sent back through WR lines when mode changes.

Lines between sensors and effectors correspond to reflex actions. Line E provides high level instructions to effectors according to action plan which must be decoded on specific effector actions. The fundamental function of the command and control system is the selection of the mode. All behavioural modes are mutually exclusive; the units also receive less processed information from all sensors, generating control signals and settings filters for all external inputs. According to McCulloch the command and control computer decides what must be sensed and then what is important. This command computer must have a neuron-like structure.

Each computing unit make a diagnosis of inputs and all of them must reach a consensus about the diagnostics which requires a large cooperation. The result of this intercommunication is the convergence into one single mode of behaviour, so that the system behaves as a whole.

This cooperative structure allows to:

a) Accelerate the decission process in order to select the behaviour mode.
b) Provides a high reliability, that is, to get a reliable neural net.

The Representation of the Environment requires a multisensorial mapping. There are two ways of structuring multisensorial information:

- Integrated representation, in low level acquisition and in high level sensorial processing.
- Step representation in which integration only occurs at high levels in symbolic structures.

These two ways admit different levels of representation from geometric to symbolic; at least for artificial systems. Planification takes place in a scenario of representation correponding to a selected mode of behaviour.

The generation of plans for solving problems involve artificial intelligence methods which implies two domains. The first domain is similar to a robot that moves avoiding obstacles, and is based on a geometric scenario. The second domain can contain symbolic scenarios.

3 Lastres Approach

Now the objective is to converge towards an approach for the granular decomposition of an artificial perception system or subsystem (id est, a vision system and specifically the classification stage) by means of neurons. This constitutes an illustration of McCulloch´s Artificial Program-II.

We begin with some considerations about McCulloch and Rosenblatt neurons. Let be a non feedback network of N computational units i and M external input lines j. Input j value at instant t is x_j (t). We make the following definitions:

Definition 1

An elemental neuron (EN) is a computational unit which stores a table of natural numbers α_{ij} with threshold ϑ_i computing:

- Input j to neuron i: $X_i(t)= \Sigma\, \alpha_{ij}\, x_j(t)$

- Single Output $Y_i(t+\Delta t) = \begin{cases} \text{McCulloch} = \begin{cases} 1 \text{ if } X_i(t) \geq \theta_i \\ 0 \text{ otherwise} \end{cases} \\ \text{Rosenblat} = \begin{cases} X_i(t) \text{ if } X_i(t) \geq \theta_i \\ 0 \text{ otherwise} \end{cases} \end{cases}$

Definition 2

An autoprogrammable elemental neuron (AEN) is a computational unit with two states, S_1 and S_2, controlled by an additional input line in such a way that:

- State S_1: Takes input data x_j during Δt generating α_{ij} table.
- State S_2: Computes the former EN equation.

The network is strictly programmable if, during S_1, neurons generate the weights α_{ij} and the function to perform.

Definition 3

An autoprogrammable neuron (AN) is a computing unit along with two states:

- State 1: Takes input data xj, perform parameter computing, generates the table and stores it.
- State 2: Computes a function F giving an output: $y_i(t+\Delta t) = F[\alpha_{ij}, x_j(t)]$

4 PEN Lastres Theorem

PEN are unfit for computing their own weights. Yet one or more PEN layer constitues an extraordinarily poor classifier unless specialized PEN layers are introduced. This situation is well illustrated by demonstrating the theorem as follows.

We introduce two types of specialized PEN:

- PEN1 During S_1, the neuron is desconnected from the net and in S_2 computes Σ δ_i p_i being δ_i descriptor i of presented unknown pattern. As can be seen is a neuron of weights p_i.

- PEN2 During S_1 takes α_{ij} as weight and during S_2 is desconnected from the net and computing:

$$\sum \alpha_{ij} \cdot \alpha_{ij}$$

Theorem

The square of the euclidean distance d_j can be computed by a PEN.

In effect:

$$d_j^2 = \sum_i (\delta_i - \alpha_{ij})^2 = \sum_i \delta_i^2 + \sum_i \alpha_{ij}^2 - 2\sum_i \alpha_{ij}\delta_i$$

The first term is realizable by means of a specialized PEN1 network; the second term is realizable by means of one layer specialized PEN2 network; third term is realizable by a one layer conventional PEN network. The sum is realizable by means of a PEN neuron with fixed weights +1, +1 and –2 demonstrating the theorem.

Let us consider the problem of pattern recognition for isolated shapes for which we adjunct the following figure:

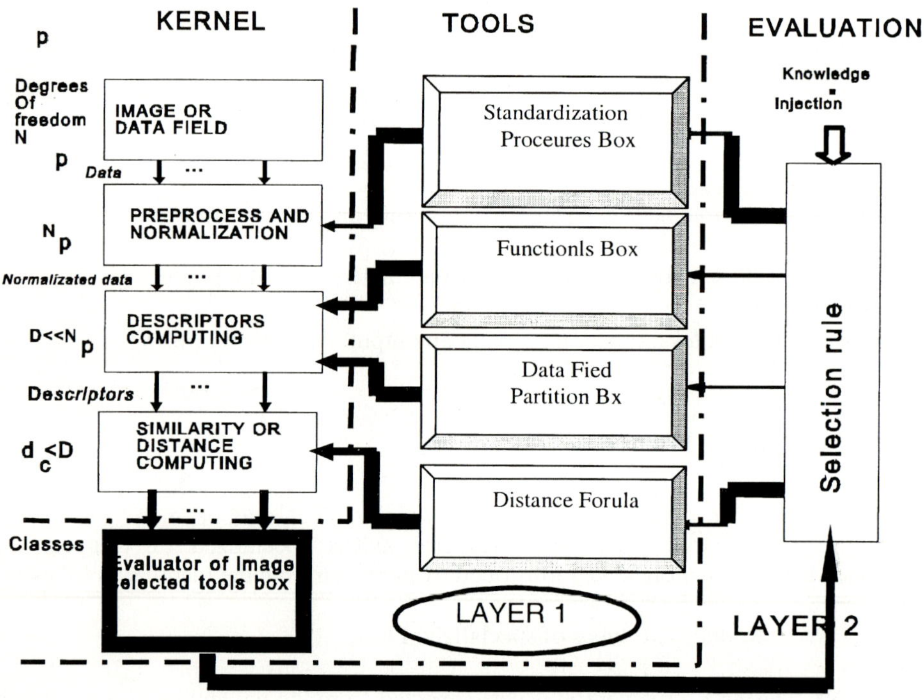

We are interested in the Lastres Approach in order to obtain an autoprogrammable neural net relative to the kernel of a pattern recognition system.

Lastres Conjecture

There is a reliable parallel distributed network comprising the kernel of a pattern recognition system.

Lastres Problem

Should it exist, which is a reliable parallel distributed net of autoprogrammable neurons solving Lastres Conjecture?

Lastres Approach

There exists an AN layer net solving Lastres Problem for a family of classifiers. The net can be build up through an AEN layer having certain distance formulations.

Focusing Lastres Approach, the central idea is that each autoprogrammable neuron (AN) computes a distance or a class similarity. For our case a C units NA network is built (one for each class), computing (during state S_1) a weighted center of gravity with each unit connected to all descriptor lines, coming from descriptor computing layer. Figure illustrates Lastres Procedure for a pattern recognition system:

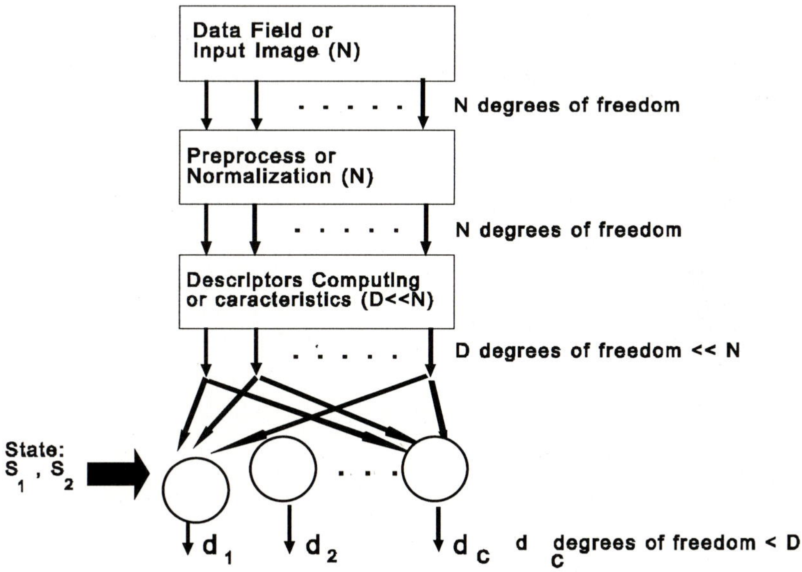

Class i sampling forms are presented for which S_1 is activated in order to compute α_{ij} ; this operation is repeated for each i. Then all autoprogrammable neurons AN go to state S_2 computing a fix distance function d_i (α_{ij}, D_j) previously selected.

Conclusions

We present an original formulation of McCulloch´s Artificial Program II obtaining a general methodology in order to build layer distributed granular computing machines being fast, secure and reliable.

In this context we have applied [4] Program-II to the classification stage of a pattern recognition system thus obtaining Lastres Approach identifying the Conjecture, the Problem and the Procedure and presenting the Euclidean Distance Compatibility Theorem for recognition by means of a Programmable Elemental Neuron Network. This constitutes the generalization of McCulloch Program-II for Artificial Perception Systems.

References

1. McCulloch, W., Pitts, W. : A Logical Calculus of the Ideas Immanent in Nervous Activity. Bulletin of Mathematical Biophysics, 5 (1943) 115-133.
2. Mira, J., Moreno-Díaz, R.: Un marco teórico para interpretar la función neuronal a altos niveles. Editorial Siglo XXI (1984) 149-171.
3. Mira, J.: System Behaviour and Computing Structure. Lecture Notes in Computer Sciences. Springer-Verlag (1989) 267-283.
4. Rovaris, E.:Images, Sampling and Neuron-like Distributed Computing: Towards an Unified Approach. Doctoral Thesis. University of Las Palmas (1984).

A Medical Ontology for Integrating Case-Based Reasoning, Rule-Based Reasoning, and Patient Databases*

M. Taboada[1], J. Des[2,3], M. Arguello[1], J. Mira[2], and D. Martínez[4]

[1] Dpto. de Electrónica e Computación. Universidad de Santiago de Compostela. 15706 Santiago de Compostela. Spain.
Tel.: 34-981-563100, Fax: 34-981-599412
chus@dec.usc.es

[2] Dpto. de Inteligencia Artificial. UNED. Madrid. Spain

[3] Servicio de Oftalmología. Hospital Comarcal de Monforte. Monforte, Spain

[4] Dpto. de Física Aplicada. Universidad de Santiago de Compostela. Spain

Abstract. Ontologies provide an explicit and shared specification of the domain knowledge in some field. With the objective of facilitating the integration of case-based reasoning, rule-based reasoning and patient databases, we propose a medical ontology, which is inherent to the ophthalmology domain, but it can be reused by another medical domains with the same representation requirements. In order to achieve this degree of reusing, the ontology was designed making a difference between core and peripheral concepts, domain-specific and method-specific concepts and task(method)-relevant and task(method)-specific concepts.

1 Introduction

Case-Based Reasoning (CBR) is an Artificial Intelligence approach based on the idea of recalling previous problems and reusing their solutions in order to evaluate and solve new cases [1,10,11]. In spite of the fact that CBR is becoming an important reasoning paradigm, there is a lack of methodologies for building and validating case-based systems [13].Currently, CBR systems are built in an ad-hoc manner, and the results obtained are not as good as could be expected. We propose to develop case-based systems by applying the current methodologies available for building Knowledge-Based Systems (KBS). These methodologies are based on the *knowledge-level hypothesis* introduced by Allen Newell in the 1980s [12]: the behaviour of an intelligent agent can be described at a level that is independent from its symbolic representation. Some of the most representative current approaches for knowledge-level modelling are Components of expertise [17], Generic Task [3], Task Structures [4], KADS [19], CommonKADS

* This work has been funded by the Comisión Interministerial para la Ciencia y la Tecnología (CICYT), through the research project TIC97-0604, and by the Secretaría Xeral de Investigación e Desenvolvemento da Xunta de Galicia, through the research project XUGA20601A98

F. Pichler, R. Moreno-Díaz, and P. Kopacek (Eds.): EUROCAST'99, LNCS 1798, pp. 521–527, 2000.

[16,15] and Protege-II [14]. All of these approaches promote the reuse of knowledge elements by providing, in many cases, libraries of knowledge types (mainly, problem-solving methods). Currently, there are many researchers working on the reuse of ontologies [5,7,18,8], as they provide an explicit and shared specification of the knowledge structure.

In this paper, we propose a medical ontology with the objective of obtaining an explicit specification of the expert knowledge. This specification facilitates the integration of case-based reasoning, rule-based reasoning and patient databases, by providing a support to validate and to maintain the system. In section 2, the building of our ontology for reusing in other clinical domains is commented on. In the following sections, the different parts of our ontology are described. Lastly, the discussion of this work is presented.

2 Building Ontologies for Reusing

Building, integration and validating ontologies is still an open field for researching [7,18]. The interaction problem [2] states that the nature of an ontology depends on the tasks and methods that use their contents. With the objective of making easy the building of our ontology in an independent manner of problem-solving methods, we have distinguished between:

- Core and peripheral ontology [18]: In the core part, the definitions are very general, whereas the peripheral part describes application specific concepts.
- Domain-specific and method-specific concepts [18]: These attributes are introduced with the objective of promoting the reuse of peripheral parts of ontologies. The domain-specificity attribute indicates to what domain a concept applies, and so, which sub-domains can reuse this concept.
- Relevant and specific concepts [8]: These attributes provide a way of reusing task or method specific concepts, and they are based on the following idea: 'a concept may be relevant for a task (or method) but not necessarily specific'.

During the building of our ontology, we have borne in mind that the knowledge was going to be used by a case-based system, but, in a near future, it could be reused by another knowledge-based systems. So, the minimal of our work should be to:

- keep apart the case representation and the expert knowledge model (in order to reuse this model), and
- represent some expert knowledge that facilitates the retrieval and matching phases during the case-based reasoning

In order to achieve these objectives, the ontological analysis was carried out in two stages. First, we acquired the knowledge independently of problem-solving method and we obtained a medical ontology with a structure very similar to a data model in a database. So, this ontology part was named Medical Data Ontology. Second, we extended Medical Data Ontology, taking into account the medical diagnosis task and the case-based method. This analysis gave rise to distinguish another three ontology parts: Medical Case Ontology, Medical Diagnosis Ontology and Medical Context Ontology. We made a difference among these four ontology parts in order to promote their reuse. Next sections deal these ontology parts in more detail.

3 Medical Data Ontology

This ontology part has been developed by reusing some theories in the core library described in [6], and later, by extending these theories with more specific concepts to our clinical domain. In this way, we have obtained a standard representation vocabulary that has been modelled using CML [15]. Fig. 1 shows a small part of the Medical Data Ontology. The higher levels in this hierarchical structure corresponds to the core ontological part (denoted in Fig. 1 by Medical Central Ontology, MCO), as it includes general categories of medical knowledge, such as 'Anamnesis', 'Findings', 'Clinical state abstractions', ...

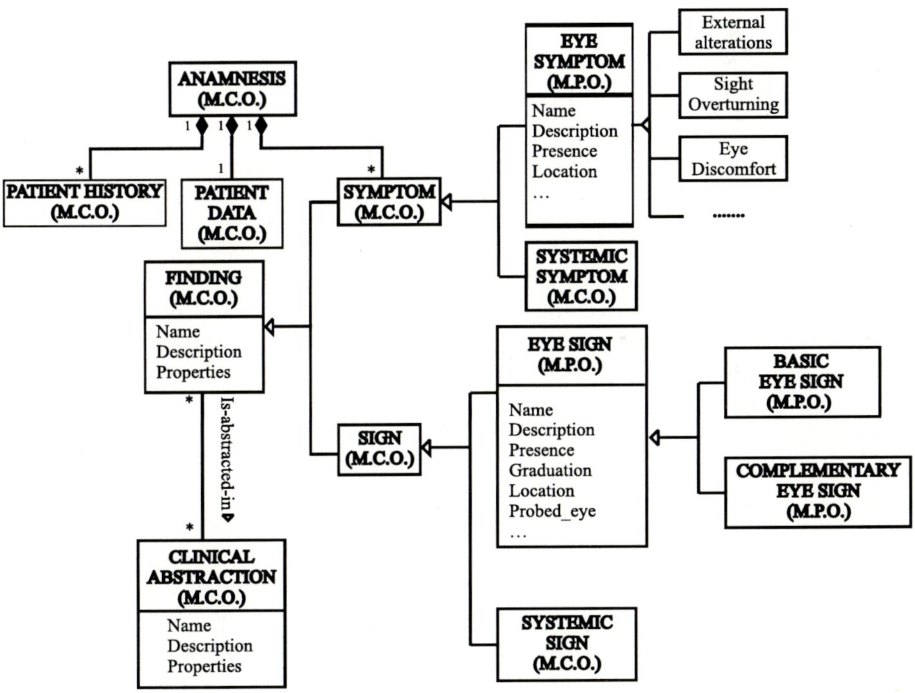

Fig. 1. A Small Part of Medical Data Ontology

As the hierarchical structure is traversed, the medical concepts are more specific to our clinical domain, giving rise to the peripheral ontological part, that contains concepts related to the ophthalmology domain (denoted by Medical Peripheral Ontology, MPO).

4 Medical Case Ontology

Medical Case Ontology has been developed with the objective of specifying the three components of a case:

- the description of the problem: includes the greater part of the concepts specified in Medical Data Ontology.
- the description of the solution to the problem: contains concepts about proposed medical diagnosis, applied treatments and some information about the reason for the choice of a treatment, taking into account the proposed diagnosis.
- the outcome as a result of applying the solution: incorporates information about the follow-up of a patient, that is, 1) expected results from the applied therapy to the patient, 2) obtained observations and 3) performed therapeutic actions as a response to the differences between the expected results and the obtained observations.

5 Medical Context Ontology

Mainly, there are two basic ways of structuring a case library [10]:

- as a flat memory, where each case is represented as a set of attribute/value pairs
- as a hierarchical memory, which clusters together similar cases, with the objective of making a more efficient retrieval phase. There are basically three approaches: shared feature networks, discrimination networks and redundant discrimination networks. All of these approaches requires complex procedures for the addition of new cases in the case library. Moreover, it is difficult to 1) keep the network optimal when new cases are added and 2) deal with missing information.

In our approach, the case library is structured in a flat memory and stored in database records. Each case consist of a simple feature/value list (the description is in Medical Case Ontology), but as there is a structured description of the domain knowledge (Medical Context and Diagnosis Ontology), we can view the representation of a case as structured. Then, the domain knowledge is organized in two ontology parts:

- Medical Context Ontology: contains knowledge used during case retrieval
- Medical Diagnosis Ontology: incorporates knowledge about case matching

Medical Context Ontology explicitly specifies a set of medical context, which are used for retrieving only cases highly relevant to the new case. A context can be defined as a set of attributes relevant for a given retrieval [9], that is, in our approach, a set of constraints on the patient clinical state. The advantages of defining and using clinical contexts can be summarize in:

- model the expert heuristic knowledge, and
- focus diagnostic task, reducing the set of relevant cases

An example of medical context in the ophthalmology domain is the named 'Ocular Pain', which can be described by the next constraints on the following symptoms:

$$'Ocular\ Pain' = \{< redness : yes, no >, < lessening\ of\ visual\ acuteness : yes, no >, < ocular\ pain : light, acute >, \quad < lacrimosity : moderate, no >, < secreations : yes, no >\}$$

In Medical Context Ontology, medical contexts are organized in a network similar to a redundant discrimination network, where

- each internal node (not leaf) is a question about a patient clinical state (that is, a question about a finding or a clinical abstraction),
- each successor node is a different answer to the question on the higher node
- each leaf contains a medical context, that is, a set of constraints on the set of findings and clinical abstractions of new case.

The advantages of our approach, that is, of using a flat memory combined with a hierarchy of medical contexts, are the following:

- retrieval phase is more efficient
- the addition of a new case is easy, just as a pure flat memory, and
- the choice of a context is intuitive for the expert, as consists of traversing a network

On the contrary, the disadvantages are the same of a hierarchical approach:

- it is necessary to maintain updated the network
- it is not confirmed that the best case will be retrieved
- it is necessary to define redundant networks in order to confirm the selection of a context

6 Medical Diagnosis Ontology

Medical Diagnosis Ontology has been developed with the objective of incorporating knowledge that facilitates the case matching phase during the case-based reasoning. Medical Diagnosis Ontology contains several types of causal relationships between 1) pathologies and findings, 2) patient history and pathologies, 3) pathologies and pathologies, and 4) findings and findings. This relationships includes attributes such as sensitivity, specificity or prognosis degrees. These relationships are used by a rule-based component in order to evaluate the amount of evidence of the new case with respect to each diagnostic hypothesis. Medical Diagnosis Ontology also contains a set of meta-rules that control the evaluation of evidence phase, providing an ophthalmology domain independent mechanism. The application of these meta-rules consists of filtering the set of possible diagnostic hypothesis before the application of case matching phase. Finally, the causal relationships in Medical Diagnosis Ontology are used during the case matching and ranking phases, as only the findings expected to each pathology are taking into account (the findings of the new case not expected are noted as 'unexplained').

7 Conclusions

In this work, we propose an explicit medical ontology for facilitating the integration of case-based reasoning, rule-based reasoning and patient databases. We started building our ontology by reusing some theories that define general categories of medical domain knowledge, such as described in [6], and later, by extending these theories with more specific concepts to our clinical domain. In this way, we have obtained the named Medical Data Ontology, which is inherent to the ophthalmology clinical domain, but independent of the medical diagnosis task and the case-based method. The specification of this ontology part has given rise to the design of patient database. In a second phase of ontological analysis, we extended our ontology taking into account the case-based medical diagnosis, and we built another three ontology parts: 1) Medical Case Ontology, which is specific to case-based method (and has given rise to the design of case library), 2) Medical Diagnosis Ontology, which is specific to medical diagnosis, but only relevant to case-based method (this ontology part defines the knowledge rules and meta-rules applied during rule reasoning), and 3) Medical Context Ontology, which is specific to medical diagnosis and can be only relevant to case-based method by a small adaptation of its structure (this ontology part describes the discrimination network applied during retrieval phase). So, we have obtained a medical ontology, which is inherent to the ophthalmology domain, but it can be reused by another medical domains with the same representation requirements. As some authors have emphasized [5], the development of explicit ontologies can be the basis for the generation of knowledge representation languages specific to a domain.

References

1. Aamodt, A., Plaza, E.: Case-Based Reasoning: fundamental issues, methodological variations, and system approaches. Artif. Intell. Commun. **7(1)** (1994) 39-59
2. Bylander, T., Chandrasekaran, B.: Generic tasks in knowledge-based reasoning: The right level of abstraction for knowledge acquisition. In: Gaines and Boose (eds.): Knowledge Acquisition for Knowledge Based Systems. Academic Press, London (1988) 65–77
3. Chandrasekaran, B.: Generic task in knowledge-based reasoning: High level building blocks for expert system design. IEEE Expert. **1(3)** (1986) 23-30

4. Chandrasekaran, B., Johnson, T.R.: Generic task and task structures: History, critique and new directions. In: David, Krivine and Simmons (eds.): Second Generation Expert Systems. Springer-Verlag, Berlin Heidelberg New York (1993) 232–272
5. Chandrasekaran, B., Josephson, J.R., Benjamins, R.: What are ontologies, and why do we need then? IEEE Intelligent Systems and their applications **14(1)** (1999) 20–26
6. Falasconi, S., Stefanelli, M.: A library of implemented ontologies. Proc. of the ECAI Workshop on Comparison of Implemented Ontologies, Amsterdam (1994) 81–91
7. Gómez-Pérez, A.: Knowledge sharing and reuse. In: Liebawitz, J. (eds.): The Handbook of Applied Expert Systems. CRC Press LCC, Boca Raton (1998) 10-1–10-35
8. Guarino, N.: Understanding, building and using ontologies. Int. J. Human-Computer Studies. **46** (1998) 1–24
9. Jurisica, I., Mylopoulos, J., Glasgow, J., Shapiro, H., Casper, R.: Case-based reasoning in IVF: prediction and knowledge mining. Artificial Intelligence in Medicine. **12** (1997) 293–310
10. Kolodner, J.: Case-based Reasoning. Morgan Kaufmann, San Mateo (1993)
11. Kolodner, J., Mark, W.: Guest Editors' Introduction. Case-Based Reasoning. IEEE Expert. **7(5)** (1992) 5-6
12. Newell, A.: The knowledge level. Artificial Intelligence. **18** (1982) 87-127
13. Ong, L., Narasimhalu, A.: Case-Based Reasoning. In: Liebawitz, J. (eds.): The Handbook of Applied Expert Systems. CRC Press LCC, Boca Raton (1998) 11-1–11-16
14. Puerta, A., Edgar, J., Tu, S.W., Musen, M.A.: A multiple-method knowledge-acquisition shell for the automatic generation of knowledge-acquisition tools. Proc. of the 6th Knowledge Acquisition for Knowledge-Based System Workshop, Banff (1991)
15. Schreiber, A., Akkermans, J.M., Anjewierden, A.A., de Hoog, A., Val de Velde, W., Wielinga, B.: Knowledge Engineering and Management. The CommonKADS Methodology. MIT Press.
16. Schreiber, A., Wielinga, B., Hoog, R., Akkermans, H., Val de Velde, W.: CommonKADS: A comprehensive Methodology for KBS development. IEEE Expert. **9(6)** (1994) 28-37
17. Steels, L.: Components of expertise. AI Magazine. **11(2)** (1990) 29-49
18. van Heijst, G., Schreiber, A., Wielinga, B.: Using explicit ontologies in KBS development. Int. J. Human-Computer Studies. **45** (1997) 183–292
19. Wielinga, B.J., Schreiber, A.T., Breuker, J.A.: KADS: A modelling approach to knowledge engineering. Knowledge Acquisition. **4(1)** (1992) 5–53

Uncertain Variables in the Computer Aided Analysis of Uncertain Systems[1]

Zdzislaw Bubnicki

Institute of Control and Systems Engineering
Wroclaw University of Technology
Wyspianskiego 27, 50-370 Wroclaw, POLAND
phone: +48 71 320 33 28 , +48 71 21 62 26 ; fax: +48 71 320 38 84
bubnicki@ists.pwr.wroc.pl

Abstract. The paper is concerned with static uncertain systems described by a function or by a relation. Unknown parameters in the mathematical models are considered as so called uncertain variables described by certainty distributions given by an expert. Two versions of the uncertain variables based on two versions of uncertain logics are defined. In the second part of the paper the formulations of the analysis and decision making problems adequate to the description of uncertainty are presented and the general procedures of the problem solving are described. Simple examples and an algorithm of the decision problem solving for a discrete case illustrate the computational aspects of the approach based on the uncertain variables and the possibility of the application to computer-aided analysis and decision making.

1 Introduction

There exist a great variety of definitions and formal models of uncertainties and uncertain systems (e.g. [8, 9, 10]). The purpose of this paper is to present definitions and basic properties of so called uncertain variables (introduced in a brief form in [2, 4, 5]) and to show how they may be applied to the analysis and decision making in a class of systems with unknown parameters in their mathematical descriptions. The unknown parameters will be assumed to be uncertain variables and the systems with uncertain parameters will be called uncertain systems.

The uncertain variables, related to random variables and fuzzy numbers, are described by their certainty distributions given by an expert and evaluating his opinion on approximate values of the uncertain variable. Two versions of the uncertain variables are introduced in Section 3, based on two versions of uncertain logics defined in Section 2. The definitions contain not only the formal description but also their interpretation, which is of much importance. The uncertain variable in the first version may be formally considered as a very special case of the fuzzy number (exactly speaking – the possibilistic number) with a specific interpretation of the

1 This work was supported by the State Committee for Scientific Research under grant no. 8 T11C 012 16.

F. Pichler, R. Moreno-Díaz, and P. Kopacek (Eds.): EUROCAST'99, LNCS 1798, pp. 528–542, 2000.

membership function. Nevertheless for the sake of simplicity and unification it is better to introduce it independently and not as a special case of much more complicated formalism with different semantics. In Section 4 the applications of the uncertain variables to analysis and decision making problems are presented for the system described by a function (functional system) and described by a relation (relational system). In the second case the static system is described by a relation between input and output vectors and the analysis consists in finding the output property (i.e. the property concerning the output vector or the set to which the output vector belongs) for the given input property. The decision making is an inverse problem (see [3]). For the system with uncertain parameters the modified versions of these problems adequate to the description of uncertainty are presented. Simple examples and an algorithm given in Section 5 show how the methods and procedures described in the paper may be applied to computer-aided analysis and decision making for the uncertain systems under consideration.

2 Uncertain Logics

Our considerations are based on multi-valued logic. To introduce terminology and notation employed in our presentation of uncertain logic and uncertain variables, let us remind that multi-valued (exactly speaking – infinite-valued) propositional logic deals with propositions ($\alpha_1, \alpha_2, \ldots$) whose logic values $w(\alpha) \in [0,1]$ and

$$\left.\begin{aligned} w(\neg\alpha) = 1-w(\alpha)\,, \quad w(\alpha_1 \vee \alpha_2) &= \max\{w(\alpha_1), w(\alpha_2)\}\,, \\ w(\alpha_1 \wedge \alpha_2) &= \min\{w(\alpha_1), w(\alpha_2)\}\,. \end{aligned}\right\} \tag{1}$$

Multi-valued predicate logic deals with predicates $P(x)$ defined on a set X, i.e. properties concerning x, which for the fixed value of x form propositions in multi-valued propositional logic, i.e.

$$w[P(x)] \stackrel{\Delta}{=} \mu_p(x) \in [0,1] \quad \text{for each} \quad x \in X\,. \tag{2}$$

For the fixed x, $\mu_p(x)$ denotes *degree of truth*, i.e. the value $\mu_p(x)$ shows to what degree P is satisfied. If for each $x \in X$ $\mu_p(x) \in \{0,1\}$ then $P(x)$ will be called here a *crisp* or a *well-defined property*, and $P(x)$ which is not well-defined will be called a *soft property*. The crisp property defines a set

$$D_x = \{x \in X\colon w[P(x)] = 1\} \stackrel{\Delta}{=} \{x \in X\colon P(x)\}\,. \tag{3}$$

Consider now a universal set Ω, $\omega \in \Omega$, a set X which is assumed to be a metric space, a function $g\colon \Omega \to X$, and a crisp property $P(x)$ in the set X. The property P and the function g generate the crisp property $\Psi(\omega, P)$ in Ω: "For the value $\bar{x} = g(\omega) \stackrel{\Delta}{=} \bar{x}(\omega)$ assigned to ω the property P is satisfied", i.e.

$$\Psi(\omega, P) = P[\bar{x}(\omega)]\,.$$

Let us introduce now the property $G(\bar{x}, x) =$ "$\bar{x} \cong x$" for $\bar{x}, x \in X$, which means: "$\bar{x}$ is approximately equal to x". The equivalent formulations are: "x is the approximate value of $\bar{x}$" or "x belongs to a small neighbourhood of $\bar{x}$" or "the value of the metric $d(x, \bar{x})$ is small". Note that $G(\bar{x}, x)$ is a reflexive, symmetric and transitive relation in $X \times X$. For the fixed ω, $G[\bar{x}(\omega), x] \triangleq G_\omega(x)$ is a soft property in X. The properties $P(x)$ and $G_\omega(x)$ generate the soft property $\overline{\Psi}(\omega, P)$ in Ω: "the approximate value of $\bar{x}(\omega)$ satisfies P" or "$\bar{x}(\omega)$ approximately satisfies P", i.e.

$$\overline{\Psi}(\omega, P) = G_\omega(x) \wedge P(x) = [\bar{x}(\omega) \cong x] \wedge P(x) \tag{4}$$

where x is a free variable. The property $\overline{\Psi}$ may be denoted by

$$\overline{\Psi}(\omega, P) = \text{"}\bar{x}(\omega) \mathrel{\tilde{\in}} D_x\text{"} \tag{5}$$

where D_x is defined by (3) and "$\bar{x} \mathrel{\tilde{\in}} D_x$" means: "the approximate value of $\bar{x}$ belongs to D_x" or "$\bar{x}$ approximately belongs to D_x". Denote by $h_\omega(x)$ the logic value of $G_\omega(x)$:

$$w[G_\omega(x)] \triangleq h_\omega(x), \qquad \bigwedge_{x \in X} (h_\omega(x) \geq 0), \tag{6}$$

$$\max_X h_\omega(x) = 1. \tag{7}$$

Definition 1 (uncertain logic): The uncertain logic is defined by a universal set Ω, a metric space X, crisp properties (predicates) $P(x)$, the properties $G_\omega(x)$ and the corresponding functions (6) for $\omega \in \Omega$. In this logic we consider soft properties (4) generated by P and G_ω. The logic value of $\overline{\Psi}$ is defined in the following way

$$w[\overline{\Psi}(\omega, P)] \triangleq v[\overline{\Psi}(\omega, P)] = \begin{cases} \max\limits_{x \in D_x} h_\omega(x) & \text{for } D_x \neq \varnothing \\ 0 & \text{for } D_x = \varnothing \end{cases} \tag{8}$$

and is called a degree of certainty or *certainty index*. The operations for the certainty indexes are defined as follows:

$$v[\neg\overline{\Psi}(\omega, P)] = 1 - v[\overline{\Psi}(\omega, P)], \tag{9}$$

$$v[\Psi_1(\omega, P_1) \vee \Psi_2(\omega, P_2)] = \max\{v[\Psi_1(\omega, P_1)], v[\Psi_2(\omega, P_2)]\}, \tag{10}$$

$$v[\Psi_1(\omega, P_1) \wedge \Psi_2(\omega, P_2)] = \begin{cases} 0 & \text{if for each } x \; w(P_1 \wedge P_2) = 0 \\ \min\{v[\Psi_1(\omega, P_1)], v[\Psi_2(\omega, P_2)]\} & \text{otherwise} \end{cases} \tag{11}$$

where Ψ_1 is $\overline{\Psi}$ or $\neg\overline{\Psi}$, and Ψ_2 is $\overline{\Psi}$ or $\neg\overline{\Psi}$. □

Using the notation (5) we have

$$v\,[\,\overline{x}(\omega)\,\tilde{\notin}\,D_x\,] = 1 - v[\,\overline{x}(\omega) \in D_x\,]\ , \tag{12}$$

$$v[\,\overline{x}(\omega)\,\tilde{\in}\,D_1 \vee \overline{x}(\omega)\,\tilde{\in}\,D_2\,] = \max\{\,v[\,\overline{x}(\omega)\tilde{\in}\,D_1],\,v[\,\overline{x}(\omega)\tilde{\in}\,D_2\,]\}\ , \tag{13}$$

$$v[\,\overline{x}(\omega)\,\tilde{\in}\,D_1 \wedge \overline{x}(\omega)\,\tilde{\in}\,D_2\,] = \min\{\,v[\,\overline{x}(\omega)\,\tilde{\in}\,D_1\,],\,v[\,\overline{x}(\omega)\,\tilde{\in}\,D_2\,]\} \tag{14}$$

for $D_1 \cap D_2 \neq \varnothing$ and 0 for $D_1 \cap D_2 = \varnothing$ – where $\tilde{\in}\,D_1$ and $\tilde{\in}\,D_2$ may be replaced by $\tilde{\notin}\,D_1$ and $\tilde{\notin}\,D_2$, respectively. From (7) and (8) $v[\,\overline{x}\,\tilde{\in}\,X\,] = 1$. One can note that $G_\omega(x) =$ "$\overline{x}(\omega)\cong x$" is a special case of $\overline{\Psi}$ for $D_x = \{\,x\,\}$ (a singleton) and

$$v[\,\overline{x}(\omega) \cong x\,] = h_\omega(x)\ , \qquad v[\,\overline{x}(\omega) \not\cong x\,] = 1 - h_\omega(x)\ . \tag{15}$$

From (8) one can immediately deliver the following property: if $P_1 \rightarrow P_2$ for each x (i.e. $D_1 \subseteq D_2$) then

$$v[\overline{\Psi}(\omega,P_1)] \le v[\overline{\Psi}(\omega,P_2)] \ \text{ or } \ v[\,\overline{x}(\omega)\,\tilde{\in}\,D_1\,] \le v[\,\overline{x}(\omega)\,\tilde{\in}\,D_2\,]\ . \tag{16}$$

Theorem 1:

$$v[\,\overline{\Psi}(\omega,\,P_1 \vee P_2)\,] = v[\,\overline{\Psi}(\omega,\,P_1) \vee \overline{\Psi}(\omega,P_2)\,]\ , \tag{17}$$

$$v[\,\overline{\Psi}(\omega,P_1 \wedge P_2)\,] \le \min\{\,v[\overline{\Psi}(\omega,P_1)],\,v[\overline{\Psi}(\omega,P_2)\,]\}\ . \tag{18}$$

$$v[\,\overline{\Psi}(\omega,\neg P)\,] \ge v[\neg\overline{\Psi}(\omega,P)]\ . \tag{19}$$

Proof: From (8) and (10)

$$v[\overline{\Psi}(\omega,P_1) \vee \overline{\Psi}(\omega,P_2)] = \max\{\max_{x\in D_1} h_\omega(x),\, \max_{x\in D_2} h_\omega(x)\} =$$

$$= \max_{x\in D_1\cup D_2} h_\omega(x) = v[\overline{\Psi}(\omega,P_1 \vee P_2)]\ .$$

Inequality (18) follows immediately from $D_1 \cap D_2 \subseteq D_1$, $D_1 \cap D_2 \subseteq D_2$ and (16). Let $P_1 = P$ and $P_2 = \neg P$ in (17). Since $w(P \vee \neg P) = 1$ for each x ($D_x = X$ in this case),

$$1 = v[\overline{\Psi}(\omega,P) \vee \overline{\Psi}(\omega,\neg P)] = \max\{v[\overline{\Psi}(\omega,P)],\ v[\overline{\Psi}(\omega,\neg P)]\}$$

and

$$v[\overline{\Psi}(\omega,\neg P)] \ge 1 - v[\overline{\Psi}(\omega,P)] = v[\neg\overline{\Psi}(\omega,P)]\ . \qquad \square$$

Inequality (19) may be written in the form

$$v[\bar{x}(\omega) \,\tilde{\in}\, \overline{D}_x] \geq v[\bar{x}(\omega) \,\tilde{\notin}\, D_x] = 1 - v[\bar{x}(\omega) \,\tilde{\in}\, D_x] \tag{20}$$

where $\overline{D}_x = X - D_x$.

As was said in Section 1, the definition of uncertain logic should contain two parts: a mathematical model (which is described above) and its interpretation (semantics). The semantics is here the following: the uncertain logic operates with crisp predicates $P[\bar{x}(\omega)]$, but for the given ω it is not possible to state whether $P(\bar{x})$ is true or false because the function $\bar{x} = g(\omega)$ and consequently the value $\bar{x}$ corresponding to ω is unknown. The exact information, i.e. the knowledge of g is replaced by $h_\omega(x)$ which for the given ω characterizes the different possible approximate values of $\bar{x}(\omega)$. If we use the terms: knowledge, information, data etc., it is necessary to determine the subject (who knows ?, who gives the information ?). In our considerations this subject is called *an expert*. So the expert does not know exactly the value $\bar{x}(\omega)$, but "looking at" ω he obtains some information concerning $\bar{x}$, which he does not express in an explicit form but uses it to formulate $h_\omega(x)$. Hence, the expert is the source of $h_\omega(x)$ which for particular x evaluates his opinion that $\bar{x} \cong x$. That is why $h_\omega(x)$ and consequently $v[\overline{\Psi}(\omega, P)]$ are called degrees of certainty. E.g. Ω is a set of persons, $\bar{x}(\omega)$ denotes the age of ω and the expert looking at the person ω gives the function $h_\omega(x)$ whose value for the particular x is his degree of certainty that the age of this person is approximately equal to x. The predicates $\overline{\Psi}(\omega, P)$ are soft because of the uncertainty of the expert.

Definition 2 (C-uncertain logic): The first part is the same as in Definition 1. The certainty index of $\overline{\Psi}$ and the operations for the certainty indexes are defined as follows:

$$v_c[\overline{\Psi}(\omega,P)] = \frac{v_p[\overline{\Psi}(\omega,P)] + v_n[\overline{\Psi}(\omega,P)]}{2} = \frac{1}{2}[\max_{x \in D_x} h_\omega(x) + 1 - \max_{x \in \overline{D}_x} h_\omega(x)], \tag{21}$$

$$v_c[\neg \overline{\Psi}(\omega,P)] = v_c[\overline{\Psi}(\omega, \neg P)], \tag{22}$$

$$v_c[\overline{\Psi}(\omega,P_1) \vee \overline{\Psi}(\omega,P_2)] = v_c[\overline{\Psi}(\omega,P_1 \vee P_2)], \tag{23}$$

$$v_c[\overline{\Psi}(\omega,P_1) \wedge \overline{\Psi}(\omega,P_2)] = v_c[\overline{\Psi}(\omega,P_1 \wedge P_2)]. \tag{24}$$

□

The operations may be rewritten in the following form

$$\bar{x} \,\tilde{\notin}\, D_x = \bar{x} \,\tilde{\in}\, \overline{D}_x \ , \tag{25}$$

$$v_c[\,\bar{x}(\omega) \,\tilde{\in}\, D_1 \vee \bar{x}(\omega) \,\tilde{\in}\, D_2] = v_c[\,\bar{x}(\omega) \,\tilde{\in}\, D_1 \cup D_2] \ , \tag{26}$$

$$v_c[\,\bar{x}(\omega) \,\tilde{\in}\, D_1 \wedge \bar{x}(\omega) \,\tilde{\in}\, D_2] = v_c[\,\bar{x}(\omega) \,\tilde{\in}\, D_1 \cap D_2\,] \ . \tag{27}$$

From (8) and (21)

$$v_c[\,\bar{x}(\omega) \,\tilde{\in}\, X\,] = 1, \qquad v_c[\,\bar{x}(\omega) \,\tilde{\in}\, \varnothing\,] = 0 \, . \tag{28}$$

Using (21), we obtain the following property: If for each x $P_1 \rightarrow P_2$ (i.e. $D_1 \subseteq D_2$) then

$$v_c[\overline{\Psi}(\omega, P_1)] \le v_c[\overline{\Psi}(\omega, P_2)] \quad \text{or} \quad v_c[\,\bar{x}(\omega) \,\tilde{\in}\, D_1\,] \le v_c[\,\bar{x}(\omega) \,\tilde{\in}\, D_2\,] \ . \tag{29}$$

Theorem 2:

$$v_c[\overline{\Psi}(\omega, P_1 \vee P_2)] \ge \max\{v_c[\overline{\Psi}(\omega, P_1)],\ v_c[\overline{\Psi}(\omega, P_2)]\} \ , \tag{30}$$

$$v_c[\overline{\Psi}(\omega, P_1 \wedge P_2)] \le \min\{v_c[\overline{\Psi}(\omega, P_1)],\ v_c[\overline{\Psi}(\omega, P_2)]\} \ , \tag{31}$$

$$v_c[\neg\overline{\Psi}(\omega, P)] = 1 - v_c[\overline{\Psi}(\omega, P)] \ . \tag{32}$$

The proof is analogous to that of Theorem 1.

Till now it has been assumed that $\bar{x}(\omega), x \in X$. The considerations can be extended for the case $\bar{x}(\omega) \in \overline{X}$ and $x \in X \subset \overline{X}$. It means that the set of approximate values X evaluated by an expert may be a subset of the set of the possible values $\bar{x}(\omega)$. In a typical case $X = \{x_1, x_2, \ldots, x_m\}$ (a finite set), $x_i \in \overline{X}$ for $i \in \overline{1,m}$.

3 Uncertain Variables

The variable $\bar{x}$ for a fixed ω will be called an uncertain variable. Two versions of uncertain variables will be defined. The precise definition will contain: $h(x)$ given by an expert, the definition of the certainty index $w(\,\bar{x} \,\tilde{\in}\, D_x\,)$ and the definitions of $w(\,\bar{x} \,\tilde{\notin}\, D_x\,)$, $w(\,\bar{x} \,\tilde{\in}\, D_1 \vee \bar{x} \,\tilde{\in}\, D_2)$, $w(\,\bar{x} \,\tilde{\in}\, D_1 \wedge \bar{x} \,\tilde{\in}\, D_2)$.

Definition 3 (uncertain variable): The uncertain variable $\bar{x}$ is defined by the set of values X, the function $h(x) = v(\,\bar{x} \cong x\,)$ (i.e. the certainty index that $\bar{x} \cong x$, given by an expert) and the following definitions:

$$v(\bar{x} \tilde{\in} D_x) = \begin{cases} \max\limits_{x \in D_x} h(x) & \text{for } D_x \neq \varnothing \\ 0 & \text{for } D_x = \varnothing , \end{cases} \tag{33}$$

$$v(\bar{x} \tilde{\notin} D_x) = 1 - v(\bar{x} \tilde{\in} D_x) , \tag{34}$$

$$v(\bar{x} \tilde{\in} D_1 \vee \bar{x} \tilde{\in} D_2) = \max\{ v(\bar{x} \tilde{\in} D_1), v(\bar{x} \tilde{\in} D_2)\} , \tag{35}$$

$$v(\bar{x} \tilde{\in} D_1 \wedge \bar{x} \tilde{\in} D_2) = \begin{cases} \min\{ v(\bar{x} \tilde{\in} D_1), v(\bar{x} \tilde{\in} D_2)\} & \text{for } D_1 \cap D_2 \neq \varnothing \\ 0 & \text{for } D_1 \cap D_2 = \varnothing . \end{cases} \tag{36}$$

The function $h(x)$ will be called a *certainty distribution*. □

The definition of the uncertain variable is based on the uncertain logic. Then the properties (15), (16), (17), (18), (20) are satisfied. The properties (17) and (18) may be presented in the following form

$$v(\bar{x} \tilde{\in} D_1 \cup D_2) = \max\{ v(\bar{x} \tilde{\in} D_1), v(\bar{x} \tilde{\in} D_2)\} , \tag{37}$$

$$v(\bar{x} \tilde{\in} D_1 \cap D_2) \leq \min\{ v(\bar{x} \tilde{\in} D_1), v(\bar{x} \tilde{\in} D_2)\} . \tag{38}$$

Definition 4 (*C*-uncertain variable): *C*-uncertain variable $\bar{x}$ is defined by the set of values *X*, the function $h(x) = v(\bar{x} \cong x)$ given by an expert, and the following definitions:

$$v_c(\bar{x} \tilde{\in} D_x) = \frac{1}{2}[\max_{x \in D_x} h(x) + 1 - \max_{x \in \overline{D}_x} h(x)] , \tag{39}$$

$$v_c(\bar{x} \tilde{\notin} D_x) = 1 - v_c(\bar{x} \tilde{\in} D_x) , \tag{40}$$

$$v_c(\bar{x} \tilde{\in} D_1 \vee \bar{x} \tilde{\in} D_2) = v_c(\bar{x} \tilde{\in} D_1 \cup D_2) , \tag{41}$$

$$v_c(\bar{x} \tilde{\in} D_1 \wedge \bar{x} \tilde{\in} D_2) = v_c(\bar{x} \tilde{\in} D_1 \cap D_2) . \tag{42}$$

□

The definition of *C*-uncertain variable is based on *C*-uncertain logic. Then the properties (28), (47) are satisfied. According to (22) and (32)

$$v_c(\bar{x} \tilde{\notin} D_x) = v_c(\bar{x} \tilde{\in} \overline{D}_x) . \tag{43}$$

Inequalities (30) and (31) may be presented in the following form

$$v_c(\bar{x} \tilde{\in} D_1 \cup D_2) \geq \max\{v_c(\bar{x} \tilde{\in} D_1), v_c(\bar{x} \tilde{\in} D_2)\}, \tag{44}$$

$$v_c(\bar{x} \tilde{\in} D_1 \cap D_2) \leq \min\{v_c(\bar{x} \tilde{\in} D_1), v_c(\bar{x} \tilde{\in} D_2)\}. \tag{45}$$

The function $v_c(\bar{x} \cong x) \stackrel{\Delta}{=} h_c(x)$ expressed by (29) may be called a *C-certainty distribution*. The formula (39) may be presented in the following way

$$v_c(\bar{x} \tilde{\in} D_x) = \begin{cases} \frac{1}{2} \max\limits_{x \in D_x} h(x) = \frac{1}{2} v(\bar{x} \tilde{\in} D_x) & \text{if } \max\limits_{x \in \overline{D}_x} h(x) = 1 \\ 1 - \frac{1}{2} \max\limits_{x \in \overline{D}_x} h(x) = v(\bar{x} \tilde{\in} D_x) - \frac{1}{2} v(\bar{x} \tilde{\in} \overline{D}_x) & \text{otherwise}. \end{cases} \tag{46}$$

In particular, for $D_x = \{x\}$

$$h_c(x) = \begin{cases} \frac{1}{2} h(x) & \text{if } \max\limits_{\breve{x} \in X - \{x\}} h(\breve{x}) = 1 \\ 1 - \frac{1}{2} \max\limits_{\breve{x} \in X - \{x\}} h(\breve{x}) & \text{otherwise}. \end{cases} \tag{47}$$

For the further considerations we assume $X \subseteq R^k$ (k-dimensional real number vector space) and we shall consider two cases: the discrete case with $X = \{x_1, x_2, \ldots, x_m\}$ and the continuous case in which $h(x)$ is a continuous function. It is easy to see that in the continuous case $h_c(x) = \frac{1}{2} h(x)$.

Note that the certainty distribution $h(x)$ is given by an expert and C-certainty distribution may be determined according to (47). The C-certainty distribution does not determine the certainty index $v_c(\bar{x} \tilde{\in} D_x)$. To determine v_c, it is necessary to know $h(x)$ and to use (46). The formula (46) shows the relation between the certainty indexes v and v_c: if $D_x \neq X$ and $D_x \neq \emptyset$ then $v_c < v$. In comparison with uncertain variable, C-uncertain variable has two advantages: In the definition of $v_c(\bar{x} \tilde{\in} D_x)$ the values of $h(x)$ for $\overline{D}_x$ are also taken into account and the logic operations (negation, disjunction and conjunction) correspond to the operations in the family of subsets D_x (complement, union and intersection). On the other hand, the certainty indexes for disjunction and conjunction are not determined by the certainty indexes $v_c(\bar{x} \tilde{\in} D_1)$, $v_c(\bar{x} \tilde{\in} D_2)$, i.e. they cannot be reduced to operations in the set of certainty indexes $v_c(\bar{x} \tilde{\in} D_x)$. We can define a *mean value* $M(\bar{x})$ in the similar way as for the random variable. In the discrete case

$$M_x = \sum_{i=1}^{m} x_i \bar{h}(x_i), \qquad \bar{h}(x_i) = \frac{h(x_i)}{\sum\limits_{j=1}^{m} h(x_j)}. \tag{48}$$

For C-uncertain variable M_{xc} is defined in the same way, with h_c in the place of h. In the continuous case $h_c(x) = \frac{1}{2}h(x)$, then $\overline{h}_c(x) = \overline{h}(x)$ and $M_c = M$. In the discrete case $M_{xc} \approx M_x$, if the dispersion in the certainty distribution is great.

To compare uncertain variables with fuzzy numbers, let us remind three basic definitions of the fuzzy number in a wide sense of the word, i.e. the definitions of the fuzzy set based on the number set $X = R^1$:

a. The fuzzy number $\hat{x}(d)$ for the given fixed value $d \in X$ is defined by X and the membership function $\mu(x, d)$ which may be considered as a logic value (*degree of truth*) of the soft property "if $\hat{x} = x$ then $\hat{x} \cong d$ ".

b. The linguistic fuzzy variable $\hat{x}$ is defined by X and a set of membership functions $\mu_i(x)$ corresponding to different descriptions of the size of $\hat{x}$ (small, medium, large, etc.). E.g. $\mu_1(x)$ is a logic value of the soft property "if $\hat{x} = x$ then $\hat{x}$ is small".

c. The fuzzy number $\hat{x}(\omega)$ (where $\omega \in \Omega$ was introduced at the beginning of this section) is defined by X and the membership function $\mu_\omega(x)$ which is a logic value (*degree of possibility*) of the soft property "it is possible that the value x is assigned to ω ".

In the first two definitions the membership function does not depend on ω, in the third case there is a family of membership functions (a family of fuzzy sets) for $\omega \in \Omega$. The difference between $\hat{x}(d)$ or the linguistic fuzzy variable $\hat{x}$ and the uncertain variable $\overline{x}(\omega)$ is quite evident. The variables $\hat{x}(\omega)$ and $\overline{x}(\omega)$ are formally defined in the same way by the fuzzy sets $\langle X, \mu_\omega(x) \rangle$ and $\langle X, h_\omega(x) \rangle$, respectively, but the interpretations of $\mu_\omega(x)$ and $h_\omega(x)$ are different. In the case of the uncertain variable there exists a function $\overline{x} = g(\omega)$, the value $\overline{x}$ is determined for the fixed ω but is unknown to an expert who formulates the degree of certainty that $\overline{x}(\omega) \cong x$ for the different values $x \in X$. In the case of $\hat{x}(\omega)$ the function g may not exist. Instead we have a property of the type "it is possible that $P(\omega, x)$ " (or shortly speaking "it is possible that the value x is assigned to ω ") where $P(\omega, x)$ is such a property concerning ω and x for which it makes sense to use the words “it is possible”. Then $\mu_\omega(x)$ for the fixed ω means the degree of possibility for the different values $x \in X$, given by an expert. From the point of view presented above $\overline{x}(\omega)$ may be considered as a special case of $\hat{x}(\omega)$ (when the relation $P(\omega, x)$ is reduced to the function g), with a specific interpretation of $\mu_\omega(x) = h_\omega(x)$. The further difference is connected with the definitions of $w(\overline{x} \,\tilde{\in}\, D_x)$, $w(\overline{x} \,\tilde{\notin}\, D_x)$, $w(\overline{x} \,\tilde{\in}\, D_1 \vee \overline{x} \,\tilde{\in}\, D_2)$ and $w(\overline{x} \,\tilde{\in}\, D_1 \wedge \overline{x} \,\tilde{\in}\, D_2)$. The function $w(\overline{x} \,\tilde{\in}\, D_x) \triangleq m(D_x)$ may be considered as a *measure* defined for the family of sets $D_x \subseteq X$. Two measures have been defined in the definitions of the uncertain variables: $v(\overline{x} \,\tilde{\in}\, D_x) \triangleq \overline{m}(D_x)$ and $v_c(\overline{x} \,\tilde{\in}\, D_x) \triangleq m_c(D_x)$. Taking into account the properties of $\overline{m}$ and m_c, and comparing them with known cases of fuzzy measure (belief measure, plausibility measure and possibility measure which is a special case of plausibility measure – see e.g. [8]), it is easy to show that $\overline{m}$ is a possibility measure and m_c is neither belief nor plausibility measure.

4 Analysis and Decision Making Problems for Uncertain Systems

Let us consider a static system with input vector $u \in U$ and output vector $y \in Y$, where U and Y are real number vector spaces. When the system is described by a function $y = F(u)$, the analysis problem consists in finding the value y for the given value u. The problem may be extended to the system described by a relation $u \;\rho\; y \triangleq R(u, y) \subset U \times Y$ in the following way (see [1, 3]): for the given R and $D_u \subset U$ where $u \in D_u$ is a given input property, find the smallest set D_y such that the implication $u \in D_u \rightarrow y \in D_y$ is satisfied. Then

$$D_y = \{ y \in Y : \bigvee_{u \in D_u} (u, y) \in R \}. \tag{49}$$

The relation R may have the form of a set of equalities and inequalities concerning u and y. Consider now the functional system described by $y = F(u, x)$ and the relational system described by $R(u, y, x) \subset U \times Y \times X$ where $x \in X$ is an unknown vector parameter which is assumed to be a value of an uncertain variable $\overline{x}$ with $h_x(x)$ given by an expert. Then y is a value of an uncertain variable $\overline{y}$ and for the fixed u, $\overline{y}$ is the function of $\overline{x}$: $\overline{y} = F(u, \overline{x})$.

Analysis problem for the functional system may be formulated as follows: for the given F, $h_x(x)$ and u find $h_y(y)$ for $\overline{y}$. Having $h_y(y)$ one can determine M_y and

$$y^* = \arg\max_{y \in Y} h_y(y), \quad \text{i.e. } h_y(y^*) = 1.$$

Using (33) one obtains

$$h_y(y;u) = v(\overline{y} \cong y) = \max_{x \in D_x(y;u)} h_x(x) \tag{50}$$

where $D_x(y;u) = \{x \in X : F(u,x) = y\}$. If F as a function of x is one-to-one mapping and $x = F^{-1}(u,y)$ then

$$h_y(y;u) = h_x[F^{-1}(u,y)]$$

and $y^* = F(u,x^*)$ where $x^* = \arg\max h_x(x)$. From the definition of the certainty distributions h and h_c it is easy to note that in both continuous and discrete cases $y^* = y_c^*$ where $y_c^* = \arg\max h_c(y)$.

Analysis problem for the relational system may be formulated as follows: for the given R, $h_x(x)$, D_u (i.e. the property "$u \in D_u$" is satisfied) and $\Delta_y \subset Y$, find $v(\bar{y} \mathrel{\tilde{\in}} \Delta_y)$.

Using (49) we can determine $D_y(x)$ for $R(u,y,x)$. Then

$$v(\bar{y} \mathrel{\tilde{\in}} \Delta_y) = v[\bar{x} \mathrel{\tilde{\in}} D_x(\Delta_y)] = \max_{x \in D_x(\Delta_y)} h_x(x) . \tag{51}$$

where

$$D_x(\Delta_y) = \{x \in X : D_y(x) \subseteq \Delta_y\} . \tag{52}$$

In the case when $\bar{x}$ is considered as C-uncertain variable it is necessary to find v (51) and

$$v(\bar{y} \mathrel{\tilde{\in}} \overline{\Delta}_y) = v[\bar{x} \mathrel{\tilde{\in}} \overline{D}_x(\Delta_y)] = \max_{x \in \overline{D}_x(\Delta_y)} h_x(x) . \tag{53}$$

Then, according to (39) with y in the place of x,

$$v_c(\bar{y} \mathrel{\tilde{\in}} \Delta_y) = \frac{1}{2}[v(\bar{y} \mathrel{\tilde{\in}} \Delta_y) + 1 - v(\bar{y} \mathrel{\tilde{\in}} \overline{\Delta}_y)] . \tag{54}$$

Example 1: Let $u, x \in R^2$, $y \in R^1$, $y = x^{(1)}u^{(1)} + x^{(2)}u^{(2)}$, $x^{(1)} \in \{3,4,5,6\}$, $x^{(2)} \in \{5,6,7\}$ and the corresponding values of h_{x1}, h_{x2} given by an expert are (0.3, 0.5, 1, 0.6) for $x^{(1)}$ and (0.8, 1, 0.4) for $x^{(2)}$. Assume that $\bar{x}^{(1)}$ and $\bar{x}^{(2)}$ are independent, i.e. $h_x(x_i^{(1)}, x_j^{(2)}) = \min\{h_{x1}(x_i^{(1)}), h_{x2}(x_j^{(2)})\}$. Then for $x = (x^{(1)}, x^{(2)}) \in$ {(3,5), (3,6), (3,7), (4,5), (4,6), (4,7), (5,5), (5,6), (5,7), (6,5), (6,6), (6,7)} the corresponding values of h_x are (0.3, 0.3, 0.3, 0.5, 0.5, 0.4, 0.8, 1, 0.4, 0.6, 0.6, 0.4). Let $u^{(1)} = 2$, $u^{(2)} = 1$. The values of $y = 2x^{(1)} + x^{(2)}$ corresponding to the set of pairs $(x^{(1)}, x^{(2)})$ are the following: {11, 12, 13, 13, 14, 15, 15, 16, 17, 17, 18,

19}. Then $h_y(11) = h_x(3,5) = 0.3$, $h_y(12) = h_x(3,6) = 0.3$, $h_y(13) = \max\{h_x(3,7), h_x(4,5)\} = 0.5$, $h_y(14) = h_x(4,6) = 0.5$, $h_y(15) = \max\{ h_x(4,7), h_x(5,5) \} = 0.8$, $h_y(16) = h_x(5,6) = 1$, $h_y(17) = \max\{ h_x(5,7), h_x(6,5) \} = 0.6$, $h_y(18) = h_x(6,6) = 0.6$, $h_y(19) = h_x(6,7) = 0.4$. For $h_y(y)$ we have $y^* = 16$, and using (48) for $\overline{y}$ we obtain $\overline{h}_y = 5$, $M_y = 15.4$. Using (47) we obtain the corresponding values of $h_{yc}(y)$: (0.15, 0.15, 0.25, 0.25, 0.4, 0.6, 0.3, 0.3, 0.2), $y_c^* = y^* = 16$, $v_c(\overline{y} \cong 16) = 0.6$, $\overline{h}_{yc} = 2.6$, $M_{yc} = 15.43 \approx M_y$. □

For the functional system $y = F(u)$ the basic decision problem consists in finding the decision $\hat{u}$ for the given desirable value $\hat{y}$. It may be extended to the relational system in the following way (see [1, 3]): for the given $R(u, y)$ and $D_y \subset Y$ where $y \in D_y$ is a desirable output property, find the largest set D_u such that the implication $u \in D_u \rightarrow y \in D_y$ is satisfied. Then

$$D_u = \{ u \in U : D_y(u) \subseteq D_y \} \tag{55}$$

where

$$D_y(u) = \{ y \in Y : (u, y) \in R(u, y) \}. \tag{56}$$

Consider now the system with the unknown parameter x.

Decision problem for the functional system: for the given $F(u,x)$, $h_x(x)$ and the desirable value $\hat{y}$ find the decision $\hat{u}$ maximizing $v(\overline{y} \cong \hat{y})$. To solve the problem one should determine $h_y(y; u)$ according to (50) and

$$\hat{u} = \arg \max_{u \in U} h_y(\hat{y}; u).$$

We can obtain $\hat{u}$ by solving the equation $F(u, x^*) = \hat{y}$ where $x^* = \arg\max h_x(x)$. In Example 1 we have $x^* = (5,6)$ and the set of solutions $\hat{u} = (u^{(1)}, u^{(2)})$ satisfying the equation $5u^{(1)} + 6u^{(2)} = \hat{y}$. Since $y^* = y_c^*$, then $\hat{u} = \hat{u}_c$ which means that the result for the *C*-uncertain variables is the same as for the uncertain variable. We can consider another version of the decision problem consisting in finding $\hat{u}$ such that $M_y(\hat{u}) = \hat{y}$. Now $\hat{u}_c$ may differ from $\hat{u}$ if $M_{yc} \neq M_y$.

Decision problem for the relational system: for the given $R(u, y, x)$, $h_x(x)$ and D_y (where $y \in D_y$ is a desirable output property) find $\hat{u}$ maximizing $v(\overline{y} \,\tilde{\in}\, D_y)$.

To solve the problem one should determine $D_y(u,x)$ and $D_u(x)$ according to (56) and (55) with $(u,y,x) \in R(u,y,x)$ in the place of $(u,y) \in R(u,y)$. Then for the fixed u

$$v(\overline{y} \,\tilde{\in}\, D_y) \stackrel{\Delta}{=} v(u) = v[u \,\tilde{\in}\, D_u(\overline{x})] = v[\overline{x} \,\tilde{\in}\, D_x(u)] = \max_{x \in D_x(u)} h_x(x) \tag{57}$$

where

$$D_x(u) = \{x \in X : u \in D_u(x)\} \tag{58}$$

and $\hat{u} = \arg\max v(u)$. When $\overline{x}$ is considered as *C*-uncertain variable it is necessary to find v (57) and

$$v(\overline{y} \,\tilde{\in}\, \overline{D}_y) = v[\overline{x} \,\tilde{\in}\, \overline{D}_x(u)] = \max_{x \in \overline{D}_x(u)} h_x(x) \ .$$

Then, according to (39) with y in the place of x,

$$v_c(\overline{y} \,\tilde{\in}\, D_y) \stackrel{\Delta}{=} v_c(u) = \frac{1}{2}[v(\overline{y} \,\tilde{\in}\, D_y) + 1 - v(\overline{y} \,\tilde{\in}\, \overline{D}_y)] \tag{59}$$

and $\hat{u}_c = \arg\max v_c(u)$. Let us note that in both cases (functional and relational) the solution $\hat{u}$ (or $\hat{u}_c$) may not exist or may be not unique (as in the deterministic system without x).

Example 2: Let $u, y, x \in R^1$, the relation R is given by inequality $xu \le y \le 2xu$, $D_y = [y_1, y_2]$, $y_1 > 0$, $y_2 > 2y_1$. Then $D_u(x) = [\frac{y_1}{x}, \frac{y_2}{2x}]$, $D_x(u) = [\frac{y_1}{u}, \frac{y_2}{2u}]$. Assume that x is a value of an uncertain variable $\overline{x}$ with triangular distribution $h_x(x)$ determined by $(0, \frac{1}{2}, 1)$. It is easy to note that $\hat{u}$ is any value from $[\ 2y_1,\ y_2\ ]$ and $v(\hat{u}) = 1$. Using (59) we obtain

$$v_c(u) = \begin{cases} \dfrac{y_2}{2u} & \text{when} \quad u \ge y_1 + 0.5y_2 \\ 1 - \dfrac{y_1}{u} & \text{when} \quad y_1 \le u \le y_1 + 0.5y_2 \\ 0 & \text{when} \quad u \le y_1 \ . \end{cases}$$

It is easy to see that $\hat{u}_c = y_1 + 0.5y_2$ and $v_c(\hat{u}_c) = \dfrac{y_2}{2y_1 + y_2}$. E.g. for y_1=2, y_2=12 the results are the following: $\hat{u} \in [4, 12]$ and $v = 1$, $\hat{u}_c = 8$ and $v_c = 0.75$. □

5 Computational Aspects

The application of C-uncertain variables (i.e. v_c instead of v) means better using the expert's knowledge, but may be connected with much greater computational difficulties. In the discrete case, when the number of possible values of x is small, it may be acceptable to determine all possible values v_c. Let us explain it for the decision problem and relational plant. Assume that $X = \{x_1,..., x_m\}$, $U = \{u_1,...,u_p\}$, $Y = \{y_1,..., y_k\}$. Now the relation $R(u, y, x)$ is reduced to the family of sets $D_y(u^{(i)}, x^{(j)}) \subset Y$, $i \in \overline{1,p}$, $j \in \overline{1,m}$, and the algorithm for the determination of $\hat{u}$ is the following:

1. For $u^{(i)}$ $(i = 1, \ldots, p)$ prove if

$$D_y(u^{(i)}, x^{(j)}) \subseteq D_y \ , \quad j = 1, 2, \ldots, m. \tag{60}$$

If yes then $x^{(j)} \in D_x(u^{(i)})$. For $j = m$ we obtain the set $D_x(u^{(i)})$.

2. Determine v_{ci} according to (59):

$$v_{ci} = \begin{cases} 1 - \dfrac{1}{2} \max\limits_{x \in \overline{D}_x(u^{(i)})} h_x(x) & \text{if} \quad x^* \in \overline{D}_x(u^{(i)}), \\ \dfrac{1}{2} \max\limits_{x \in D_x(u^{(i)})} h_x(x) & \text{otherwise} \end{cases}$$

where $x^* \in X$ is such that $h_x(x^*) = 1$.

3. Choose $i = i^*$ such that v_{ci} is the maximum value in the set of v_{ci} determined in the former steps. Then $u^* = u_i$ for $i = i^*$.

Special forms of this algorithm have been elaborated for two special cases of $R(u, y, x)$ and $D_y = [\underline{y}, \overline{y}]$:

1. $c^{\mathrm{T}}u \le y \le d^{\mathrm{T}}u$ where c, d are subvectors of x.
2. $0 < y < u^{\mathrm{T}}Qu$ where $Q = \operatorname{diag} x$, $x = (x^{(1)},..., x^{(s)})$, $x^{(r)} > 0$, $r \in \overline{1,s}$.

For example, in the first case, the second step with (60) is reduced to testing if $c_j^{\mathrm{T}}u^{(i)} \ge \underline{y}$ and $d_j^{\mathrm{T}}u^{(i)} \le \overline{y}$.

For these cases computer programs have been elaborated and used for simulations.

6 Conclusions

Two versions of uncertain logics have been defined and used in the definitions of two versions of uncertain variables. The certainty distributions seemed to be the most natural, simplest and practically available description of the uncertainty evaluating an expert's opinion on approximate values of an unknown parameter. The methods of the analysis and decision making presented in the paper were used to elaborating the computer programs for special cases. The simulations showed a significant influence of the parameters in certainty distributions on the final results. The uncertain variables may be applied to uncertain systems described by a logical knowledge representation [2, 6] and may be combined with an idea of learning for knowledge-based systems [4, 7].

References

1. Bubnicki, Z.: Logic-algebraic method for a class of knowledge based systems. In: Pichler, F., and Moreno-Diaz, R. (eds.): Computer Aided Systems Theory. Lecture Notes in Computer Science, Vol. 1333. Springer Verlag, Berlin (1997) 420–428
2. Bubnicki, Z.: Logic-algebraic approach to a class knowledge based fuzzy control systems. In: Proc. of European Control Conference, Brussels, Belgium, Vol.1 (1997) TU-E-G2
3. Bubnicki, Z.: Logic-algebraic method for knowledge-based relation systems. Systems Analysis Modelling and Simulation, Vol. 33 (1998) 21–35
4. Bubnicki, Z.: Uncertain variables and learning algorithms in knowledge-based control systems. Artificial Life and Robotics, Vol. 3, 3 (1999)
5. Bubnicki, Z.: Uncertain variables and logic-algebraic method in knowledge-based systems. In: Hamza, M.H. (ed): Intelligent Systems and Control. Proc. of IASTED International Conference, Halifax, Canada. Acta Press, Zurich (1998) 135–139
6. Bubnicki, Z.: Learning processes and logic-algebraic method in knowledge-based control systems. In: Tzafestas, S.G., and Schmidt, G. (eds.): Progress in System and Robot Analysis and Control Design. Lecture Notes in Control and Information Sciences, Vol. 243. Springer Verlag, London (1998) 183–194
7. Bubnicki, Z.: Learning control systems with relational plants. In: Proc. of European Control Conference, Karlsruhe, Germany (1999)
8. Dubois, D., Wellman, M. P., D'Ambrosio, B., and Smets, P.: Uncertainty in Artificial Intelligence. Morgan Kaufmann, San Mateo CA (1992)
9. Klir, G. J., and Folger, T. A.: Fuzzy Sets, Uncertainty, and Information. Prentice-Hall, Englewood Cliffs NJ (1988)
10. Ranze, K. C., and Stuckenschmidt, H.: Modelling Uncertainty in Expertise. In: Cuena, J. (ed.): Proc. of the XV IFIP World Computer Congress, Vienna, Austria, and Budapest, Hungary. Osterreichische Computer Gesellschaft, Vienna (1998) 105–116

Variable-Structure Learning Controllers

Antonio Sala, Pedro Albertos, and Manuel Olivares

Dept. de Ing. de Sistemas y Automática.
Universidad Politécnica de Valencia
Apdo. 22012, E-46071 Valencia (Spain)
asala@isa.upv.es, pedro@aii.upv.es

Abstract. In this paper, a direct sliding control learning strategy is combined with a fuzzy system with variable granularity to achieve different precision requirements in different zones of the state-space; non-local basis functions are also added. A coordinate transformation gives a clearer meaning over performance evaluation and the fuzzy system operates on this transformed space.

1 Introduction

There are many control design techniques, being its basic role to act on a system according to some goals and information from the system and the environment. It is almost impossible to fully model the behaviour of the system, the actual scenario and the detailed requirements. Thus, there have been many attempts to provide some kind of learning to the controller, in order to cope with new situations not fully considered at the initial design stage, from mere parameter adaptation to a full-scale decision system.

One of the approaches to robust control of partly-known nonlinear systems is to use variable structure (sliding mode) controllers [8,1], if the system model can be expressed in control-affine form $\dot{x} = f(x) + g(x)u$ and a pseudo-output $s = h(x)$ can be defined so that it has unity relative degree and $s(t) = 0$ implies asymptotic convergence of the output to zero. For inverse-stable systems, the sliding variable $s(t)$ is usually formed as a linear combination of output derivatives $s = \sum \alpha_i y^{(i)}$ such that $s = 0$ is a stable linear differential equation.

Sliding controller produce a discontinuous control, synthesised based on known model error bounds. Even if stability is guaranteed, a high control activity results if those bounds are overestimated (the limit case is a switching control law). Furthermore, commutation at sampling frequency (chattering) and errors due to discretization are present. The high frequency components can excite unmodelled dynamics so robustness can be degraded. Filtering, dead-zones and small sampling periods are used to solve these issues [3].

The sliding mode idea is well suited to learning and adaptation because the temporal credit assignment problem has an easy solution: the blame for unsatisfactory performance can be credited to the immediately past control action.

A learning scheme can act upon the overall system by improving a learned model, thus reducing the uncertainty bounds (to be estimated) or directly acting

F. Pichler, R. Moreno-Díaz, and P. Kopacek (Eds.): EUROCAST'99, LNCS 1798, pp. 543–549, 2000.

on the control actions by increasing or reducing the control activity based on performance evaluation.

Some continuous-time adaptive approaches to this problem are reported in the literature if initial error bounds are not known[7]. Other learning algorithms in the literature (for example, [6]) can be thought of as variations of the sliding strategy, in particular many of those that define *performance zones* on a phase plane.

In the following, a model-free direct discrete-time adaptive controller is developed, so that a pre-specified approach to the sliding surface is achieved. Different precision requirements are needed in different zones of the state space. Hence, the controller divides the state space in coarser partitions as the sliding variable increases. Interpolative fuzzy models will be used to model a controller $u(k) = f(x, r)$.

In the coarsest zone, even a saturating control action could be appropriate (for stable processes) in the sense that it produces no chattering. In the fine zone, control actions have to be modulated to keep the discrete-time system into a *quasi-sliding* mode [3]: the true sliding regime is not attainable due to the finite sampling period jointly with the approximation error of the fuzzy system used as a controller.

2 Sliding-Mode Control

Sliding-mode (variable-structure) control [8] is now a classical strategy to control a class of systems with the following characteristics:

- The system dynamics can be described by $\frac{d^n y}{dt^n} = f(x) + g(x) * u$
- The system state is measurable, and the zero dynamics is stable.

Let a reference signal $r(t)$ be defined, as well as a reduced-order stabe target dynamics for the closed loop error $e(t) = r(t) - y(t)$:

$$\mathbf{s}(e(t)) = \frac{d^{n-1}e}{dt^{n-1}} + a_{n-2}\frac{d^{n-2}e}{dt^{n-2}} + \cdots + a_0 e = 0 \tag{1}$$

This target dynamics is called *sliding mode* regime, chosen by the designer according to control specifications. It is a linear combination of system state variables and reference derivatives, assumed a smooth enough reference. The usual choice is the critically damped $(\frac{d}{dt} + 1)^n$ dynamics.

By derivation with respect to time of (1) and substituting the process equation, an expression such as $\frac{ds(e(t))}{dt} = P(r, \mathbf{x}) - f(\mathbf{x}) - g(\mathbf{x})u$ can be obtained, where $\mathbf{x} = (\frac{d^{n-1}y}{dt^{n-1}}, \ldots, \frac{dy}{dt}, y)$ and $P(r, \mathbf{x})$ can be calculated from measurements.

The controller is synthesised by calculating the interval of control actions u such that a convergence condition is met, such as:

$$\text{sign}(s)\frac{ds}{dt} \leq -\eta \tag{2}$$

that will ensure convergence to $s = 0$ in finite time.

The control action is discontinuous, as it depends on sign(s) and the uncertainty in f and g. This discontinuity is called *chattering*. Dead zones on s can diminish this high frequency control activity.

When using discrete-time versions of sliding controllers (such as the ones obtained by numerical discretisation of continuous-mode ones) then chattering around the sliding regime becomes an oscillation with frequency $\omega_s/2$. If this oscillation has bounded amplitude Δ it is named a quasi-sliding mode [5].

In the discrete case, fulfilment of reaching laws such that $|S_{k+1}| < |S_k|$ ensure closed-loop stability. Usual reaching laws are $|S_{k+1}| = 0$ or $S_{k+1} = \alpha S_k$ (equivalent to pole-assignment in perfectly known systems). Unfortunately, uncertainty makes impossible to design a discrete controller fulfilling the reaching law all over the state space, specially for small $|S(k)|$. That's the reason quasi-sliding behaviour is present in the discrete case.

3 A Learning Sliding Controller

The controller will control a continuous-time sliding function via a discrete controller implemented by a universal function approximator such as fuzzy systems. Many standard fuzzy and neurofuzzy techniques [4,2] used in intelligent learning control can be cast into the form:

$$f(p) = \sum \mu_i(p)\bar{f}_i$$

so that for each μ_i (membership of basis function) there exists a point p_i such that $\mu_i(p_i) = 1$. That point is called the "vertex" or prototype center point of the basis functions.

Some considerations will be made to apply them to the sliding control framework:

The first one is that the sliding variable $s(t)$ is not a state variable because it directly depends on the reference, but a state variable can be formed by adding to it the appropriate reference and reference derivatives. From now on, it will be assumed that derivatives of the reference are zero (i.e., the system is to track constant references). Under that assumption, a linear transformation exist so that the operation space $(\mathbf{x}, r)$ (composed by the plant state and the reference) is mapped to $(s, \mathbf{s}^{\perp}, r)$, where $\mathbf{s}^{\perp}$ is a $n-1$ dimensional basis orthogonal to s, being n the plant order.

The first distinctive feature here proposed is that the controller will be a non-linear fuzzy interpolator given by:

$$u(x, r) = \sum \mu_i(s, \mathbf{s}^{\perp}, r)\bar{u}_i$$

i.e., the control surface will be defined on the transformed space, instead of the common setup $u(x, r) = \sum \mu_i(x, r)\bar{u}_i$. In that way, different precision requirements in the controller are easier to incorporate: when satisfactory closed loop performance will be achieved, the controller will operate nearly always in the "small s" region. For big values of s sliding control can be approximated even with bang-bang actions so precision requirements are far less.

Based on the previous idea, a special node arrangement is set up, in which precise steady state control is ensured via a fine granularity near the $(s, \mathbf{s}^{\perp}) \approx 0$ region, but a much coarser granularity is gradually used away from that points. In that way, the number of nodes is reduced hence giving a regulator with fewer parameters.

In the referred arrangement, the distance between node centers increases geometrically as s and $s^{\perp}$ get away from the origin (see figure 1). Partitions are set up along each dimension so that the nodes p_k on them verify: $d(p_k, p_{k+1}) = \beta d(p_{k-1}, p_k)$. Notwithstanding, the final global setup is not the cartesian product of those partitions, unlike many usual fuzzy system configurations. Some additional operations will be made.

The geometric multiplication factors β_i are different for each of the dimensions: precision is needed when s is small, whatever the value of $s^{\perp}$ is, so scattering along $s^{\perp}$ is more uniform (a uniform scattering corresponds to $\beta = 1$). Anyway, high values of $s^{\perp}$ indicate that the system isn't near the steady-state point, even if the sliding mode has been reached ($s \approx 0$), so actual precision requirements are less and nonuniform partitions along that dimension is hence justified.

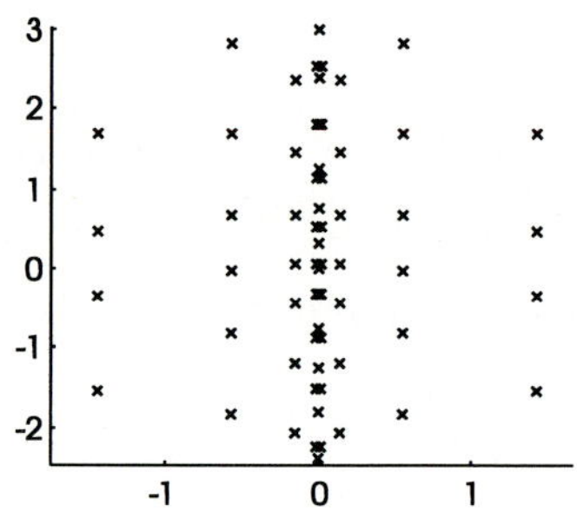

Fig. 1. Centroid spacing

In this way, a precise approximation inside a small user-defined dead zone ϵ around $s = 0$ is achieved without sacrifice of memory and generalisation capabilities. The sliding variable values in which nodes are placed are $\{0, \pm\epsilon, \pm\beta\epsilon, \pm\beta^2\epsilon, \dots\}$. A similar approach is taken in the $\mathbf{s}^{\perp}$ coordinates, but with a ϵ^* value that increases with s, so the final arrangement is not a cartesian-product one, as previously mentioned. Appropriate straightforward modifications to usual fuzzy-interpolation routines are made to deal with the nonuniform sampling points (fig 1).

In this way, a fine control is learnt at or near the quasi-sliding surface, and a coarser control law is applied if far from it. The fine control will allow reducing the chattering magnitude, and the coarser one will enable better generalisation.

A totally saturated coarse action $u = u_{max}\text{sign}(s)$ is taken when either $|s| > s_0$ or $|\mathbf{s}^{\perp}| > s_1$.

3.1 Learning Algorithm

A desired $\hat{S}_{k+1}$ is calculated according to a pre-specified reaching law, given S_k. After applying the controller action, error $e_{k+1} = S_{k+1} - \hat{S}_{k+1}$ is obtained.

After determination of a variable dead zone (function of closed-loop precision requirements and the increasing approximation error due to decreased granularity of the fuzzy system) in the form $d(s) = d_0 + d_1 * |s|$, the error is suitably reduced to e^d_{k+1}.

The change in parameters follow the law $\Delta\bar{u}_i = \eta\mu_i e^d_{k+1}$, where η is a learning rate. Knowledge of the sign of the plant gain is assumed. If parameter increment produces an *a posteriori* control action above saturation limits, an antiwindup-like strategy is applied, in such a way that parameter increments are scaled down, thus avoiding drift.

Learning speed depends on η and the unknown process gain. At reduced learning rate, reduced sampling time is similar to the discretisation of analog sliding-mode adaptive control [7]. Very high learning rates would lead to a $\pm u_{sat}$ switching worst-case performance. In any case, that would lead to chattering similar to a binary switching controller, but it will not destabilise a stable process. Chattering may appear if the number of nodes is too reduced to properly approximate the needed controller for small values of the sliding variable.

To improve generalisation and allow a more reduced number of overall nodes, some of the regressors $\mu_i(x, r)$ have been set to non-local functions, in particular to r, s, s^2, $\tanh(s/\epsilon)$, and $s^{\perp}$, with a reduced learning rate.

Example: The previously outlined algorithm has been tested with a simulated nonlinear spring second-order mechanical system (figure 2) whose equations are:

$$l = \sqrt{h^2 + (u - y)^2} \tag{3}$$

$$f = (u - y)/l * k * (atan(kp * (l - l_0)) + kh) \tag{4}$$

$$\frac{d^2x}{dt^2} = f - 0.125 * v - 0.06 * sign(v) \tag{5}$$

with physical parameters being $h = 1$,$l_0 = 0.85$, $k = 2$, $kp = 1$, $kh = 0$.

Figure 3 shows the behaviour in the first iterations and figure 4 illustrates the final results. In the simulations, the reaching law was $S_{k+1} = 0.62S_k$ for $|s| > d_0$ and $S_{k+1} = 0.9S_k$ for $|s| \leq d_0$, where d_0 is an user-defined deadzone where quasi-sliding mode behaviour is assumed satisfactory, set to 0.01. The granulation had 6 nodes over the s and r variables and a maximum of 10 nodes in $s^{\perp}$ (for small s).

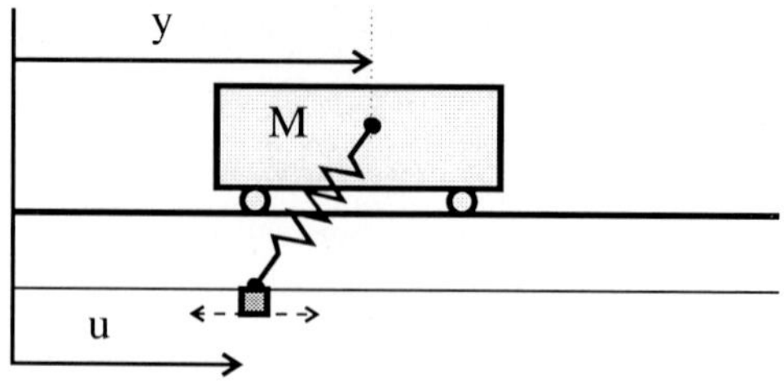

Fig. 2. Example system

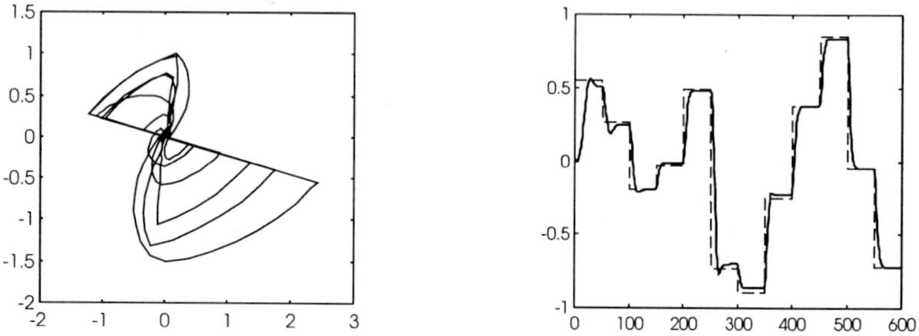

(a) **Initial phase plane on** s, $s^{\perp}$ **domain** (b) **Initial Output**

Fig. 3. Initial behaviour

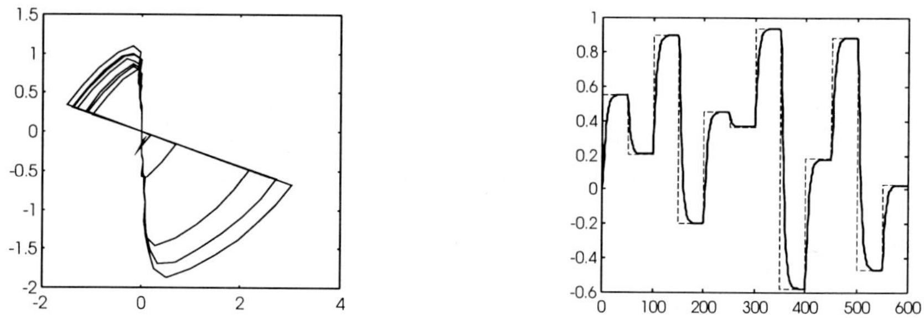

(a) Final phase plane **(b) Output response**

Fig. 4. Learning results

4 Conclusions

In this paper, a fuzzy system approximator plus nonlocal basis functions is used for learning direct sliding control, learning how to reach the sliding regime according to a pre-specified reaching law.

As precision requirements are different depending on different regions of the state space, a progressive granularity reduction is made for big values of the sliding variable. The coarse-fine control transition is efficiently implemented via a change of coordinates that transform the state-reference space into another one with the r,s and $s^{\perp}$ variables as their axes. Partitions on this space have a clearer meaning towards performance evaluation. Granularity reduction and nonlocal basis functions allow for a more reduced number of controller parameters, thus enhancing generalisation.

References

1. Special issue on sliding control. *Int. Journal of Control*, 57(5), 1993.
2. M. Brown and C.J. Harris. *Neurofuzzy Adaptive Modelling and Control.* Prentice Hall, Englewood Clifffs, NJ, 1994.
3. F. Naranjo. *Sliding mode control of sampled data systems (in Spanish).* PhD thesis, Universidad Politecnica Valencia (Spain), 1998.
4. W. Pedrycz. *Fuzzy Control and Fuzzy Systems.* J.Wiley and Sons, New York, NY, second edition, 1993.
5. S.Z. Sarpturk. On the stability of discrete-time sliding mode control systems. *IEEE Trans. Automatic Control*, 32(10):930–932, 1987.
6. S. Shenoi, K. Ashenayi, and M. Timmerman. Implementation of a learning fuzzy controller. *IEEE Control Systems*, pages 73–80, 1995.
7. J.J.E Slotine and W. Li. Adaptive manipulator control: a case study. In *Proc. IEEE Intl. Conf. Robotics and Autom.*, pages 1392–1401, 1987.
8. V. I. Utkin. *Sliding Modes in Control and Optimization.* Springer-Verlag, 1992.

An Identification Algorithmic Toolkit for Intelligent Control Systems

Kirill Chernyshov and Feodor Pashchenko

Institute of Control Sciences
Profsoyuznaya 65, 117806 Moscow, Russia
myau@ipu.rssi.ru

Abstract. The paper presents a set of new non-parametric and parametric identification algorithms oriented to using within the input/output system description. A purpose of elaborating the identification techniques is to involve as broad as possible, to some extend, classes of stochastic system descriptions, both linear and nonlinear ones, assuming that the lack of knowledge with respect to the system may vary from unknown system parameters to unknown system structure at all.

1 Introduction

A control problem solution considerably depends on a choice of the investigated system model to be used within the control process. An approach to analytical description of various systems is based on input/output model description. Within this, deriving an explicit dependence between input and output of the investigated system plays an important role. Such a dependence should ensure an adequate approximation of the system considered by an effective, from a practical point of view, manner. In other words, the identification problem takes an important place within a control process, being a necessary preliminary step when the investigated system model is unknown to some extent. Conventionally, identification problems are classified as structure identification, nonparametric identification, and parametric identification. Such a classification is motivated by body of knowledge available with respect to the investigated system model. In turn, body of knowledge about the model may vary from lack of information on the model structure at all to uncertainties in values of the model parameters. Thus, aim of the paper is to present an algorithmic toolkit suitable within the conditions outlined and to be used for intelligent control systems design.

2 Structure/Nonparametric Identification

When considering nonlinear stochastic systems, the most general identification approaches are based on using non-parametric methods. In turn, solving non-parametric problems is considerably influenced by a choice of a measure of stochastic

F. Pichler, R. Moreno-Díaz, and P. Kopacek (Eds.): EUROCAST'99, LNCS 1798, pp. 550–564, 2000.

dependence of random processes. Among the measures, the ordinary product correlation functions as well as the disperssional functions [1]

$$\theta_{yx}(v) = \mathbf{M}\left(\mathbf{M}\left\{ {y(t)}/{x(s)} \right\} - \mathbf{M}\,y(t)\right)^2, \quad v = t - s$$

are commonly used. Throughout the paper, symbols $\mathbf{M}(\bullet)$, $\mathbf{D}(\bullet)$, $\mathbf{M}\left\{ {\bullet}/{\bullet} \right\}$, and $\mathbf{cov}\,(\bullet,\bullet)$ will respectively stand for the mathematical expectation, variance, conditional expectation, and covariance. However, the functions are known to be able to provide restricted magnitudes of actual stochastic dependence, especially within nonlinear problems. In particular, the product correlation and disperssional functions may vanish provided that there exists a deterministic dependence between the input and output variables of a system [1-3]. Thus, the most suitable way within the nonlinear system identification is based on using consistent, following to Kolmogorov's terminology, measures of dependence. Consistency of measure of dependence $\mu(x,y)$ between random variables x and y means that $\mu(x,y)=0$ if and only if x and y are stochastically independent, with $\mu(x,y)=1$ if there exist a deterministic functional dependence between x and y . Among the consistent measures, the maximal correlation function [4] is, say, of a "covariance nature":

$$S_{yx}(v) = \sup_{\{\mathrm{B}\},\{\mathrm{C}\}} \mathbf{cov}\left(\mathrm{B}(y(t)),\mathrm{C}(x(s))\right), \quad v = t - s, \tag{1}$$

$$\mathbf{M}\left(\mathrm{B}(y(t))\right) = \mathbf{M}\left(\mathrm{C}(x(s))\right) = 0,$$

$$\mathbf{D}\left(\mathrm{B}(y(t))\right) = \mathbf{D}\left(\mathrm{C}(x(s))\right) = 1.$$

In contrast to the correlation and disperssional functions, the maximal correlation function is a complete characteristic of link between random processes. It also should be noted that there exists an example [4] when actual dependence between the input and output variables is nonlinear even provided that the regression of a variable onto another one and vise versa is linear. For the example, such a dependence is properly handled ultimately by the maximal correlation.

A problem statement naturally leading to using the maximal correlation function of input and output processes of investigated systems is as follows. The model's operator is searched for as a linear dynamic mapping $\mathcal{A}$ of a nonlinear input transformation $\mathcal{C}$ into a nonlinear output transformation $\mathcal{B}$:

$$\mathrm{B}\; y(t) = \mathrm{A}\;\; \mathrm{C}\; x(s). \tag{2}$$

Within the problem, the all three components are subject to identification in accordance with minimization of the mean squared error value:

$$J(\mathrm{A}, \mathrm{B}, \mathrm{C}) = \mathbf{D}\left(\mathrm{B}\; y(t) - \mathrm{A}\;\; \mathrm{C}\; x(s)\right).$$

Formally, one should find the operator triplet $\left(\mathrm{A}^*, \mathrm{B}^*, \mathrm{C}^*\right)$ meeting the condition

$$\left(\mathcal{A}^*,\mathcal{B}^*,\mathcal{C}^*\right) = \arg\inf_{\{\mathcal{A}\},\{\mathcal{B}\},\{\mathcal{C}\}} J(\mathcal{A},\mathcal{B},\mathcal{C}),$$

$$\mathbf{D}\left(\mathcal{B}\; y(t)\right) = \mathbf{D}\left(\mathcal{C}\; x(s)\right) = 1 .$$

A structural scheme corresponding to the problem statement is presented at fig. 1 where $N_1(\bullet)$ and $N_2(\bullet)$ are some nonlinear transformations, $L(\bullet)$ is a linear dynamic operator. Within the notations, above operator $\mathcal{B}$ is considered as inverse, in some sense, to transformation $N_2(\bullet)$.

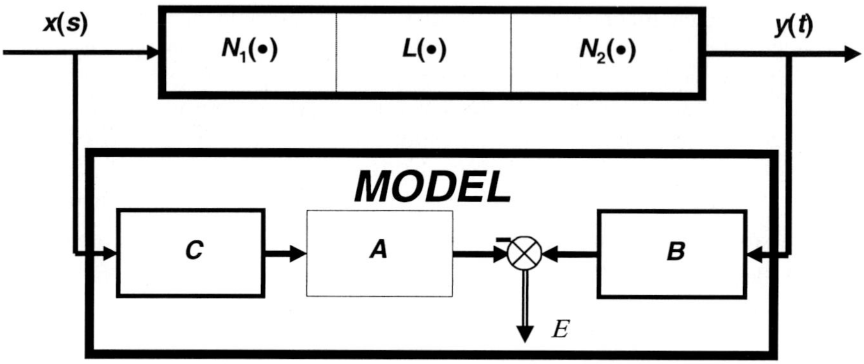

Fig. 1. A structure scheme of the system model identification based on the maximal correlation function approach

Obviously, such a representation covers a broad class of systems. If $\mathcal{B} = \mathcal{C} = \mathcal{I}$, with $\mathcal{I}$ being the identity transformation, relationship (2) corresponds to conventional linear system representation

$$y(t) = \mathcal{A}\; x(s) .$$

If $\mathcal{B} = \mathcal{I}$ and $\mathcal{C}$ is a nonlinear static one $f(\cdot)$ then (2) describes a class of nonlinear systems which is referred as Hammerstein systems:

$$y(t) = \mathcal{A}\; f\left(x(s)\right).$$

The systems are obtained by nonlinear static gain followed by linear dynamic one.

If $\mathcal{C} = \mathcal{I}$ and $\mathcal{B}$ is a nonlinear static transformation, relationship (2) represents systems which may be thought of to be of the Wiener type. The Wiener models are obtained by sequential linking linear dynamic and nonlinear static $f(\cdot)$ gains, with the transformation $\mathcal{B}$ being considered to be inverse to the nonlinear characteristic of the Wiener model:

$$y(t) = f(\mathsf{A}\ x(s)).$$

For a case when the operators $\mathscr{B}$ and $\mathscr{C}$ are conditional mathematical expectations,

$$\mathsf{B}\ y(t) = \mathbf{M}\left\{{z_1}/{y(t)}\right\},\ \mathsf{C}\ x(s) = \mathbf{M}\left\{{z_2}/{x(s)}\right\},$$

where z_1, z_2 are some random elements, representation (2) gives rise to the disperssional type models [1]

$$\mathbf{M}\left\{{z_1}/{y(t)}\right\} = \mathsf{A}\ \mathbf{M}\left\{{z_2}/{x(s)}\right\}.$$

A concrete form of the above model is determined by corresponding choice of elements z_1 and z_2. Say, for the case $z_1 = z_2 = y(t)$ the former expression corresponds to the linear in mean models [1], i.e. the models which are linear in the conditional mathematical expectation of the output process with respect to the input one, i.e.

$$y(t) = \mathsf{A}\ \mathbf{M}\left\{{y(t)}/{x(s)}\right\}.$$

In accordance with the above problem statement, the identification scheme is separated onto two stages. At the first one, the only nonlinear transformations of the input and output processes are determined in accordance with the condition of the maximal arithmetization [5] of probability distribution given by the joint distribution density of the input and output random processes. Within the context, the maximal arithmetization assumes determining a pair of transformations of the input and output processes, which provide maximization of the correlation function of the processes in accordance with expression (1).

From another hand side, such a pair of transformations is just the pair of the first eigenfunctions corresponding to the largest (except unity) eigenvalue of the stochastic kernel $\mathrm{K}(y,x,v)$ given by the joint $p(y,x,v)$ and marginal $p(y), p(x)$ distribution densities of the input and output processes [5-7]:

$$\mathrm{K}(y,x,v) = \frac{p(y,x,v)}{\sqrt{p(y)p(x)}}.$$

In turn, this largest eigenvalue is the maximal correlation. Thus, within the approach described, choice of the nonlinear transformations of the input and output processes is completely formalized. Such a choice does not require any heuristics, restrictive assumptions on distributions of the random processes, or assumptions that the nonlinear transformations belong to a parametric family.

At the second stage of the identification scheme, following to conventional techniques in accordance with the mean square error criterion, the linear mapping is determined by use of the nonlinear transformations obtained at the preceding stage.

Also, provided that the all three components of the identification problem are obtained, the identification scheme may be supplemented by a procedure of determining a statistical equivalent transformation which is inverse to the output transformation. This finally leads to the conventional input/output description based on a relationship which is solved with respect to the output process.

In the fullness of time [1], to express quantitatively nonlinearity of a system, a notion of degree of nonlinearity has been introduced. Within the disperssional identification, such a measure is derived by a comparison of the ordinary product correlation function and the disperssional function of the input and output processes as follows

$$\eta_{disp}(v) = \sqrt{1 - \frac{K_{yx}^2(v)}{\theta_{yx}^2(v)}},$$

with $K_{yx}(v)$, $\theta_{yx}(v)$ standing for the product correlation and cross-disperssional function as defined above.

At the same time, the approach presented, enables one to improve the characteristic of nonlinearity by introducing a corresponding comparison between the ordinary product correlation and the maximal correlation functions of the input and output processes of a system, i.e.

$$\eta_{\max corr}(v) = \sqrt{1 - \frac{K_{yx}^2(v)}{S_{yx}^2(v)}},$$

with $\eta_{\max corr}(v)$ vanishing if and only if all the transformations (B,C) are linear ones. Obviously,

$$\eta_{\max corr}(v) \geq \eta_{disp}(v).$$

In addition, using the maximal correlation enables one to introduce another measure of nonlinearity, the degree of nonlinearity in mean. Such a measure is based on comparison of the disperssional and the maximal correlation functions:

$$\eta_{mean}(v) = \sqrt{1 - \frac{\theta_{yx}^2(t,s)}{S_{yx}^2(t,s)}}.$$

Correspondingly, vanishing such a measure means the system under study is linear in mean, that is linear in conditional mathematical expectation of the output process with respect to the linear one.

Some examples may be presented which demonstrate usefulness of such quantitative characteristics introduced. Consider a simple (static) input/output system for which the joint distribution density (not known in reality, of course, within an identification problem statement) of the input and output variables has the following form [4]

$$p(x,y)=\frac{1}{3\pi\sqrt{3}}\left\{e^{-\frac{3}{2}\left(x^2+y^2+xy\right)}+2e^{-\frac{2}{3}\left(x^2+y^2-xy\right)}\right\}=$$

$$=\frac{e^{-(x^2+y^2)/2}}{2\pi}\left\{1+\sum_{k=1}^{\infty}\frac{2+(-1)^k}{3\cdot 2^k}H_k(x)H_k(y)\right\}=$$

$$=\frac{e^{-(x^2+y^2)/2}}{2\pi}\left\{1+\sum_{k=1}^{\infty}c_k H_k(x)H_k(y)\right\}$$

where

$$H_k(x)=\frac{(-1)^k}{\sqrt{k!}}e^{x^2/2}\frac{d^k}{dx^k}e^{-x^2/2}=\frac{1}{\sqrt{k!}}\left\{x^k-\frac{k(k-1)}{1!\cdot 2}+\ldots\right\},$$

with $H_k(x)$ being the Hermite polynomials.

For the system, the degree of nonlinearity based on the disperssional function is equal to zero, while that of based on the maximal correlation does not: $\eta_{\max corr}=\sqrt{5}/3$.

For another system, let the joint distribution density of the input and output variables be as follows [3]

$$p_{-\frac{1}{2},\lambda}(x,y)=\frac{1}{4\pi\sqrt{1-\lambda^2}}\exp\left[-\frac{x^2+y^2-2\lambda xy}{2\left(1-\lambda^2\right)}\right]+$$

$$+\frac{1}{4\pi\sqrt{1-\lambda^2}}\exp\left[-\frac{x^2+y^2+2\lambda xy}{2\left(1-\lambda^2\right)}\right],\ \text{with } |\lambda|<1.$$

Then, for such a case, the corresponding disperssional characteristic is undefined, while the degree of nonlinearity and degree of nonlinearity in mean both are equal to unity.

Thus, the technique proposed enables one:

- to split the nonlinear system identification scheme onto simpler sequential stages,
- to achieve completely formalized choice of nonlinear input and output transformations without any heuristics, and a priori assumptions on distributions of the random processes, or that the transformations to belong to a parameterized family,
- to use a consistent measure of dependence of random processes, the maximal correlation function, which properly reflects the actual inherent stochastic linking between the processes,

– to derive a nonlinearity measure of the system under study, with the measure being more accurate in comparison with those of based on ordinary product correlation functions, or disperssional functions.

3 Parametric Identification

An opposite case to the nonparametric approach considered above, is concerned with unknown system parameters. When the system model is unknown up to values of parameters, the parameterized description of the models are commonly used as a prediction $\hat{y}(t,\theta)$ of future values of the output process $y(t)$ [8] where (for the linear-in-parameters systems)

$$y(t) = \theta^{*T} \varphi(t) + v(t) .$$

Here θ stands for the model parameter vector, with θ^{*} standing for the truth system parameter vector. Within such a case, the main attention is focused on deriving recursive parameter estimation algorithms. Conventionally, recursive parametric identification involves least square algorithms, extended and generalized least squares, maximum likelyhood, and instrumental variables. In turn, choosing a technique among them is based on the available information on disturbances $v(t)$ affecting the system.

For instance, known optimal algorithms and optimal on a class algorithms derived for identification of ARMAX (Auto Regressive Moving Average with eXogeneous inputs) models considerably use the disturbance model representation as a moving average process of some known order. Deriving such algorithms is based, in entity, on using the maximum likelyhood method. Similar assumptions on the external disturbances model structure are used within a number of optimal instrumental variables algorithms. And some general enough approaches, which are based on considering the disturbance model as an autoregressive moving average process, require information on orders of degrees of the corresponding polynomials of the disturbance filter, that is the disturbance model structure is also assumed to be known.

From another hand side, unknown disturbance model structure is just the natural limitation of an identification problem, while deriving optimal algorithms is based on accounting such a model. In turn, a number of identification problems may require both obtaining system parameter estimation and determining disturbance model. Another case of a problem statement may be concerned ultimately with identification of the system parameters. Within the former case, using an optimal algorithm seems to be not necessary while using an algorithm, which does not require involving disturbance model, would be acceptable. Thus, the consideration has been focused on recursive parametric identification of dynamic systems of the above form affected by a disturbance having a completely unknown model structure.

Under unknown disturbance model structure, the only way to obtain unbiased estimates of the system parameters is using the instrumental variable methods. Among them, the ovedetermined extended instrumental variable method originally proposed by Stoica and Soderstrom [9-11] is the most general one. In entity, identification criterion corresponding to such a technique may be expressed as a condition of coincidence of a generalized covariance function of the output system process and the instrumental variable vector, from one hand side, and the predicted output process and the same instrumental variable vector, from another hand side:

$$\hat{\theta} = \arg\min_{\theta} I(\theta) , \tag{3}$$

$$I(\theta) = \left\| K_{yz} - K_{\hat{y}z}(\theta) \right\|_Q^2 .$$

In the above relationship,

$$K_{yz} = \mathbf{M}\left(Z(t) F(q^{-1}) y(t) \right),$$

$$K_{\hat{y}z}(\theta) = \mathbf{M}\left(Z(t) F(q^{-1}) \hat{y}(t,\theta) \right)$$

are the cross-covariance functions of the instrumental variable vector $Z(t)$, $\dim Z(t) \geq \dim\theta$, and, correspondingly, observed output $y(t)$ and predicted output $\hat{y}(t,\theta)$, $\hat{y}(t,\theta) = \theta^T \varphi(t)$, with the outputs $y(t)$ and $\hat{y}(t,\theta)$ being transformed by an asymptotically stable filter $F(q^{-1})$. Here Q is a positively defined weighting matrix, and, conventionally, for a column-vector X, $\|X\|_Q^2 = X^T Q X$.

Conditions, which the instrumental variables are to meet to, are obvious and have the form

$$\mathbf{M}\left(Z(t) v(s) \right) = 0 \ \ \forall \ t,s \ , \ \ rank \mathbf{M}\left(Z(t) \varphi^T(t) \right) = \dim \varphi(t) .$$

By ergodicity

$$\lim_{t\to\infty} \frac{1}{t} \sum_{k=1}^{t} Z(k) F(q^{-1}) y(k) = K_{yz} \ \text{ a.s.,}$$

$$\lim_{t\to\infty} \frac{1}{t} \sum_{k=1}^{t} Z(k) F(q^{-1}) \hat{y}(k,\theta) = K_{\hat{y}z}(\theta) \ \text{ a.s.,}$$

and criterion (3) may be rewritten in the form

$$\hat{\theta} = \arg\min_{\theta} I(\theta) ,$$

$$I = \frac{1}{t^2} \left\| \sum_{k=1}^{t} Z(k) F(q^{-1}) \left(y(k) - \hat{y}(k,\theta) \right) \right\|_Q^2 . \tag{4}$$

Conventional recursive instrumental variables algorithms are constructed in analogy to those of recursive least squares, which have been widely used in problem of linear-in-parameters model identification. As well known, the basic idea of the RLS algorithm is to obtain parameter estimates by minimizing the sum of the squared errors between the observed and estimated output variables of the model. In the RLS algorithm, the initial covariance matrix should be chosen properly to ensure the existence of all the estimates in the estimation chain, especially in the ill-conditioned situations. It is well known that the covariance matrix may affect the convergence rate of the estimates of the RLS algorithm, hence applications to problems requiring a fast convergence rate or involving ill-conditioning situations will be limited [12]. Some modified versions of the RLS algorithm, being applicable for the ill-conditioned situations, are known. These, however, are oriented for a restricted model description. In turn, conventional recursive instrumental variables algorithms inherit some recursive least square features. Among them, both the condition number of the identification criterion Hessian and autocorrelation disturbance properties play an important role. These are well known to be able to influence the identification processes and lead to unacceptable results.

In contrast, within the proposed approach, the Hessian inversion is eliminated by representation of the recursive identification algorithms as a linear combination of the estimate obtained at the preceding algorithm step and the current observation vector:

$$\theta(t) = \alpha(t)\theta(t-1) + \beta(t)\varphi(t) .$$

Coefficients $\alpha(t)$, $\beta(t)$ of such a combination are to be chosen to meet a condition suitable within a specified identification problem statement. Following to the considered problem assumptions, that is the condition of colour disturbances having completely unknown model structure, choosing the above coefficients is implemented by described criterion (4) corresponding to the overdetermined extended instrumental variables. As a result, such an approach enabled one to obtain a strongly consistent recursive identification algorithm possessing increased stability of current estimates with respect to sample data.

A number of examples demonstrate a good efficiency of the algorithm presented under various characteristics of systems subject to identification. Figures 2 to 7 represent the behavior of the current Euclidean identification error square norm $\eta^2(t) = \left(\theta(t) - \theta^*\right)^T \left(\theta(t) - \theta^*\right)$ corresponding to the algorithm obtained, the solid line Olc_t, and to the conventional recursive algorithm of the extended instrumental variables [13], the dotted line $Roiv_t$. Stability of the algorithm behavior is clearly manifested both with respect to the condition number of criterion (4) Hessian (example 1 (fig. 2) where the condition number is of order 10^2, and example 2 (fig. 3 and fig. 4) where the condition number is of order 10^4), and with respect to the external disturbance structure (example 2 (fig. 3 and fig. 4) where the condition number is of order 10^4 under colour disturbances, and example 3 (fig. 5) where the condition number is of order 10^4 under white-noise disturbances; example 4 (fig. 6) where the condition number is of order 10^5 under colour disturbances, and example 5 (fig. 7) where the condition number is of order 10^5 under white-noise disturbances).

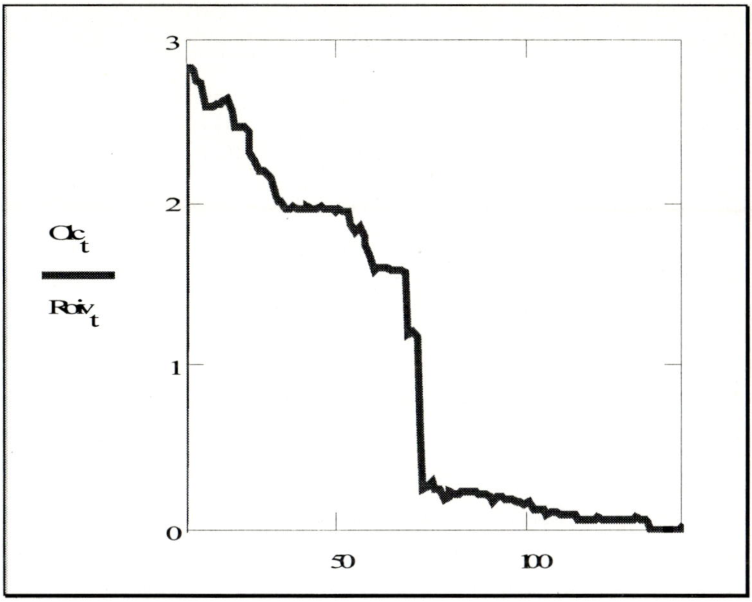

Fig. 2. Example 1: behavior of the current Euclidean identification error square norm when criterion (4) Hessian condition number is of order 10^2 under colour disturbances

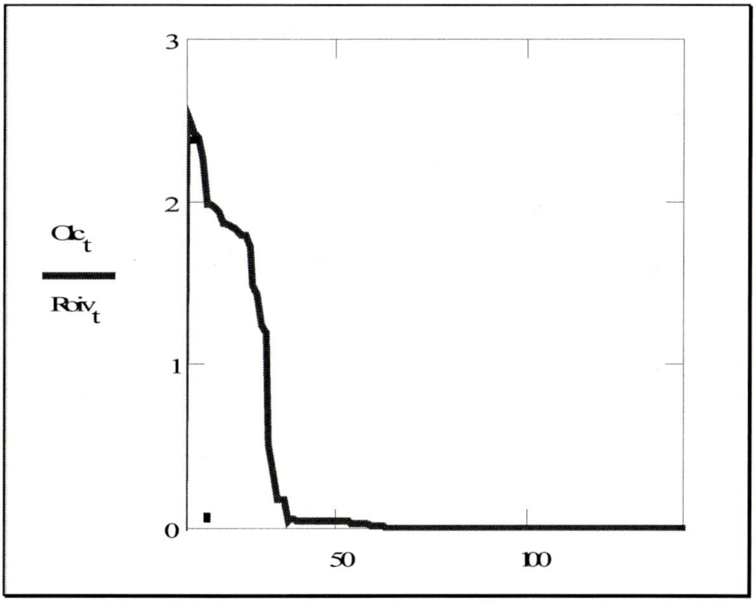

Fig. 3. Example 2 (fine scale): behavior of the current Euclidean identification error square norm when criterion (4) Hessian condition number is of order 10^2 under colour disturbances

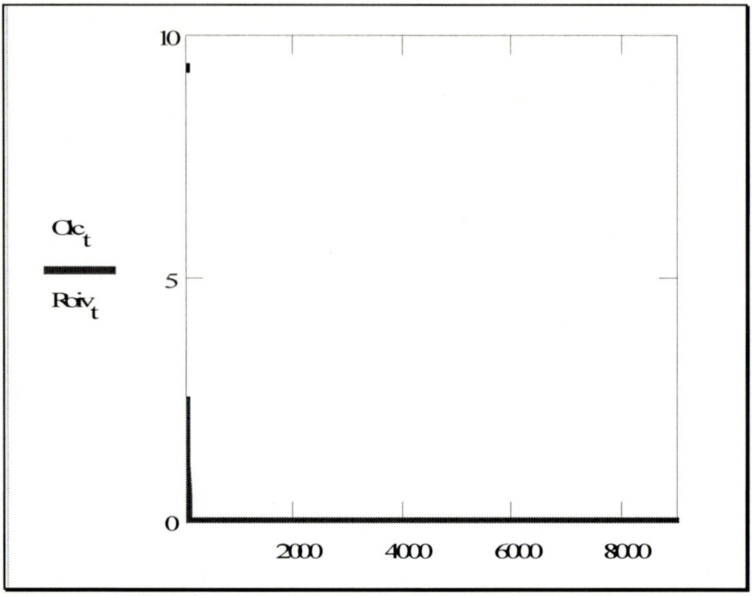

Fig. 4. Example 2 (rough scale): behavior of the current Euclidean identification error square norm when criterion (4) Hessian condition number is of order 10^2 under colour disturbances

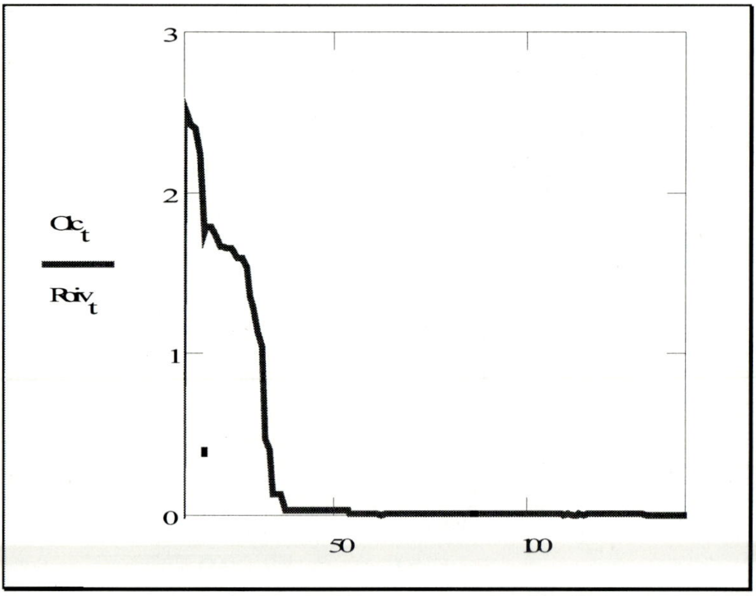

Fig. 5. Example 3: behavior of the current Euclidean identification error square norm when criterion (4) Hessian condition number is of order 10^4 under white-noise disturbances

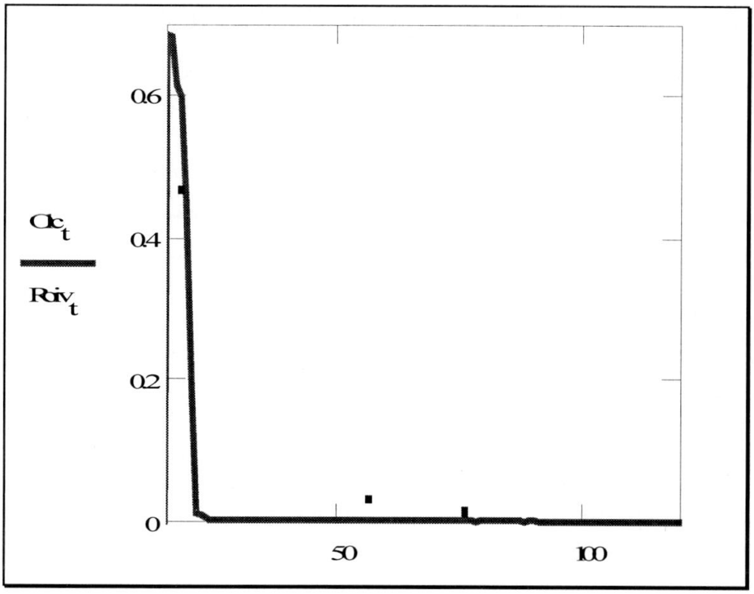

Fig. 6. Example 4: behavior of the current Euclidean identification error square norm when criterion (4) Hessian condition number is of order 10^5 under colour disturbances

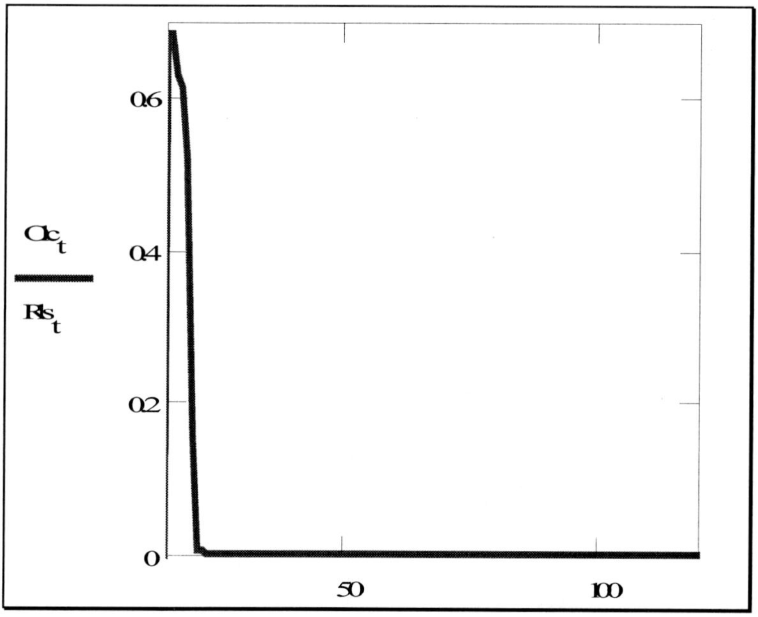

Fig. 7. Example 5: behavior of the current Euclidean identification error square norm when criterion (4) Hessian condition number is of order 10^5 under white-noise disturbances

From another hand side, the fact, that the identification criterion Hessian is ill-posed, is not necessary an obstacle for convergence of the recursive schemes based on direct minimization of criteria of form (4) (examples 3 and 5). Provided that Hessian is ill-posed, just auto-correlation nature of the external disturbances should be considered as a significant factor which considerably affects the sample covariances forming the identification criterion Hessian components and, finally, worsening convergence properties of the conventional identification schemes (examples 2 and 4).

As a branch of practical implementation of the approach presented, an example may be considered referring to the fault detection problem based on system model parameter identification. So, figures 8 and 9 demonstrate, as above, the behavior of current Euclidean square norm of deviation of the current system parameters from the nominal ones $\eta_{dev}^2(t) = (\theta(t) - \theta_{nom})^T (\theta(t) - \theta_{nom})$ of the systems from example 1 (fig. 8) and example 2 (fig. 9) correspondingly. Within the example, before the 2000-th time step the vector of the truth system parameters had been corresponding to the nominal system parameters, while after the 2000-th time step an abrupt change has been appeared. For the case, loose of efficiency of the conventional recursive instrumental variable algorithm is clearly manifested, the dotted line in example 2 (fig 9). This leads to false indications on a possible fault. From another hand side, stable behavior of the algorithm presented is expressed both under nominal system parameters and under a fault appeared.

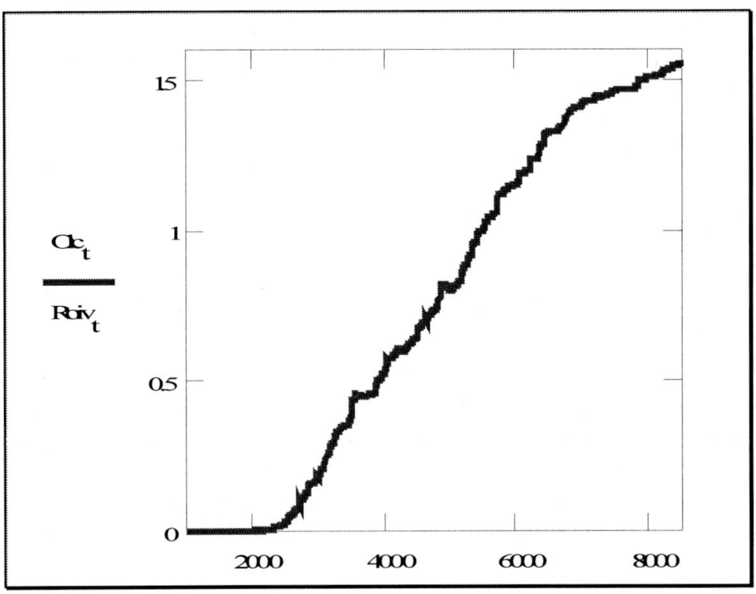

Fig. 8. Fault detection in example 1 system

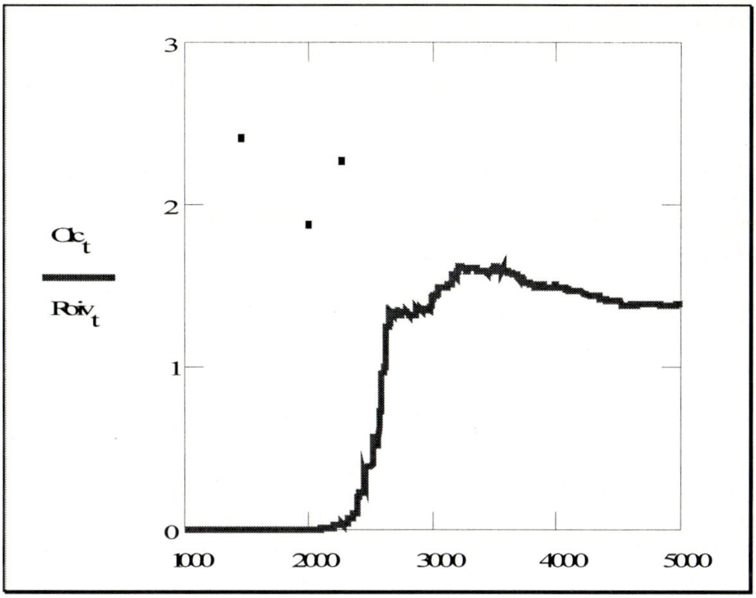

Fig. 9. Fault detection in example 2 system

It is interesting to note that, at fig. 9, the Hessian condition number of the "faulty" system, i.e. after the 2000-th time step, is of order 10. This confirms convergence properties of the conventional recursive instrumental variable algorithm in accordance with the above inference.

4 Conclusions

Algorithmic identification tools have been presented covering general enough classes of input/output stochastic system descriptions. When applied to nonlinear system identification, the algorithms are based on nonparametric approach requiring minimal body of a priori assumptions with respect to the considered system. When applied to linear-in-parameters systems, the corresponding approach enables one to derive computationally efficient recursive algorithms under broad assumptions with respect to external disturbances affecting the considered system. As a basic mathematical tools both within the nonparametric and parametric approaches the *functional* correlation function, i.e. correlation (covariance) of some transformations of the considered processes, has been used. In dependence of a type of investigated system, the corresponding transformations may be chosen in accordance with appropriate criteria. Such a choice leads to corresponding partial types of the functional correlation: the maximal correlation function, the generalized covariance functions.

Both theoretical and computational examples have been presented illustrating the results obtained.

References

1. Rajbman, N.S.: Extensions to nonlinear and minimax approaches. In: Eykhoff, P. (ed.): Trends and Progress in System Identification. Pergamon Press, Oxford (1981) 185-237
2. Renyi, A.: On measures of dependence. Acta Math. Hung. **10** (1959) 441-451
3. Sarmanov, O.V.: Pseudonormal correlation and its various generalizations. Dokl. AN SSSR. **132** (1960) 299-302 (in Russian)
4. Sarmanov, O.V., Bratoeva, Z.N.: Probabilistic properties of bilinear expansions of Hermite polynomials. Theor. Probability Appl. **12** (1967) 470-481
5. Sarmanov, O.V., Zakharov, E.K.: Measures of dependence between random variables and spectra of stochastic kernels and matrices. Matematicheskiy Sbornik. **52(94)** (1960) 953-990 (in Russian)
6. Sarmanov, O.V.: The maximal correlation coefficient (nonsymmetric case). Sel. Trans. Math. Statist. Probability. **4** (1963) 207-210
7. Sarmanov, O.V.: Investigation of stationary Markov processes by the method of eigenfunction expansion. Sel. Trans. Math. Statist. Probability. **4** (1963) 245-269
8. Ljung, L.: System Identification: Theory for the User. 2nd edn. Prentice Hall (1999)
9. Stoica, P., Soderstrom, T.: Optimal instrumental variable estimation and approximate implementations. IEEE Trans. Autom. Control. **AC-28** (1983) 757-772
10. Stoica, P., Soderstrom, T.: Optimal instrumental variable methods for identification of multivariable linear systems. Automatica. **19** (1983) 425-429
11. Soderstrom, T., Stoica, P.: On the generic consistency of instrumental variable estimates. In: Proceedings of the Ninth Triennial World Congress of IFAC. Budapest, Hungary, 2-6 July 1984. Pergamon Press, Oxford (1985) 603-607
12. Ljung, L., Soderstrom, T.: Theory and Practice of Recursive Identification. M.I.T. Press, Cambridge, MA (1983)
13. Friedlander, B.: The overdetermined recursive instrumental variable estimation method. IEEE Trans. Autom. Control. **AC-29** (1984) 353-356

Non Selective Gas Sensors and Artificial Neural Networks – Determination of Gas Mixtures

B.W. Licznerski, P.M. Szecówka, A. Szczurek, and K. Nitsch

Wrocław University of Technology, Institute of Microsystem Technology,
Wrocław 50-370, Wybrzeże Wyspiańskiego 27, Poland

Abstract. The paper presents examples of artificial neural networks approach for analysis of gas sensors responses. The research focused on quantitative analysis of gas mixtures appearing in dry and humid air. Despite difficulties in development of selective gas sensors, application of neural networks as self tuning signal processors provide construction of sensor systems capable of reliable measurements as well as analysis of gas mixtures with reasonable accuracy. Possibility of implementation of neural processing in low-cost devices enables eventual fabrication of microsystems integrating gas sensor matrices with intelligent data processing devices.

1 Introduction

Recent development of adaptive neural networks derivates, to a high degree, from the theory of Wiener's filter. This paper presents practical application of neural networks for identification of gas compounds appearing in mixtures.

Authors first focused on detection of dangerous concentrations of methane and carbon monoxide, which reveal respectively explosive and poisonous properties. Both these gases may appear in houses or boiler-rooms because of leaky installations or improper combustion in furnaces. Facing strong need for alarming or security systems appropriate low-cost solutions should be proposed. The authors came to conclusion that most attractive proposition would be chemical sensors based on metal oxide semiconductors, especially SnO2. Although these sensors reveal poor selectivity and strong dependence on the properties of the atmosphere, especially humidity, still according to the literature their lifetime is the longest and cost of fabrication very low. Thus a few variants of systems containing combinations of methane, carbon monoxide and humidity sensors were proposed. Both qualitative and quantitative analysis systems were considered.

Similar approach to volatile organic compound mixtures analysis was found to be slightly more difficult task. Sensor matrix was composed of commercial TGS 800 series sensors made by Figaro. Long-term experiments involved investigation of sensor matrix reaction for mixtures of compounds changing together with humidity level. Large amount of data collected provided appropriate patterns for development of several variants of neural networks. Eventually quantitative analysis of reasonable accuracy was found possible. Implementation of developed neural network structures in single-chip microcontroller and dedicated digital integrated circuits was considered.

F. Pichler, R. Moreno-Díaz, and P. Kopacek (Eds.): EUROCAST'99, LNCS 1798, pp. 565–572, 2000.

2 Gas Sensors

Gas sensors are classified among chemical sensors. Methane, carbon monoxide and vapours of volatile organic compounds reveal combustible properties. Their interaction with semiconductor SnO2 sensors relies on oxidation – burning of molecules of these gases on the surface of sensor, which then changes its electrical conductance:

$$G = G_0 + ap^n \tag{1}$$

where:

G_0 – conductance of SnO2 in clean atmosphere
a – constant characteristic for the gas (for oxygen it is negative)
p – partial pressure (concentration) of combustible gas
n – exponent, characteristic for particular reaction (e.g. for oxygen could be from 1/6 to 1/2, for methane 1/3, for carbon monoxide 1/2)

Generally the sensors are not selective, although appropriate catalysts added to SnO2 semiconductor together with selected temperature of operation set, provide increase of sensitivity to desired gas. Thus precise control of sensor operating temperature becomes very important issue [1].

Contemporary gas sensors are fabricated with technologies commonly used in microelectronic industry. In this case thick film technology was applied. All sensor elements are screen-printed on ceramic substrate and fired in appropriate high temperatures (Figure 1).

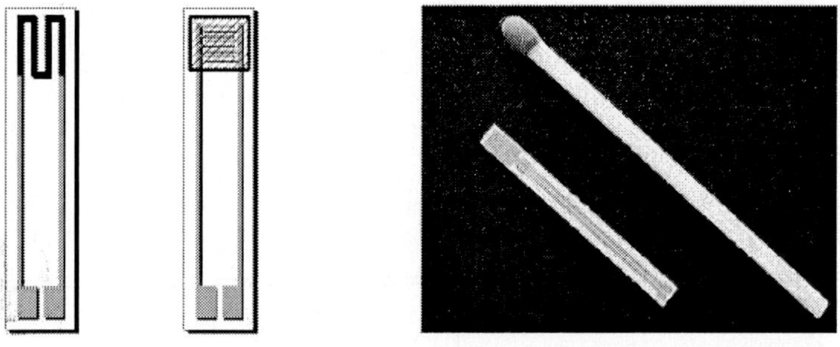

Fig. 1. Design of thick film gas sensor

Before the application sensors are measured with special gas installation providing transport of desired gas mixture to the chamber with sensors (Figure 2) [2].

Typical characteristics of the sensors show sensitivity to selected gas dependence on temperature (Figure 3a) or conductance versus gas concentration for chosen fixed temperature (Figure 3b).

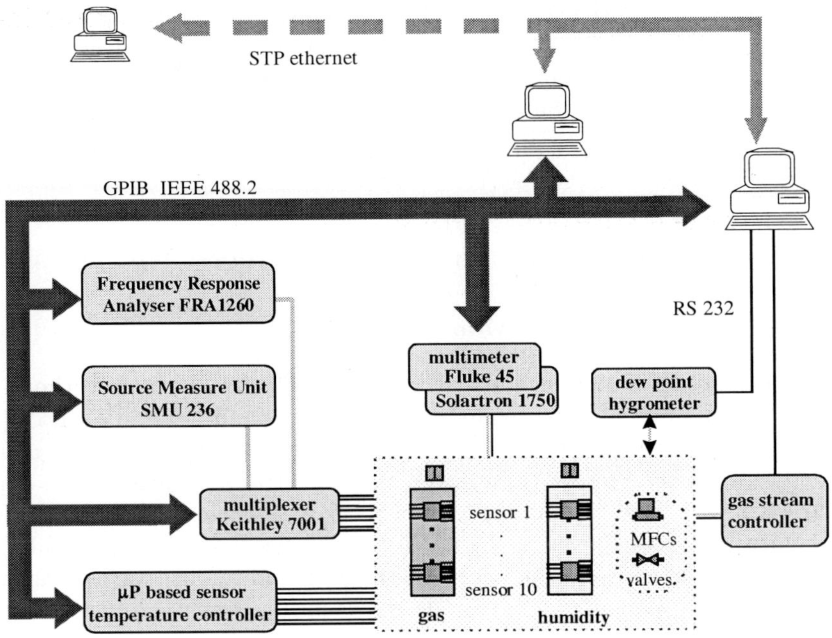

Fig. 2. Automatic test system for gas and humidity sensor testing

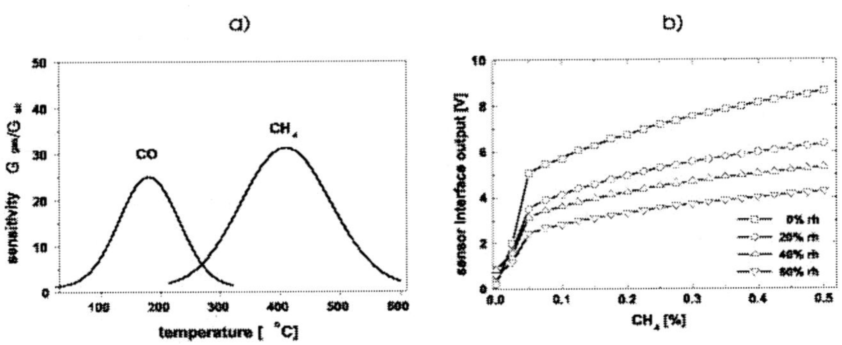

Fig. 3. Responses of the SnO2 thick film sensors, temperature sensitivity – (a), sensor interface output – (b)

Considering the chart presented in Figure 3b it is visible that methane sensor would react for carbon monoxide and carbon monoxide sensor for methane. Additionally both sensors were found to be sensitive to humidity.

In opposition to single gas influence on sensor response, strictly described by equation (1), derivating from theory, there is no adequate formula describing sensors reaction for mixture of compounds. Extensive investigations led by several research groups revealed only complex character of interaction between gas molecules on the surface of the sensor. Similar problem was found with explanation of water molecules impact, i.e. the influence of humidity. Either reverse formulas – providing calculation of gas concentrations from sensor matrix responses have not been developed yet for the general case. These reasons cause growing interest in neural networks application in sensor systems for both qualitative [3], [4], [5] and quantitative [6], [7], [8] analysis.

3 Neural Networks for Analysis of Sensors Responses

3.1 Methane and Carbon Monoxide

For the purpose of methane and carbon monoxide measurements the sensor matrix was composed of appropriate two gas sensors and humidity sensor [8].

The network contained three input units, two hidden layers with 22 neurons in each and three neurons in output layers. Sigmoid transfer function was applied in all the neurons. The training process involving classical error backpropagation (BP) algorithm [9] was performed with Neural Works II software (NeuralWare). After 150 thousand of iterations satisfying local minimumwas reached. Accuracy of neural network responses was tested on 460 patterns containing responses of the three sensors for different mixtures of methane and carbon monoxide, with four humidity levels (half of this data was used directly in training process).

The results of testing are presented in Figures 4 and 5. The two charts show appropriate neural network outputs responses versus real concentrations of the two gases. In both cases the samples contained also another gas. For methane output the inaccuracy does not exceed 7% of the range whilst for carbon monoxide it does not exceed 13%.

3.2 Volatile Organic Compounds

Similar methodology was applied for analysis of different volatile organic compound mixtures containing alcohols and aromatic compounds – benzene, toluene and xylene. Six commercial TGS 800 series sensors made by Figaro were characterised in appropriate mixtures. The experiments revealed similar reactions of all the sensors for all compounds and high sensitivity to humidity. None of these sensor applied alone would be capable of reliable measurements of any vapour if to assume that compounds would appear together or e.g. the humidity would change. This paper focuses on mixtures of butanol and toluene, which are commonly used in organic solvents. Both these compounds are known to impact human health.

After initial analysis of data four sensors were chosen for butanol/toluene mixtures analysis. In-house developed software tool was applied for construction of several variants of neural networks providing translation of sensors responses to continuous values determining concentrations of the two compounds. The best results were obtained for the network containing four input units, two hidden layers with 20 and 30 neurons respectively and output layer with two regular

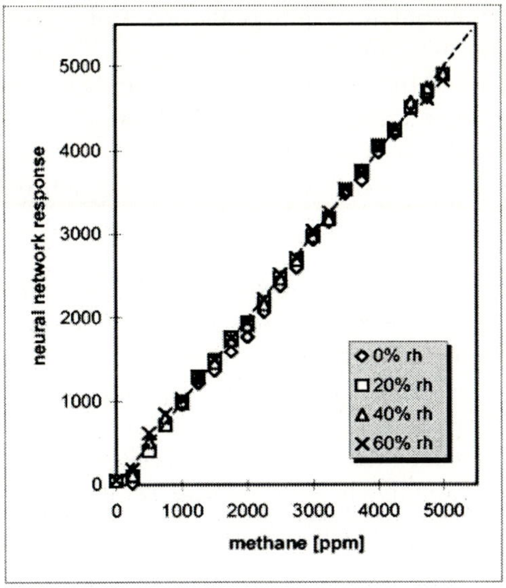

Fig. 4. Neural network responses for methane in mixture

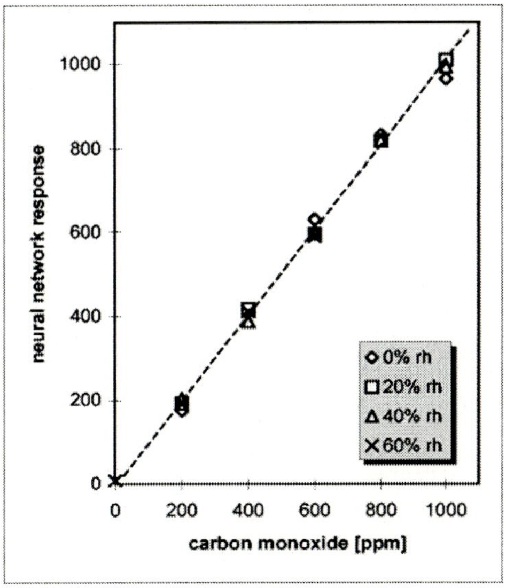

Fig. 5. Neural network responses for carbon monoxide in mixture

neurons. Simple summation of weighted signals and sigmoid transfer function were applied in all the neurons excluding ones from input (dummy) layer. 101 patterns finely distributed in the space of compounds concentrations and humidity were used for network training and whole data set (ca. 150 patterns) for the tests of evaluated structures. Output vectors were scaled to the range of [0.1, 0.9], whilst the input vectors (sensors responses) were left in physical range (0 to 12V). Learning strategy was based on BP algorithm again, with uniform noise and random schedule of patterns presentation. Additional mechanism was implemented for tracking of the training process. Periodic memorising of the best actually reached result provided comfortable way of long-term network evaluation with significant reduction of overtraining danger.

The learning ratio was initially set to 0.1 and then decreased to 0.001. Strong momentum ratio (0.8) was applied. Evaluation of the network took 7.8 million iterations. Such a long time was found as a kind of optimum by the mentioned tracking procedure, although there were also smaller structures of not much worse quality obtained after e.g. 200 thousand iterations.

The test results of the best solution reached are presented in Figures 6 and 7. Horizontal axes denote real concentrations of each compound in the mixture, while vertical axes stand for the responses of each output unit of the neural network (responses are scaled again to physical range). The neural network provides responses with inaccuracy lower than 14% of range for toluene output (average 2.3%) and lower than 7.6% for butanol output (average 1.5%).

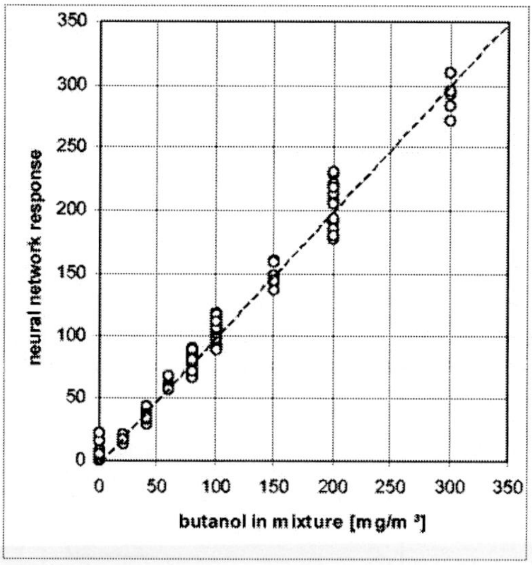

Fig. 6. Butanol dedicated output responses for butanol and toluene mixtures

These parameters relate to each output for all assumed concentrations of the other ("noisy") component and all tested levels of humidity.

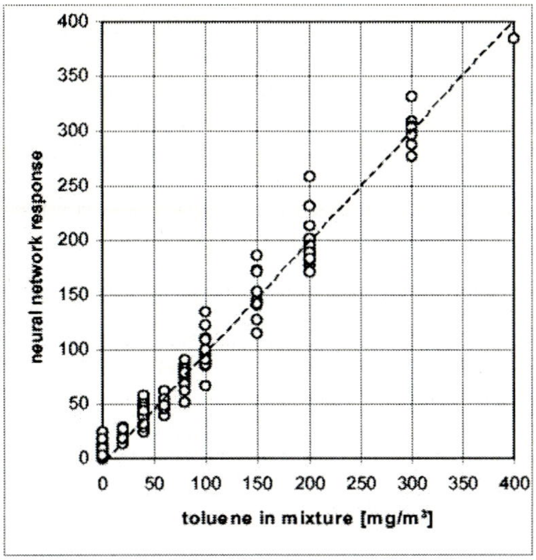

Fig. 7. Toluene dedicated output responses for butanol and toluene mixtures

3.3 Neural Networks Implementation

Considering off-computer implementation of the neural networks, additional software was created, providing automatic generation of appropriate function code in standard C language. The neural network for methane/carbon monoxide mixtures analysis was successfully implemented in Intel 8051 compatible microcontroller [11].

Experiments on similar tool based on VHDL language were also performed, aiming in ASIC realisation of neural networks for the purposes of microsystem technology.

4 Conclusions

Two sensor systems for quantitative analysis of gas mixtures were presented. Reasonable accuracy of responses make them potential replacement of traditional devices, usually providing better quality but extremely expensive and troublesome in off-laboratory operation.

It was shown that tin oxide based semiconductor gas sensors may be successfully applied for the tasks which are far beyond the vendors expects, however somewhat sophisticated methods of data processing seem to be necessary. On the other hand, in such applications, poor selectivity of sensors reverses to significant advantage.

References

1. Licznerski B.W., Nitsch K., Teterycz H., Szecówka P.M., Wioniewski K., Humidity Insensitive Thick Film Methane Sensor, Proc. 12th European Conference on Solid State Transducers Eurosensors XII, Southampton, UK, Sept. 13-16, 1998, pp. 493-496.
2. Teterycz H., Licznerski B. W., Nitsch K., Wioniewski K., Golonka L. J., Anomalous behaviour of new thick film gas sensitive composition, Sensor and Actuators B. Chemical 47, 1998, 152-156.
3. Di Natale C., Macagnano A., Mantini A., Davide F., D'Amico A., Paolesse R., Boschi T., Faccio M. and Ferri G., Advances in Food Analysis by Electronic Nose, Proc. IEEE International Symposium on Industrial Electronics, Guimaraes, Portugal 1997, SS122-SS127.
4. Gardner J. W., Pearce T. C., Friel S., Bartlett P. N. and Blair N., A multisensor system for beer avour monitoring using an array of conducting polymers and predictive classifiers, Sensors and Actuators B 18-19, 1994, 240-243.
5. Tan T., Loubet F., Labreche S. and Amine H., Quality Control of Coffee Using the FOX4000 Electronic Nose, Proc. IEEE International Symposium on Industrial Electronics, Guimaraes, Portugal 1997, SS140- SS145.
6. Gutierez F. J., Ares L., Robla J., Horillo M. C., Sayago I., Getino J. and Garcia C., Integrated Sensors for Monitoring Contaminant Gases in Atmospheres and Soils, Proc. IEEE International Symposium on Industrial Electronics, Guimaraes, Portugal 1997, SS113-115.
7. Ulmer H., Mitrovics J., Noetzel G., Weimar U. and Gopel W., Odours and avours identified with hybrid modular sensor system, Sensors and Actuators B 43 1997, 24-33.
8. Huyberechts G., Szecówka P. M., Roggen J. and Licznerski B. W., Simultaneous quantification of carbon monoxide and methane in humid air using a sensor array and artificial neural network, Sensors and Actuators B, 45, 1997, 123-130.
9. Rumelhart D. E., Hinton G. E. and Williams R. J., Learning Representations by Back-Propagating Errors, Nature 322, 1986, 533-536.
10. Figaro Gas Sensors, (Figaro Engineering Inc., 1-5-3 Senbanishi, Mino, Osaka 562, Japan), catalogue.
11. Janiczek J., Stepien S., Licznerski B. W., Szecówka P. M. and Huyberechts G., Implementation of gas sensors responses processing neural network on Siemens 8xC515 microcontroller, Proc. of the Third Conference Neural Networks and Their Applications, Kule-Czestochowa, Poland 1997, 570-575.

The Supervision of Hybrid Control Systems – A Layered Architecture

Virginia Ecaterina Oltean, Theodor Borangiu, and Mitică Manu

"Politehnica" University of Bucharest, Faculty of Control and Computers
Spl. Independentei 313, sector 6, 77206 Bucharest, Romania
olteana@aii.pub.ro, {borangiu, mitica}@icar.cimr.pub.ro

Abstract. The scenario treated in this paper concerns the discrete event control and supervision of two decoupled continuous systems. The theoretical framework is based on the hybrid control system architecture, developed by P.J. Antsaklis and his co-workers and on the supervised control concept, proposed by the Sylodi Group from Grenoble. The main contribution of this paper is an algorithm for building the discrete event approximation of a continuous system with unknown, constant and constrained parameters. The Antsaklis formalism is extended to a disturbed continuous first order system.

1 Introduction

This contribution proposes a scenario and a layered architecture for the supervision of two hybrid control systems (HCS) that can work independently without supervision. The motivation of this study is an introductory discussion concerning some problems arising in the modeling and design of complex systems, implying combined continuous and discrete approaches.

The supervised control of discrete event systems (DES) has been defined by the researchers of the Sylodi Group from the Automation Laboratory of Grenoble, as an extension of the Ramadge-Wonham supervisory theory of DES [1], [2]. In fig.1, the events from Σ_{pr} are generated by the process and the events from Σ_{co} are generated by the logic controller. The controller forces some events in the process to occur, while the supervisor prevents some events from Σ_{co} to occur. The control and supervision tasks are separated.

The HCS structure considered in this paper is a variant of the framework proposed by Antsaklis and his co-workers from the ISIS Group, and it comprises a continuous plant that is controlled, through an interface, by a DES (fig.2) [3]. The plant and the interface are first abstracted to a DES, called the DES-plant. Then the controller is built as a Moore machine, by adapting the techniques from the Ramadge-Wonham DES control theory [4], [5].

The example presented below combines these two approaches. The plant comprises the undisturbed levels dynamics in a two tanks system and the disturbed temperature dynamics in the first tank. The intuitive description of the plant is presented in section 2. The HCS for the levels and temperature are developed in

F. Pichler, R. Moreno-Díaz, and P. Kopacek (Eds.): EUROCAST'99, LNCS 1798, pp. 573–587, 2000.

sections 3 and 4 respectively. In section 5, the supervisor will solve an additional logical restriction, imposed to the modular extended hybrid process represented by the two HCS.

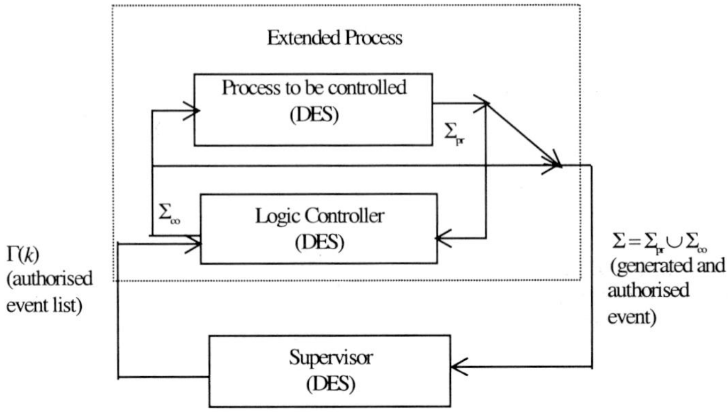

Fig. 1. The supervised control of a DES [1]

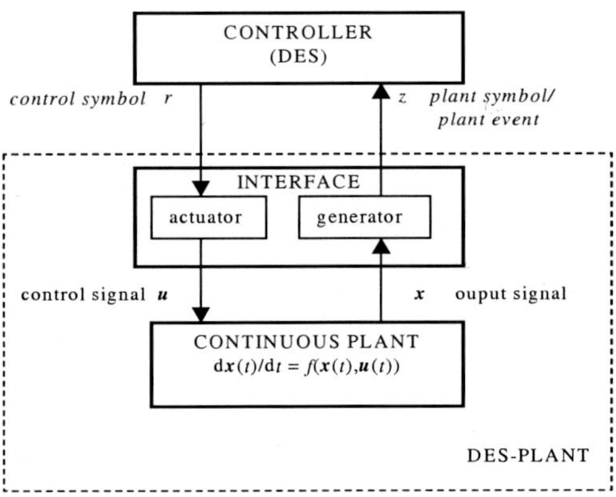

Fig. 2. The architecture of a HCS [3]

2 The Continuous Plant – An Intuitive Description

The plant is represented by a two tanks filling process and an additional first order system, describing the evolution law of a property of the liquid in the first tank, for example the temperature (fig.3). The plant is equipped with level and temperature threshold sensors. u_1 and u_2 are the control signals (1 = on, 0 = off) for the valves V1 and V2. V3 remains open. u_3 is the control signal for the temperature actuator TE and it can switch between the values –D, 0 or D, with D > 0 a given value. The continuous filling process and the temperature evolution are independent.

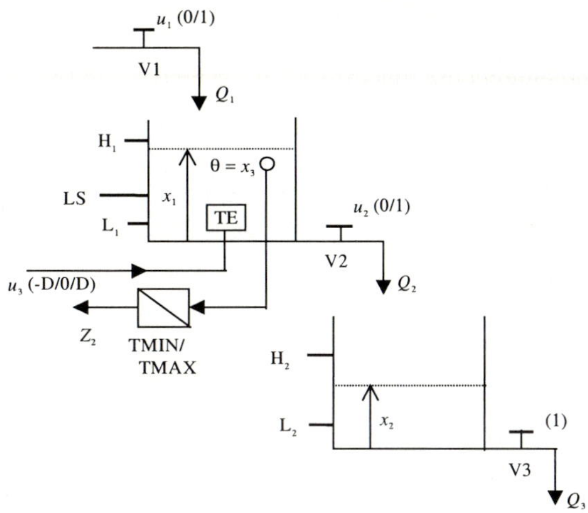

Fig. 3. The plant equipped with threshold sensors

3 The Unsupervised Discrete Event Control of the Levels

The HCS associated to the levels dynamics has the structure depicted in fig.2. The generator, the DES-plant model and the controller have to be synthesized, starting from a primal control objective.

3.1 The State Equations of the Levels Dynamics

In fig.3, x_1 and x_2 are the liquid levels in tank 1 and 2 respectively and Q_1, Q_2, Q_3 are the flows. Denote |R the set of real numbers. The levels dynamics is described by the differential system

$$\dot{\boldsymbol{x}} = f(\boldsymbol{x},\boldsymbol{u}), \text{ with } f(\boldsymbol{x},\boldsymbol{u}) = [-\mathrm{a}x_1u_2 + \mathrm{b}u_1 \quad \mathrm{a}x_1u_2 - \mathrm{c}x_2]^T, \tag{1}$$

where $\boldsymbol{x} = [x_1 \quad x_2]^T \in |\mathrm{R}^2$ is the state vector, $\boldsymbol{u} = [u_1 \quad u_2]^T \in |\mathrm{R}^2$ is the control vector, a, b, c $\in$ |R are parameters, $Q_1 = \mathrm{b}u_1$, $Q_2 = \mathrm{a}x_1u_2$ and $Q_3 = \mathrm{c}x_2$.

The parameters a, b, c together with the threshold limits L_1, L_2, H_1 and H_2 (fig.3) satisfy the restriction

$$(Res): \mathrm{a} = \mathrm{c} > 0,\ 0 < \mathrm{L}_1 = \mathrm{L}_2 < \mathrm{H}_1 = \mathrm{H}_2,\ \mathrm{aL}_1 < \mathrm{b} < \mathrm{aH}_1. \tag{2}$$

3.2 The Actuator

The control vector $\boldsymbol{u}$ takes values in the set $U_1 = \{[1\ \ 0]^T, [0\ \ 1]^T, [0\ \ 0]^T, [1\ \ 1]^T\}$. Consider the alphabet of control symbols

$$R_1 = \{r_1, r_2, r_3, r_4\}. \tag{3}$$

The *actuator* implements the function $\gamma_1 : R_1 \to U_1$ defined by

$$\gamma_1(r_1) = [1\ \ 0]^T,\ \gamma_1(r_2) = [0\ \ 1]^T,\ \gamma_1(r_3) = [0\ \ 0]^T,\ \gamma_1(r_4) = [1\ \ 1]^T. \tag{4}$$

Denote $k \in |\mathrm{N} = \{0, 1, 2, \ldots\}$ the logical time variable. A sequence of control symbols $w_r = r(0),r(1),\ldots,r(k), \ldots, r(k) \in R_1, \forall k \in |\mathrm{N}$, generates a piecewise constant control signal

$$\boldsymbol{u}(t) = [u_1(t)\ \ u_2(t)]^T = \sum_{k\geq 0} \gamma_1(r(k)).I(t,t_c(k),t_c(k{+}1)), \tag{5}$$

where $t_c(k) \in |\mathrm{R}$ is the moment when $r(k)$ is received from the DES controller, $t_c(k) < t_c(k+1)$, $\forall k \in |\mathrm{N}$ and $I : |\mathrm{R} \times |\mathrm{R} \times |\mathrm{R} \to \{0, 1\}$ is a characteristic function defined by $I(t,t_1,t_2) = 1$, if $t_1 \leq t < t_2$ and $I(t,t_1,t_2) = 1$ if else.

3.3 The Generator

The primal control objective. *The primal control objective of the levels dynamics* is to drive the state vector $\boldsymbol{x} = [x_1\ \ x_2]^T$, by means of a control signal $\boldsymbol{u}(.)$ (5), in order to satisfy the *string of restrictions* $S_{c1} = c(0),c(1),c(2)$, $\forall \boldsymbol{x}(0) \in c(0)$, where $c(0)$, $c(1)$ and $c(2)$ are open regions in $|\mathrm{R}^2$ defined as follows:

$$\begin{aligned} &c(0) : 0 < x_1 < \mathrm{L}_1, 0 < x_2 < \mathrm{L}_2\ ;\ \ c(1) : \mathrm{L}_1 < x_1 < \mathrm{H}_1, 0 < x_2 < \mathrm{L}_2\ ; \\ &c(2) : \mathrm{L}_1 < x_1 < \mathrm{H}_1, \mathrm{L}_2 < x_2 < \mathrm{H}_2. \end{aligned} \tag{6}$$

The state space partition. Based on (6), consider the smooth functionals $h_i : |\mathrm{R}^2 \to |\mathrm{R}$, $i = 1, 2, 3, 4$, defined as follows:

$$h_1(\boldsymbol{x}) = x_1 - \mathrm{L}_1,\ h_2(\boldsymbol{x}) = -x_1 + \mathrm{H}_1,\ h_3(\boldsymbol{x}) = x_2 - \mathrm{L}_2,\ h_4(\boldsymbol{x}) = -x_2 + \mathrm{H}_2. \tag{7}$$

Denote $S_h^4 = \{h_i : |\mathrm{R}^2 \to |\mathrm{R} \mid i = 1, 2, 3, 4\}$. $\forall h_i \in S_h^4$, h_i separates $|\mathrm{R}^2$ into two open halfspaces and $\mathrm{Ker}(h_i) = \{\boldsymbol{x} \in |\mathrm{R}^2 \mid h_i(\boldsymbol{x}) = 0\}$ (fig.4) is a nonsingular hypersurface. The gradients of the hypersurfaces are constant, so they don't depend on the current $\boldsymbol{x}$:

$$\mathrm{grad}(h_1) = [1\ \ 0]^T,\ \mathrm{grad}(h_2) = [-1\ \ 0]^T,\ \mathrm{grad}(h_3) = [0\ \ 1]^T,\ \mathrm{grad}(h_4) = [0\ \ -1]^T. \tag{8}$$

Define $\mathrm{sgn} : |\mathrm{R} \to |\mathrm{R}$, $\mathrm{sgn}(y) = -1$, if $y < 0$, $\mathrm{sgn}(y) = 0$, if $y = 0$ and $\mathrm{sgn}(y) = 1$, if $y > 0$. The *quality function* $b : |\mathrm{R}^2 \to \{-1, 0, 1\}^4$ is defined by

$$b(\boldsymbol{x}) = [\mathrm{sgn}(h_1(\boldsymbol{x}))\ \mathrm{sgn}(h_2(\boldsymbol{x}))\ \mathrm{sgn}(h_3(\boldsymbol{x}))\ \mathrm{sgn}(h_4(\boldsymbol{x}))]. \tag{9}$$

The value $b(\boldsymbol{x})$ is *consistent* if and only if $\mathrm{sgn}(h_i(\boldsymbol{x})) \neq 0$, $\forall h_i \in S_h^4$ and *inconsistent* if else [5]. Consider $DX = |\mathrm{R}^2 \setminus Fr$, with $Fr = \bigcup_{i=1}^4 \mathrm{Ker}(h_i)$.

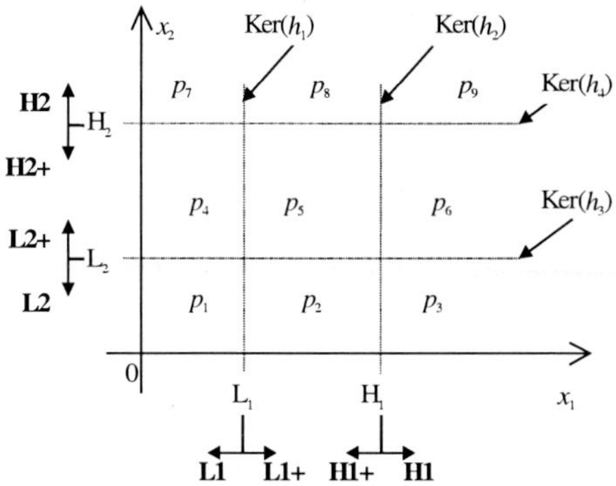

Fig. 4. The state space partition of the filling process in fig.3 and the associated plant symbols

Table 1. The significance of the discrete states of the levels DES-plant model

The alphabet P_1	The cellular space $C = DX_{/Rel}$	The set of consistent quality values B
p_1	$c_1 = \{x \in \|R^2\| h_1(x) < 0, h_2(x) > 0, h_3(x) < 0, h_4(x) > 0\}$	$b_1 = [-1\ +1\ -1\ +1]$
p_2	$c_2 = \{x \in \|R^2\| h_1(x) > 0, h_2(x) > 0, h_3(x) < 0, h_4(x) > 0\}$	$b_2 = [+1\ +1\ -1\ +1]$
p_3	$c_3 = \{x \in \|R^2\| h_1(x) > 0, h_2(x) < 0, h_3(x) < 0, h_4(x) > 0\}$	$b_3 = [+1\ -1\ -1\ +1]$
p_4	$c_4 = \{x \in \|R^2\| h_1(x) < 0, h_2(x) > 0, h_3(x) > 0, h_4(x) > 0\}$	$b_4 = [-1\ +1\ +1\ +1]$
p_5	$c_5 = \{x \in \|R^2\| h_1(x) > 0, h_2(x) > 0, h_3(x) > 0, h_4(x) > 0\}$	$b_5 = [+1\ +1\ +1\ +1]$
p_6	$c_6 = \{x \in \|R^2\| h_1(x) > 0, h_2(x) < 0, h_3(x) > 0, h_4(x) > 0\}$	$b_6 = [+1\ -1\ +1\ +1]$
p_7	$c_7 = \{x \in \|R^2\| h_1(x) < 0, h_2(x) > 0, h_3(x) > 0, h_4(x) < 0\}$	$b_7 = [-1\ +1\ +1\ -1]$
p_8	$c_8 = \{x \in \|R^2\| h_1(x) > 0, h_2(x) > 0, h_3(x) > 0, h_4(x) < 0\}$	$b_8 = [+1\ +1\ +1\ -1]$
p_9	$c_9 = \{x \in \|R^2\| h_1(x) > 0, h_2(x) < 0, h_3(x) > 0, h_4(x) < 0\}$	$b_9 = [+1\ -1\ +1\ -1]$

The *equivalence relation* induced by S_h^4 is $Rel \subset DX \times DX$, defined by

$$[x_a \;\; x_b]^T \in Rel \Leftrightarrow h_i(x_a)h_i(x_b) > 0, \forall h_i \in S_h^4. \tag{10}$$

The *cellular space* or the *state space partition* $DX_{/Rel} = C$ is the set of all *classes of equivalence* of the relation *Rel*. Define $Q = \text{card}(C) = 9$ and $I_Q = \{1, 2, \ldots, 9\}$. The *alphabet of discrete states* of the DES-plant is a set of Q distinct indexed symbols (see fig.4 and Table 1)

$$P_1 = \{p_1, \ldots, p_9\}. \tag{11}$$

The map $et : C \to P_1$, $et(c_q) = p_q$, $\forall q \in I_Q$ is the *label function* of C. $\forall c_q \in C$, $b(x)$ is constant and consistent, $\forall x \in c_q$. Denote $b(x) = b_q = [b_q^1 \;\; b_q^2 \;\; b_q^3 \;\; b_q^4]$, $\forall x \in c_q \in C$ and define $B = \{b_1, \ldots, b_9\}$ the set of all consistent quality values (Table 1). The bijective maps $ech : C \to B$, $ech(c_q) = b_q$, $\forall q \in I_Q$ and $et_B : B \to P$, $et_B(b_q) = p_q$, $\forall q \in I_Q$ have the property $et_B \circ ech = et$ [5].

The discrete states p_q, $p_s \in P_1$ are *adjacent* if $\exists h_i \in S_h^4$ s.t. the vectors $\boldsymbol{b}_q = et_B^{-1}(p_q)$ and $\boldsymbol{b}_s = et_B^{-1}(p_s)$ satisfy the relations $b_q^i b_s^i = -1$ and $b_q^j b_s^j = 1$, $\forall j \neq i, j \in \{1, 2, 3, 4\}$[3]. The *open adjacency frontier* is $A(h_i,p_q,p_s) = \{\boldsymbol{x} \in \mathbb{R}^2 \mid \text{sgn}(h_i(\boldsymbol{x})) = 0 \text{ and } \text{sgn}(h_j(\boldsymbol{x})) = b_q^j, \forall j \neq i, j \in \{1, 2, 3, 4\}\} = A(h_i,p_s,p_q) \subseteq \text{Ker}(h_i)$ [5].

The speed of the state vector of the system (1) can have four distinct expressions:

$$\begin{aligned} f_1(\boldsymbol{x}) &\equiv f(\boldsymbol{x},\gamma_1(r_1)) = [\text{b} \quad \text{-cx}_2]^T; \quad f_2(\boldsymbol{x}) \equiv f(\mathbf{x},\gamma_1(r_2)) = [\text{-ax}_1 \quad \text{ax}_1\text{-cx}_2]^T \\ f_3(\boldsymbol{x}) &\equiv f(\boldsymbol{x},\gamma_1(r_3)) = [0 \quad \text{-cx}_2]^T; \quad f_4(\boldsymbol{x}) \equiv f(\boldsymbol{x},\gamma_1(r_4)) = [\text{-ax}_1\text{+b} \quad \text{ax}_1\text{-cx}_2]^T. \end{aligned} \tag{12}$$

Property 1. $\forall p_q$, $p_s \in P_1$ *that are* adjacent *on* $\text{Ker}(h_i)$, $h_i \in S_h^4$ *and* $\forall r_m \in R_1$, *the following relation is satisfied*: $\text{sgn}(f_m^T(\underline{\boldsymbol{x}}).\text{grad}(h_i)) = const.$, $\forall \underline{\boldsymbol{x}} \in A(h_i,p_q,p_s)$.

The above property can be directly tested and it holds because *Res* (2) is satisfied.

Property 2 [4]. $\forall p_q$, $p_s \in P_1$, *if* p_q *and* p_s *are* adjacent *on* $\text{Ker}(h_i)$, $h_i \in S_h^4$, *then* p_s *is the* only *discrete state that is adjacent on* $\text{Ker}(h_i)$ *to* p_q.

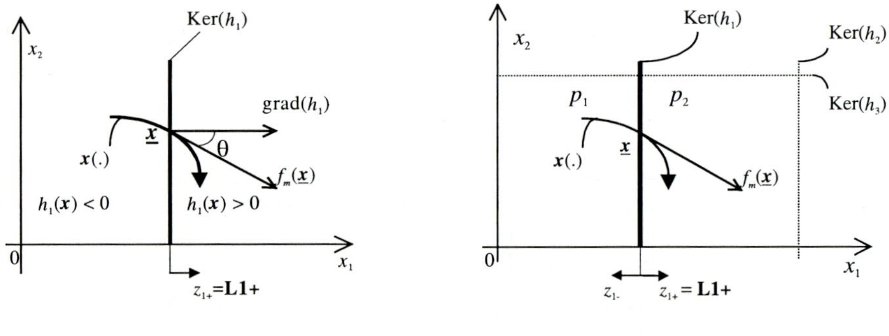

Fig. 5. Examples : a) the plant event (1+) occurs if $f_m(\underline{\boldsymbol{x}})^T.\text{grad}(h_1) > 0$, with $\underline{\boldsymbol{x}} \in \text{Ker}(h_1)$; b) the discrete state transition $p_1 \rightarrow p_2$ is observed as the occurrence of the plant symbol $z_{1+} =$ **L1+**

Plant events and plant symbols. Consider a functional $h_i : \mathbb{R}^2 \rightarrow \mathbb{R}$, $h_i \in S_h^4$ and $\boldsymbol{x}(.)$ a continuous evolution of the plant model (1).

According to [3], the *plant-event* (*i*+) (or (*i*-)) occurs at $t_e \in \mathbb{R}$ if the following conditions are satisfied: a) $h_i(\boldsymbol{x}(t_e)) = 0$; b) $\exists \text{d}_1 > 0$ s.t. for all ε, $0 < \varepsilon < \text{d}_1$, $h_i(\boldsymbol{x}(t_e+\varepsilon)) > 0$ (respectively $h_i(\boldsymbol{x}(t_e+\varepsilon)) < 0$)) ; c) $\exists \text{d}_2 > 0$ s.t. for all ε, $0 < \varepsilon < \text{d}_2$, $h_i(\boldsymbol{x}(t_e-\text{d}_2)) < 0$ and $h_i(\boldsymbol{x}(t_e-\varepsilon)) \leq 0$ (respectively $h_i(\boldsymbol{x}(t_e-\text{d}_2)) > 0$ and $h_i(\boldsymbol{x}(t_e-\varepsilon)) \geq 0$). Eight distinct plant events can be defined with respect to the functionals from S_h^4. Denote the set of plant events $E_a = \{(1+), (1-), (2+), (2-), (3+), (3-), (4+), (4-)\}$. The associated *plant symbols* $z_{1+} =$ **L1+**, $z_{1-} =$ **L1**, $z_{2+} =$ **H1+**, $z_{2-} =$ **H1**, $z_{3+} =$ **L2+**, $z_{3-} =$ **L2**, $z_{4+} =$ **H2+**, $z_{4-} =$ **H2** (fig.4) form the alphabet

$$Z_1 = \{z_{1+}, z_{1-}, z_{2+}, z_{2-}, z_{3+}, z_{3-}, z_{4+}, z_{4-}\} \tag{13}$$

s.t $\forall i \in \{1, 2, 3, 4\}$, z_{i+} labels uniquely $(i+) \in E_a$ and z_{i-} labels uniquely $(i-) \in E_a$.

Hypothesis 1. $\forall h_i \in S_h^4$ and $\forall \boldsymbol{x}(.)$ a continuous trajectory of the plant (1), if $\exists t_e \in \mathbb{R}$ s.t. $h_i(\boldsymbol{x}(t_e)) = 0$ and $\dfrac{\mathrm{d}h_i(\boldsymbol{x}(t_e))}{\mathrm{d}t} = 0$, then t_e is a *local extremum* of $h_i(\boldsymbol{x}(.))$ and $\boldsymbol{x}(.)$ does *not* cross $\mathrm{Ker}(h_i)$ at t_e.

Proposition 1 [5]. *Consider* $h_i \in S_h^4$ *and assume that* hypothesis 1 *is true. The state trajectory* $\boldsymbol{x}(.)$ *of* (1), *controlled by* $\boldsymbol{u}(.)$ (5), *can produce at* $t_e \in \mathbb{R}$ *the plant event :*

A) *(i+)* $\Leftrightarrow$ *(a1)* $h_i(\boldsymbol{x}(t_e)) = 0$ *and* $\dfrac{\mathrm{d}h_i(\boldsymbol{x}(t_e))}{\mathrm{d}t} > 0$

$\Leftrightarrow$ *(a2)* $\exists \underline{\boldsymbol{x}} \in \mathrm{Ker}(h_i)$ *and* $r_m \in R_1$ *s.t.* $\boldsymbol{x}(t_e) = \underline{\boldsymbol{x}}$, $\boldsymbol{u}(t) = \gamma_1(r_m)$, $\forall t \in (t_e^-, t_e^+)$ *and* $f_m^T(\underline{\boldsymbol{x}}).\mathrm{grad}(h_i) > 0$ *or*

B) *(i-)* $\Leftrightarrow$ *(b1)* $h_i(\boldsymbol{x}(t_e)) = 0$ *and* $\dfrac{\mathrm{d}h_i(\boldsymbol{x}(t_e))}{\mathrm{d}t} < 0$

$\Leftrightarrow$ *(b2)* $\exists \underline{\boldsymbol{x}} \in \mathrm{Ker}(h_i)$ *and* $r_m \in R_1$ *s.t.* $\boldsymbol{x}(t_e) = \underline{\boldsymbol{x}}$, $\boldsymbol{u}(t) = \gamma_1(r_m)$, $\forall t \in (t_e^-, t_e^+)$ *and* $f_m^T(\boldsymbol{x}).\mathrm{grad}(h_i) < 0$,

where $f_m(\underline{\boldsymbol{x}}) = f(\boldsymbol{x}(t_e),\boldsymbol{u}(t_e))$ *is the speed of the continuous state of* (1) *at* t_e (fig.5(a)).

Proof (of proposition 1). a) Only the basic ideas of the proof are presented. Consider $\boldsymbol{x}(.)$ a continuous-time evolution of (1), controlled by a signal $\boldsymbol{u}(.)$ (5) and $t_e \in \mathbb{R}$ s.t. $\boldsymbol{u}(t) = \gamma_1(r_m)$, $\forall t \in (t_e^-, t_e^+)$), i.e. $\boldsymbol{u}(.)$ does *not* switch at t_e. Consequently, the function $(h_i \circ \boldsymbol{x})(.) = h_i(\boldsymbol{x}(.))$ is smooth on (t_e^-, t_e^+) and $\forall t \in (t_e^-, t_e^+)$, the following relation holds:

$$\frac{\mathrm{d}h_i(\boldsymbol{x}(t))}{\mathrm{d}t} = \sum_{j=1}^{2} \frac{\partial h_i}{\partial x_j}(\boldsymbol{x}(t)) \frac{\mathrm{d}x_j(t)}{\mathrm{d}t} = (\dot{\boldsymbol{x}}(t))^T \mathrm{grad}(h_i) = f^T(\boldsymbol{x}(t),\boldsymbol{u}(t)).\mathrm{grad}(h_i)\,. \tag{14}$$

Hence *(a1)* $\Leftrightarrow$ *(a2)*. *(a1)* $\Rightarrow$ (*i*+), according to the general definition of a plant event. (*i*+) $\Rightarrow$ *(a1)* because *hypothesis 1* holds, which means that if (*i*+) occurs, then *(a1)* is satisfied. If $h_i(\boldsymbol{x}(t_e)) = 0$ and $\dfrac{\mathrm{d}h_i(\boldsymbol{x}(t_e))}{\mathrm{d}t} = 0$ then (*i*+) cannot occur because, in *hypothesis 1*, $\boldsymbol{x}(.)$ does *not* cross $\mathrm{Ker}(h_i)$ and it returns to the halfspace $H_i^- = \{\boldsymbol{x} \in \mathbb{R}^2 \mid h_i(\boldsymbol{x}) < 0\}$.

b) The proof is similar. □

Hypothesis 2. The plant events do *not* occur simultaneously.

Consider $Fr = \bigcup_{i=1}^{4} \mathrm{Ker}(h_i)$ and define the set of all intersection points of the hypersurfaces of S_h^4, $Int = \{\boldsymbol{x} \in Fr \mid \exists h_i, h_j \in S_h^4, h_i \neq h_j, \text{ s.t. } \boldsymbol{x} \in \mathrm{Ker}(h_i) \cap \mathrm{Ker}(h_j)\}$.

In *hypothesis 2*, the *generator* implements the function $gev : Fr \setminus Int \to Z_1$, defined s.t. $gev(\boldsymbol{x}(t)) = z_{i+}$ (or z_{i-}) if (*i*+) (or (*i*-), respectively) occurs at $t \in \mathbb{R}$.

A sequence of plant events is denoted $w_e = e(1),e(2),\ldots,e(k), \ldots$ with $e(k) \in E_a$, $\forall k \in \mathbb{N}$, $k \neq 0$, and $t_e(k) \in \mathbb{R}$ the moment when $e(k)$ occurs, s.t. $t_e(k) < t_e(k+1)$, $\forall k \in \mathbb{N}$, $k \neq 0$. $t_e(0)=0$ is the moment when the external event "initialization" occurs.

w_e produces a sequence of plant symbols $w_z = z(1),z(2),\ldots,z(k), \ldots$ s.t. $z(k) = gev(\boldsymbol{x}(t_e(k))) \in Z_1$, $\forall k \in \mathbb{N} \setminus \{0\}$.

3.4 The DES-Plant Automaton G_{p1} and the DES Controller G_{c1}

In this section, the DES-plant model G_{p1} is built first, *without integrating* the differential system (1). The DES controller for the levels dynamics is synthesized next, based on G_{p1}.

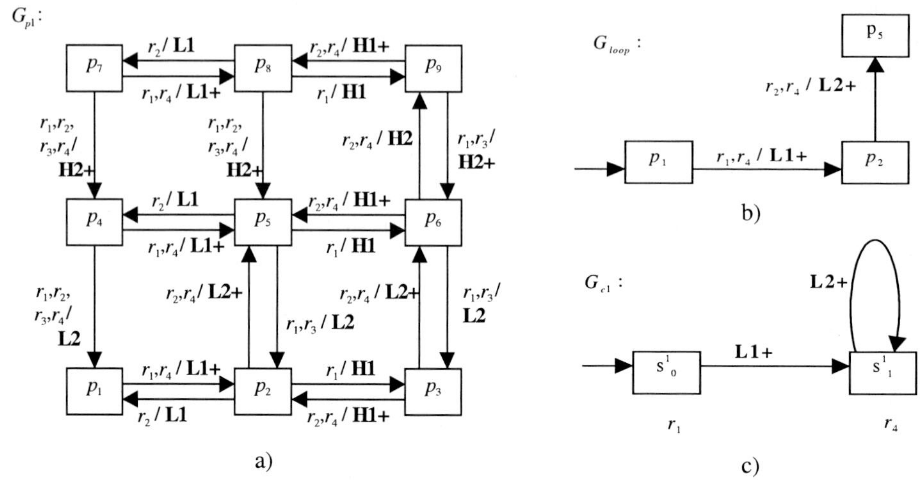

Fig. 6. a) The DES-plant automaton associated to the levels dynamic in fig.3; b) the DES model $G_{loop} \subset G_{p1}$ with the desired evolution $p_1p_2p_5$; c) the DES controller of the levels dynamics

The *DES-plant model* associated to the state equations (1) is a nondeterministic automaton $G_{p1} = \{P_1, R_1, f_{p1}, Z_1, g_{p1}\}$, where P_1 (10) is the set of discrete states, R_1 (3) is the input alphabet of control symbols, Z_1 (13) is the output alphabet of plant symbols, $f_{p1} : P_1 \times R_1 \rightarrow 2^{P_1}$ is the state transition function and $g_{p1} : P_1 \times P_1 \rightarrow Z_1$ is the output function. The dynamic equations are

$$p(k+1) \in f_{p1}(p(k),r(k)),\ g_{p1}(p(k),p(k+1)) = z(k+1), \tag{15}$$

where $p(k) \in P_1$, $z(k+1) \in Z_1$, $r(k) \in R_1$ and $p(k) \neq p(k+1)$, $\forall k \in \mathbb{N}$ [5].

Consider an evolution of G_{p1}, $w_{p1} = p(0),p(1),\ldots,p(k), \ldots$, with $p(k) \in P_1$, $\forall k \in \mathbb{N}$, and $t_e(k+1) \in \mathbb{R}$ the moment when the transition $p(k) \rightarrow p(k+1)$ occurs, observed as $z(k+1) \in Z_1$ (fig.5(b)). G_{p1} is in $p(k+1)$ if $\lim_{\varepsilon \to 0} \boldsymbol{x}(t_e(k+1)+\varepsilon) \in et^{-1}(p(k+1))$, with $\boldsymbol{x}(t_e(k+1)) \in \mathrm{Ker}(h_i)$ and $h_i \in S_h^4$ defining the *adjacency frontier* between $p(k)$ and $p(k+1)$. w_{p1} is *controlled* by a sequence of control symbols $w_{r1} = r(0),r(1),\ldots,r(k), \ldots$, $r(k) \in R_1$, $\forall k \in \mathbb{N}$, and is *observed* as a sequence of plant symbols w_{z1} = $z(1),z(2),\ldots,z(k+1), \ldots, z(k+1) \in Z_1$, $\forall k \in \mathbb{N}$. It is assumed that $t_c(k) = t_e(k)$, $\forall k \in \mathbb{N}$.

Continuous masked evolutions in G_{p1}. Assume that $p_s \in f_{p1}(p_q,r_m)$, with $p_q, p_s \in P_1$, $r_m \in R_1$ *and also* that $\exists \boldsymbol{x}_0 \in c_q = et^{-1}(p_q)$ with the property $F_m(\boldsymbol{x}_0,t) \in c_q$, $\forall t \geq t_0$, where t_0 is the moment when r_m has occurred and $F_m(\boldsymbol{x}_0,t) = \boldsymbol{x}_0 + \int_{t0}^{t} f(\boldsymbol{x}(\tau),\gamma_1(r_m))\mathrm{d}\tau$. Then the trajectory $\boldsymbol{x}(.) \equiv F_m(\boldsymbol{x}_0,.)$ of (1) is called a *continuous masked evolution* associated to the pair (p_q, r_m) [5]. The presence or absence of continuous masked evolutions in G_{p1} can be analyzed only by inspecting the phase portraits of the system (1), with the control vector value $\boldsymbol{u} = \gamma_1(r_m)$, for each $r_m \in R_1$, respectively.

Building the DES-plant automaton G_{p1}. The purpose of this section is to build G_{p1} *without integrating* the state equations (1). Assume that the following hypothesis is true.

Hypothesis 3. $\forall p_q \in P_1$ and $\forall r_m \in R_1$, there are *no* masked evolutions associated to the pair (p_q, r_m).

Criterion 1 [5]. *Assume that* hypotheses 1, 2 *and* 3 *are true. Consider* $p_q, p_s \in P_1$ *and* $r_m \in R_1$.

A) $p_s \in f_{p1}(p_q,r_m)$ *in* G_{p1} *if and only if*

(*a1*) $\exists h_i \in S_h^4$ *s.t.* p_q *and* p_s *are* adjacent *on* $\mathrm{Ker}(h_i)$ and

(*a2*) $b_q^i.(f_m^T(\underline{\boldsymbol{x}}).\mathrm{grad}(h_i)) < 0,\ \forall \underline{\boldsymbol{x}} \in A(h_i,p_q,p_s) \subseteq \mathrm{Ker}(h_i)$,

where b_q^i *is the* i^{th} *component of the quality value* $\boldsymbol{b}_q = ech(p_q)$ *and* $f_m(.)$*is the speed of the continuous state associated to* r_m *according to* (12).

B) *Assume that* $p_s \in f_{p1}(p_q,r_m)$. *Then*

(*b1*) $g_{p1}(p_q,p_s) = z_{i+} \in Z_1$ *if and only if* $b_q^i = -1$, or

(*b2*) $g_{p1}(p_q,p_s) = z_{i-} \in Z_1$ *if and only if* $b_q^i = 1$.

The proof follows directly from proposition 1 and from property 1 and is given in [5]. The algorithm for the extraction of G_{p1} (fig.6(a)) is based on criterion 1 and is presented in the Appendix. Note that G_{p1} is *observable* [3], [5].

The DES controller. G_{c1} is a deterministic Moore machine that evolves according to the received plant symbols and sends the adequate control symbols to G_{p1} [3], [4], [5].

The primal control objective (6), presented in subsection 3.3, is modeled by the subautomaton $G_{loop} \subset G_{p1}$, with the desired evolution $w_{p1}^D = p(0),p(1),p(2) = p_1p_2p_5$ (fig.6.(b)). w_{p1}^D is *controllable* [4], [5], because there is *a sequence of control symbols* $w_{r1}^D = r(0),r(1) = r_1r_4$ s.t. i) $f_{p1}(p_1,r(0)) = \{p_2\}$, ii) $f_{p1}(p_2,r(1)) = \{p_5\}$ and iii) $f_{p1}(p_5,r(1))$ is *not* defined, i.e. there is no spontaneous undesired transition from p_5 when $r(1) = r_4$ is active.

The levels DES control problem is to design the DES controller G_{c1} (that produces w_{r1}^D) s.t. G_{p1}, coupled from $p(0) = p_1$ to G_{c1}, evolves like G_{loop}. This is a *desired states and transitions problem*, with the solution G_{c1} depicted in fig.6(c).

4 The Unsupervised Discrete Event Control of the Temperature

The HCS associated to the temperature dynamics comprises a disturbed continuous plant that is controlled, through an interface, by a DES. The plant coupled to the interface will be modeled as a disturbed DES-plant and the DES controller is a Moore machine.

The temperature dynamics is described by the differential equation

$$\dot{x}_3 = -x_3 + \mathrm{T_D} + u_3 + d\ , \tag{16}$$

where $x_3 \in \mathbb{R}$ is the state, $\mathrm{T_D} > 0$ is the desired medium temperature, $u_3 \in \mathbb{R}$ is the control and d is a scalar disturbance that takes values within any of the mutual exclusive ranges listed in Table 2. Each Boolean variable in Table 2 takes the value 1 (TRUE) if and only if d resides within the corresponding range. $0_d = 1$ (TRUE) if and only if (16) is *not* disturbed. Denote $Dist = \{d_1, d_1^-, d_2, d_2^-, 0_d\}$ the set of all Boolean variables associated to d.

Table 2. The logical modeling of the temperature disturbance

The possible ranges of the disturbance d		The associated Boolean variables
(1)	$0 < d < \mathrm{D}$	d_1
(2)	$-\mathrm{D} < d < 0$	d_1^-
(3)	$\mathrm{D} < d < 2\mathrm{D}$	d_2
(4)	$-2\mathrm{D} < d < -\mathrm{D}$	d_2^-
(5)	$d = 0$	0_d

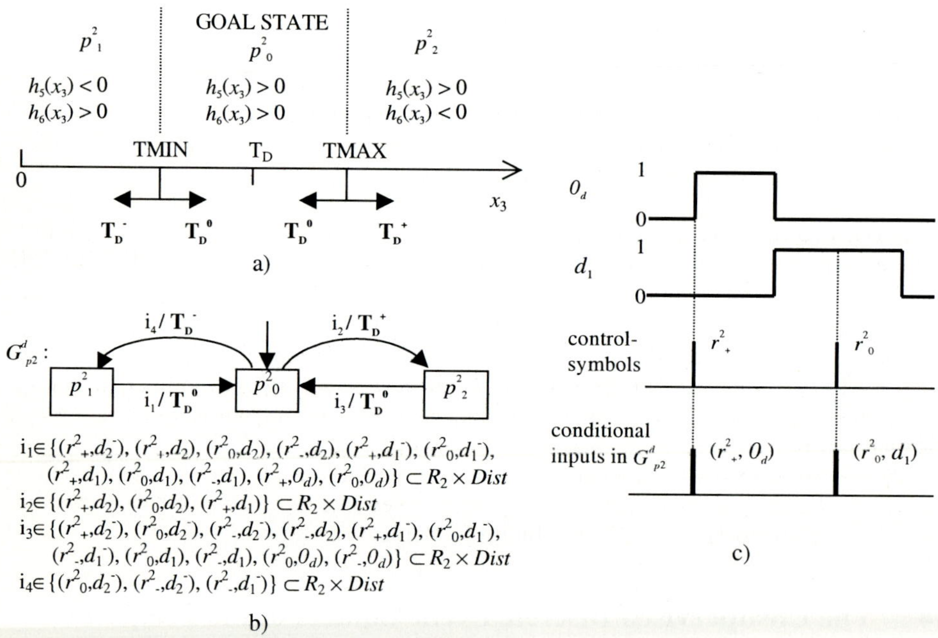

Fig. 7. a) The significance of the alphabets P_2 (19) and Z_2 (20); b) the disturbed DES-plant G^d_{p2} associated to the temperature dynamics; c) example: $(r^2_+, 0_d)$, $(r^2_0, d_1) \in R_2 \times Dist$ are conditional inputs in G^d_{p2}

The primal control objective of the temperature dynamics is to drive the continuous variable x_3 within the interval (TMIN,TMAX), and to maintain it there in the presence of the external disturbance d, where TMIN = T_D - D > 0 and TMAX = T_D + D.

U_2 = {-D, 0, D} is the set of control values. The *alphabet of control symbols* is

$$R_2 = \{r^2_-, r^2_0, r^2_+\}. \tag{17}$$

The *actuator* implements the function $\gamma_2 : R_2 \to U_2$, with the values

$$\gamma_2(r^2_-) = -D,\ \gamma_2(r^2_0) = 0,\ \gamma_2(r^2_+) = D. \tag{18}$$

The smooth functionals $h_5, h_6 : \mathbb{R} \to \mathbb{R}$, $h_5(x_3) = x_3$ - TMIN, $h_6(x_3) = -x_3$ + TMAX, are chosen starting from the primal control objective. The *alphabet of discrete states*

$$P_2 = \{p^2_0, p^2_1, p^2_2\} \tag{19}$$

and the alphabet of plant symbols

$$Z_2 = \{\mathbf{T_D^-}, \mathbf{T_D^0}, \mathbf{T_D^+}\} \tag{20}$$

have the significances respectively described in fig.7(a).

The disturbed DES-plant model associated to the temperature evolution (16) is the automaton $G^d_{p2} = \{P_2, R_2 \times Dist, f_{p2}, p^2_0, Z_2, g_{p2}\}$ with $f_{p2} : P_2 \times (R_2 \times Dist) \to P_2$ the state transition function, $g_{p2} : P_2 \times P_2 \to Z_2$ the output function and p^2_0 the initial state. A control symbol $r_m \in R_2$ occurs when a Boolean variable $d_j \in Dist$ is or becomes TRUE, i.e. $d_j = 1$, so G^d_{p2} receives a conditional input $(r_m, d_j) \in R_2 \times Dist$ (fig.7(b),(c)).

The temperature DES control problem is to find a Moore machine G_{c2} s.t. G^d_{p2}, coupled from the *goal state* p^2_0 (that corresponds to (TMIN,TMAX) in fig.7(a)) returns to p^2_0, whenever G^d_{p2} leaves p^2_0, due to the external disturbance d. The solution G_{c2} (fig.8(a)) is synthesized intuitively, based on G^d_{p2}.

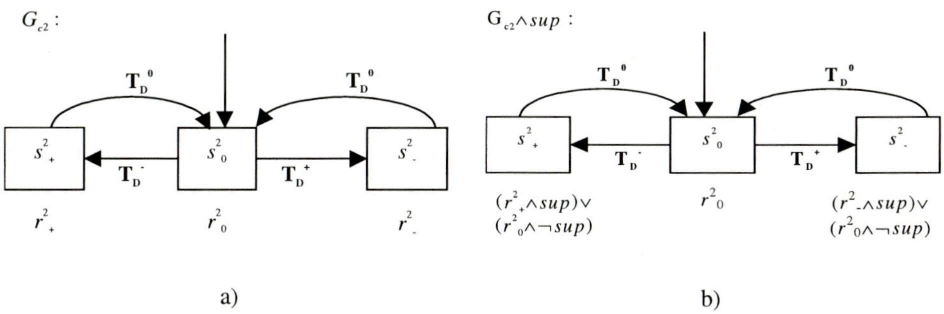

Fig. 8. a) The unsupervised DES controller of the temperature x_3; b) the supervised DES controller of the temperature x_3, included in the layered architecture depicted in fig.9.

5 The Supervision Architecture

The supervision objective is to prohibit the switching of u_3 to the values D or -D, no matter the temperature evolution, until the level x_1 in tank 1 reaches the extra limit LS, with $L_1 < LS < H_1$. This imposes an additional level sensor (LS) (fig.3). The outputs of the sensor (LS) are the symbols **LS+** and **LS-**, which are sent at a moment $t \in \mathbb{R}$ if and only if $x_1(t) = LS$ and $\dot{x}_1(t) > 0$ or if $x_1(t) = LS$ and $\dot{x}_1(t) < 0$, respectively. Define

$$Z_s = \{\mathbf{LS\text{-}}, \mathbf{LS+}\}. \tag{21}$$

The supervisor can potentially *observe* the occurrence of the symbols $\Sigma = \Sigma_{pr} \cup \Sigma_{co}$, where $\Sigma_{pr} = Z_s \cup Z_1 \cup Z_2$ and $\Sigma_{co} = R_1 \cup R_2$. The symbols from Σ_{co} are *potentially controllable* [1]. The control symbols that have to be *prevented to occur* when **LS-** is observed in the plant are $\Sigma_c = \{r^2_-, r^2_+\} \subset \Sigma_{co}$.

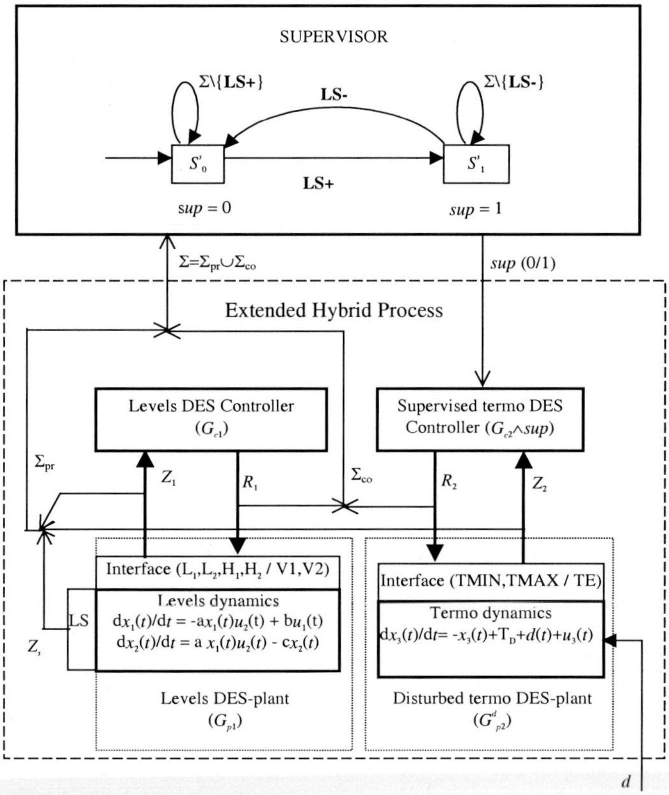

Fig. 9. The layered architecture for the supervision of the two HCS for the levels and temperature dynamics respectively.

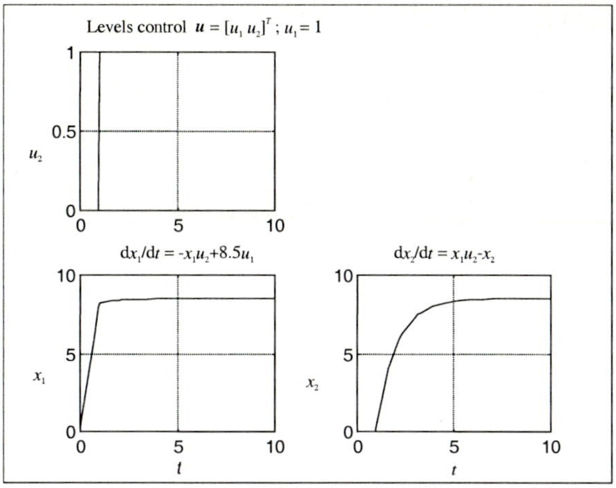

a) $u_2 = 0.5(\mathrm{sgn}(x_1 - L_1)+1)$

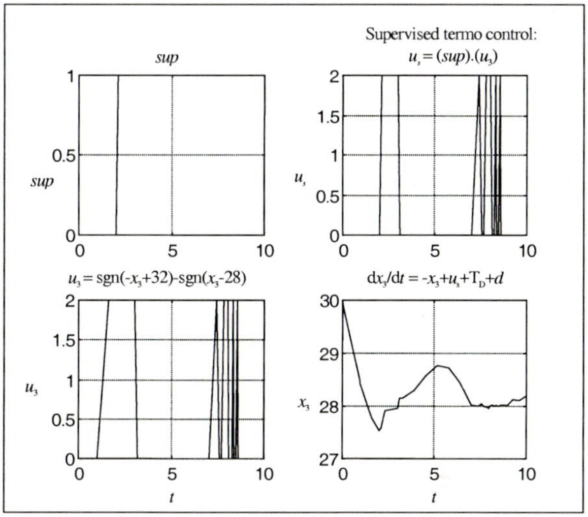

b) $sup = 0.5(\mathrm{sgn}(x_1 - LS)+1)$

Fig. 10. Simulation result of the controlled and supervised continuous plant depicted in fig.13, with: $L_1 = L_2 = 8$, $H_1 = H_2 = 12$, LS = 8.4, TMIN = 28, TMAX = 32, a = c = 1, b = 8.5, $T_D = 30$, $d(t) = -2\text{-}\sin(t)$. MATLAB integration routine: ODE45, with tol = 0.0001

The supervisor is synthesized intuitively, as a Moore machine that produces the logical variable *sup* (fig.9 and fig.10): *sup* = 0 if $x_1 < LS$ and *sup* = 1 if $x_1 > LS$. The supervised temperature DES controller $G_{c2} \wedge sup$ (fig.8(b)) produces conditional outputs: the control symbols from Σ_c are prevented to occur if *sup* = 0 and are permitted if else. The levels DES controller G_{c1} (fig.6(c), fig.9) is *not* influenced by *sup*.

6 Conclusions

In the proposed scenario, each DES controller is first synthesized as a solution of a distinct DES control problem. The main original contribution of this paper is the algorithm presented in the Appendix, for building the levels DES-plant, *without integrating* the state equations (1) and assuming that *hypotheses 1*, *2* and *3* are (generically) true. In section 4, the Antsaklis HCS formalism was adapted in order to treat also disturbed plants. The supervised control concept was intuitively extended to a group of two independent HCS (fig.9). A general approach of these two design problems is still open. Each continuous system in fig.9 perceives the supervised control part as a *switching control law* [4], [5], which permits the numerical simulation of the continuous evolution in MATLAB (fig.10).

References

1. Charbonnier, Fr., Alla, H., David, R.: The Supervised Control of Discrete Event Dynamic Systems: A New Approach. In: Proc. of the 34th Conf. on Decision & Control, New Orleans, LA (1995) 913-920
2. Ramadge, J.G., Wonham, W.M.: The Control of Discrete Event Systems. In: IEEE Proc., **77**(1) (1989) 81-98
3. Stiver J.A., Antsaklis P.J., Lemmon M.D.: A Logical DES Approach to the Design of Hybrid Control Systems. Technical Report of the ISIS Group at the University of Notre Dame, ISIS-94-011 (1994)
4. Oltean, V.E., Cârstoiu, D.: The Controllability Concept for a Class of Hybrid Control Systems – an Example. In: Bajić V. (ed.): Advances in Systems, Signals, Control and Computers, (SSCC'98, Durban, South Africa), published by IAAMSAD and ANS, Vol. II (1998) 442-446
5. Oltean, V.E.: Contributions to the Modeling, Analysis and Design of Hybrid Control Systems. PhD Thesis, "Dunărea de Jos" University of Galați, Romania (1998)

Appendix : Algorithm for Building the Levels DES-Plant Model G_{p1}

Input: a) h_i (7) and grad(h_i) (8), $i = 1,4$; b) P_1 (11) and B (Table 1); c) R_1 (3) and $f_m(\boldsymbol{x}) = f(\boldsymbol{x}, \gamma_1(r_m))$, $m = 1,4$ (12); d) Z_1 (13). *Output*: $f_{p1} : P_1 \times R_1 \rightarrow 2^{P_1}$, $g_{p1} : P_1 \times P_1 \rightarrow Z_1$

begin

1. Build the *adjacency matrix*[1] associated to P_1, $AD = (AD(p_q,p_s))_{1 \le q,s \le 9}$ according to:

   Rule 1: for each $p_q, p_s \in P_1$, with $\boldsymbol{b}_q = ech(p_q)$, $\boldsymbol{b}_s = ech(p_s)$ do

   if $\exists i \in \{1, 2, 3, 4\}$ s.t. $b_q^i b_s^i = -1$ and $b_q^j = b_s^j$, $\forall j \neq i, j \in \{1, 2, 3, 4\}$
   then $AD(p_q,p_s)$:= "h_i"
   else $AD(p_q,p_s)$:= "-"

2. for i := 1 to 4 do

   build the *frontier matrix* $H_i = (H_i(p_q,r_m))_{1 \le q \le 9,\ 1 \le m \le 4}$, according to:

   Rule 2: for each $p_q \in P_1$ and $r_m \in R_1$ do

[1] In this example, the matrix AD can be built directly from fig.4. For example, $AD(p_1,p_2)$ = "h_1".

if $\exists p_s \in P_1$ s.t. $AD(p_q,p_s) = "h_i"$
then begin
  $S = \text{sgn}(b_q^{\,i}.(f_m^{\,T}(\boldsymbol{x}).\text{grad}(h_i)),\ \forall \boldsymbol{x} \in A(h_i,p_q,p_s)$ % see *properties 1, 2*
  if $S = -1$
  then $H_i(p_q,r_m)$:= "I" % Ker(h_i) is input frontier for (p_q, r_m)
  else if $S = 1$
    then $H_i(p_q,r_m)$:= "E" % Ker(h_i) is exit frontier for (p_q, r_m)
    else $H_i(p_q,r_m)$:= 0 % Ker(h_i) is not an exit or input
      % frontier for (p_q, r_m)
  end
else $H_i(p_q,r_m)$:= "-" % p_q has no frontiers on Ker(h_i)

3. for q := 1 to 8 do % building the state transition function $f_{p1} : P_1 \times R_1 \rightarrow 2^{P_1}$
  for s := q+1 to 9 do
    if $\exists i \in \{1, 2, 3, 4\}$ s.t. $AD(p_q,p_s) = "h_i"$
    then for m := 1 to 4 do
      if $H_i(p_q,r_m) \neq$ "-"
      then if $H_i(p_q,r_m)$ = "E"
        then $p_s \in f_{p1}(p_q,r_m)$ and $p_q \notin f_{p1}(p_s,r_m)$ % $p_q \rightarrow p_s$
        else $p_q \in f_{p1}(p_s,r_m)$ and $p_s \notin f_{p1}(p_q,r_m)$ % $p_s \rightarrow p_q$
      else $p_s \notin f_{p1}(p_q,r_m)$ and $p_q \notin f_{p1}(p_s,r_m)$
    else $\forall r_m \in R_1$, $p_s \notin f_{p1}(p_q,r_m)$ and $p_q \notin f_{p1}(p_s,r_m)$ % p_q and p_s are not adjacent

4. for q := 1 to 8 do % building the output function $g_{p1} : P_1 \times P_1 \rightarrow Z_1$
  for s := q+1 to 9 do
    if $\exists r_m \in R_1$ s.t. $p_s \in f_{p1}(p_q,r_m)$ and $AD(p_q,p_s) = "h_i"$
    then if $b_q^{\,i} = -1$
      then $g_{p1}(p_q,p_s)$:= z_{i+} % (i+) has occurred when $p_q \rightarrow p_s$
      else $g_{p1}(p_q,p_s)$:= z_{i-} % (i-) has occurred when $p_q \rightarrow p_s$
    else $g_{p1}(p_q,p_s)$ is not defined
end

$$H_1 = \begin{bmatrix} E & I & 0 & E \\ I & E & 0 & I \\ - & - & - & - \\ E & I & 0 & E \\ I & E & 0 & I \\ - & - & - & - \\ E & I & 0 & E \\ I & E & 0 & I \\ - & - & - & - \end{bmatrix} \quad H_2 = \begin{bmatrix} - & - & - & - \\ E & I & 0 & E \\ I & E & 0 & I \\ - & - & - & - \\ E & I & 0 & E \\ I & E & 0 & I \\ - & - & - & - \\ E & I & 0 & E \\ I & E & 0 & I \end{bmatrix} \quad H_3 = \begin{bmatrix} I & I & I & I \\ I & E & I & E \\ I & E & I & E \\ E & E & E & E \\ E & I & E & I \\ E & I & E & I \\ - & - & - & - \\ - & - & - & - \\ - & - & - & - \end{bmatrix} \quad H_4 = \begin{bmatrix} - & - & - & - \\ - & - & - & - \\ - & - & - & - \\ I & I & I & I \\ I & I & I & I \\ I & E & I & E \\ E & E & E & E \\ E & E & E & E \\ E & E & E & E \end{bmatrix}$$

Fig. 11. The frontier matrices built at step 2. For example, $H_1(p_1,r_1)$ = "E" means that the cell c_1 associated to p_1 has a frontier on Ker(h_1) and, if the control $\gamma_1(r_1)$ is applied to the system (1) with $\boldsymbol{x}_0 \in c_1$, then the continuous trajectory $\boldsymbol{x}(.)$ will leave c_1 by crossing Ker(h_1) and will pass into a next adjacent cell. $H_1(p_1,r_2)$ = "I" means that if the control $\gamma_1(r_2)$ is applied to the system (1) with $\boldsymbol{x}_0$ on the frontier of c_1 which is common with Ker(h_1), then $\boldsymbol{x}(.)$ will enter the cell c_1.

Automatic Players for Computer Games

Werner DePauli-Schimanovich-Göttig

University of Vienna
Institute for Statistics and Decision Support
Universitätsstraße 5, A-1010 Vienna
werner.schimanovich@univie.ac.at
http://www.univie.ac.at/bvi

1 Introduction

One of the most important fields of research in computer science is the investigation of intelligent agents for the web which should search in the net for informations. Usually these agents are cooperating and form a working group of "multi agents". But these groups or different agents can also stay in competition to each other. In both cases we have to model the behavior of the agents; i. e. to program a set of rules (consisting of preconditions and actions) which tells every agent what he has to do.

The present rapid development of the web (and other networks and information systems) gives us the need to develop proper methods of modeling the intelligent agents. Sometimes these methods already exist more or less in principle, even when they are not yet in the final applicable form. An example for such a preliminary method is the design of automatic players for games. A lot of research has been done on this field, and the experiences and insights gained from this research can partially be used for the modelling of agents.

The following report presents an overview on different concepts and methods to program automatic players for games. The main focus in this presentation is on the methods developed by the author during his research and teaching on this field at the University Vienna.

2 Search Methods for Computer Chess

Norbert Wiener wrote in his famous book Cybernetics (first edition 1948) after the main chapters 1 to 8 an appendix about a chess playing machine. He quoted there the article "Theorie der Gesellschafts-Spiele" (i.e. theory of games) of Johann von Neumann. In this article von Neumann developed for the first time his theory of finite games with the main theorem: "Every finite game has an optimal strategy." Later Oskar Morgenstern applied the von Neumann theory of games to economics and together they published the famous book "Game Theory and Economic Behavior" (cf Neumann & Morgenstern). In the theory of games developed by von Neumann we distinguish between games in extensional form and games in normal form. For our investigation here we consider only the games in extensional form.

F. Pichler, R. Moreno-Díaz, and P. Kopacek (Eds.): EUROCAST'99, LNCS 1798, pp. 588–600, 2000.

Extensional games are mainly 2-person games. Key words for these games are: Game tree, Mini-Max, Full information, etc. We are interested here in programming "Optimal players".

By the theorem of von Neumann, they can be constructed with mini-max methods. In practise this, however, is only possible for small games, e.g: for the endgames of chess. (To allow more than 6 pieces is not possible for chess, because of the combinatorial explosion.) For the construction of the optimal strategy, the nodes must be asigned to real optimal values. Therefore we need first a computation of the optimal values. The reader interested in this further should consult the author´s "Calculus of Winning" and Ken Thompson's "Retrograde Analysis of Certain Endgames".

Since optimal players are not feasable for large games like chess, the socalled "Heuristic Players" have been developed. 1948 Norbert Wiener was the first one who recognized that, and he designed an heuristic player for chess. After him, Claude Shannon made a first classification of searching algorithms into the following 3 types:

- Type A (already Wiener's suggestion):
  (1) Look ahead n plys (= half moves, or moves of only 1 player)
  (2) Estimate the value for every node: evalu8 (suggested too by Wiener)
  (3) Compute the value back to the predecessor node by mini-max (or similar methods)!

  Improvements of Type A:
  (4) The great danger of type A is the "horizon effect": this algorithm cannot see a check mate in n+1 plys!
  (5) Better dynamic search: like minimax, but alpha-beta search. A short explanation is in the book of Nilsson. or of Knuth & Moore. Alpha-Beta allows also foward pruning with cut-offs, i.e. without throwing away the best move.

- Type B (suggested by Shannon):
  (1) Forward pruning: ordering and selection of moves by the value of the nodes.
  (2) Throw away the stupid moves! Search only from nodes with a good value.
  (3) But it can happen that a Type B algorithm throws away also some good moves (when evalu8 was not correct!)

  Conclusions for Type B:
  (4) Therefore this algorithm will not necessarily find the best solutions.
  (5) Type B was at the beginning of chess programming the only chance to realize it on weak computers (with only little computability-power.) Chess 4.x switched to type A (as a pioneer at this time!)

- Type C (also suggested by Shannon):
  (1) Problem-oriented forward pruning: analysis of chess-oriented features of a position, selection of goals and moves leading to this goals.

(2) Modelling the human player: e.g. quiescence search (i.e. searching selective deeper, when it is necessary, because of the horizon effect). When you cannot evaluate the real value of a position by a static evaluation, you have to compute a dynamical value by searching.

(3) Since 1980 those programs had been available for chess endgames.

Chess is certainly the most important game for which automatic players have been developed. But there exists a number of other Computer Games than Chess:

Hans Berliner's Backgammon-Program BKG9.8 won 1981 against the world champion. Around 1985 Nicolas Findler developed programs for Bridge and Poker. About 1995 Ralph Gasser (from the ETH Zürich) showed by decompositional methods that 9-Men's-Morris is drawn, and Martin Müller developed an endgame-database for Go (by decomposition of endgames with methods of Sprague-Grundy theory and combinatorial game theory).

3 Computer Games: Own Developments

My work on computer games can be summarized in this way: My students and I programmed some automatic players based on heuristic strategies for the usual wellknown games, some optimal and semi-optimal players for the diminuative versions of these games, and other optimal players for small games. In particular I investigated the following commonly known games: Kalaha (6 and 4), De Bono's L-game, Hexi (= an application of Ramsey Theory, click on http://www.dbai.tuwien.ac.at/proj/ramsay/ or /staff/slany), Colour-Triangle (7, 5 and 3), 9-Men-Morris (Draughts) in restricted form, 3 in a Row (with gravitation), 4 in a Row (with gravitation), 5 in a Row (without gravitation), and others.

Furthermore I invented some other games too and programmed automatic players for them. Let me mention the more important ones:

Fig. 1. A starting position of Bear-Dog-Hunt

Bear-Dog-Hunt (BDH):

BDH is played on a 5 x 5 board with 5 dogs and 2 bears. All these 7 pieces can move alternately in one of the 8 directions of a square (i.e. N, S, E, W and NE, NW, SE, SW). In the starting position the 5 dogs are positioned on the left side of the board in a zip-line; the 2 bears are on the right side. The starting directions are created randomly and marked with a star or an arrow. The dogs-player (who starts) can change the directions of 2 dogs. After that all 5 dogs move 1 field ahead simultaneously. On the boarders they are sent back like a billard ball plus 1 field more. The bears-player can change only the direction of 1 bear. After that both bears are moving ahead. If 2 dogs meet 1 bear: the bear is killed. If a bear meets only 1 dog: the dog is killed. 2 dogs or 2 bears kill themselves! When there is only 1 dog left, the bears win. When there is no bear left, of course the dogs win. If there is only 1 bear and 2 dogs left, after 10 moves the game is drawn.

Several modifications of that game are possible. E.g: as videogame on a board with 10 x 20 squares (with black spots between). At the beginning 20 bears and 80 dogs, move alternatly every 3 seconds or 5 seconds resp. (the dogs). Players can change direction only in clockwise steps with the joystick. BDH will be soon available on my homepages.

Poker-TicTacToe (PTTT):

PTTT is a paper-and-pencil game similar to TicTacToe played on a 4 x 4 board, but with additional rules! The player who gets first 4 signs of his own in a row or a diagonal line will of course win. But this is not the regular case, because the game is drawn (as its value by optimal play) like TTT. Therefore the board will be filled with naughts and crosses in the regular play. After that you count the 3's in a row of each player. The one who has more 3's wins. If the number of the 3's are equal, you must count the pairs that are not already part of a "drull" (i.e. a 3). The player with more pairs wins. If the number of pairs is also equal, the game is drawn. (My student Marcus Wagner and I computed the full database of the values of the positions and programmed an optimal player.)

Cyclop and CyclopeD:

The simplest version of the game is played by moving 2 pieces A and B respectively (e. g. coins) one field ahead in horizontal or vertical direction, on a board with 2 x 3 squares (Let´s mark them 11, 12, 13, 21, 22, 23.) The starting position is A = 11 and B = 23. The player who first moves into a repetitional position (i.e. a position that occured already in the game before) looses! This diminuative version of the game is for the experienced player too simple. Therefore in the advanced version Cyclop-D diagonal moves are also allowed.

Loss-Drawn-Win Marienbad-game:

This is a special threefold Marienbad-game with 3 trees. Remember: The Marienbad-game is a NIM-game (constituted by 4 heaps of 1, 3, 5 and 7 sticks). Now we have to consider 3 Marienbad-games side by side with different winning rules: If the last stick of this 3-fold Marienbad-game is taken from the left tree, the player who has to take it looses the whole game. If the last stick of the 3-fold game is taken from the middle tree, the whole game ends with drawn. Only if a player can take the last stick (of the whole 3-fold Marienbad-game) from the right tree, he wins the whole game. (As a variation of these rules one can also design an unbalanced payoff-function where the game need not to be a zero-sum-game.)

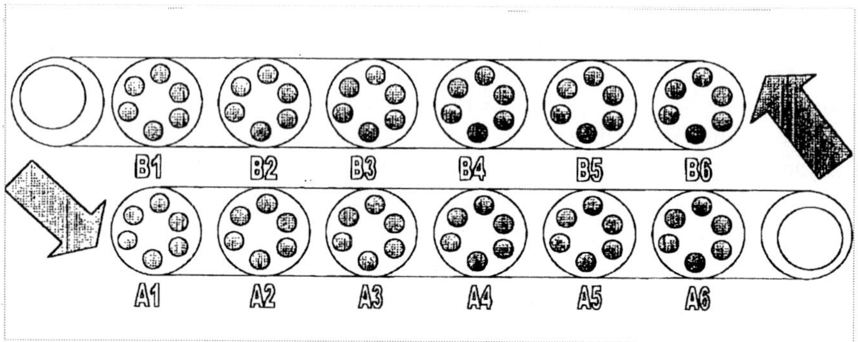

Fig. 2. Starting position of Kalaha

Kalah4:
This game is similar to Kalaha, which contains 6 holes for each player with 6 stones in each hole at the starting position, and 1 kalah-hole (See Figure 2. Kalaha will be described later.) Kalah4 has only 4 holes with 4 stones, and 1 (empty) kalah-hole (for both players at the beginning). My student Wolfgang Wiesbauer and I programmed an optimal player for Kalah4. With this diminuative player we could test and improve the evaluation function of Kalah6. Therefore also our heuristic player for Kalaha was unbeatable for a long time.

4 Computer Games: Experiences

In the following sections I want to discuss by the above mentioned games the work we did some years ago. Let´s start with Kalaha! In the initial board configuration are 6 stones in every hole. The two players alternately do empty an own hole and distribute the stones anti-clockwise (one by one) in the 12 holes and the own kalah (surpassing the opponents kalah). Depending on the position of the hole, where the last stone had been placed, we can distinguish 3 sorts of moves: (1) An Ordinary Move happens when the last stone is placed into a hole of the opponent or into a not-empty own hole (of course except the kalah which is not considered as a hole). (2) A Kalah Move is when the last stone falls into the (own) kalah. In this case the move is possibly not finished yet: the player can empty another (own) hole. (3) Robbery

Move: If the last stone falls into an empty own hole the player can rob the stones in the adjacent hole of the opponent and put them into his own kalah.

The goal of the play is to collect (at least) 36 stones in the own Kalah. The player who succeeds, wins. But be careful: A player who does not have any stones in his holes, cannot move and looses even if his opponent does not have 36 stones yet. (Sometimes Kalaha is played with the additional condition that in this case the opponent is allowed to put his own stones into his Kalah-hole. Then the game can end drawn too!) There are also other slightly different rules of the game in use, e.g. that in the case of a Kalah Move the player has to empty another hole! (In this case the player can loose also, when he has already 36 stones.)

Wiesbauer (1980) computed the whole database for the optimal values of the positions of Kalah4 (= the smaller version with only 4 holes per player and 4 stones in each hole and corresponding rules, e.g. collect 16 stones for a win!) With this database we could program an optimal player for Kalah4 and test heuristic evaluation functions. The insights (we won by this test) we used to tune and optimize the heuristic evaluation function of the large game. Therefore our automatic player was unbeatable in that time, where these games could only be played on large mainframe computers.

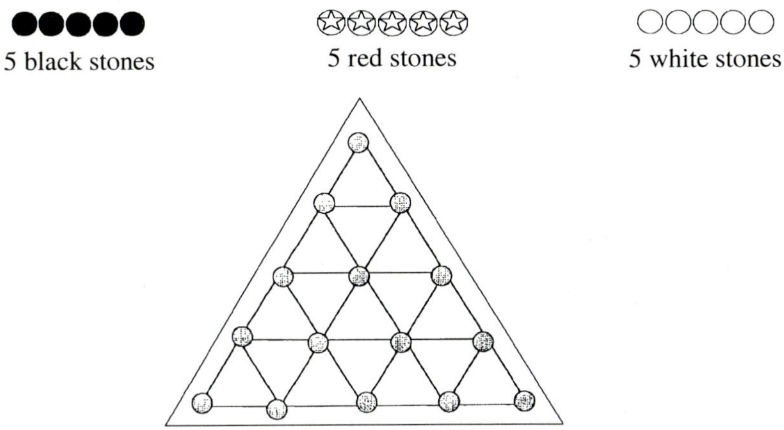

Fig. 3. Empty Board of Colour-Triangle 5

The next game I want to describe is Colour-Triangle. The normal version consists of a triangle with 7 circular dots at its 3 vertices/edges (including the corners). Parallel to each edge, 5 lines are drawn connecting the 2 corresponding dots at the other 2 edges. At the intersection of every 3 lines, new dots are placed: 4 one at the 1st parallel line after the bottom vertice/edge, 3 one at the next, etc. Therefore the board will be covered with 28 dots (= 7 + 6 + 5 + 4 + 3 + 2 + 1).

At each of these dots, a stone can be placed. At the beginning we have the empty board, and a pool of 28 stones; always 7 of one of the 4 colours (= white, black, red and blue). 2 players alternately take a stone (no matter which colour) from the pool and place it on the board. If a player can set a stone this way, that it forms together with 2 other stones of the same colour on the board the corners of a regular (=60 degrees) triangle, he wins!

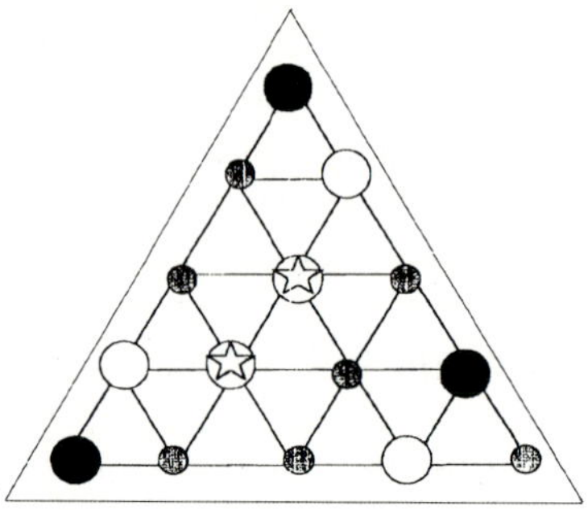

Fig. 4. A win for the player who put the last white stone onto the board

The smaller version Colour-Triangle 5 has only a board with 15 dots/points (shown in Figure 3). Figure 4 shows an end position with a win for the player who put the last white stone onto the board. Hermann Kaindl and I solved this game and programmed an optimal player for it. (See Kaindl.)

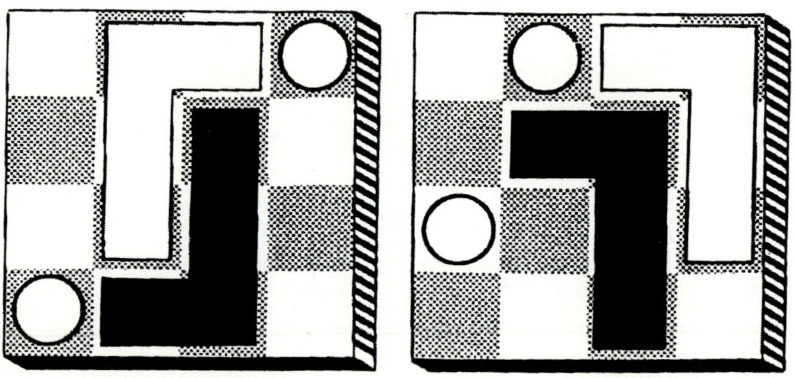

Fig. 5. Starting position and an end position of De Bono's L-Game

Another game which is an excellent example for our methods is De Bono´s L-Game. It is played on a 4 x 4 grid, where the 2 L´s (of the 2 players) and 2 neutral pieces are placed on the board. The starting position is shown in Figure 5. A move consists of 2 parts: first the player has to change the position of his L and second he can change the position of 1 neutral piece (but he does not need to). The game ends if a player cannot move and looses therefore. (E.g. White in Figure 5.)

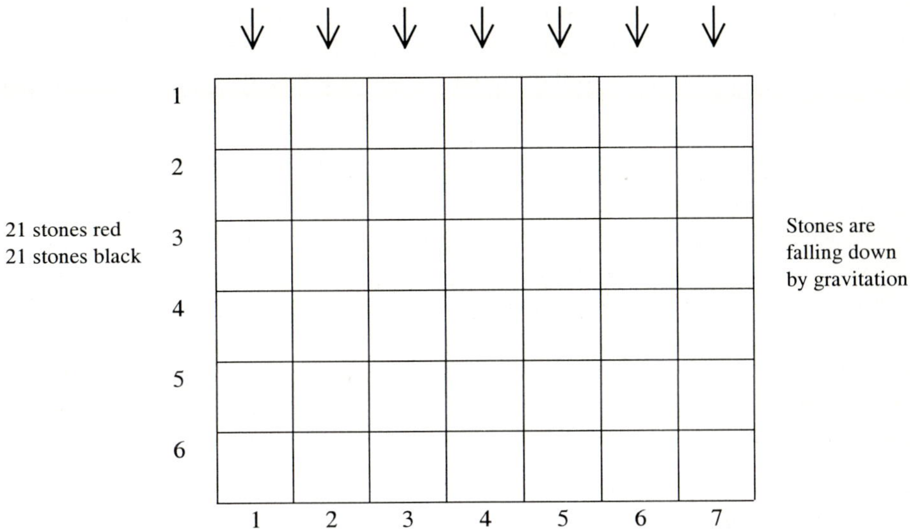

Fig. 6. The empty comb of 4-in-a-Row

The game "4 in a row" is well known. A vertical board or (better to say:) a standing comb with 6 rows and 7 columns shall be filled alternately by each of the 2 players with their own stones (black or red. See Figure 6.) The stones are always falling down to the bottom or onto the stone below in that column. This is producing a pattern of black and red dots that form sometimes also lines. If a line with 4 red stones can be formed (horizontal, vertical or diagonal), the red player wins. The same is valid for the black player.

5 Computational Methods

After the description of different games, I want to explain the methods by which they can be treated computationally. The most wellknown method to construct heuristic players is the Alpha-Beta-Search, explained in several textbooks about Artificial Intelligence. (E.g. consult Nielsson.) A static evaluation function computes heuristic values of the positions which are computed backward by a mini-max like dynamical search. The feasability of this methods depends of the branching factor: the average number of moves a player can perform from a position. Alpha-beta is very good for chess (branching factor = 45), bad for GO (BF = 400) or "4 in a Row" with gravitation (BF = only 6, but horizon effect = 20 plys or more). Alpha-Beta search has also a theoretical disadvantage. Dana Nau could show the paradoxon: "The deeper you search, the worse the result!" (See Judea Pearl.)

However, this paradoxon gives reason to chose different approaches than in classical AI. One of it is based on Sprague-Grundy Theory which is wellknown for solving NIM-games (similar to the Marienbad game.) The Sprague-Grundy Theory (=SGT) was created about 1935. It allows to compute (for a special class of games) natural numbers as absolute values of the positions of a given game (in contrary to the heuristic values, which are usual when we are constructing automatic players with the computer).

The values of a special Sprague-Grundy function are defined at the start by the terminal positions concerning to the rules of the game (0 for win, 1 for loss), and after that for the non-terminal positions backward by computation:
$f(\mathrm{p}) := \min (\mathrm{N} \setminus \{ f(\mathrm{q}) \ (\in \mathrm{N}) \mid : \exists$ move from p to q $\})$. Here should be noted that $0 \in \mathrm{N}$. The set $\{ f(\mathrm{p}) = 0 \}$ forms the kernel K of all those positions from which the moving player can only loose (or has lost already in the terminal case). Games with drawn (and other) values und puzzles (=1-person games) and 3- or more-person games cannot be treated with SGT.

Let me shortly note the following facts:
"p Move q" means: some player moves from p to q.
Then the following implications can be proven:
p Move q & $\mathrm{q} \in \mathrm{K} \Rightarrow \mathrm{p} \notin \mathrm{K}$.
$\mathrm{q} \in \mathrm{K}$ & q Move r $\Rightarrow \mathrm{r} \notin \mathrm{K}$.
(If $\mathrm{r} \notin$ Stop:) $\mathrm{r} \notin \mathrm{K} \Rightarrow \exists \mathrm{s}$: r Move s & $\mathrm{s} \in \mathrm{K}$.
Furthermore: The SG-function makes the games decomposable into partial games, with
$f(\mathrm{G}_1 + \mathrm{G}_2) = f(\mathrm{G}_1) + f(\mathrm{G}_2)$. Because:
$\mathrm{p} \in \mathrm{K} \cap \mathrm{G}_1$ & $\mathrm{q} \in \mathrm{K} \cap \mathrm{G}_2 \Rightarrow (\mathrm{p}+\mathrm{q}) \in \mathrm{K} \cap (\mathrm{G}_1 + \mathrm{G}_2)$.
A further development is the Combinatorial Game Theory. (See Berlecamp, Conway and Guy.)

6 Calculus of Winning

Based on this SGT, I developed the Calculus of Winning (=CW). See Schimanovich (1978, 1980, 1981, 1982) and DePauli-Schimanovich (1999, 2000).

The Calculus of Winning is the counterpart (and an extension) of Sprague-Grundy theory. It is an axiom system in first order logic with a number of theorems that gives us an insight into the structure and the internal relations of the specific games. Its precise language allows also a finer typization of games (in the extended form, concerning to its essential properties) and a precise classification of positions (win, loss, drawn). Eventually the Calculus of Winning delivers simple algorithms to construct the optimal strategy for a computer as an automatical player. (E.g. for the endgames of chess we get an algorithm that ignores all drawn or not yet classified positions, computing only the win or loss positions.) With these optimal players for small games you can tune the evaluation function of the heuristic players for large games. In this aspect the Calculus of Winning is more related to Computer Games and Artificial Intelligence than to the Theory of Games (which treats mainly games in normal form).

The Calculus of Winning is a very sophisticated logical system, where one can derive a lot of theorems, like in other theories of predicate logic. This delivers a good insight into the structure of the game. Some games (the so called Drawn-Circle games or sometimes called "loopy games"), where a position can be moved to again and the game ends with drawn, can be treated with the CW excellently. It allows very simple algorithms for these games to construct the database of absolute values of the positions (ignoring all drawn and not yet classified positions!) This counts for the endgames of Chess, for De Bono´s L-game, for 9-Men´s-Morris or Bear-Dog-Hunt.

The other important group of games are the so-called stratified and strict stratified games, like Kalaha, Colour-Triangle or 4-in-a-row. They do not allow a repetition of positions (or restrict them like in the case of Go.) For those games the CW delivers also good classification algorithms, but not so excellent ones as for the drawn-circle games. Also for the last group of games (the win-circle and the loss-circle games) some feasable algorithms can be established. But most games (e.g. like chess) are a mixture of different types.

In addition to the common Artificial Intelligence methods, and to Sprague-Grundy Theory, Combinatorial Game Theory, and the Calculus of Winning, Ranan Banerji developed the "Homomorphism Method" for games: given the solution of a game (i.e. the absolute value classification of its positions), construct the solution of another game. We demonstrate this by the 9-penny game: You have 9 different coins with all numbers between 1 and 9 pennies. 2 players choose alternately 1 coin. The one who has first 3 coins that add up exactly to 15 pennies, wins. Solution: Play this game on a 3 x 3 magic square, because it is isomorphic to Tic-Tac-Toe. Also the game of race (from Newell and Simon) is homomorphic to Tic-Tac-Toe.

7 Puzzles or 1-Person Games

At the end I want to mention the puzzles (= 1-person games) on which I worked with my students: Solitaire (in restricted form), Rock em up, Master Mind, Rubik´s cube (2 x 2 x 2, 3 x 3 x 3, 4 x 4 x 4 and n x n x n), Tower of Hanoi.

The Tower of Hanoi is usually used as an example for recursion in the textbooks. But it has a simple explicit moving function, that computes the next move of a given position without recursion. This moving function can be easily explained by the "thumb-rule strategy" for the Tower of Hanoi: At the beginning of the game put the thumb (of your right hand) on the 2nd stick in the middle, and then move the smallest disc (which is at the begin on the top of the other discs on the 1st (= left) stick) to the

3rd (= right) stick. After this always move the thumb clockwise (2nd, 3rd, 1st stick, then start again with the 2nd) and make the corresponding move with the disc. (Note: concerning to the rule that no larger disc can be put ahead of a smaller one, there is only one move possible!) After some clockwise moves of the thumb the pile will have been shifted from the left to the right side, and the puzzle is solved. In Buneman & Levi you can find this thumb-rule that I discovered independently too. (See also Cull & Ecklund.)

Finaly, I shortly want to explain my method of "Relative Problem Solving" on the example of the Rubik´s Cube. Some years ago puzzling was a mania under children and students and they knew how to solve the Cube (i.e. the 3 x 3 x 3 Cube.) With this knowledge of the solving procedures it is easy to solve the baby cube (= 2 x 2 x 2). Some people can only solve the fool´s cube (= 1 x 1 x 1) or the military cube (all sides are red). But the interesting question for us is: How to solve the 4 x 4 x 4 Cube with a minimum of new algorithms?

When we treat that whole 4 x 4 x 4 Cube like the baby cube, it will lead to nothing. However we bring the 4 inner squares of each side to one unique colour first and then form such way a "super cube". This can be done with the knowledge of the baby cube. Therefore the only additional knowledge we need is an algorithm to assimilate the 2 inner neighbor-cubies of each edge of the 4 x 4 x 4 cube. (I.e. to give them the same colour side on side on both fronts.) This should work somehow without destroying the 6 super cubies. Now we see immediately that the 4 x 4 x 4 cube can be treated as 3 x 3 x 3 cube, and can be solved with the minimal additional knowledge of the super-cubies algorithm.

Following these lines Erwin Sulzgruber established a general theory to solve the n x n x n Cube!

8 Conclusion

The preceding explanations are of course only a snapshot of the large field of computer games, and of methods used there to program automatic players. With this article I wanted to guide young researchers only to the magic triangle of A.I., theory of games, and cybernetics. For the actual task of developing methods to support intelligent agents to navigate through the web and gain information, those results of the pioneer time of computer science can bring a lot of hints how to get insights, especially in respect to the goal how the design of agents should be handled properly. Systems analysis and modelling with mathematical and logical tools (like it has been done for games with the Sprague-Grundy theory or the Calculus of Winning) can successfully be applied to intelligent agents too!

References

Banerji R. (1980) Artificial Intelligence: A Theoretical Approach. North-Holland PublComp, Amsterdam.

Bell A.G. (1972) Games Playing with Computers. George Allen & Unwin, London.

Berlecamp E., Conway J., Guy A. (1982) Winning Ways (for your mathematical plays). Vol. 1 + 2, Academic Press, London, N.Y.

Buneman P., Levy L. (1980) The Towers of Hanoi Problem. Information Proceeding Letters 10, pp. 243 – 244.

Butter S., Payer K., Pokorny L., Schimanovich W. (1981) Minimühle. Konstruktion eines optimalen Spieles für die Zieh- und Sprungphase. Technical Report of the Institute of Statistics and Computer Science, University Vienna.

Casti J (1990) Searching for Certainty. What Scientists Can Know About the Future. William Morrow & Comp., N.Y.

Casti J., DePauli W. (2000) Kurt Gödel. A mathematical Legend. Perseus Books. Reading MA.

Cull P., Ecklund E. (1985) Towers of Hanoi and Analysis of Algorithms. American Mathematical Monthly, Vol 92, No 6, June-July.

DePauli-Schimanovich W. (1999) Programming Chess & Computer Games (Part I + II). In: Wiener's Cybernetics: 50 years of evolution. Cybernetics'99, Extended Abstracts, ISBN-84-8416-950-2.

DePauli-Schimanovich W. (2000) The Calculus of Winning. In: Yearbook 1999 of the Kurt Gödel Society, TU-Vienna.

Feda M., Bartl R. (1985) Pente: 5er-Reihe mit 5-Pärchen-Klau. Technical Report of the Institute of Statistics and Computer Science, University Vienna.

Grob K. (1978) Ein Automatischer Spieler für Master Mind. Diploma Thesis at the Institute of Statistics and Computer Science, University of Vienna.

Hartmann G. (1980) Konstruktion der Optimierungsstrategie und Programmierung eines automatischen Spielers für das L-Spiel von Edward de Bono. Diploma Thesis at the Institute for Statistics and Informatics, University of Vienna.

Horacek H. (1982) Ein Modell des menschlichen Problemlösens an Hand von Schach-Bauern-Endspielen. Dissertation at the Institute for Logistic, University of Vienna.

Jauernik R. (1990) Entwurf eines heuristischen Spielers für das BEAR-DOG-HUNT-Spiel. Diploma Thesis at the Institute for Statistics and Informatics, University of Vienna.

Kaindl H., Schimanovich W. (1978) Catrin, mathematisches Brettspiel für 2 Partner. In: Technical Report Nr.7 at the Institute of Statistics and Computer Science, University of Vienna.

Kaindl H. (1983) Neue Wege im Computerschach. Dissertation at the Institute for Information Systems, Technical University Vienna.

Laufer H., Schimanovich W. (1981) Peggy: Arbeitsweise eines Optimalen Dialogspieles. Technical Report of the Institute of Statistics and Computer Science, University Vienna.

Laufer H. (1982) SOLITAIRE. Arbeitsweise eines optimalen Dialogspielers. Diploma Thesis at the Institute for Statistics and Informatics, University of Vienna.

Levi L. (1980) Computer Games. Springer Verlag Inc., N.Y.

Neumann J. von (1928) Theorie der Gesellschafts-Spiele. Mathematische Annalen vol. 100.

Neumann J. von, Morgenstern O. (1944) Game Theory and Economic Behavior. Princeton University Press.

Nielsson N. (1971) Problem Solving Methods in AI. McGraw-Hill Book Comp., N.Y.

Nielsson N. (1980) Principles of AI. Tjoga Publ.Press, or Springer Inc., N.Y.

Österreicher K., Sulzgruber D., Schimanovich W. (1980) Tertio en Raya. Konstruktion eines optimalen Spielers. Technical Report of the Institute for Statistics and Computer Science, University of Vienna.

Payer W. (1983) Ein Optimaler Spieler für die Ovalmühle. Klassifikation der Stellungen und Konstruktion eines optimalen Dialogspielers. Diploma Thesis at the Institute for Statistics and Informatics, University of Vienna.

Pearl J. (1983) Search and Heuristics. North-Holland Publ. Comp., Amsterdam.

Pearl J. (1984) Heuristics. Intelligent Search Strategies for Computer Problem Solving. Addison-Wesley Publ. Comp., Reading MA.

Schimanovich W. (1978) Der Gewinnkalkül. Abstract für den Workshop "Berichte aus Informatik-Instituten", September 1978.

Schimanovich W. (1980) Prädikatenlogische Theorie zur Behandlung von Positionsspielen. Proceedings of the 4th International Wittgenstein Symposium 1979, Hölder-Pichler-Tempsky-Verlag, Vienna.

Schimanovich W. (1980) Spiele und Intelligenz. In: Wissenschaft Aktuell, Heft 3 (Juni) und Heft 4 (September), Wien.

Schimanovich W. (1981) An Axiomatic Theory of Winning Mathematical Games. In: Cybernetics and Systems 12: pp. 1-19.

Schimanovich W. (1982) An Axiomatic Theory of Winning Mathematical Games and Economic Actions. Proceedings of the 5th European Meeting on Cybernetics and System Research 1980. Hemisphere Publishing Corp., N.Y.

Schimanovich W. (1984) Relative Problem Solving and the Role of Logic in AI. (Invited Lecture of AIMSA, Varna.) Technical Report of the Institute for Statistics and Computer Science, University of Vienna.

Shannon C. (1950) Programming a Computer for Playing Chess. In: Phil. Magazine, Band 41, S. 256-275.

Slagle J. (1971) Artificial Intelligence: The Heuristic Programming Approach. McGraw-Hill Book Comp., N.Y. London.

Slany W. (1984) Hexi: Happy Perfect Hexagon Player. Technical Report of the Institute for Statistics and Computer Science, University of Vienna.

Thompson K. (1986) Retrograde Analysis of Certain Endgames. In: ICCA Journal, September.

Trubel F., Schibl A. (1982) Rock em up: Beschreibung des automatischen Spielers und des Programmes samt Lösungs-Algorithmus. Technical Report of the Institute for Statistics and Computer Science, University of Vienna.

Wagner M., Schimanovich W. (1980) Laufzeituntersuchung und Optimierung des Schachprogrammes Chess.5. Technical Report of the Institute for Statistics and Computer Science, University of Vienna.

Wiesbauer W. (1981) Kalah4: Ein optimaler Dialog-Spieler. Diploma Thesis at the Institute for Statistics and Informatics, University of Vienna.

Wiener N. (1948) Cybernetics. John Wiley, N. Y.

Author Index